T0180784

Lecture Notes in Computer Science

Lecture Notes in Computer Science

Edited by G. Goos and J. Hartmanis

319

J. H. Reif (Ed.)

VLSI Algorithms and Architectures

3rd Aegean Workshop on Computing, AWOC 88
Corfu, Greece, June 28 – July 1, 1988
Proceedings

Springer-Verlag

New York Berlin Heidelberg London Paris Tokyo

Editor

John H. Reif
Department of Computer Science, Duke University
202 North Building, Durham, NC 27706, USA

CR Subject Classification (1987): F.2, B.7

ISBN 0-387-96818-0 Springer-Verlag New York Berlin Heidelberg
ISBN 3-540-96818-0 Springer-Verlag Berlin Heidelberg New York

Printing and binding: Druckhaus Beltz, Hemsbach/Bergstr.
2145/3140-543210

FOREWORD

The papers in this volume were presented at the Aegean Workshop on Computing: VLSI Algorithms and Architectures (AWOC 88), organized by the Computer Technology Institute in Patras in cooperation with ACM, EATCS, IEEE and the General Secretariat of Research and Technology (Ministry of Industry, Energy & Technology of Greece). They were selected from 119 abstracts submitted in response to the program committee's call for papers and to additional invitations from John Reif.

AWOC 88 took place in Corfu, Greece, June 28-July 1, 1988. AWOC 88 is the third meeting in the International Workshop on Parallel Computing & VLSI series; the first meeting took place in Amalfi, Italy, 1984, and the second at Loutraki, Greece, 1986.

F. Preparata and J.H. Reif were the program chairmen of AWOC 88.

The program committee consisted of:

G. Bilardi (USA),
U. Lauther (FRG),
T. Lengauer (FRG),
F. Makedon (USA),
G. Miller (USA),
T. Papatheodorou (GREECE),
J. van Leeuwen (THE NETHERLANDS),
M. Yannakakis (USA),
S. Hambrusch (USA),
T. Leighton (USA),
F. Luccio (ITALY),
K. Mehlhorn (FRG),
C. Papadimitriou (GREECE),
P. Spirakis (GREECE), and
U. Vishkin (ISRAEL)

The very enjoyable local arrangements were made by:

F. Makedon
P. Spirakis, and
T. Papatheodorou

TABLE OF CONTENTS

Parallel Tree Contraction

Simulation and Embedding of Parallel Networks

Session Chair: T. Leighton (M.I.T., USA)

Compaction and Channel Routing

Session Chair: K. Mehlhorn (Univ. des Saarlandes, W. Germany)

VLSI Layout

Session Chair: M. Yannakakis (Bell Labs, USA)

VLSI Testing and Derivation

Session Chair: F. Preparata (Univ. of Ill-Urbana, USA)

Distributed Computing

Session Chair: Jan van Leeuwen (Univ. of Utrecht, The Netherlands)

Distributed Computing

Session Chair: P. Vitányi (Centrum voor Wiskunde en Informatica,
The Netherlands

Parallel Routing and Sorting

Late Paper

PARALLEL ALGORITHMS FOR EVALUATING SEQUENCES OF SET-MANIPULATION OPERATIONS
(Preliminary Version)

Mikhail J. Atallah[1]

Dept. of Computer Science, Purdue Univ., West Lafayette, IN 47907

Michael T. Goodrich

Dept. of Computer Science, Johns Hopkins Univ., Baltimore, MD 21218

S. Rao Kosaraju[2]

Dept. of Computer Science, Johns Hopkins Univ., Baltimore, MD 21218

Abstract

Given an off-line sequence S of n set-manipulation operations, we investigate the parallel complexity of evaluating S (i.e. finding the response to every operation in S and returning the resulting set). We show that the problem of evaluating S is in NC for various combinations of common set-manipulation operations. Once we establish membership in NC (or, if membershp in NC is obvious), we develop techniques for improving the time and/or processor complexity.

1 Introduction

The evaluation of operation sequences is something that is fundamental in the design of algorithms. Given a sequence S of set-manipulation operations the evaluation problem is to find the response to every operation in S and return the set one gets after evaluating S. There are a host of problems that are either instances of an evaluation problem or can be solved by a reduction to an evaluation problem. For example, sorting a set $S = \{x_1, x_2, ..., x_n\}$ can easily be reduced to the problem of evaluating the sequence $(I(x_1), I(x_2), ..., I(x_n), E, E, ..., E)$, where $I(x)$ stands for "Insert x," E stands for "ExtractMin," and there are n E's. The answers to all the "ExtractMin" operations immediately give us a sorting of the items in S.

Given an off-line sequence S of n set-manipulation operations, we investigate the parallel complexity of evaluating S. This is a well studied problem in sequential computation, but surprisingly little is known about its parallel complexity. Our motivation, then, comes from a desire to begin a systematic treatment of this im-

portant area from a parallel perspective. In addition, because of the foundational aspect of off-line evaluation problems, we are also interested in these problems for their possible applications. We already know of applications to such areas as processor scheduling, computational geometry, and computational graph theory (we discuss some of these below).

This paper contains two types of results:

(i) We show that the problem of evaluating S is in the class NC for various combinations of operations appearing in S. That is, it can be evaluated in $O(\log^c n)$ time using $O(n^d)$ processors, for constants c and d.

(ii) Once we establish membership in NC (or, if membershp in NC is obvious), we develop techniques for improving the time and/or processor complexity.

The following is a list of the results presented in this paper. We give the results in the following format: First we mention the operations that may appear in S, then we give a pair $T(n), P(n)$ representing (respectively) the time and processor complexities of our algorithms, to within constant factors.

1. $Insert(x)$, $Delete(x)$, $ExtractMin$: $\log^2 n$, n. Showing that this problem is in NC, let alone that it can be solved using a linear number of processors, is perhaps our most surprising result.

2. $Insert(x)$, $ExtractMin(x)$: $\log^2 n$, $n^3/\log n$. An $ExtractMin(x)$ operation returns and simultaneously removes the smallest element larger than or equal to x.

3. $Insert(x)$, $ExtractMin(x)$, in which the $ExtractMin(x)$'s have non-decreasing arguments: $\log n$, n^3. The technique we use here is very different from our solution to problem 2. We give an application of this problem to maximum matching in a convex bipartite graph.

4. $Insert(x)$, $ExtractMin(x)$: $\log n$, n. This $T(n) * P(n)$ product is optimal.

[1]Research supported by the Office of Naval Research under Grants N00014-84-K-0502 and N00014-86-K-0689, and the National Science Foundation under Grant DCR-8451393, with matching funds from AT&T.

[2]Research supported by the National Science Foundation under Grant DCR-8506361.

5. $Insert(x)$, $Delete(y)$, and "tree-search" queries: $\log n$, n. The queries include any which do not modify the set and could be performed sequentially in logarithmic time if the set were stored in a binary tree which has $O(1)$ labels associated with each node v (which could be computed by applying associative operations to the decendents of v).

6. $Insert(x, A)$, $Delete(x, A)$, $Min(A)$, $Union(A, B)$, $Find(x, A)$: $\log n$, n. Here A and B are set names. When we allow operations such as these, that manipulate sets as well as their elements, then the methods of result 5 no longer hold. In this case we show that the problem can still be solved optimally, however.

We briefly outline the conventions we use in this paper. The computational model we use is the CREW-PRAM model, unless otherwise specified. Recall that this is the shared-memory model of where the processors operate synchronously and can concurrently read any memory cell, but concurrent writes are not allowed. (Some of our results are for the weaker EREW PRAM, when no concurrent memory accesses are allowed.) Another convention we use is that sets are actually multisets (i.e. multiple copies of an element are allowed), so that whenever we say "element x" we are actually referring to a particular copy of x. It is straightforward to modify our results for the case when having multiple copies of an element would be erroneous.

The details for each result are given in what follows, one per section. We conclude with some final remarks in Section 8.

2   Insert(x),Delete(x),ExtractMin

If A is a set and S a sequence of set manipulation operations, then AS denotes applying the sequence S to a set whose initial value is A. In addition to the responses to the operations in S, an evaluation of AS also returns the set "left over" after S is evaluated. In this notation, we are trying to solve $\emptyset S$ where the operations in S come from the set $\{I(x), D(x), E\}$.

Let S_1 (resp., S_2) be the sequence consisting of the first (resp., last) $n/2$ operations in S. Recursively solve $\emptyset S_1$ and $\emptyset S_2$ in parallel. The recursive call for $\emptyset S_1$ returns (i) the correct responses for the operations in it (i.e. the same as in $\emptyset S$), and (ii) the set just after $\emptyset S_1$ terminates (let L_1 denote this set). The recursive call for $\emptyset S_2$ returns responses and a final set that may differ from the correct ones, because it applies S_2 to $\emptyset$ rather than to L_1. The main problem that we now face is how to incorporate the effect of L_1 into the solution returned by the recursive call for $\emptyset S_2$, obtaining the solution to $L_1 S_2$.

Notation: If R is a subsequence of S, then $S - R$ denotes the sequence obtained by removing every element in R from S.

Convention: Throughout, our algorithms adopt the following conventions. A $D(x)$ executed when there are many copies of x in the set removes the copy that was inserted latest. Similarly, an E executed when there are many copies of the smallest element in the set removes the copy that was inserted latest. These conventions cause no loss of generality because they do not change any response. However, they do simplify our correctness proofs.

Let us first make some observations about $\emptyset S_2$. Let L_2 be the set resulting from $\emptyset S_2$ (i.e. the set after $\emptyset S_2$ terminates). Consider an $I(x)$ for which x is not removed by any E in $\emptyset S_2$, i.e. it either ends up in L_2 or gets removed by a $D(x)$ (in the latter case we say that the $D(x)$ *corresponds* to $I(x)$). Let S' be the sequence obtained from S_2 by removing every such $I(x)$ and its corresponding $D(x)$ (if any). In other words the only $I(x)$ operations in S' are those whose x was removed by an E in $\emptyset S_2$, and the only $D(x)$ operations in S' are those whose response in $\emptyset S_2$ was "x not in set". Obviously the response to any operation in S' is the same in $\emptyset S'$ as in $\emptyset S_2$. However, the following also holds.

Lemma 2.1 The responses to the operations in S' are the same in $L_1 S'$ as in $L_1 S_2$. The set resulting from $L_1 S_2$ equals L_2 plus the set resulting from $L_1 S'$.

Proof. The proof is based on a careful case analysis of the types of operations that can appear in S, which, for space reasons is included only in the full paper. ■

Lemma 2.1 has reduced the problem of solving $L_1 S_2$ to that of solving $L_1 S'$, so we now focus on obtaining the responses and final set for $L_1 S'$. The next lemma will further reduce the problem to one in which a crucial *suffix property* holds, as is later established in Lemma 2.3.

Lemma 2.2 Let $\hat{S}$ be obtained from S' by moving every $I(x)$ to just before the E whose response it was in $\emptyset S_2$ (such an E must exist by definition of S'). Then the responses to the operations in S' are the same in $L_1 \hat{S}$ as in $L_1 S_2$. The set resulting from $L_1 S_2$ equals L_2 plus the set resulting from $L_1 \hat{S}$.

. . .

Proof. Included only in full paper. ∎

Since we already know the responses to $\emptyset S_2$ (they were returned by one of the two parallel recursive call), a simple parallel prefix computation easily identifies the set S' (and hence $S_2 - S'$ and $\hat{S}$), in $O(\log n)$ time and with $O(n/\log n)$ processors. (Recall that the parallel prefix technique is to reduce a problem to that of computing all prefix partial sums of a sequence of integers, for some associated summation operator.) The responses in $L_1 S_2$ to the operations in $S_2 - S'$ are now trivially known: the response to an $I(x)$ is "x inserted" by the definition of $I(x)$, and the response to a $D(x)$ is "x deleted" by the definition of S'. The main problem we face is obtaining the responses in $L_1 S_2$ to the operations in S', and the final set resulting from $L_1 S_2$. Lemma 2.2 has reduced this problem to that of solving $L_1 \hat{S}$, so we now focus on obtaining the responses and final set for $L_1 \hat{S}$. The rest of this section shows that they can be obtained in $O(\log n)$ time and with $O(n)$ processors, thus implying for the overall problem an $O(\log^2 n)$ time and $O(n)$ processor bounds.

Let $\hat{S} = O_1, O_2, ..., O_m$, $m \leq n/2$. For every j, $1 \leq j \leq m$, let $S(j)$ be the sequence of operations obtained from $O_1, ..., O_j$ by removing the E's from it. Note that $S(j)$ contains only two kinds of operations: (i) $I(x)$ for which x was a response to an E in $\emptyset S_2$, and (ii) $D(x)$ whose response was "x not in set" in $\emptyset S_2$. Let $L(j)$ denote the set resulting from $L_1 S(j)$. Let $L(0)$ denote L_1. Recall that, by convention, element x is better than element y iff either (i) $x < y$, or (ii) $x = y$ and x was inserted later than y.

Lemma 2.3 For every j such that O_j is a $D(x)$ or an E, $1 \leq j \leq m$, there is an integer $f(j)$, $1 \leq f(j) \leq |L(j)|$, such that the set resulting from $L_1 O_1, ..., O_j$ consists of the $f(j)$ worst (i.e. largest) elements in $L(j)$.

Proof. It suffices to prove that the $D(x)$'s and E's (in $L_1 O_1, ..., O_j$) remove the k best elements in $L(j)$, for some integer k (this would establish the lemma, with $f(j) = |L(j)| - k$). The proof is by contradiction: suppose to the contrary that some O_i, $i < j$, removes an element x of $L(j)$ and that some element y of $L(j)$, where y is better than x, is not removed by any operation (in $L_1 O_1, ..., O_j$). We distinguish two cases.

Case 1. O_i is an E (call it E_1). Since y is better than x, y could not have been present when E_1 removed x, and therefore y was inserted by an $I(y)$ which comes after E_1 and before O_j. Such an $I(y)$ is (by definition) in $\hat{S}$, and therefore (by

the definition of $\hat{S}$) it is immediately followed in $\hat{S}$ by an E (call it E_2) which is after E_1 and not after O_j (possibly $E_2 = O_j$, since $I(y) = O_{j-1}$ is possible). By hypothesis, y is not removed in $L_1 O_1, ..., O_j$ and hence E_2 must have removed a z which is better than y. Since z is better than x, z could not have been present when E_1 removed x and therefore z was inserted by an $I(z)$ which comes in between E_1 and E_2. Such an $I(z)$ is (by definition) in $\hat{S}$, and therefore (by the definition of $\hat{S}$) $I(z)$ is immediately followed in $\hat{S}$ by an E (call it E_3) which is in between E_1 and E_2 ($E_3 \neq E_2$ because it is $I(y)$ and not $I(z)$ which occurs just before E_2 in $\hat{S}$). Now, repeat the argument with E_3 playing the role of E_2. Eventually, after (say) q iterations of this argument, a contradiction is reached (when there is no E in between E_1 and E_q). Thus O_i cannot be an E.

Case 2. O_i is a $D(x)$. Using an iterative argument similar to that in Case 1 we can arrive at a contradiction for this case as well. ∎

Lemma 2.4 All the $f(j)$'s can be computed in $O(\log n)$ time with $O(n)$ processors. The responses and final set for $L_1 \hat{S}$ can be computed within the same time and processor bounds.

Proof. Using the array-of-trees technique, described below in Section 6, we can compute an implicit representation of each of $L(0), L(1), ..., L(m)$ as being stored in a binary tree. Let $L(i, k)$, $1 \leq i \leq m$, $1 \leq k \leq |L(i)|$, denote the set consisiting of the worst k elements of $L(i)$. Note that $L(0, |L_1|) = L_1$. If $L(i, k) O_{i+1}$ results in $L(i+1, p)$, then we say that $L(i+1, p)$ is the successor of $L(i, k)$. Note that if $L(i+1, p)$ is the successor of $L(i, k)$, then $|p - k| \leq 1$. An $L(i, k)$ has at most one successor. It is straightforward to compute what the successor function is for each $L(i, k)$ based on O_{i+1}, the value of k, and the number of elements in $L(i)$ greater than the argument of O_{i+1} (which can be computed in $O(\log n)$ time using the array-of-trees representation). The successor function defines a forest whose $O(n^2)$ nodes are the $L(i, k)$'s and such that the parent of $L(i, k)$ is its successor. In the full version of this paper we show how to use the cascading flow technique, outlined below in Section 5.1, to construct an implicit representation of this graph in $O(\log n)$ time using $O(n)$ processors. From this representation we can then compute the path of successors from $L(0, |L_1|)$ in these same bounds. We complete the algorithm by using the k-index of each $L(i, k)$ descendent of $L(0, |L_1|)$ to look up the k-th largest element in the data structure for $L(i)$, thus giving us the response for operation i. ∎

The following theorem summarizes the main result of this section.

Theorem 2.5 *Given any sequence of n $I(x)$, $D(x)$ and E operations, a CREW-PRAM can determine the responses and resulting set of $\emptyset S$ in $O(\log^2 n)$ time with $O(n)$ processors.*

3 Insert(x), ExtractMin(x)

Let the operation $ExtractMin(x)$ ($E(x)$ for short) return and simultaneously remove from the set the smallest element $\geq x$ (if there are many copies of it then, by convention, the one inserted latest gets removed). This section concerns itself with the case where the operations appearing in S are $I(x)$ and $E(x)$. For expository reasons, we prove membership in NC by giving a rather inefficient algorithm that runs in $O(\log^2 n)$ time with $O(n^5/\log n)$ processors, and in the full version of this paper show how to reduce the number of processors to $O(n^3/\log n)$. The next lemma reduces the problem to that of determining which $E(x)$'s have an empty response.

Lemma 3.1 Let S be a sequence of n $I(x)$ and $E(x)$ operations. Let O be any one of the $E(x)$ operations in S, and let $TEST(S, O)$ be any algorithm which solves the problem of determining whether O has an empty response in $\emptyset S$. Let $T(n)$ and $P(n)$ be the time and processor complexities of $TEST(S, O)$. Then determining the actual responses to all the $E(x)$ operations in $\emptyset S$ can be done in time $O(T(n) + \log n)$ with $O(n^2 P(n))$ processors.

Proof. To every operation O which is an $E(x)$, assign $P(n)$ processors which perform $TEST(S, O)$ to determine whether it has a nonempty response in $\emptyset S$. If $TEST(S, O)$ determines that the response to O in $\emptyset S$ is empty, then that is the correct response for O. However, if $TEST(S, O)$ determines that O has a nonempty response in $\emptyset S$, then O gets assigned $nP(n)$ processors whose task it will be to determine the actual response of O. We now show how these $nP(n)$ processors can find the (nonempty) response of such an O in time $O(T(n) + \log n)$. We need only consider the prefix of S that ends with O, i.e. if $S = O_1, O_2, ..., O_n$ and $O = O_j = E(x)$ then we need only look at $\emptyset S_j$ where S_j is $O_1, O_2, ..., O_j$. Let $(x_1, x_2, ..., x_q)$ be the elements inserted in S_j that are $\geq x$, sorted from worst to

best (and hence $x_1 \geq x_2 \geq ... \geq x_q \geq x$). In other words, if there are, in S_j, q insertions of elements $\geq x$, then the sequence $(x_1, x_2, ..., x_q)$ is the sorted version of

$$\{y : I(y) \in S_j \text{ and } y \geq x\}$$

One of these x_i's is the correct response to O. To determine which one it is, we create q sub-problems where the the kth sub-problem is that of determining whether O has a nonempty response in

$$\emptyset O_1, ..., O_{j-1}, E(x_1), ..., E(x_k), O_j, \qquad (3.1)$$

i.e. the kth sub-problem is obtained by putting just before O_j in $\emptyset S_j$ the sequence $E(x_1), E(x_2), ..., E(x_k)$. Each such kth sub-problem is solved in $T(n)$ time with $P(n)$ processors using the $TEST$ procedure (there are enough processors for this because O_j has $nP(n)$ processors assigned to it). We claim that the response of O_j in $\emptyset S_j$ is then x_s, where s is the maximum k such that the response of O_j in the kth sub-problem is not empty. We now show that x_s is indeed the response of O_j in $\emptyset S_j$. Let r_k be the response to O_j in the kth sub-problem (possibly r_k is an empty response, i.e. $r_k = $"set empty"). Observe that the sequence $r_1, r_2, ..., r_q$ is initially monotonically decreasing, then at some threshold index, consists of "set empty" responses (this monotonicity follows from the way the q sub-problems are defined). Let x_t be the response to O_j in $\emptyset S_j$. Then surely the response to O_j is still x_t in every kth subproblem for which $k < t$ (because the k $E(y)$ operations just before O_j in that sub-problem remove elements about which O_j "doesn't care" because they are worse than its own response x_t). On the other hand, if $k \geq t$, then surely the response to O_j in the kth sub-problem is empty, because otherwise that response is better than x_t, a contradiction (the response to O_j in any kth sub-problem cannot be better than its response in $\emptyset S_j$). Therefore $t = s$, completing the proof (the additive $\log n$ term in the time complexity comes from the max operation needed for computing s). ∎

Next, we focus on describing a procedure $TEST(S, O)$ that has a $T(n) = O(\log^2 n)$ and a $P(n) = O(n^3/\log n)$. This will imply a total number of processors which is $O(n^5/\log n)$. In the full version of this paper we bring down the processor complexity to $O(n^3/\log n)$ by exploiting similarities between the n^2 copies of the $TEST$ing problem that are created in the proof of Lemma 3.1.

WLOG, we may describe $TEST(S, O)$ assuming that O is the last operation in S, i.e. $S = O_1, O_2, ..., O_n$ where $O = O_n$.

We begin with the observation that solving $TEST(S, O)$ amounts to determining the cardinality of a maximum *up-left matching* problem [10]. Create n distinct points in the plane, as follows: for every operation O_i in S, create a corresponding planar point whose x-coordinate is i and whose y-coordinate is the parameter of O_i (i.e. z if $O_i = I(z)$ or $O_i = E(z)$). The points corresponding to $E(z)$'s are called *pluses*, those corresponding to $I(z)$'s are called *minuses*. The responses to the $E(z)$'s in $\emptyset S$ can be viewed as being the result of the following matching procedure: scan the pluses in left to right order (i.e. by increasing x coordinates), matching the currently scanned plus with the lowest unmatched minus that is to the left of it and not below it. The correspondence between the matching so produced and the responses in $\emptyset S$ should be obvious. Furthermore, one can show [10] that this greedy left-to-right matching procedure produces a matching of maximum cardinality among all possible up-left matchings. These remarks imply that in order to determine whether O has a response in $\emptyset S$, it suffices to compare the cardinality c of a maximum matching for the configuration of pluses and minuses corresponding to S, with the cardinality c' of a maximum matching for the configuration of pluses and minuses corresponding to $S - O = O_1, O_2, ..., O_{n-1}$. If $c = c'$ then the presence of O does not make a difference and hence its response in $\emptyset S$ is empty, while $c = c' + 1$ implies that it has a nonempty response.

If $p = (a, b)$ is a plus, then $Region(p)$ is the region $(-\infty, a] \times [b, +\infty)$, i.e. the closure of the region of the plane that is to the left of p and above it. If P is a set of pluses, then $Region(P) = \cup_{p \in P} Region(p)$.

The *deficiency* of any region of the plane is the number of pluses in it minus the number of minuses in it. The deficiency of a set of pluses P is that of $Region(P)$ and is denoted by $def(P)$.

Lemma 3.2 [10]. Let Π denote the set of pluses. The cardinality of a maximum up-left matching is then equal to

$$|\Pi| - \max\{def(P) : P \subseteq \Pi\}.$$

Proof. A straightforward application of Hall's Theorem (see [10] for details). ∎

The above lemma implies that one can compute $TEST(S, O)$ in $O(\log^2 n)$ time using

$O(n^3/\log n)$ processors provided we can compute the quantity $\max\{def(P) : P \subseteq \Pi\}$ within those same bounds. This is what we show how to do next.

Let $G(S)$ be the weighted directed acyclic graph whose vertex set is the set of pluses and two new special vertices s and t, and whose edge set is defined as follows. For every two vertices p and q, there is an arc from p to q if and only if one of the following conditions (i)-(iii) holds:

(i) $p = s$ and $q \neq t$.

(ii) $p \neq s$ and $q = t$.

(iii) p is a point (a, b) and q a point (c, d) such that $a \leq c$ and $b \leq d$ (i.e. q is to the right and above p).

In case (i) the cost of the arc (s, q) is equal to $def(q)$. In case (ii) the cost of the arc (p, t) is zero. In case (iii) the cost of the arc (p, q) is the deficiency of the region $[a, c] \times [d, +\infty)$. It is not hard to see that the cost of a longest s-to-t path in $G(S)$ is precisely equal to the quantity $\max\{def(P) : P \subseteq \Pi\}$. Since $G(S)$ is acyclic, computing its longest s-to-t path is trivial to do in $O(\log^2 n)$ time with $O(n^3/\log n)$ processors.

The above $O(\log^2 n)$ time, $O(n^3/\log n)$ processor algorithm for $TEST(S, O)$ immediately implies (by Lemma 3.1) an $O(\log^2 n)$ time, $O(n^5/\log n)$ processor algorithm for evaluating sequence S. In the full version of this paper we reduce the processor complexity to $O(n^3/\log n)$.

4 A Special Case of $I(x)$ and $E(x)$

In this section we study the case when the $E(x)$ operations in S are such that the sequence of x's is in non decreasing order. As a consequence of this result, we can obtain an $O(\log n)$ time algorithm for finding a maximum matching in a convex bipartite graph, an improvement by a factor of $\log n$ on the Dekel-Sahni parallel algorithm for this problem [6].

Let m denote the number of $E(x)$ operations in S, and let $E(y_i)$ denote the i-th such operation. Note that, by hypothesis, we have $y_1 \leq y_2 \leq ... \leq y_m$. Let A_i denote the set of elements inserted by the $I(x)$ operations between $E(y_{i-1})$ and $E(y_i)$, so that the sequence S can be written $S = A_1 E(y_1) A_2 E(y_2)...A_m E(y_m)$ (some of the A_i's may be empty). Without loss of generality, we assume that no A_i contains an element less than y_i (such an element would be useless anyway).

The longest-paths characterization of Section 3 apparently does not result in an $O(\log n)$ time algorithm: that $y_1 \leq y_2 \leq \dots \leq y_m$ implies that the pluses form an increasing chain, but this in itself does not give an $O(\log n)$ time algorithm for the resulting longest-path problem. Our solution actually avoids the characterization of Section 3. Instead, we replace the problem with a polynomial number of sub-problems each of which is such that the first $E(x)$ occurs after the last $I(x)$. The next lemma observes that this type of problem is solvable in $O(\log n)$ time.

Lemma 4.1 If $S = AE(y_1)E(y_2)E(y_m)$, then all the responses can be computed in $O(\log n)$ time and with $O(n)$ processors.

Proof. Let $L(0) = A$ and, for $1 \leq i \leq m$, let $L(i)$ be $\{x \in A : x \geq y_i \}$. Let $L(i, k)$, $1 \leq i \leq m$, $0 \leq k \leq |L(i)|$, denote the set consisiting of the largest k elements of $L(i)$. Note that $L(0, |A|) = A$. Let the *successor* of $L(i, k)$ be the set obtained by removing all the elements that are less than y_i from the set resulting from $L(i, k)E(y_i)$. It is easy to see that the successor of $L(i, k)$ is equal to $L(i+1, p)$ for some integer $p < k$. An $L(i, k)$ with $i < m$ has exactly one successor and hence the successor function defines a tree whose $O(n^2)$ nodes are the $L(i, k)$'s and such that the parent of $L(i, k)$ is its successor. The root of this tree is $L(m, q)$ for some integer q. The successor function can easily be computed in $O(\log n)$ time with $O(n^2)$ processors. In the tree defined by the successor function, consider the path originating from the leaf $L(0, |A|)$ and terminating at the root $L(m, q)$. Tracing this path is trivial to do in time $O(\log n)$ with $O(n^2)$ processors. This path constitutes a complete description of the responses to the $E(y_i)$'s, as follows. If $L(i, k)$ is on this path and $k > 0$ then the response to $E(y_i)$ is the smallest element in $L(i, k)$. If $L(i, k)$ is on this path and $k = 0$ then $E(y_i)$ has a "set empty" response. The way the above algorithm is implemented with a linear rather than quadratic number of processors is very similar to the proof of Lemma 2.4 and hence is omitted. ∎

We now show how to solve the problem for $S = A_1 E(y_1)A_2E(y_2)A_mE(y_m)$ by solving a polynomial number of problems each of which is of the type considered in Lemma 4.1.

Definitions. Let $A_{ij} = A_i \cup A_{i+1} \cup \dots \cup A_j$. Let r_{ij} be the response to $E(y_j)$ in $A_{ij}E(y_i)E(y_{i+1})\dots E(y_j)$. Let z_{ij} be the response to $E(y_j)$ in $A_iE(y_i)A_{i+1}E(y_{i+1})\dots A_jE(y_j)$.

Note that in this notation the response to $E(y_j)$ in S is z_{1j}. Also note that the r_{ij}'s

can be computed in $O(\log n)$ time because of Lemma 4.1. The following theorem establishes a crucial link between the r_{ij}'s and the z_{1j}'s and implies that the z_{1j}'s can also be computed in $O(\log n)$ time.

Theorem 4.2 *For every* j, $1 \leq j \leq m$, $z_{1j} = \min_{1 \leq i \leq j} r_{ij}$.

Proof. The proof is based on a number of monotonicity properties of the sequence S, but is quite involved. For space reasons we include it only in the full version of this paper. ∎

We are now ready to state the main result of this section.

Theorem 4.3 *The responses to the* $E(y_i)$'s *in a sequence* $S = A_1E(y_1)A_2E(y_2)A_mE(y_m)$ *where* $y_1 \leq y_2 \leq \dots \leq y_m$, *can all be computed in* $O(\log n)$ *time with* $O(n^3)$ *processors.*

Proof. Assign n processors to every pair i and j, $i \leq j$, and use them to compute r_{ij} in $O(\log n)$ time. Then assign $n/\log n$ processors to every $E(y_i)$ and use them to compute $z_{1j} = \min_{1 \leq i \leq j} r_{ij}$. The overall time complexity is clearly $O(\log n)$ using $O(n^3)$ processors. ∎

4.1 Application: Maximum Matching in a Convex Bipartite Graph

Many problems can be formulated as an off-line sequence of set manipulation operations. One such problem is that of computing a maximum matching in a convex bipartite graph. An $O(\log^2 n)$ time algorithm for solving this problem on an EREW-PRAM model was given by Dekel and Sahni [6]. In this section, by formulating the problem as a sequence of $I(x)$ and $E(y)$ operations where the y's are nondecreasing, we make use of Theorem 4.3 to obtain an $O(\log n)$ time EREW-PRAM algorithm, an improvement by a factor of $\log n$. We now sketch how to formulate the problem in this manner. First recall that a convex bipartite graph is such that its vertex set can be written as $A \cup B$ where $A = \{a_1, ..., a_p\}$ and $B = \{b_1, ..., b_q\}$, where (i) every edge has one endpoint in A and the other endpoint in B, and (ii) if (a_i, b_j) and (a_i, b_{j+k}) are edges then so is (a_i, b_{j+s}) for every $1 \leq s < k$. Let l_i (r_i) be the smallest (largest) j such that (a_i, b_j) is an edge. Glover's algorithm [8] for finding a maximum matching in such a graph works as follows: Consider the vertices of B one by one, starting at b_1. When b_j is considered, match it against a remaining a_k that is adjacent to it and whose r_k is smallest, and then delete a_k from the graph. It is Glover's algorithm that we formulate as a sequence of $I(x)$ and $E(x)$ operations, as follows.

WLOG, we assume that the a_i's are re-named so that $r_1 \leq \ldots \leq r_p$. Let L_j (R_j) denote the set which contains every a_i whose l_i (r_i) equals b_j. Then Glover's algorithm is equivalent to the problem of evaluating the sequence S created by considering the vertices of B one by one, starting at b_1 with $S = \emptyset$ and $\beta = -\infty$. When b_j is considered, we append to the end of S an $I(a_i)$ for every $a_i \in L_j$, followed by an $E(\beta)$. Then (before moving to b_{j+1}) we set β equal to the max of its old value and the largest element in R_j. If, in S, the response to the j-th $E(y)$ is (say) a_i, then the edge (a_i, b_j) is in the maximum matching. It is easy to see that this procedure results in exactly the same matching as Glover's algorithm. Since the resulting sequence S satisfies the conditions of Theorem 4.3, we have the following:

Theorem 4.4 *The maximum matching problem for convex bipartite graphs can be solved in $O(\log n)$ time with $O(n^3)$ processors on a CREW-PRAM.*

5 Insert(x) and ExtractMin

The main idea of our algorithm for this off-line priority queue problem is to combine some observations from [5] with a new cascading merge technique. In the next subsection we describe this technique, and in the subsequent subsection we give our algorithm for the off-line priority queue problem.

5.1 Cascading in a Flow Graph

In this subsection we describe a generalization of the cascading divide-and-conquer technique [2], which we call the *cascading flow* technique. The generic problem is defined as follows. We are given a directed graph G which has constant degree, a sorted list of elements $A_0(v)$ for each node v in (some of which may be empty), and definitions for sets $A_{t+1}(v)$ for each node v (at time $t + 1$) which have the following form:

$$A_{t+1}(v) = \bigcup_{w \in In(v)} F_{w,v}(A_t(w)), \quad (5.1)$$

where $F_{w,v}(A_t(w))$ denotes a set which can be constructed from $A_t(w)$ in $O(1)$ time using a number of processors proportional to its size (i.e.,

$O(|F_{w,v}(A_t(w))|)$ processors). We use $In(v)$ to denote $\{v : (w, v) \in E\}$, and use $Out(v)$ to denote $\{v : (v, w) \in E\}$. All sets are stored as sorted lists.

Definitions. Given an element a and sorted list B we use $rank(a)$ to denote the *rank* of a in B, i.e., the number of elements $\leq a$. If for every element in a sorted lists A we know its rank in B we say A is *ranked* in B. If for any two consecutive elements in A their ranks in B differ by at most a constant c, we say A is a *c-cover* for B.

The cascading flow problem is the following: given a specific time $t^* > 0$ and p processors, construct $A_{t^*}(v)$ for each v in G in $O(t^*)$ time using p processors. The following lemma specifies gives two simple conditions which imply a solution to this problem.

Lemma 5.1 (The Cascading-Flow Lemma) If the following conditions hold, then one can construct $A_{t^*}(v)$ for each $\overline{v}$ in G in $O(t^*)$ time using p processors.

1. For any $t > 0$, $F_{w,v}(A_{t-1}(w))$ is a c-cover for $F_{w,v}(A_t(v))$ and is ranked in $F_{w,v}(A_t(v))$ for each edge (v, w) in G.

2. One can give an initial processor distribution and a distribution function for the p processors, so that if $\phi_t(v)$ denotes the number of processors assigned to v in G at time t and $f_t(v, w)$ denotes the number of processors transfered from v to w at the end of time t, then

$$\sum_{v \in G} \phi_0(v) = p, \quad (5.2)$$

$$\phi_{t+1}(v) = \phi_t(v) + \sum_{w \in In(v)} f_t(w, v)$$
$$- \sum_{w \in Out(v)} f_t(v, w), \quad (5.3)$$

$$|A_{t+1}(v)| \leq d\phi_t(v), \quad (5.4)$$

where c and d are constants. In addition, these conditions imply that $A_t(v)$ will always be a c-cover of $A_{t+1}(v)$ and that the computation of $A_{t+1}(v)$ will automatically give us a ranking of $A_t(v)$ in $A_{t+1}(v)$.

Proof: Included only in full paper. ∎

Intuitively, the Cascading Flow lemma allows us to completely specify a global computation in terms of local computations being performed at the nodes of G. Typically, the function $F_{w,v}(A_t(w))$ will simply be a b-sample of

$A_t(w)$, that is, the list consisting of every b-th element of $A_t(w)$. Cole [4] has shown that if all the F's in each $A_{t+1}(v)$ are simply b-samples of the $A_t(w)$'s, then the c-cover property (Condition 1) holds automatically. In Cole's scheme (for the sorting problem) the graph is a complete binary tree, all elements are initially only stored at the leaves. He has $b = 4$ for every node v at every stage, so long as $A_t(v)$ doesn't contain all the elements stored in its leaf descendents. When $A_t(v)$ is "full," he sets $b = 4$; then in the next stage has $b = 2$ for this node; then sets $b = 1$ in the stage after that (at which time the parent of v is full). Thus, one can use his algorithm to sort n elements in $O(\log n)$ time using $O(n)$ processors.

5.2 The Off-line Priority Queue

Let $S = (O_1, O_2, ..., O_n)$ be a sequence of $Insert(x)$ and $ExtractMin$ operations. Construct a complete binary tree T "on top" of S so that each leaf of T is associated with a single operation O_i (listed from left to right). For each node v let $e(v)$ denote the number of $ExtractMin$ operations stored in the descendent leaves of v. It is trivial to compute $e(v)$ for each v in T in $O(\log n)$ time using $O(n)$ processors by a simple bottom-up summation computation. For every leaf of T corresponding to an E operation we replace that leaf with an node v with two leaf-node children such that its left child corresponds to an $I(\infty)$ operation and its right child corresponds to an E. This allows us to assume that each E has a response (which could be ∞).

For each v in T let $S(v)$ denote the substring of S which corresponds to the descendents of v. For each v in T we will compute two sets $A(v)$ and $L(v)$: $A(v)$ will be the sorted list of answers to all the E's in $\emptyset S(v)$, and $L(v)$ will be the sorted list of elements left after we perform all the E's in $\emptyset S(v)$. We let $Prefix_m(B)$ denote the list consisting of the first m elements in B. Similarly, we let $Suffix_m(B)$ denote the list consisting of the last $|B| - m$ elements in B. The following lemma establishes the foundation of our method.

Lemma 5.2 Let v be a node in T with left child x and right child y. Then we have the following relationships:

$$A(v) = A(x) \cup Prefix_{e(y)}(L(x) \cup A(y))$$
$$L(v) = Suffix_{e(y)}(L(x) \cup A(y)) \cup L(y)$$

Proof: Omitted. ∎

Using this lemma and the cascading flow technique we construct $A(v)$ for every v in T. After performing this computation, $A(root)$ will store the set of answers to $\emptyset S$. This does not tell us the response to each specific $ExtractMin$ in S, however. For that we will have to perform one more cascading operation, this time going down T. Let $L'(v)$ (resp., $L''(v)$) denote the set of elements that is left over after performing the operations in S (starting with $\emptyset$) up to, but not including, (resp., including) the operations in $S(v)$. In addition, let $A'(v)$ denote the set of elements that are actually the answers to the E's in $S(v)$. In order to find the true answers to all the E's in S we must compute $A'(v)$ for each node v' whose right child is a leaf $ExtractMin$ node. The following lemma gives us the main idea for performing the downward sweep.

Lemma 5.3 Let v be a node in T with left child x and right child y. Suppose we have $L'(v)$ and $L''(v)$ at v, $A(x)$ at x, and $A(y)$ at y. Then

$$L'(x) = L'(v)$$
$$L''(x) = Suffix_{e(x)}(L'(x) \cup A(x)),$$
$$L'(y) = L''(x),$$
$$L''(y) = L''(v).$$

Proof: Omitted. ∎

We can use these definitions to define a top-down computation to construct all the possible "left-over" sets. The response of an E operation at leaf-node v is simply the first element in the left-over set for the node just left of v in T. This approach is not enough to give us an efficient algorithm, however. As it is expressed now it would be impossible to construct the necessary left-over sets in $O(\log n)$ time using $O(n)$ processors. This is because for each level of the tree we are essentially doubling the amount of space we need to represent all the left-over sets. We can get around this problem, however, by noting that for any node v there can be at most $n_v/2$ ExtractMin operations stored in the descendent leaves of v, where n_v denotes the total number of descendent leaves of v. Thus, for any node v we only need to consider the first $n_v/2$ elements of any left-over set we are computing with. Given these observations and the equations of the above lemmas it is then straightforward to set up the computation of the L' and L'' sets as a cascading flow problem. The operations we define for going from one node v to its children will include only b-samples and prefix computations, thus, the resulting algorithm runs in $O(\log n)$ time and $O(n \log n)$ space using $O(n)$ processors.

6 The Array of Trees

In this section we study the problem of evaluating sequences of $I(x)$'s, $D(x)$'s, and "tree-search" queries. By the name "tree-search" query we mean any query that could be performed in $O(\log n)$ time if the elements in our set were stored in a height-balanced binary tree where each node v of this tree could store up to $O(1)$ labels, each label possibly being the value of some associative operation computed for all the elements stored in descendents of v. Examples of such tree-search queries include selecting the minimum element, finding the k-th smallest element, computing the number of elements in a certain range, etc.

Suppose we are given a sequence S of n operations of the from described above. WLOG, assume that every element x which is the arguemnt of an $Inset(x)$ is unique (if not, then we can think of the element as a pair (x, i) where i is the position of the insert operation in S). Let Y be the sorted list of all elemetns x which are the argument of an $Insert(x)$ operation. Clearly, Y can be constructed in $O(\log n)$ time using $O(n)$ processors in the EREW PRAM model [4]. We define the *array of trees* data structure as follows. It is a directed acyclic graph G with n "access nodes" $v_1, v_2, ..., v_n$. The subgraph of G which is accessible from any particular access node v_i forms a binary tree containing exactly those elements of Y which would be present in the set resulting from performing the first i operations in S.

For each element x in Y let $i(x)$ denote the position in S where $Insert(x)$ occurs, and let $d(x)$ denoted the position in S where $Delete(x)$ occurs. If there is no $Delete(x)$, then $d(x)$ is 0. We can define $G = (V, E)$ in a recursive fashion as follows. The basis case is when Y Y contains only one item. In this case $V = \{v, i(x), d(x)\}$ (or $\{v, i(x)\}$ if $d(x) = 0$) and $E = \{(i(x), v), (d(x), \text{nil})\}$. The list of access nodes is simply $A = (i(x), d(x))$ (or just $(i(x))$ if $d(x) = 0$). For the general case, let Y_1 (resp., Y_2) denote the list consisting of the first (resp., last) $n/2$ elements of Y. We recursively construct sorted lists A_1 and A_2, where A_1 (resp., A_2) is the list of indices of $i(x)$'s and $d(x)$'s for the x's in Y_1 (resp., Y_2). Associated with each index position i in A_1 (resp., A_2) are a left pointer and a right pointer (edges) which point to the children of the node for position i. We merge A_1 and A_2 by index numbers and for each index i we build left and right pointers (edges) to its predecessor in A_1 and A_2 respectively. Since this set A will be the A_1 or A_2 for the next level up in

the recursion, this computation can be done in a cascading fashion. The entire computation takes $O(\log n)$ time and $O(n \log n)$ space using $O(n)$ processors. When it completes we can perform a search query for any index position i by finding the interval in A containing i and then doing a simple search in the "tree" for that position. The computation can be implemented in $O(n)$ space in the CREW PRAM model if all the queries are the values of associative operations (see [2] for some applications).

7 Using Set Names

The methods of the previous sections only apply when the set-manipulation operations all are for the same set. In this section we study sequences of operations which can take set names as arguments in addition to specific elements. In particular, we address the problem of evaluating a sequence of operations from the set $\{Insert(x, A), Delete(x, A), Min(A), Union(A, B), Find(x, A)\}$. Membership in the class NC is not obvious for this problem, mainly because the tree defined by the union operations need not be a binary tree, and can have $O(n)$ depth and/or some $O(n)$ degree nodes.

We begin by describing the semantics associated with each of the five operations. Initially, every set named in the sequence S exists and is empty.

1. $Union(A, B)$: Union the elements of A and B into the set B, destroying A.

2. $Find(x, A)$: find the name of the set that x currently belongs to, where x was inserted using an $Insert(x, A)$ operation.

3. $Insert(x, A)$: Insert x into the set A, provided A still exists.

4. $Delete(x, A)$: Delete an element x which was inserted by some $Insert(x, A)$ operation.

5. $Min(A)$: return the value of the minimum element currently in A. Here, "minimum" can be replaced by any semi-group operation.

We outline below how to solve this evaluation problem in $O(\log^2 n)$ time using $O(n/\log n)$ processors. In the final version of this paper we show how to use cascading to implement this algorithm in $O(\log n)$ time using $O(n)$ processors.

We begin by creating the union tree U (there is an edge from A to B for each $Union(A, B)$ operation). Let $S(A)$ denote the subsequence of S made up of all the operations $Insert(x, A)$, $Delete(x, A)$, and $Find(x, A)$. With each operation we store its position index in S (call this the operation's *time of evaluation*). Using the techniques of the previous section, consider all the operations in $S(A)$ as applying only to A and construct a vector $m(A)$ which contains the value of the minimum element for each time of evaluation in $S(A)$.

The problem left is to construct for each node A of U the vector $M(A)$ which contains the value of the true minimum element in A (including all the sets which were unioned to form A) for each time of evaluation that the minimum element in A changes. Take U and divide it into two trees U_1 and U_2, each having at most $(2/3)|U|$ nodes. Recursively construct $M(A)$ for each node in U_1 and U_2. Let r_1 denote the root of U_1 and r_2 denote the root of U_2. WLOG, r_1 is also the root of U. Thus, the $M(A)$ vector for each node in U_2 is correct. In addition, all the nodes A in U_1 have a correct $M(A)$ vector unless A is on the path from r_2 to r_1 in U. To compute the correct $M(A)$ vector for each of these nodes we must must update the old $M(A)$ for each such A to reflect the fact that A contained more elements than was considered in the recursive call. This is accomplished by merging $M(r_2)$ with each $M(A)$ such that A is on the path from r_2 to r_1. For each overlapping interval in $M(r_2)$ and $M(A)$ we can then update $M(A)$ to contain the minimum value for each time unit of elements that affect A. Actually, we need only store that part of $M(A)$ that spans the time until the evaluation time of $Union(A, B)$. These observations allow us to use parallel merging [3] and parallel prefix [11,12] to update all these $M(A)$'s in $O(\log n)$ time using $O(n/\log n)$ processors. This implies a total time for the algorithm of $O(\log^2 n)$ using $O(n/\log n)$ processors.

8 Final Remarks

The problem of efficiently evaluating an off-line sequence of data structure operations has been extensively studied for sequential models of computation. However, surprisingly little work had previously been done on the parallel complexity of such problems. This paper provides a first step in the study of the parallel complexity of these problems. Here we focussed primarily on problems whose membership in NC was nonobvious, because of the behaviour of $ExtractMin$ and $ExtractMin(x)$ operations. The main open question that remains is whether the problem is in NC when S contains $I(x)$, $D(x)$ and $E(x)$ operations.

References

[1] A.V. Aho, J.E. Hopcroft and J.D. Ullman, *The Design and Analysis of Computer Algorithms*, 1974.

[2] M. Atallah, R. Cole, and M. Goodrich, "Cascading Divide-and-Conquer: A Technique for Designing Parallel Algorithms," *28th IEEE Symp. on Found. of Comp. Sci.*, 151–160 (1987).

[3] G. Bilardi and A. Nicolau, "Adaptive Bitonic Sorting: An Optimal Parallel Algorithm for Shared Memory Machines," TR 86-769, Dept. of Comp. Sci., Cornell Univ., August 1986.

[4] R. Cole, "Parallel Merge Sort," *27th IEEE Symp. on Found. of Comp. Sci.*, 1986, 511–516.

[5] E. Dekel and S. Sahni, "Binary Trees and Parallel Scheduling Algorithms", *IEEE Trans. on Computers*, 307–315, March 1983.

[6] E. Dekel and S. Sahni, "A Parallel Matching Algorithm for Convex Bipartite Graphs and Applications to Scheduling", *J. of Par. and Dist. Comp.*, 185–205, 1984.

[7] P.W. Dymond and S.A. Cook, "Hardware Complexity and Parallel Comp.," *21st IEEE Symp. on Found. of Comp. Sci.*, 1980, 360–372.

[8] F. Glover, "Maximum Matching in a Convex Bipartite Graph", *Naval Res. Logist. Quart.*, 313–316 (1967).

[9] M.T. Goodrich, "Efficient Parallel Techniques for Computational Geometry," Ph.D. thesis, Dept. of Comp. Sci., Purdue Univ., August 1987.

[10] T. Leighton and P. Shor, "Tight Bounds for Minimax Grid Matching, With Applications to the Average Case Analysis of Algorithms," *18th ACM Symp. on Theory of Comp.*, 91–103 (1986).

[11] C.P. Kruskal, L. Rudolph, and M. Snir, "The Power of Parallel Prefix," *1985 Int. Conf. on Parallel Processing*, 180–185.

[12] R.E. Ladner and M.J. Fischer, "Parallel Prefix Computation," *J. ACM*, October 1980, 831–838.

[13] W.L. Ruzzo, "On Uniform Circuit Complexity," *J. of Comp. and Sys. Sci.*, Vol. 22, No. 3, June 1981, 365–383.

Fast Parallel and Sequential Algorithms For Edge-Coloring Planar Graphs (extended abstract)

Marek Chrobak
Dept. of Mathematics
University of California
Riverside, CA 92521

Moti Yung
IBM Research Division
Almaden Research Center
San Jose, CA 95120

Abstract

Vizing [21,22,8] proved that any graph G of maximal degree $\Delta(G)$ is either Δ or $\Delta + 1$ colorable. For edge-coloring of *planar graphs*, he showed that all graphs G with $\Delta(G) \geq 8$ are Δ-colorable. Based on the proof of Vizing's theorem Gabow *et al.* [10] designed an $O(n^2)$ algorithm for Δ-coloring planar graphs when $\Delta(G) \geq 8$. Boyar and Karloff [2] proved that this problem belongs to NC if $\Delta(G) \geq 23$. Their algorithm runs in time $O(\log^3 n)$ and uses $O(n^2)$ processors on a CRCW PRAM.

We present the following algorithms for Δ-edge coloring planar graphs:

- A linear time sequential algorithm for graphs G with $\Delta(G) \geq 19$.

- An efficient parallel EREW PRAM algorithm for graphs G with $\Delta \geq 19$. This algorithm runs in time $O(\log^2 n)$ and uses $O(n)$ processors.

- An $O(n \log n)$ time sequential algorithm for graphs G with $\Delta(G) \geq 9$.

- An efficient parallel EREW PRAM algorithm for graphs with $\Delta \geq 9$. This algorithm runs in time $O(\log^3 n)$ and uses $O(n)$ processors.

1 Introduction

In this paper we consider the problem of edge-coloring planar (and constant-genus) graphs, that is, the problem of assigning colors to the edges of a graph in such a way that edges with a common endpoint have different colors. This problem has various applications in scheduling and routing (see, for example, [8]).

Let $\chi_1(G)$ denote the chromatic index of G, that is the minimum number of colors necessary to color the edges of G. Vizing [21,8] proved that $\chi_1(G)$ is either $\Delta(G)$ or $\Delta(G) + 1$ for each graph G, where $\Delta(G)$ denotes the maximum degree of a vertex in G.

The problem of deciding whether a given graph is $\Delta(G)$-colorable is NP-complete even when restricted to cubic graphs [14,18]. Therefore it is natural to look for approximate algorithms which use $\Delta(G) + 1$ colors for each G. A straightforward implementation of the proof of Vizing's theorem yields such an algorithm with complexity $(O(m+n)^2)$. Gabow et al. [10] describe more efficient algorithms for this problem.

Parallel $\Delta(G)+1$-coloring is one of the major open problems in the theory of parallel algorithms. So far, it is only known that, as was shown by Karloff and Shmoys [15], the problem is in NC for graphs G with $\Delta(G) = O(\log^k n)$, where k is fixed.

In view of the difficulty of the general edge coloring problem, special classes of graphs have been investigated. Much research has been done on bipartite graphs. A classical theorem of König and Hall states that all bipartite graphs are $\Delta(G)$-colorable. Efficient algorithms for optimal edge coloring bipartite graphs, both sequential and parallel, are given in [4,9,19].

In this paper we concentrate on another class of graphs, the class of planar (and fixed-genus) graphs. In this case, the situation is rather peculiar. Vizing [22] proved that all planar graphs G with $\Delta(G) \geq 8$ are $\Delta(G)$-colorable. He also conjectured that this can be extended to $\Delta(G) = 6, 7$ (still an open problem [8]). If $2 \leq \Delta(G) \leq 5$ then G can belong to either of the classes. If G is cubic then G is $\Delta(G)$-colorable providing that it does not have bridges (if it has bridges, it is not). This is a corollary of the 4-Color Theorem [1], since the problem was shown by Tait in 1880 (see [8]) to be equivalent to the 4-Color Theorem. Indeed, the problem is already non-trivial for this case. For $\Delta(G) = 3, 4, 5$ the complexity of the decision problem is open.

Coloring of planar graphs, especially vertex coloring, has been extensively investigated. The proof of the 4-Color Theorem [1] yields a sequential $O(n^2)$ algorithm for 4-coloring every planar graph. The parallel complexity of this problem is open. There are, however, many algorithms for coloring planar graphs with 5 colors. Sequential algorithms in [3,7,23, 20] achieve linear-time. Recent parallel algorithms solve it in time $O(\log n \log^* n)$ [11,13]. The complexity of edge-coloring planar graphs has also been considered before. Gabow et al [10] give an $O(n^2)$ algorithm which colors each planar graph G such that $\Delta(G) \geq 8$ with $\Delta(G)$ colors. This algorithm is based on a modified proof of Vizing's theorem for planar graphs. Boyar and Karloff [2] prove that this problem is in NC if $\Delta(G) \geq 23$. Their algorithm runs in time $O(\log^3 n)$ and uses $O(n^2)$ processors on a CRCW-PRAM.

In this paper we improve these results by presenting the following algorithms for Δ-edge coloring of planar graphs:

(1.1) A linear time sequential algorithm for graphs G with $\Delta(G) \geq 19$.

(1.2) An efficient parallel EREW PRAM algorithm for graphs G with $\Delta \geq 19$. This algorithm runs in time $O(\log^2 n)$ and uses $O(n)$ processors.

(2.1) An $O(n \log n)$ time sequential algorithm for graphs G with $\Delta(G) \geq 9$.

(2.2) An efficient parallel EREW PRAM algorithm for graphs with $\Delta \geq 9$. This algorithm runs in time $O(\log^3 n)$ and uses $O(n)$ processors.

In the sequential case, our results substantially improve the time complexity of the problem. In the parallel case, we extend the class of planar graphs for which the problem is known to be in NC, our algorithms miss optimality only by a poly-logarithmic factor, and we use the weakest PRAM model of EREW.

EREW is a shared memory PRAM model, in which read as well as write access to the memory cells is exclusive; no conflicts are allowed. Usually, when parallel algorithms are developed, if it is the case that they do not use simultaneous access to memory cells, they are claimed to be EREW. In this work, to achieve EREW algorithms systematically, we apply a data-structure technique called the *conflict graph* . It is used by the processors to coordinate their access to the memory. Given a set of processor activities, they are being connected into a graph which represents (by edges) possible memory conflicts. The conflict graph construction itself can be done without conflicts. Then, a Maximal Independent Set (MIS) algorithm can be combined with the data structure to choose an independent set of activities. A similar idea has been used in [13].

The design of MIS algorithms is an important case study in the development of parallel algorithms [16,17,12,5,11] as a seemingly inherent sequential process, but can also serve as an important sub-program in other algorithms. In our cases the graphs are of constant-degree and the very fast $O(\log^* n)$-time MIS techniques of Cole and Vishkin, and Goldberg Plotkin and Shannon [5,11] can be applied to find a linear fraction of the activities which are non-conflicting and therefore can access the memory simultaneously. This guarantees that all the activities can be performed in logarithmic number of MIS steps on an EREW PRAM.

The technique we use is based on refinements of Euler's and Vizing's theorems for planar graphs. The original proofs were not sufficient for our purpose. Roughly, in the proof of Vizing's theorem it is shown that each planar graph G with $\Delta(G) \geq 8$ contains an edge e with the property that if we remove e and color the remaining graph with $\Delta(G)$ colors, then the obtained coloring can be extended to e without using more colors. Such edges will be called *reducible*. We strengthen this result for $\Delta(G) \geq 9$ by showing that in this case the number of such edges is $\Theta(n)$, where n is the number of vertices of G. We also have an example that this is not true if $\Delta(G) = 8$; in this case it may happen that G will have only $O(1)$ such edges.

2   Combinatorial Results: Degree Distribution

Let $G = (V, E)$ be a planar graph. For $u \in V$ by $\deg(u)$ we denote the degree of u, and $\Delta(G) = \max_{u \in V} \deg(u)$. By n, and respectively m, we denote the number of vertices and edges of G. By n_d we denote the number of vertices of degree d. By $\deg_v^*(u)$ we denote the number of neighbors of u of degree $\Delta(G)$ different than v.

In this section we will present some results about the distribution of the vertex degrees in a planar graph (omitting the proofs). The results are derived from Euler's formula. In particular, we will use the fact that for a planar graph $m \leq 3n$, and for bipartite graph Euler's formula implies that $m \leq 2n$. The distribution is used to show that certain objects occur a linear number of times in a planar graph. In our algorithms we apply reduction techniques to this objects. A similar analysis was done for parallel vertex-coloring algorithms on planar graphs [5,13]. In our case, however, this analysis is more involved. Below we state the results, the proofs will appear in the journal version. Lemma 2 is needed for the correctness and analysis of the algorithms for large (≥ 19) maximal degree graphs (1.1, 1.2), while Lemma 3 is used in the algorithms for smaller (≥ 9) maximal degree (2.1, 2.2).

The following lemma can be derived from Euler's formula.

Lemma 1 *If G is a planar graph then $m \leq 3n - 2n_1 - n_2$.*

Let K and L be constants which will be specified later.

Lemma 2 *Let G be a planar graph, and $\phi(G) = n_1 + \max(n_2 - \sum_{d \geq L} n_d, 0) + \sum_{d=3}^{6} n_d'$, where n_d' is the number of vertices of degree d which have a neighbor of degree at most K. Then for $K = 12$ and $L = 19$ we have $\phi(G) = \Theta(n)$.*

For the algorithms (2.1) and (2.2), we define an edge $(u, v) \in E$ to be *reducible* if either $\deg(u) + \deg_u^*(v) \leq \Delta(G)$ or $\deg_v^*(u) + \deg(v) \leq \Delta(G)$.

Lemma 3 *Let $G = (V, E)$ be a planar graph and R the set of reducible edges in G. If $\Delta(G) \geq 9$ then $|R| \geq \frac{1}{36}n$.*

The above lemmas are optimal in the sense that they are not true for $\Delta(G) = 18$ and 8. (We have constructed counter-examples to the lemmas in such graphs which we will give in the journal version).

3 Algorithms for Large Maximal Degree

3.1 The sequential algorithm (1.1)

We present the sequential linear-time algorithm which colors a planar graph $G_0 = (V_0, E_0)$ with at most $\max(\Delta(G_0), 19)$ colors. For a vertex $v \in V_0$, $\deg_0(v)$ denotes the degree of v in the original graph. The graph G_0 will be changed in the course of the execution of the algorithm, so by $G = (V, E)$ we will denote the current graph, and $\deg(v)$ is the current degree of a vertex v. We will always have that $V \subseteq V_0$ and $E \subseteq E_0$, so $\deg(v) \leq \deg_0(v)$ for each $v \in V$. D is a constant, $D = \max(L, K + 5) = 19$. The graph is represented as a list of vertices, each vertex is associated with an adjacency list, the two copies of each edge are doubly connected (such a representation can be constructed efficiently from a given input, in the sequential as well as the parallel case.)

We will describe the idea of algorithm (1.1). It is divided into two stages. In the first stage we delete the edges of the graph until $E = \emptyset$, keeping the information about the deleted edges on a stack. In the second stage the edges are added consecutively to the graph in reverse order. Each added edge is colored with an appropriate color.

Three types of edges are being reduced. An edge (u, v) is of the *type* (1) if $\deg(u) + \deg(v) \leq D + 1$. It is of *type* (2) if it is not of type (1) and one of u or v has degree 1. In the first phase of the reduction stage we delete edges of types (1) and (2) as long as such edges still exist. (We use a queue of edges of types (1) and (2), we put and get edges of these types until the queue is empty.) Each edge manipulation costs constant time. Once there are no edges of type (1) and (2) the graph must contain a linear number of vertices in the so called *bunches*. A *bunch* is a set of degree-2 vertices with the same common neighbors u and v where u and v have degree at least D. It is denoted by $B(u, v)$. Edges joining u

and v with the vertices in $B(u, v)$ are of *type* (3). The second phase of the reduction step is the bunch deletion. In this phase the cost of deletion is linear in the size of the graph, and we are guaranteed to delete a constant fraction of the edges. After the bunches are deleted we perform the reductions again from start, and so on.

We explain now why these edges are called reducible. If we have an edge (u, v) of type (1), that is $\deg(u) + \deg(v) \leq D + 1$, then after removing (u, v) and coloring the rest of the graph, we have at least one free color for (u, v) (because the total number of colors used at u and v is at most $\deg(u) - 1 + \deg(v) - 1 \leq D - 1$).

Edges of type (2) are very easy to color, because if we remove such an edge (u, x), for $\deg(x) = 1$, and color the rest of the graph, then there is at least one free color at u, and all colors are free at x.

Finally, let us consider the hardest case, the edges of type (3). We do not know how to reduce a *single* edge of this type. However, if we have a bunch of size at least two, there is a simple method to do it. Consider a bunch $B(u, v)$, and assume for simplicity that $B(u, v) = \{x, y, \}$. Again, remove x and y, color the rest of the graph and now extend the coloring to the edges joining u, v with x and y as follows. Take two colors c_1 and c_2, which are free at u, and set $\mathrm{col}(u, x) := c_1$, $\mathrm{col}(u, y) := c_2$. Let c_3 and c_4 be the colors missing at v, and choose them such that $c_3 \neq c_1$ and $c_4 \neq c_2$. This is always possible. And then set $\mathrm{col}(v, x) := c_3$, $\mathrm{col}(v, y) := c_4$. For larger bunches, this coloring scheme is similar.

It is difficult to maintain the information about bunches in constant time after each reduction, so we use search. Also, we do not remove all bunches; the bunches which contain only one vertex have to stay in the graph since it may not be possible to color them in the coloring stage. (Two data structures are used in the bunch deletion phase. If u is a vertex whose bunches are currently being searched for, then `Buddies` is a list of the vertices which share with u a common neighbor of degree 2 that have been found so far, and for $v \in$ `Buddies`, `Bunch`(v) stores vertices in $B(u, v)$ which have been already discovered.)

The fact that we delete a linear number of edges is shown by applying Lemma 2 to the graph $G' = (V', E')$ where $V' = \{u \in V : \deg(u) \geq L\}$ and $E' = \{(u, v) : u, v \text{ have a common neighbor of degree 2 }\}$. Then $|E'| \leq 3|V'| \leq 3 \sum_{d \geq L} n_d$ is a bound on the number of bunches in G. Since we delete all vertices of degree 2 (all of them are in bunches), except when a bunch is a singleton, we have that $\phi(G)$ is a lower bound on the number of vertices in the bunches which are deleted in Phase 2. By Lemma 2 we have $\phi(G) = \Theta(n)$.

Once the whole graph is deleted, we add the deleted edges one by one in the reverse order. For an edge of type (1) we can find an available color in constant time. At the moment when the degree of a vertex v becomes large (greater than D), we construct the list of colors available at v, `Free_Col`(v). Initially, `Free_Col`(v) contains all colors among $\{1, 2, \ldots, \deg_0(v)\}$, which are not yet used on the edges incident to v. The cost of generating these lists in the entire algorithm is $O(n_0)$. Then finding a free color of edges of type (2) costs also a constant time. The costs of coloring a bunch can be made in time proportional to its size.

Theorem 1 *Algorithm colors the edges of a planar graph G_0 with $\max(\Delta(G_0), 19)$ colors, and it runs in time $O(n_0)$.*

Remark: It is important, we think, to note a property of planar graphs which was implicitly shown in our algorithm: If we want to color the edges of a planar graph G_0 with

$\max(\Delta(G_0), D)$ colors, $D = 19$, then the colors of the vertices v with $\deg_0(v) > D$ can be chosen from the set $\{1, 2, \ldots, \deg_0(v)\}$, except at most D colors. This property makes our algorithm run in linear time, because when the degree of a vertex v becomes greater than D we can fix a set of at most $\deg_0(v)$ colors which are free at v, and then the choice of colors can be made locally. Also the coloring of the edges incident to a vertex v with $\deg_0(v) \leq D$ has similar "local" property. Note that in the coloring stage, at most 2 edges incident to v can have type (2) or (3). These edges can be assigned large colors. But after these edges are colored, all other edges which are incident to v and are being returned to G are of type (1), so their colors can be chosen from the set $\{1, 2, \ldots, \deg_0(v)\}$.

3.2 The parallel algorithm (1.2)

We present an EREW-PRAM algorithm which colors the edges of a planar graph G with $\max(\Delta(G), 19)$ colors.

First we describe the idea of the algorithm. As in the sequential case, the graph is represented as a list of vertices with their adjacency lists, and the algorithm consists of three stages: graph representation construction, reduction, and coloring. The most important stage is the reduction stage which is divided into $O(\log n)$ phases. In each phase of the algorithm a set $\mathcal{I}$ of reductions is executed. This set is required to have some independent properties, and to be appropriately large, i.e., $\mathcal{I} = \Theta(n)$. In the third stage the algorithm undoes the reductions performed in the second stage, in the reverse order; the returned edges are colored with available colors. In order to execute the reductions in parallel avoiding memory conflict, the conflict graph data structure is used.

We identify three types of reductions:

type 1: $\rho = \{(u, v)\}$; where $\deg(u) + \deg(v) \leq D + 1$,

type 2: $\rho = \{(u, v)\}$; where one of u or v has degree 1, and (u, v) is not of type 1,

type 3: $\rho = \{(u, x), (x, v), (u, y), (y, v)\}$; where $\deg(u), \deg(v) \geq D$ and $\deg(x) = \deg(y) = 2$. (x and y is a pair of vertices from $B(u, v)$)

A set $\mathcal{R}$ of reductions is called *proper* if:

(p1) $\forall \rho_1, \rho_2 \in \mathcal{R}, \rho_1 \cap \rho_2 = \emptyset$,

(p2) $\mathcal{R}$ contains all reductions of type (1) and type (2),

(p3) $\mathcal{R}$ contains $\lfloor B(u, v)/2 \rfloor$ reductions of type (3) from each $B(u, v)$.

We will have to make sure that the reductions performed in each phase are independent in the sense that both the edge deletion, and later the coloring of the deleted edges can be made locally without causing memory access conflicts between processors.

Let $\mathcal{A}(\rho)$ be the set of entries in the adjacency lists which have to be either read from or written into when executing the reduction ρ. It is important to note that $|\mathcal{A}(\rho)| = O(1)$.

We define *conflicts* between reductions. If ρ_1 and ρ_2 are of type (1) then they are in conflict if $\rho_1 = \{e_1\}, \rho_2 = \{e_2\}$ and e_1, e_2 have a common endpoint. If ρ_1 and ρ_2 are of type (2) or (3) then they are in conflict if $\mathcal{A}(\rho_1) \cap \mathcal{A}(\rho_2) \neq \emptyset$. Note here that there are no conflicts between reductions of type (1) and these of types (2) or (3).

The data structure we use for choosing the set of reductions is the *reduction conflict graph* $\mathcal{H} = (\mathcal{R}, \mathcal{F})$, where $\mathcal{R}$ is a proper set of reductions and $\mathcal{F} = \{(\rho_1, \rho_2) | \rho_1$ and $\rho_2 \in \mathcal{R}$ and are in conflict $\}$. It is easy to observe that $\Delta(\mathcal{H}) = O(1)$. On this graph we find the reduction set $\mathcal{I}$ by finding a maximal independent set (MIS) of vertices of $\mathcal{H}$ in parallel [CV,GPS].

Below we give an outline of the algorithm, followed by the technical details.

Algorithm 1.2:

Stage 1: Initialization

    $G = (V, E) := G_0 = (V_0, E_0)$; construct the representation of G;

    $k := 0$;

Stage 2: Reductions

    <u>while</u> $E \neq \emptyset$ <u>do</u>

    <u>begin</u>

        $k := k + 1$;

(1)   Find a proper set of reductions $\mathcal{R}$;

(2)   Construct the conflict graph $\mathcal{H} = (\mathcal{R}, \mathcal{F})$;

(3)   Find a maximal independent set $\mathcal{I}$ in $\mathcal{H}$;

        <u>for each</u> $\rho \in \mathcal{I}$ <u>pardo</u>

        <u>begin</u>

(4)   execute ρ;

            $\mathrm{nr}(\rho) := k$

        <u>end</u>

    <u>end</u>;

    $k_0 := k$;

Stage 3: coloring

    <u>for</u> $k := k_0$ <u>downto</u> 1 <u>do</u>

    <u>for each</u> ρ with $\mathrm{nr}(\rho) = k$ <u>pardo</u>

    <u>begin</u>

(5)   undo ρ;

(6)   color the edges in ρ;

    <u>end</u>

We assign a processor to each vertex and each edge of the graph G_0. In step (1) a proper set of reduction is chosen and a processor is assigned to each reduction. Reductions of type (1) are identified by edge processors which recognize that a corresponding edge has two small endpoints. Reduction of type (2) and (3) are identified by edge processors assigned to edges joining vertices of degree 1 and 2 with vertices of degree greater than D. Let $\deg(u) \geq D$. Then the processors on edges (u, x), for $\deg(x) = 1, 2$, form a list of u's neighbors of degree 1 and a list of u's neighbors of degree 2, and then the latter list is cut into bunch lists.

Then, the list containing $B(u, v)$ is ranked [6] and organized in pairs (It is done in the vertex $x = \max(u, v)$). Each pair is identified and completed to a reduction of type (3). All operations in this step take $O(\log n)$ time.

In step (2) we construct the conflict graph $\mathcal{H}$. The processors assigned to reductions find possible conflicts with other processors. A reduction of type (1) can only conflict with other type (1) reduction; it is found by the processors in the endpoint vertices in constant time. Conflicts between other types of reduction are identified on the adjacency lists, a reduction can conflict with a constant number of other reductions and it is easy to mark the reduction and then to coordinate the identification of conflicts in constant time, avoiding read- and write- conflicts of the processors.

In step (3) we find independent reductions by computing a maximal independent set $\mathcal{I}$ in $\mathcal{H}$. This can be done in $O(\log^* n)$ [CV,GPS]. Then in step (4) each reduction processor in constant time performs the deletion of an edge, and it stores on a local stack the ρ and its number $\mathrm{nr}(\rho)$ which indicates the number of the round in which it is deleted.

The function $\phi(G)$ introduced in Lemma 2 is a lower bound on the number of edges in any proper set $\mathcal{R}$ of reductions in G. $\phi(G)$ evaluates the number of vertices of degree one plus vertices of degree from 3 to 6 with a neighbor of degree less than K, each of these corresponds to a deletion of an edge. It also adds to the above a bound on the number of vertices of degree 2 that we delete (each corresponds to a deletion of two edges). To calculate this it evaluates the number of vertices of degree 2 minus the bound on the number of bunches, since for each bunch we may leave in the graph one member. From Lemma 2 which gave a bound on the value of $\phi(G)$ we conclude that $\mathcal{R} = \Theta(n)$. Also, since $\Delta(\mathcal{H}) = O(1)$ and $\mathcal{I}$ is a MIS in $\mathcal{H}$ we conclude that $\mathcal{I} = \Theta(n)$.

After the whole graph is deleted the coloring stage is done backwards. In each round the corresponding reductions are undone. The algorithm assigns colors to restored edges. A type (1) reduction $\{(u, v)\}$ chooses the first free color in u and v. Processors assigned to reductions of type (2) and (3) choose colors from the available colors lists in the vertex (the assignment of colors is done in $O(\log n)$ time by standard techniques, such as sorting and ranking). Then each reduction colors the edges it contains in constant time.

Theorem 2 *Algorithm (1.2) colors the edges of a planar graph G_0 with $\max(\Delta(G_0, 19))$ colors in time $O(\log^2 n_0)$, using $O(n_0)$ processors, and can be implemented on an EREW PRAM.*

4 Algorithms for Small Maximal Degree

4.1 Fan sequences

Next we refine Vizing's analysis [22,8], adjusting it to enable fast parallel algorithm. Assume some of G's edges are already colored. We use colors from the set $\{1, 2, \ldots, \Delta\}$. Let $\mathrm{col}(x, y)$ denote the color of the edge (x, y). For $x \in V$ we define Used(x) to be the set of colors used at x and Free(x) to be the set of colors free at x. Clearly, for each $x \in V$, Used$(x) \cup$ Free$(x) = \{1, 2, \ldots, \Delta\}$.

Let $(u, v) \in E$, and suppose that all edges $(u, x) \in E$ except (u, v) are already colored. By a *fan sequence* centered at u and starting at v we denote a sequence of u's neighbors $F = [v = x_0, x_1, \ldots, x_k]$, where all the x_i are different, and

(s) $\mathrm{col}(u, x_{i+1}) \in$ Free(x_i) for $i = 0, 1, \ldots, k - 1$.

By $\mathcal{S}_{u,v}$ we denote the family of all fan sequences centered at u and starting at v.

If, additionally to (s), a fan sequence $F = [x_0, x_1, \ldots, x_k] \in \mathcal{S}_{u,v}$ satisfies

(l) Free$(x_k) \cap$ Free$(u) \neq \emptyset$,

then F is called a *local-fan*. $\mathcal{L}_{u,v}$ is the family of all local fans in $\mathcal{S}_{u,v}$. If $F \in \mathcal{L}_{u,v}$ then we can extend the coloring to (u, v) by simply rearranging the colors around u.

If c, d are different colors, then by a (c, d)-*path* we denote an alternating path colored with colors c, d. By *recoloring* this path we mean exchanging the colors c, d of the edges along this path.

Lemma 4 *Let $F \in \mathcal{S}_{u,v}$, $S = [x_0, x_1, \ldots, x_k]$, and $c \in \text{Free}(u)$, $d \in \text{Free}(x_k)$. Suppose that the (c, d)-path P starting from u (possibly empty) avoids x_k. Then, after recoloring P, there is a local fan $F_0 \in \mathcal{L}_{u,v}$.*

F_0 is constructed as follows: **if** $c \in \text{Free}(x_j)$ for some $0 \leq j \leq k$ **then** $F_0 :=$ $[x_0, x_1, \ldots, x_j]$; otherwise, let i be smallest such that $d \in \text{Free}(x_i)$ and x_i does not belong to P; then recolor P and $F_0 := [x_0, x_1, \ldots, x_i]$

The proof of the lemma shows that F_0 is indeed a local fan (properties (s) and (1) hold).

Let $F_1, F_2 \in \mathcal{S}_{u,v}$, $F_1 \neq F_2$, where $F_1 = [x_0, x_1, \ldots, x_k]$, and $F_2 = [y_0, y_1, \ldots, y_l]$. Clearly, $x_0 = y_0 = v$. Then the pair $F = (F_1, F_2)$ is called a *double fan* if $x_k \neq y_l$ and

(d) $\text{Free}(x_k) \cap \text{Free}(y_l) \neq \emptyset$.

$\mathcal{D}_{u,v}$ is the family of all double fans in $\mathcal{S}_{u,v}{}^2$. We show how a double fan $F = (F_1, F_2)$ can be transformed into a local fan by the following procedure, (thus enabling us to extend the coloring to (u, v)):

let $c \in \text{Free}(u)$, $d \in \text{Free}(x_k) \cap \text{Free}(y_l)$, and let P be the (c, d)-path starting from u; then, exchange the colors on P; **if** P contains x_k **then** $F_0 := F_2$ **else** $F_0 := F_1$.

Using the above lemma, we immediately obtain that this procedure is correct (that is F_0 is a local fan), because P must avoid either x_k or y_l.

If F is a double fan, as defined above, and c, d are colors, then F is called a (c, d)-*fan* if $c \in \text{Free}(u)$ and $d \in \text{Free}(x_k) \cap \text{Free}(y_l)$.

Let $G = (V, E)$ be a planar graph with $\Delta = \Delta(G) \geq 8$. Vizing [22] and Gabow *et al.* [10] proved that G must have a reducible edge (see Lemma 3). Let (u, v) be such an edge, satisfying $\deg_v^*(u) + \deg(v) \leq \Delta$. We will show that after removing (u, v) and coloring the remaining edges with Δ colors we can extend the coloring to (u, v). This will prove that the procedure uses $\Delta(G)$ colors.

Let $c \in \text{Free}(v)$. We will attempt to color (u, v) by constructing a fan sequence F_c as follows. The first vertex in F_c is $x_0 = v$. If $c \in \text{Free}(u)$ then we can color (u, v). Else, find an edge (u, x_1) such that $\text{col}(u, x_1) = c$ and extend F_c to $[x_0, x_1]$. Suppose that we have already $F_c = [x_0, x_1, \ldots, x_k]$. If $\text{Free}(x_k) \cap \text{Free}(u) \neq \emptyset$ then F_c is a local fan, as before. Also, if $\text{Free}(x_k) \cap \text{Free}(x_j) \neq \emptyset$ for some $0 \leq j \leq k-1$ then (F_c, F_c'), for $F_c' = [x_0, x_1, \ldots, x_j]$, is a double fan and we can color (u, v). Otherwise, if $\text{Free}(x_k) \neq \emptyset$, let x_{k+1} be a vertex such that $\text{col}(u, x_{k+1}) \in \text{Free}(x_k)$ and take $F_c := [x_0, x_1, \ldots, x_{k+1}]$. Clearly, x_{k+1} cannot be equal to any x_i, $i = 0, 1, \ldots, k$. Note also that $\text{Free}(x_k) = \emptyset$ only when $\deg(x_k) = \Delta$.

Suppose that we have already constructed F_c for each $c \in \text{Free}(v)$ and we failed to color (u, v). In this case all these fan sequences F_c end in vertices of degree Δ. We have $|\text{Free}(v)| = \Delta - \deg(v) + 1$. But u has at most $\deg_v^*(u) \leq \Delta - \deg(v)$ neighbors of degree Δ different than v, so there are two fan sequences, say F_c and F_d, which end in the same vertex x of degree Δ. After removing x from F_c and F_d we obtain two fan sequences F_c' and F_d'. Then (F_c', F_d') is a double fan and we can color (u, v). This proves that $\chi_1(G) = \Delta$.

Our algorithm will be based on the above procedure. Lemma 3 says that $\Delta(G) \geq 9$ implies that the number of reducible edges in G is $\Theta(n)$, and indeed, we will reduce, and later color $\Theta(n)$ edges simultaneously. This should be done with care, especially when we recolor an alternating path. To avoid difficulties we will consider (c,d)-fans separately for different pairs (c,d), and also choose the fans to be, in some sense, independent.

4.2 Fan conflict graph

Let $\mathcal{F}$ be a family of (c,d)-fans, such that for each $F, F' \in \mathcal{F}$, if $F \in \mathcal{D}_{u,v}$ and $F' \in \mathcal{D}_{x,y}$ then the distance between the endpoints of (u,v) and (x,y) is at least 2.

We define the *fan conflict graph* for $\mathcal{F}$ as the graph $C(\mathcal{F}) = (\mathcal{F}, \mathcal{E})$, where the set of edges $\mathcal{E}$ is determined as follows. For each $F \in \mathcal{F} \cap \mathcal{D}_{u,v}$ let $P_{c,d}(D)$ be the (c,d)-path starting from u. Then $(F, F') \in \mathcal{E}$ iff either $P_{c,d}(F)$ ends in a vertex of F' or $P_{c,d}(F')$ ends in a vertex of F. (Saying that a path P ends in a fan sequence F we mean that it contains an element of this sequence, not the center).

Suppose that $|\mathcal{F}| = f$. Then, obviously, $|\mathcal{E}| \leq f$. Let $\mathcal{H} \subseteq \mathcal{F}$ be a set of vertices of degree at most 2 in $C(\mathcal{F})$. A trivial calculation shows that $|\mathcal{H}| \geq \frac{1}{3}f$. Therefore, if $\mathcal{J}$ is a maximal independent subset of $\mathcal{H}$ then $|\mathcal{J}| \geq \frac{1}{9}f$. These considerations will be applied to find the set of double fans to be recolored in our algorithm.

4.3 General Algorithmic Strategy

Both algorithms, sequential and parallel, are based on the same general strategy. Recall that $9 \leq \Delta(G) \leq 18$. The algorithms are, again, divided into *stages*: construction, reduction, and coloring. The last two consist of loop statements, a loop iteration is called a *phase*.

```
        begin
              G := G₀;
(1)           construct the representation of G;
              p := 0;
stage 1:
              while V ≠ ∅ do
              begin
                    p := p + 1;
(2)                 find a set R of reducible edges in G;
(3)                 find a maximal set Iₚ ⊆ R such that if e, e' ∈ Iₚ then the
                    minimum distance between the endpoints of e, e' is at least 2;
(4)                 E := E \ Iₚ;
(5)                 V := {v ∈ V| deg(v) ≥ 1}
              end;
              p₀ := p;
stage 2:
              for p := p₀ downto 1 do
```

```
          begin
(6)               E := E ∪ I_p;
(7)               V := V ∪ {x|(x,y) ∈ I_p for some y};
                  for each (u,v) ∈ I_p satisfying deg(v) + deg*_v(u) ≤ Δ do
(8)                   construct a fan (local or double) F_{u,v} ∈ S_{u,v};
(9)               while I_p ≠ ∅ do
                  begin
(10)                  for each local fan F_{u,v} do color (u,v);
(11)                  partition the set of remaining fans into the
                      sets A_{a,b}, where A_{a,b} contains (a,b)-fans;
(12)                  choose A_{c,d} with greatest cardinality;
(13)                  construct the fan-conflict graph C(A_{c,d});
(14)                  find the set H of vertices of degree at most 2 in C(A_{c,d});
(15)                  find a maximal independent subset J ⊆ H;
(16)                  for each F_{u,v} ∈ J do
                      begin
(17)                      recolor P_{c,d}(F_{u,v});
(18)                      transform F_{u,v} into a local fan;
(19)                      color (u,v)
                      end
                  end
              end
      end.
```

4.4 The Algorithms (2.1) and (2.2)

In the sequential algorithm (2.1), steps (2)-(5) take $O(n)$ time using searching a graph. Notice that since $|R| = \Theta(n)$, then $|I_p| = \Theta(n)$, and the number of edges in G decreases geometrically at each phase. By the choice of $A_{c,d}$ we have $|J| = \Theta(n)$. Therefore loop (9) is executed $O(\log n)$ times. So the execution time for loop (9) in one phase is $O(n \log n)$. This implies immediately the total time bound $O(n_0 \log^2 n_0)$. After taking into account that in stage 1 the size of G decreases geometrically, it is not hard to derive the bound $O(n_0 \log n_0)$. Note that only the second stage requires $O(n_0 \log n_0)$ time, the first one can be done in linear time.

The parallel algorithm (2.2) implements (3) by computing a *reduction conflict graph* (R, W), where $(e, e') \in W$ iff e, e' have a common endpoint. This costs time $O(1)$. And then let I_p be an MIS. Steps (4)-(8) and (10)-(19) (using the *fan conflict graph*) cost $O(\log n_0)$ time. The most costly step is loop (9), executed $O(\log n_0)$ times, and each execution can be done in time $O(\log^2 n_0)$. Thus stage 2 costs time $O(\log^3 n)$. Therefore the total time complexity is $O(\log^3 n_0)$.

Theorem 3 Let G be any planar graph G_0 such that $9 \leq \Delta(G_0) \leq 19$ (or any other constant). Both algorithms (2.1) and (2.2) color the edges of G with Δ colors. Algorithm

(2.1) runs in time $O(n_0 \log n_0)$. Algorithm (2.2) runs in time $O(\log^3 n_0)$, and uses $O(n_0)$ processors on an EREW PRAM.

References

[1] K.I.Appel, W.Haken, Every planar map is four colorable, Bull. Amer. Math. Soc., vol 82, pp 711-712, 1976.

[2] J.Boyar, H.Karloff, Coloring planar graphs in parallel, J. Algorithms 8 (1987) 470-479.

[3] N.Chiba, T.Nishizeki, N.Saito, A linear algorithm for five-coloring a planar graph, J. Algorithms 2 (1981) 317-327.

[4] R.Cole, J.Hopcroft, On edge coloring bipartite graphs, SIAM J. Comput 11 (1982) 540-546.

[5] R.Cole, U.Vishkin, Dermistic coin tossing and accelerating cascades: micro and macro techniques for designing parallel algorithms, 18th ACM STOC (1986) 206-219.

[6] R.Cole, U.Vishkin, Approximate and exact parallel scheduling with applications to list, Tree and Graph Problems, 27th IEEE FOCS (1986) 478-491.

[7] G.N. Frederickson, On linear-time algorithms for five-coloring planar graphs, Inform. Proc. Letters 19 (1984) 219-224.

[8] S.Fiorini, R.J.Wilson, Edge-Colourings of Graphs, Pitman, London, 1977.

[9] H.N.Gabow, O.Kariv, Algorithms for edge coloring bipartite graphs and mulitgraphs, SIAM J. Comput. 11 (1982) 117-129.

[10] H.N.Gabow, T.Nishizeki, O.Kariv, D.Leven, O.Terada, Algorithms for edge-coloring graphs, TR-41/85, Department of Computer Science, Tel Aviv University, 1985.

[11] A.V.Goldberg, S.A.Plotkin, G.E.Shannon, Parallel symmetry breaking in sparse graphs, 19th ACM STOC, New York, 1987.

[12] M.Goldberg, T.Spencer, A new parallel algorithm for the maximal independent set problem, 19th IEEE FOCS, 1987.

[13] T.Hagerup, M.Chrobak, K.Diks, Optimal parallel 5-colouring of planar graphs, 14th ICALP, 1987.

[14] I.J.Holyer, The NP-completeness of edge coloring, SIAM J. Comput 10 (1981) 718-720.

[15] H.Karloff, D.Shmoys, Efficient parallel algorithms for edge coloring problems, J. Algorithms 8 (1987) 39-52.

[16] R.M.Karp, A.Widgerson, A fast parallel algorithm for the maximal independent set problem, 16th ACM STOC, 1987.

[17] M.Luby, A simple parallel algorithm for the maximal independent set problem, 17th ACM STOC, 1985.

[18] D.Leven, Z.Galil, NP-completeness of finding the chromatic index of regular graphs, J. Algorithms 4 (1983) 35-44.

[19] G.F.Lev, N.Pippenger, L.G.Valiant, A fast parallel algorithm for routing in permutation networks, IEEE Trans. on Comput. C-30 (1981) 93-100.

[20] D.Matula, Y.Shiloah, R.Tarjan, Two linear-time algorithms for five-coloring a planar graph, Tech. Rep. No. STAN-CS-80-830, Dep. Comp. Sci., Stanford University.

[21] V.G.Vizing, On the estimate of the chromatic class of a p-graph, Diskret. Analiz 3 (1964) 25-30.

[22] V.G.Vizing, Critical graphs with a given chromatic number, Diskret. Analiz 5 (1965) 9-17.

[23] M.H.Williams, A linear algorithm for coloring planar graphs with five colors, Comput. J. 28,1 (1985) 78-81.

Optimal Parallel Algorithms on Planar Graphs[†]

Torben Hagerup

Fachbereich Informatik, Universität des Saarlandes, D–6600 Saarbrücken, West Germany

Abstract:

Few existing parallel graph algorithms achieve optimality when applied to very sparse graphs such as planar graphs. We add to the list of such algorithms by giving optimal, logarithmic-time PRAM algorithms for the connected components, spanning tree, biconnected components, and strong orientation problems. The algorithms work on classes of graphs including planar graphs and graphs of bounded genus or bounded thickness.

1. Introduction

We consider in this paper the problems below. An instance of each problem is defined by an undirected input graph $G = (V, E)$ which, except in the case of problem (1), will be assumed to be connected.

(1) The connected components problem: Compute an array $F : V \to V$ such that for all $u, v \in V$, $F[u] = F[v]$ if and only if u and v belong to the same connected component of G.

(2) The biconnected components problem: Compute an array $B : E \to E$ such that for all $e_1, e_2 \in E$, $B[e_1] = B[e_2]$ if and only if e_1 and e_2 belong to the same biconnected component of G.

(3) The undirected spanning tree problem: Compute the edge set of an undirected spanning tree of G.

(4) The directed spanning tree problem: For some directed spanning tree T of G, compute the parent of each non-root vertex in T.

(5) The strong orientation problem: Direct the edges in E in such a way that the resulting directed graph is strongly connected. For this to be possible, G must be bridgeless (i.e., each edge in G must be part of a simple cycle).

It has been known for some time that for graphs with n vertices and m edges, all of the above problems can be solved in $O(\log n)$ time on a CRCW PRAM with $O(n + m)$ processors. However, since the problems are trivially solved in linear sequential time, this is a factor of $\Theta(\log n)$ away from being optimal. Gazit [6] recently described a probabilistic algorithm that with high probability solves the connected components problem in $O(\log n)$ time with an optimal number of processors, and Cole and Vishkin [2, 4] have given deterministic logarithmic-time algorithms for the above problems which are optimal provided that the input graphs are sufficiently dense. However, the algorithms of [2, 4] are complicated, and they are not optimal when applied to very sparse graphs. We complement the results of Cole and Vishkin by giving deterministic logarithmic-time algorithms for the problems which are optimal for certain classes of very sparse graphs, namely those that are n-contractible in the sense of the following

---

† Supported by the DFG, SFB 124, TP B2, VLSI Entwurfsmethoden und Parallelität.

Definition: Let $f : \mathbb{R}_+ \to \mathbb{R}_+$ be a convex function with $\lim\limits_{x \to \infty} f(x) = \infty$. A class $\mathcal{G}$ of undirected graphs is called *$f(n)$-contractible* if

(1) For all $G = (V, E)$ in $\mathcal{G}$, $|E| = O(f(|V|))$

(2) Every subgraph and every elementary contraction of a graph in $\mathcal{G}$ is in $\mathcal{G}$. An *elementary contraction* of a graph $G = (V, E)$ is a graph obtained from G by contracting two adjacent vertices u and v to a single vertex, i.e., by replacing u and v by a new vertex z adjacent to exactly those vertices in $V \backslash \{u, v\}$ that are adjacent to at least one of u and v in G.

Examples of n-contractible graph classes are the class of planar graphs and classes of graphs of bounded genus or bounded thickness [10].

Our results are all derived from the same basic algorithm which is described in detail in Section 3 for the case of the computation of connected components. The algorithm is correct and works in logarithmic time on arbitrary undirected graphs. On n-vertex graphs drawn from an $f(n)$-contractible class of graphs, the number of processors used is $O(f(n)/\log n)$. Our computational model is the CRCW PRAM.

2. Definitions and notation

Given an undirected graph $G = (V, E)$, we may define an equivalence relation $\overset{G}{\sim}$ on V by letting

$$u \overset{G}{\sim} v \iff \text{there is a path in } G \text{ from } u \text{ to } v,$$

for all $u, v \in V$. The connected components of G are the subgraphs of G spanned by the equivalence classes of $\overset{G}{\sim}$. Given a finite set V and a function $F : V \to V$, the graph *induced* by F is the directed graph

$$G_F = (V, \{(u, F(u)) \in V \times V \mid F(u) \neq u\}).$$

If G_F is acyclic, F will be called a *forest function* on V. In this case G_F is a collection of disjoint directed trees called the trees in G_F. For $u \in V$, let $R_F(u)$ denote the root of the tree in G_F containing u, i.e., the unique vertex in the tree of outdegree 0. Given an undirected graph $G = (V, E)$ and a forest function F on V, F is said to *respect* G if

$$\forall u, v \in V : (R_F(u) = R_F(v) \Rightarrow u \overset{G}{\sim} v).$$

If F respects G, a tree in G_F is called *complete* (with respect to G) if its vertex set is identical to that of a connected component of G.

Let F be a forest function respecting G, and let $R = \{R_F(u) \mid u \in V\}$ be the set of roots of trees in G_F. The *supervertex graph* induced by F is the undirected graph

$$(R, \{\{u, v\} \in R^{(2)} \mid u \neq v \text{ and } \exists \{x, y\} \in E : R_F(x) = u \text{ and } R_F(y) = v\}),$$

where $M^{(2)}$, for any set M, denotes the set of unordered pairs of elements from M. The vertices of the supervertex graph are called *supervertices*.

An *isolated vertex* in a graph is a vertex with no incident edges. Given a graph G, the graph obtained from G by the removal of all isolated vertices will be called the *non-isolated part* of G. Such graphs can be represented in a particularly simple way. When $G = (V, E)$ is an undirected graph without isolated vertices, we will consider a representation of G to be a triple (m, H, T), where $m = |E|$ and $H, T : \{1, \ldots, 2m\} \to V$ are arrays with

$$\forall \{u, v\} \in E \; \exists i, j \in \{1, \ldots, 2m\} : H[i] = T[j] = u \text{ and } H[j] = T[i] = v.$$

In other words, each undirected edge $\{u, v\} \in E$ is represented as two directed edges (u, v) and (v, u), and G is represented by a list of all such directed edges. We consider two different representations:

(1) The *unordered* representation. Here no further restrictions are placed on H and T.

(2) The *ordered* representation. This is the representation used in [2, 4], and it essentially means that G is given as a set of adjacency lists. We stipulate that all the (directed) edges leaving a given vertex occur together, i.e.,

$$\forall i, j, k \in \{1, \ldots, 2m\} : (i \leq j \leq k \text{ and } T[i] = T[k] \Rightarrow T[j] = T[i]).$$

These definitions generalize verbatim to undirected multigraphs.

A *PRAM* (parallel RAM) is a machine consisting of a finite number p of processors (RAMs) operating synchronously on common, shared memory cells numbered $0, 1, \ldots$ We assume that the processors are numbered $1, \ldots, p$ and that each processor is able to read its own number. All processors execute the same program. We use the unit-cost model in which each memory cell can hold integers of size polynomial in the size of the input and each processor is able to carry out usual arithmetic operations including multiplication and integer division on such numbers in constant time.

Various types of PRAMs have been defined. We use here the *ARBITRARY CRCW PRAM* [5] which allows simultaneous reading as well as simultaneous writing of each memory cell by arbitrary sets of processors. When several processors try to write to the same memory cell in the same step, one of them succeeds (i.e., the value that it attempts to write will actually be present in the cell after the write step), but the algorithm must work correctly regardless of which one of the competing processors is successful.

3. The connected components algorithm

We first explain how our algorithm differs from previous algorithms for the connected components problem, and then proceed to describe the algorithm in a self-contained and more formal way.

Most parallel algorithms to compute the connected components of an undirected graph $G = (V, E)$ maintain a forest function F on V respecting G. G_F is called the *pointer graph*. Initially each vertex in the pointer graph forms a tree by itself, and the goal is to repeatedly combine trees until finally each surviving tree is complete. A tree may grow by letting its root r *hook* itself onto a vertex v in an adjacent tree (i.e., $F[r] := v$), or by being hooked to by roots of adjacent trees. Here two trees are called adjacent if their roots are adjacent in the supervertex graph induced by F. Care must be taken to prevent simultaneous hookings by several vertices from introducing cycles into the pointer graph. In order for an edge to be able to cause a hooking, the root of at least one of the trees containing its endpoints must be known. Hence hooking steps alternate with *shortcutting* steps in which the height of the trees in the pointer graph is reduced by means of pointer doubling (i.e., $F[u] := F[F[u]]$, for all $u \in V$) with the aim of turning each tree into a star, a tree of height at most 1 whose root can be determined in constant time from any vertex in the tree.

In the first algorithms of this kind [11, 1], trees in the pointer graph were reduced to stars after each hooking step. Since a hooking step potentially creates trees of height $\Theta(n)$, $\Theta(\log n)$ shortcutting steps were needed after each hooking step in order to ensure the star property. A time bound of $O((\log n)^2)$ for the whole algorithm followed from the need to perform $\Theta(\log n)$ hooking steps. Shiloach and Vishkin [13] later designed an algorithm with a running time of $O(\log n)$, which follows each hooking step by only $O(1)$ shortcutting steps.

Our algorithm also maintains a collection of trees that expand through hooking. The differences are as follows: (1) When a vertex stops being the root of its tree, its incident edges are transferred to the root of its tree. This means that unlike most earlier algorithms, our algorithm explicitly manipulates the supervertex graph, or rather its non-isolated part. Since the number of incomplete trees decreases geometrically, the non-isolated part of the supervertex graph rapidly becomes smaller (note that a supervertex is isolated exactly if the corresponding tree in the pointer graph is complete), and the result is a more economical use of the available processors. Essential in achieving this goal is a new parallel prefix sums algorithm by Cole and Vishkin [2, 3]; (2) The hooking steps are modified to create stars instead of trees of arbitrary height.

The algorithm will be cast in the framework of the following

Lemma 1: Let $f : \mathbb{R}_+ \to \mathbb{R}_+$ be a convex function with $\lim_{x \to \infty} f(x) = \infty$, and let $\mathcal{P}$ be a class of problem instances, each of which has an associated size $\in \mathbb{N}$. Suppose that

(1) Problem instances in $\mathcal{P}$ of size n can be solved in $O(\log n)$ time by $O(f(n))$ processors.

(2) There are constants $c < 1$ and $n_0 \in \mathbb{N}$ so that given any $P \in \mathcal{P}$ of size $n \geq n_0$, there is a problem instance $\Psi(P) \in \mathcal{P}$ of size at most cn (the *reduced* instance) such that $O(\log n/\log \log n)$ time and $O(f(n) \log \log n/\log n)$ processors suffice to construct the input to $\Psi(P)$ from the input to P and to deduce a solution to P from any solution to $\Psi(P)$.

Then problem instances in $\mathcal{P}$ of size n can be solved in $O(\log n)$ time by $O(f(n)/\log n)$ processors. In particular, if $f(n) = n$ and if the notion of size has been chosen reasonably, then problem instances in $\mathcal{P}$ can be solved optimally in logarithmic time.

Remark: This useful principle has been employed before [3], but it seems to not have previously been formulated explicitly.

Proof: Replacing Ψ by Ψ^K for a suitable constant $K \in \mathbb{N}$, we may assume that $c = 1/2$. Given a problem instance $P_0 \in \mathcal{P}$ of size n, execute the following algorithm (letting $\Psi(P) = P$ for all $P \in \mathcal{P}$ of size $< n_0$):

(1) $k := \lceil \log \log n \rceil$;
(2) **for** $i := 1$ **to** k
(3) **do** construct the input to $P_i = \Psi(P_{i-1})$;
(4) Solve P_k;
(5) **for** $i := k$ **downto** 1
(6) **do** obtain a solution to P_{i-1} from the solution to P_i;

By assumption (2), the size of P_k is bounded by $\max\{n_0, n/\log n\}$. Hence by assumption (1), line (4) can be executed in $O(\log n)$ time by $O(f(n/\log n)) = O(f(n)/\log n)$ processors. Using $O(f(n)/\log n)$ processors, the ith execution of line (3), for $i = 1, \ldots, k$, takes time

$$O\left(\frac{f(2^{-i}n)}{f(n)} \log n + \frac{\log n}{\log \log n}\right) = O\left(\left(2^{-i} + \frac{1}{\log \log n}\right) \log n\right),$$

and the execution time of the entire loop in lines (2)–(3) is

$$O\left(\sum_{i=1}^{k}\left(2^{-i} + \frac{1}{\log \log n}\right) \log n\right) = O(\log n).$$

The same time bound holds for the loop in lines (5)–(6), giving a total execution time of $O(\log n)$. $\blacksquare$

Now let $G_0 = (V_0, E_0)$ be an undirected input graph with n_0 vertices and m_0 edges. Since the handling of isolated vertices is trivial, we may assume that G_0 contains no isolated vertices and is given by a triple (m_0, H_0, T_0) according to the unordered representation described in Section 2. In order to apply Lemma 1, we must describe the construction of a reduced graph G_1 and indicate how to derive a solution to the connected components problem for G_0 from one for G_1. Note the recursive algorithm structure implicit in this description.

G_1 is constructed by steps 1–8 below. Most steps are supposed to be executed concurrently by several processors. This is indicated via the construction

$$\text{for } i \text{ in } \{1, \ldots, l\} \text{ pardo } S$$

which means that the statement S should be executed for $i = 1, \ldots, l$. No particular processor allocation is implied. It is thus not necessarily the case that l processors are available, and that the kth processor, for $k = 1, \ldots, l$, executes the statement with $i = k$. Indeed, the saving in the number of processors derives mainly from a careful attention to the question of processor allocation.

Note that actions that are logically associated with a vertex are executed simultaneously by one processor for each edge incident on the vertex. Although somewhat wasteful, this allows a certain simplification of the description since there is no need to introduce processors associated with the vertices.

Step 1:
    for i **in** $\{1, \ldots, 2m_0\}$ **pardo** $F_0[T_0[i]] := T_0[i]$;

Comment: The forest function F_0 is initialized by making each vertex a trivial tree.

Step 2:
    for i **in** $\{1, \ldots, 2m_0\}$ **pardo** $F'[T_0[i]] := H_0[i]$;

Comment: Concurrent writing is used to choose for each $u \in V_0$ a neighbour of u onto which it might hook itself.

Step 3:
    Let $I \subseteq V_0$ be an independent vertex set of size at least $n_0/6$ in the graph induced by F';

Comment: I is the set of vertices that will be hooked onto other vertices. It is computed as follows: Delete from $G_{F'}$ the at most $n_0/2$ vertices of indegree > 1 and let I be a maximal independent vertex set in the resulting graph, which is a collection of disjoint simple paths and simple cycles.

Step 4:
    for i **in** $\{1, \ldots, 2m_0\}$ **pardo if** $T_0[i] \in I$ **then** $F_0[T_0[i]] := F'[T_0[i]]$;

Comment: The actual hooking. Since I is independent, all trees in G_{F_0} are stars after this step.

Step 5:
    for i **in** $\{1, \ldots, 2m_0\}$ **pardo begin** $H_1[i] := F_0[H_0[i]]$; $T_1[i] := F_0[T_0[i]]$; **end;**

Comment: Each edge $\{u, v\} \in E_0$ is transformed into the edge $\{R_{F_0}(u), R_{F_0}(v)\}$ joining the roots of the trees containing u and v. Let G' be the multigraph represented by (m_0, H_1, T_1) after step 5. Steps 6–8 serve to turn G' into the simple graph G_1.

Step 6:
    for i **in** $\{1, \ldots, 2m_0\}$ **pardo** $Mark[i] := (H_1[i] = T_1[i])$;
    for i **in** $\{1, \ldots, 2m_0\}$ **pardo** $A[H_1[i], T_1[i]] := i$;
    for i **in** $\{1, \ldots, 2m_0\}$ **pardo if** $A[H_1[i], T_1[i]] \neq i$ **then** $Mark[i] := true$;

Comment: Self loops and multiple edges in G' are marked for later removal.

    $Mark : \{1, \ldots, 2m_0\} \to \{true, false\}$ and $A : V_0 \times V_0 \to \{1, \ldots, 2m_0\}$ are auxiliary arrays. Concurrent writing into A is used to select exactly one (undirected) edge between each pair of adjacent vertices in G'.

Step 7: **for** i **in** $\{1, \ldots, 2m_0\}$ **pardo** $N[i] := \#\{j \mid 1 \leq j \leq i \text{ and not } Mark[j]\}$;

Comment: For each (directed) edge, the number of unmarked edges preceding it in the list of edges is computed into an auxiliary array $N : \{1, \ldots, 2m_0\} \to \{0, \ldots, 2m_0\}$.

Step 8:
    for i **in** $\{1, \ldots, 2m_0\}$ **pardo if not** $Mark[i]$ **then**
      begin $H_1[N[i]] := H_1[i]$; $T_1[N[i]] := T_1[i]$; **end;**
    $m_1 := N[m_0]/2$;

Comment: The unmarked edges are moved to the beginning of the arrays H_1 and T_1. This effectively deletes the marked edges. m_1 becomes the number of remaining (undirected) edges.

Let $G_1 = (V_1, E_1)$ be the simple graph represented by (m_1, H_1, T_1) after step 8, and let F_1 be the result of solving the connected components problem on G_1, i.e.,

$$F_1[u] = F_1[v] \iff u \overset{G_1}{\sim} v, \quad \text{for all } u, v \in V_1.$$

A solution to the original problem may then be computed simply by executing

Step 9:
  for i **in** $\{1, \ldots, 2m_0\}$ **pardo if** $F_0[T_0[i]] \in V_1$ **then** $F_0[T_0[i]] := F_1[F_0[T_0[i]]]$;

Comment: Shortcutting through pointer doubling.

Lemma 2: After the execution of step 9,

$$u \overset{G_0}{\sim} v \iff F_0[u] = F_0[v], \quad \text{for all } u, v \in V_0.$$

Proof: Let F denote the value of F_0 at the end of step 8. For all $u, v \in V_0$, u and v belong to the same connected component of G_0 if and only if $R_F(u)$ and $R_F(v)$ belong to the same connected component of the supervertex graph G^* induced by F. Since G_1 is the non-isolated part of G^*, we have at the end of step 9,

$$u \overset{G_0}{\sim} v \iff R_F(u) = R_F(v) \text{ or } (R_F(u) \in V_1 \text{ and } R_F(v) \in V_1 \text{ and } R_F(u) \overset{G_1}{\sim} R_F(v)) \iff$$

$$F[u] = F[v] \text{ or } (F[u] \in V_1 \text{ and } F[v] \in V_1 \text{ and } F_1[F[u]] = F_1[F[v]]) \iff F_0[u] = F_0[v]. \quad \blacksquare$$

Theorem 1: On n-vertex input graphs drawn from an $f(n)$-contractible class $\mathcal{G}$ and given according to the unordered representation, the connected components and undirected spanning tree problems can be solved in $O(\log n)$ time on a CRCW PRAM with $O(f(n)/\log n)$ processors.

Proof: Use the procedure given above which was already shown to correctly determine the connected components of the input graph. The computation of a maximal independent vertex set in step 3 can be carried out in $O(\log n/\log \log n)$ time by $O(f(n) \log \log n/\log n)$ processors. Step 7 calls for a prefix sums computation which can be done within the same time and processor bounds. Both of these results are due to Cole and Vishkin [3]. All other steps can be executed trivially in $O(1)$ time by $O(f(n))$ processors, and hence in $O(\log n/\log \log n)$ time by $O(f(n) \log \log n/\log n)$ processors. The graph G_1 constructed by steps 1-8 clearly is in $\mathcal{G}$ and has at most $\frac{5}{6}n_0$ vertices. Since furthermore the connected components problem can be solved on graphs with n vertices in $O(\log n)$ time by $O(f(n))$ processors [13], the first part of the theorem follows from Lemma 1.

The connected components algorithm is easily extended to also compute an undirected spanning tree of each connected component of the input graph. As has been observed often before, it suffices to mark as a tree edge each edge of the input graph that causes a hooking. $\blacksquare$

Remark 1: If vertices are represented by integers of size $O(n)$, the result of [9] allows the algorithm to be implemented to use $O(\min\{n^2, f(n)n^\epsilon\})$ space, for any fixed $\epsilon > 0$.

Remark 2: If the ARBITRARY CRCW PRAM is replaced by the stronger Priority CRCW PRAM [8] as the model of computation, it is easy to combine the above algorithm with the near-optimal sorting algorithm of [8] to get a minimum-weight spanning tree algorithm which uses $O(\log n)$ time and $O(f(n) \log \log n/\log n)$ processors on n-vertex edge-weighted graphs drawn from an $f(n)$-contractible class, provided that edge weights are integers of size polynomial in n.

4. Derived algorithms

Theorem 2: On connected n-vertex input graphs drawn from an $f(n)$-contractible class and given according to the ordered representation, the directed spanning tree and biconnected components problems can be solved in $O(\log n)$ time on a CRCW PRAM with $O(f(n)/\log n)$ processors.

Proof: Let $G = (V, E)$ be an n-vertex input graph. Given G according to the ordered representation, any undirected spanning tree of G may be converted to a directed spanning tree of G by means of the Euler tour technique. This is described in great detail in [14], the only difference being that in order to achieve optimality, an optimal list ranking algorithm [2, 3] must be used to guide the pointer doubling. Hence the first part of the theorem follows from Theorem 1.

The algorithm to compute the biconnected components of G follows in outline the algorithm of [14], with modifications to the single steps in order to achieve optimality. The simple basic idea is to construct an undirected graph G' on the vertex set E with the property that two edges of G belong to the same biconnected component of G if and only if they are vertices in the same connected component of G', and then to apply a connected components algorithm to G'. In more detail, the algorithm proceeds as follows:

(1) Compute an undirected spanning tree $T = (V, E_T)$ of G and convert it as above to a directed spanning tree $\vec{T}$. Using the Euler tour technique, compute the preorder number $pre(u)$ and the number of descendants $nd(u)$ in $\vec{T}$ for each $u \in V$ (for some ordering of the children of each vertex). This is described in [14], and using an optimal list ranking algorithm, it can be done in $O(\log n)$ time with $O(f(n)/\log n)$ processors.

(2) By applying a prefix sums algorithm to the ordered representation of G, compute for each $u \in V$ the quantity

$$locallow(u) = \min(\{pre(u)\} \cup \{pre(v) \mid \{u, v\} \in E \backslash E_T\}).$$

Next, for all $u \in V$, compute

$$low(u) = \min\{locallow(v) \mid v \text{ is a descendant of } u \text{ in } \vec{T}\}.$$

We discuss this step separately below. By Lemma 3, $low(u)$ can be computed for all $u \in V$ in $O(\log n)$ time with $O(n/\log n)$ processors. For $u \in V$, $high(u)$ is defined and computed analogously.

(3) For each non-root vertex u in $\vec{T}$, let $p(u)$ denote the parent of u in $\vec{T}$, let

$$\Delta_2 = \{\{\{p(v), v\}, \{p(w), w\}\} \in E_T^{(2)} \mid \{v, w\} \in E \backslash E_T \text{ and } pre(v) + nd(v) \leq pre(w)\} \quad \text{and}$$

$$\Delta_3 = \{\{\{p(v), v\}, \{v, w\}\} \in E_T^{(2)} \mid 1 \neq pre(v) < pre(w), \text{ and either}$$

$$low(w) < pre(v) \text{ or } high(w) \geq pre(v) + nd(v)\}$$

and compute the undirected graph

$$G'' = (E_T, \Delta_2 \cup \Delta_3).$$

Note that G'' is isomorphic to a subgraph of G, the vertex $\{p(v), v\} \in E_T$ of G'' corresponding to the vertex $v \in V$ of G. Hence identifying the edges of G'' takes constant time with $f(n)$ processors, and an (ordered) representation of G'' may be computed by means of a parallel prefix computation in $O(\log n)$ time with $O(f(n)/\log n)$ processors.

(4) Find the connected components of G'', i.e., compute an array $B : E_T \to E_T$ such that for all $e_1, e_2 \in E_T$, $B[e_1] = B[e_2]$ if and only if $e_1 \overset{G''}{\sim} e_2$. By Theorem 1, this can be done in $O(\log n)$ time by $O(f(n)/\log n)$ processors.

(5) For all $e \in E \backslash E_T$, let $B[e] = B[\{p(w), w\}]$, where w is the endpoint of e with larger preorder number. This takes $O(1)$ time with $f(n)$ processors.

[14] proves that when the above procedure has been carried out, $B[e_1] = B[e_2]$ if and only if e_1 and e_2 belong to the same biconnected component of G, for all $e_1, e_2 \in E$. The time and processor bounds claimed in Theorem 2 follow from the above discussion. ∎

We finally fill in the last detail in the proof of Theorem 2:

Lemma 3: Given an ordered representation of a rooted tree $T = (V, E)$ on n vertices, a set R, an array $\lambda : V \to R$ and a commutative and associative function $+ : R \times R \to R$ (called the sum) which can be evaluated on specific arguments in constant time by one processor, it is possible, in $O(\log n)$ time and using $O(n/\log n)$ processors, to compute for all $u \in V$ the sum

$$s(u) = \sum_{\substack{v \text{ is a descendant} \\ \text{of } u \text{ in } T}} \lambda[v]$$

of the λ values of all descendants of u in T.

Proof: Let us call $s(u)$ the *subtree value* of u with respect to T. Our task, then, is to compute the subtree values of all vertices with respect to T.

We begin by constructing a binary tree T' that will yield the same answers as T. This is done as follows (see Fig. 1 for an example):

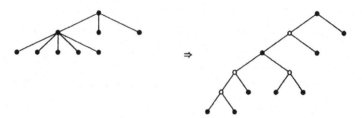

Fig. 1. The construction of a binary tree T' equivalent to T. Dummy vertices are shown white.

Each vertex u in T with more than two children is replaced by a complete binary tree with u as its root and the former children of u as its leaves. This requires the introduction of a number of dummy vertices without associated λ values. Note, however, that the number of vertices in T' is $O(n)$ and that for all $u \in V$, the subtree value of u with respect to T' is the same as the subtree value of u with respect to T. T' can be constructed from the ordered representation of T in $O(\log n)$ time by $O(n/\log n)$ processors.

We can now compute all subtree values with respect to T' by means of an optimal algorithm for expression evaluation. A very simple such algorithm was described by Gibbons and Rytter [7]. A number of inessential differences between their setting and the one used here can be dealt with as follows:

(1) Gibbons and Rytter consider expressions constructed from the usual arithmetic operators plus, times, etc. It is obvious, however, that their algorithm will work for any single associative operator.

(2) They assume the leaves of the expression tree to be numbered consecutively from left to right. This numbering can be obtained as follows: First use the Euler tour technique to construct a list of length $O(n)$ in which each leaf of T' occurs exactly once and the leaves occur in the desired order. Next use optimal list ranking to carry out a prefix sums calculation on this list, counting for each leaf the number of leaves preceding it in the list. The whole computation takes $O(\log n)$ time with $O(n/\log n)$ processors.

(3) In their problem, no values are associated with internal vertices. However, they associate a function, initially the identity function, with each internal vertex, and a different initial setting of the function associated with a given vertex is easily made to reflect the presence of a value at that vertex.

(4) Gibbons and Rytter compute only the subtree value of the root, not of all vertices. However, a simple technique described in [12] allows their algorithm to be extended to compute all subtree values. ∎

Theorem 3: On connected, bridgeless n-vertex input graphs drawn from an $f(n)$-contractible class and given according to the ordered representation, the strong orientation problem can be solved in $O(\log n)$ time on a CRCW PRAM with $O(f(n)/\log n)$ processors.

Proof: By applying techniques developed above, the algorithm for strong orientation described in [14] may be modified to use only the stated number of processors. As in the case of the biconnected components problem, the only bottleneck barring the way to a general optimal algorithm is the computation of a spanning tree. ∎

References

[1]: Francis Y. Chin, John Lam, and I-Ngo Chen: "Efficient Parallel Algorithms for Some Graph Problems". *Communications of the ACM* **25** (1982), 659–665.

[2]: Richard Cole and Uzi Vishkin: "Approximate and exact parallel scheduling with applications to list, tree and graph problems". Proceedings, 27th Annual Symposium on Foundations of Computer Science (1986), 478–491.

[3]: Richard Cole and Uzi Vishkin: "Faster Optimal Parallel Prefix Sums and List Ranking". To appear in *Information and Computation*.

[4]: Richard Cole and Uzi Vishkin: "Approximate parallel scheduling. Part II: Applications to optimal parallel graph algorithms in logarithmic time". Preprint, 1987.

[5]: Faith E. Fich, Prabhakar L. Ragde and Avi Wigderson: "Relations Between Concurrent-Write Models of Parallel Computation". Proceedings, 3rd Annual ACM Symposium on Principles of Distributed Computing (1984), 179–189.

[6]: Hillel Gazit: "An Optimal Randomized Parallel Algorithm for Finding Connected Components in a Graph". Proceedings, 27th Annual Symposium on Foundations of Computer Science (1986), 492–501.

[7]: Alan Gibbons and Wojciech Rytter: "Optimal Parallel Algorithms for Dynamic Expression Evaluation and Context-Free Recognition". Preprint, 1987.

[8]: Torben Hagerup: "Towards Optimal Parallel Bucket Sorting". *Information and Computation* **75** (1987), 39–51.

[9]: Torben Hagerup: "A Note on Saving Space in Parallel Computation". Submitted to *Information Processing Letters*.

[10]: Frank Harary: "Graph Theory". Addison-Wesley, Reading, Mass., 1969.

[11]: D. S. Hirschberg, A. K. Chandra and D. V. Sarwate: "Computing Connected Components on Parallel Computers". *Communications of the ACM* **22** (1979), 461–464.

[12]: Gary L. Miller and John H. Reif: "Parallel Tree Contraction and Its Application". Proceedings, 26th Annual Symposium on Foundations of Computer Science (1985), 478–489.

[13]: Yossi Shiloach and Uzi Vishkin: "An $O(\log n)$ Parallel Connectivity Algorithm". *Journal of Algorithms* **3** (1982), 57–67.

[14]: Robert E. Tarjan and Uzi Vishkin: "An Efficient Parallel Biconnectivity Algorithm". *SIAM Journal on Computation* **14** (1985), 862–874.

EFFICIENT PARALLEL TRICONNECTIVITY IN LOGARITHMIC TIME
(Extended Abstract)

Vijaya Ramachandran†

Coordinated Science Lab, University of Illinois, Urbana, IL 61801, and
International Computer Science Institute, Berkeley, CA

Uzi Vishkin‡

Sackler Faculty of Exact Sciences, Tel Aviv University, Tel Aviv, Israel, and
Courant Institute, New York University, New York, NY

ABSTRACT

We present two new techniques for trimming a logarithmic factor from the running time of efficient parallel algorithms for graph problems. The main application of our techniques is an improvement in running time from $O(\log^2 n)$ to $O(\log n)$ for efficient triconnectivity testing in parallel. Additional applications include almost optimal $O(\log n)$ time algorithms for recognizing Gauss codes, for testing planarity of graphs with a known Hamiltonian cycle and for testing if a permutation is sortable on two stacks.

1. INTRODUCTION

In this paper we give improved algorithms for two problems on graphs:

Problem 1 Given a graph consisting of a simple cycle C and a collection of chords on C, obtain a characterization of the interlacings between the chords.

Problem 2 Given a graph G together with a collection of disjoint subgraphs $T_i, i=1, \cdots, k$, find the bridge graph of each T_i.

Algorithms for variants of both of these problems are needed in the triconnectivity algorithm of [MiRa2]: an $O(\log^2 n)$ time efficient parallel algorithm is given in that paper for each of these problems, performing a total work of $O(m\log^2 n)$, i.e., performing $O(m\log^2 n)$ operations, where m is the number of edges and n is the number of vertices in the graph. We note that, in general, an algorithm running in $O(T)$ parallel time while performing $O(W)$ work represents an algorithm running in $O(T)$ parallel time using $O(W/T)$ processors.

In this paper we provide efficient $O(\log n)$ time parallel algorithms for both problems. For problem 1 our algorithm is optimal for all practical purposes: it performs $O((m+n)\alpha(m,n)+m+n\lambda(k,n))$ work, where α is the inverse Ackerman function, and $\lambda(k,n)$ is the inverse of a certain function at the $\lfloor k/2 \rfloor$-th level of the primitive recursive hierarchy, k being any positive integer. For problem 2 our algorithm performs $O(m\log^2 n)$ work, which is the same as the work performed by the earlier algorithm.

† Supported by Joint Services Electronics Program under N00014-84-C-0149, the Semiconductor Research Corporation under 86-12-109, and the International Computer Science Institute, Berkeley, CA.

‡ Supported by NSF grants NSF-CCR-8615337 and NSF-DCR-8413359, ONR grant N00014-85-K-0046, by the Applied Mathematical Sciences subprogram of the Office of Energy Research, U. S. Department of Energy, under contract number DE-AC02-76ER03077 and the Foundation for Research in Electronics, Computers and Communication administered by the Israeli Academy of Sciences and Humanities.

In problem 1 we consider a simple characterization of interlacings between chords on a cycle. A contribution of the present paper is in showing that such a simple characterization suffices for obtaining the 'coalesced graph' in the triconnectivity algorithm. This leads to a considerable simplification relative to the algorithm in [MiRa2]. Further our algorithm is faster and performs less work than the earlier algorithm.

Our improved algorithm for problem 2 is based on a method of pipelining the computation, which we call *streaming*, to reduce the parallel time from $O(\log^2 n)$ to $O(\log n)$, while maintaining the total work as $O(m\log^2 n)$.

In addition to the application to triconnectivity, our algorithm for problem 1 gives almost optimal $O(\log n)$ parallel algorithms for the following problems: 1) finding a planar embedding of a graph with a known Hamiltonian cycle, if such an embedding exists [RoTa]; 2) determining if a permutation is sortable on two stacks [EvIt]; and 3) determining if a sequence is a Gauss code [Ga,LoMa,Ro].

2. DETERMINING INTERLACINGS OF CHORDS ON A CYCLE

Let G be a graph with a known Hamiltonian cycle C. Then G can be viewed as the simple cycle C together with a collection of chords (v_i, v_j) on C. For simplicity we assume $C = <1, \cdots, n, 1>$.

Let $a = (u, v)$ and $b = (w, x)$ be two chords on C, where $u < v$ and $w < x$. Then a and b *interlace* on C if the vertices u, v, w, x satisfy $u < w < v < x$ or $w < u < x < v$. If a and b interlace, then they cannot be placed on the same side of C (either inside or outside) in a planar embedding (see, e.g., [Ev]). If a and b do not interlace, then they can be placed in a planar embedding on the same (opposite) side of C if and only if there exists no sequence of chords $<a = a_0, a_1, \cdots, a_{r-1} = b>$, with r odd (even) such that a_i interlaces with a_{i+1}, $0 \le i \le r-2$. If there is such a sequence with r even then a and b have *even interlacing parity* and if there is such a sequence with r odd, then a and b have *odd interlacing parity*. If no such sequence exists for r either odd or even, then a and b have *null interlacing parity*: in this case a and b can be placed either in the same side or in opposite sides of C in a planar embedding. It is possible for a and b to have both odd and even parity -- in this case, no planar embedding of G is possible.

We now present an almost optimal parallel algorithm for preprocessing the graph G so that an interlacing parity for any pair of chords can be determined in constant time. If a pair of chords a and b have both even and odd interlacing parities, then we will find only one of these parities. However, we show in section 5.1 that we can determine whether there exists a planar embedding for G. A high level description of the algorithm is as follows: We construct an auxiliary graph G' with a vertex for each chord on C. We then place some edges in G'. Each such edge connects a pair of vertices whose corresponding chords interlace. We do not put in an edge for every pair of interlacing chords, but only a subset of them, so that the size of G' is linear in the size of G. In particular, if r is the number of chords on C then G' will have at most $2r$ edges. Further if a and b are two chords for which there exists some interlacing sequence then a and b will lie in the same connected component of G'. Since each edge in G' represents an actual interlacing, we can obtain an interlacing parity for each pair of chords with an interlacing sequence by finding a spanning tree in each connected component of G' and two coloring the vertices of the spanning tree: Now two chords, whose vertices are in the same connected component in G', have odd interlacing parity if they have different colors and even parity if they have the same color. The algorithm is presented below. Since vertices in G' correspond to chords on C, we will sometimes refer to a vertex in G' as a chord; by this we mean the chord in G that this vertex represents.

Algorithm 1 Interlacing Parity

Input: Undirected graph G consisting of a simple cycle $C = <1, \cdots, n, 1>$, together with a collection of chords on C.

1. For each chord $c = (u, v)$, $u < v$, that interlaces with some other chord do

a) *left rule:* Let u_l be the minimum numbered vertex on C such that c interlaces with a chord incident on u_l. If $u_l < u$, then find the chord $l_c = (u_l, v_l)$ with maximum v_l that interlaces with c and place an edge (the 'left edge') in G' between vertices c and l_c;

b) *right rule:* Let v_r be the maximum numbered vertex on C such that c interlaces with a chord incident on v_r. If $v_r > v$, then find the chord $r_c = (u_r, v_r)$ with minimum u_r that interlaces with c and place an edge (the 'right edge') in G' between c and r_c.

2. Find a spanning tree in each connected component of G' and two-color the spanning trees. Assign a label <component number, color> to each vertex.

With this preprocessing we can determine an interlacing parity for any pair of chords c,d on C as follows: If component number of c is not equal to component number of d then c and d have null interlacing parity, otherwise they have even interlacing parity if they have the same color, and odd interlacing parity if they have different colors.

Lemma 1. If a pair of chords α,β have an interlacing parity that is not null, then α and β appear in the same connected component of G'.
Proof. By induction on the number of chords r on C.
Base: $r=2$. This is immediate.
Induction step: Assume that the claim is true for all simple cycles with up to $r-1$ chords, and let G be a graph consisting of a simple cycle C, together with r chords. Let v be the lowest numbered vertex on C that has a chord incident on it, and let $a=(v,w)$ be the chord incident on v with maximum w. Delete a from G to form G_1. By the induction hypothesis, every pair of chords that have an interlacing parity that is not null in G_1 appear in the same connected component in G'_1.

We will show two things:
A. Any pair of chords α,β that lie in the same connected component in G'_1 continue to lie in the same connected component in G'.
B. Any chord having an interlacing parity with a will be in the connected component of a in G'.

Lemma 1 follows. To see this consider any pair of chord α,β, where $\alpha \neq a$ and $\beta \neq a$. Assume α and β have an interlacing parity in G and consider an interlacing sequence. If the interlacing sequence does not include a apply claim A. If the interlacing includes a, then we may assume that a appears only once. Apply claim B to show that a is in the same connected component as its predecessor and successor in the sequence. Finally, apply claim A.

We show A. For this, observe that the only edges in G'_1 that are not present in G' are the edges introduced by the left rule for chords that interlace with a: each such edge is replaced by an edge connecting to vertex a. Let b be such a chord interlacing with a, let its left edge in G'_1 be (b,c). Its left edge in G' is (b,a). We claim that a,b and c lie in the same connected component in G'.
Case 1: Chord c interlaces with chord a. Then edges (a,c) and (a,b) are present in G' and hence a,b and c belong to the same connected component in G'.
Case 2: Chord c does not interlace with a. Consider the right edge (c,d) of c. Then chord d has its right endpoint at least as large as the right endpoint of b, and hence d interlaces with a. Hence edges $(a,d),(c,d)$ and (b,a) are present in G', i.e., a,b and c lie in the same connected component in G'.

We show claim B. For each chord b having an interlacing parity with a, consider an interlacing sequence $a_0=a,a_1,\ldots,a_k=b$. Chord a_1 and b are in the same connected component of G' by claim A. It suffices to show that chord a and a_1, which actually interlace, are in the same connected component. But, there must be an edge connecting them in G' by the left rule for a_1. Claim B and Lemma 1 follow. []

Lemma 2. Algorithm 1 correctly finds an interlacing parity for each pair of chords on C.
Proof. Since an edge (a,b) is placed in G' only if chords a and b interlace, it follows that the algorithm finds a correct interlacing parity for every pair of chords that belong to the same connected component. By Lemma 1, vertices that belong to different connected components correspond to chords that have null interlacing parity. This establishes the correctness of the algorithm.[]

To implement step 1 we determine at each vertex v, the chord with smallest attachment s_v and the chord with largest attachment l_v incident on v, and we store at v the ordered pairs $(s_v,-v)$ and (l_v,v). This can be done optimally in parallel in $O(\log n)$ time. Consider chord $c=(u,v)$, $u<v$. Applying the left rule to chord c (as required in step 1a) will be done by answering the following query: "find the minimum s_w for all w such that $u<w<v$". Similarly, for applying the right rule to chord c (as required in step 1b) we need to answer the query: "find the maximum l_w for all w such that $u<w<v$". We will have one minimum and one maximum query for each chord and a total of $2r$ queries. These queries can be answered in $O(\log n)$ time while performing $O(r+n\lambda(k,n))$ work on a CREW PRAM [AlSc], where k is any positive integer, and $\lambda(k,n)$ is the inverse of a certain function at the $\lfloor k/2 \rfloor$-th level of the primitive recursive hierarchy

("Ackerman like"). Step 2 can be implemented in $O(\log n)$ time with $O((r+n)\alpha(r,n))$ work on a CRCW PRAM using the connectivity algorithm in [CoVi]; here $\alpha(r,n)$ is the inverse Ackerman function. Thus, this gives an almost optimal parallel algorithm for the problem that runs in $O(\log n)$ time on a CRCW PRAM.

3. GENERATING BRIDGES OF A DISJOINT COLLECTION OF SUBGRAPHS

Let $G=(V,E)$ be a connected graph, and let Q be a subgraph of G. Analogous to [Ev, p. 148], we define the *bridges of Q in G* as follows: Let V' be the vertices in $G-Q$, and consider the partition of V' into classes such that two vertices are in the same class if and only if there is a path connecting them which does not use any vertex of Q. Each such class K defines a *(nontrivial) bridge $B=(V_B,E_B)$* of Q, where B is the subgraph of G with $V_B=K \cup$ {vertices of Q that are connected by an edge to a vertex in K}, and E_B containing the edges of G incident on a vertex in K. The vertices of Q which are connected by an edge to a vertex in K are called the *attachments* of B. An edge (u,v) in $G-Q$, with both u and v in Q, is a *trivial bridge* of Q, with attachments u and v. The trivial and nontrivial bridges together form the bridges of Q in G. Since trivial bridges are easy to identify, we will assume, for convenience, that all bridges are non-trivial. This would mean that Q is an induced subgraph on a subset of V in G.

Let $G=(V,E)$ be a graph, and let Q be a subgraph of G. We define the *bridge graph of Q, $S=(V_S,E_S)$* as follows: Let the bridges of Q in G be $B_i, i=1, \cdots, k$. Then $V_S=V(Q) \cup \{B_1, \cdots, B_k\}$ and $E_S=E(Q) \cup \{(v,B_i) \mid v \in V(Q), 1 \le i \le k$, and v is an attachment of $B_i\}$. Each subgraph of S consisting of one of the B_i, together with all edges incident on it, is called a *star* [MiRa2].

In this section we obtain a parallel algorithm for the following problem. Let $G=(V,E)$ be a connected undirected graph, with $|V|=n$ and $|E|=m$. Let $V=\{V_0, \cdots, V_{k-1}\}$ a collection of disjoint sets of vertices in V. Let T_i be the subgraph of G which is induced by V_i, $0 \le i \le k-1$. Our problem is to find the bridge graph of each T_i. Here, k is arbitrary and can be as large as n. A naive algorithm for this problem would be to delete each T_i in parallel and find the connected components in each resulting graph. This algorithm performs $O(nm)$ work in the worst case. A better parallel algorithm for a variant of this problem is given in [MiRa2]: this algorithm runs in $O(\log^2 n)$ time on a CRCW PRAM with $O(m)$ processors, i.e., with a total work of $O(m \log^2 n)$. The algorithm is based on decomposing the computation into $\log k$ stages of connectivity computations.

The main contribution of this section is in pipelining these stages and thereby trimming a factor of $\log k$ from the time bound. Specifically, we obtain an $O(\log n)$ parallel algorithm on a CRCW PRAM for this problem using $O(m \log k)$ processors, i.e., performing $O(m \log k \log n)$ work, as before. Below, we first describe the $O(\log k \log n)$ time algorithm and later show how to apply a type of pipelining, which we call *streaming*, to improve this to $O(\log n)$ time.

Denote by $G-T_i$ the subgraph of G induced by $V-V_i$. In order to form the bridge graph of each T_i, we find the connected components of $G-T_i$ and collapse each such component into a single vertex: there is one such collapsed vertex for each bridge of T_i. Once we determine these for each T_i, we can form the bridge graph for each T_i by deleting multiple edges of G that connect a bridge to a vertex of T_i.

3.1. Forming the bridge graphs in $O(\log n \log k)$ time:

Let H_i denote the graph obtained from $G-T_i$ by collapsing each connected component into a single vertex. (Observe that this graph has no edges.) For $k-1 \ge j > i \ge 0$, let $H_{i,j}$ denote the graph $G-(T_i \cup T_{i+1} \cup \cdots \cup T_j)$ with each connected component collapsed into a single vertex. Note that $H_i=H_{i,i}$.

We construct each H_i by constructing $H_{i,j}$ in stages for intervals of $[i,j]$ that decrease in length by half so that after $\log k$ stages we will have constructed $H_{i,i}$. The computation is guided by a complete binary tree (see Fig. 3.1.)

When referring to the binary tree it will be convenient to partition its nodes into levels, according to their distance from the root of the tree. To start with (see Step 1 in the algorithm below, and the root, or level zero, of the binary tree) we form $H_{0,k-1}$ by finding connected components in $G-(T_0 \cup \cdots \cup T_{k-1})$, and collapsing each such connected component into a single vertex. In the next stage (which already belongs to step 2 and is represented by the two children of the root, or level 1 of the tree), we form

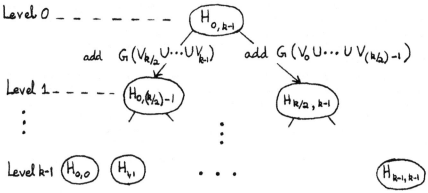

Fig. 3.1: Forming the bridge graphs

$H_{0,(k/2)-1}$ and $H_{k/2,k-1}$ in parallel. We form $H_{0,(k/2)-1}$ by adding to $H_{0,k-1}$ the subgraph of G induced by $V_{k/2} \cup \cdots \cup V_{k-1}$ and all edges of G connecting this subgraph with the subgraph represented by $H_{0,k-1}$. We then collapse each resulting connected component into a single vertex. Analogously, we form $H_{k/2,k-1}$ by adding to $H_{0,k-1}$ the subgraph of G induced by $V_0 \cup \cdots \cup V_{(k/2)-1}$ and all edges of G connecting this subgraph with the subgraph represented by $H_{0,k-1}$, and then collapse each resulting connected component into a single vertex. We proceed in this way for $\log k$ stages, at which point we will have constructed $H_{i,i}$, for each i. The algorithm is presented below (for simplicity, we assume that k is a power of 2).

Algorithm 2: Forming the Bridge Graphs in $O(\log n \log k)$ Time

Input: Undirected graph $G = (V,E)$ with a collection of disjoint (induced) subgraphs $T = \{T_0, \cdots, T_{k-1}\}$.

1. Form $H_{0,k-1}$ by collapsing connected components in $G - (T_0 \cup \cdots \cup T_{k-1})$.
2. *For* $i = 1, \cdots, \log k$ *do*

    for $j = 0$ *to* 2^i *pardo*

    (Let $a = jk, b = (j+1)k, c = (j+2)k, x = 2^i$)

    2.1. Form $H_{(a/x),(b/x)-1}$ by adding to $H_{(a/x),(c/x)-1}$ the subgraph of G induced by $V_{(b/x)} \cup \cdots \cup V_{(c/x)-1}$ and all edges of G connecting this subgraph with the subgraph represented by $H_{(a/x),(c/x)-1}$. Then collapse each connected component into a single vertex; and

    2.2. Form $H_{(b/x),(c/x)-1}$ by adding to $H_{(a/x),(c/x)-1}$ the subgraph of G induced by $V_{(a/x)} \cup \cdots \cup V_{(b/x)-1}$ and all edges of G connecting this subgraph with the subgraph represented by $H_{(a/x),(c/x)-1}$. Then collapse each connected component into a single vertex.

3. Obtain the bridge graph of T_i, for $i = 0, \ldots k-1$. For this add T_i and edges connecting T_i and $H_{i,i}$ to $H_{i,i}$, and then remove multiple edges. The resulting graph is the bridge graph of T_i.

This algorithm has $O(\log k)$ stages of connectivity computation on graphs of size $O(m+n)$, and hence, using the connectivity algorithm in [ShVi], it can be implemented to run in $O(\log k \log n)$ parallel time with a linear number of processors on a CRCW PRAM. There are several implementation considerations of Algorithm 2, e.g., 1) assigning processors to implement tasks of the algorithm; 2) recording the $H_{i,j}$ graphs; and 3) finding the initial component name for each vertex at the start of each stage. These have fairly straightforward solutions, whose details will appear in the full paper.

3.2. An $O(\log n)$ Time Algorithm Using Pipelining

We now describe a method of pipelining the above computation so that the parallel time is reduced to $O(\log n)$ while the total work done remains the same. We will continue to use the algorithm of [ShVi] to find connected components. However, we will *stream* the connected components down the levels as and when they are formed. Thus, as soon as a component being grown at a given level realizes that it is a connected component for that level and cannot grow any further, it transfers itself to the two sub-problems that use it in the next level. In general, at the ith level, the edges present in that level try to implement the connectivity algorithm at each time instant. The computation is completed when every component in the last level has been computed.

For convenience let us assume that a single stage of the connectivity algorithm in [ShVi] can be implemented in one time unit (it takes some constant number of time units whose exact number depends on the instruction set of the processors). Let us also assume that a component can test whether it can grow no further at a given level and transfer itself to the next level in one time unit. Lemma 3 below proves that the number of time units needed by the algorithm is $O(\log n)$. We assume familiarity with the connectivity algorithm in [ShVi].

Lemma 3 Consider any active hooking tree which is at level l at time t and whose height is x. Then it must contain at least $x(3/2)^{t-l}$ vertices.

Proof. By induction on t.

Base. Time $t=0$: All hooking trees are of size 1, height 1 and at level 0, and we have $(3/2)^{t-l}=1$.

Induction step. Assume inductively that the lemma is true for time up to $t-1$ and consider time t. First note that there can be no active hooking tree at a level $l>t$ since it takes at least one time unit for a component to be transferred down by one level.

Let C be a hooking tree of height x at level $l \leq t$. We show that C has at least $x(3/2)^{t-l}$ vertices.

Case 1: C was transferred from level $l-1$ to level l at time t. Then at time $t-1$, C was at level $l-1$. By the pipelining rules it had to have height 1. By the induction hypothesis C had at least $(3/2)^{t-1-(l-1)}$ vertices, i.e., $(3/2)^{t-l}$ vertices.

Case 2: All vertices in C were at level l at time $t-1$. Since C is still in level l at time t, it must consist of one of the following two at time $t-1$.

Case 2.1. At least two hooking trees C_1 and C_2, which were at level l (at time $t-1$). Let $C_1, C_2, \cdots, C_g$ all hooking trees (of level l at time $t-1$) that comprise C, and let $x_1, x_2, \ldots, x_g$ be their heights, respectively. By the induction hypothesis, C_i had at least $x_i(3/2)^{t-1-l}$ vertices, for $1 \leq i \leq g$, respectively (at time $t-1$), and hence C has at least $(x_1+x_2+...+x_g)\cdot(3/2)^{t-1-l}$ vertices at time t, i.e., at least $\dfrac{x_1+x_2+...+x_g}{(3/2)}\cdot(3/2)^{t-l}$ vertices at time t. Finally, we observe that the connectivity algorithm in [ShVi] implies that the height of C is at most $\dfrac{x_1+x_2+...+x_g}{(3/2)}$. The inductive claim follows.

Case 2.2. A single hooking tree C_1 whose height was $x_1 \geq 2$. The algorithm in [ShVi] implies that the height of C is at most $\dfrac{x_1}{(3/2)}$. The inductive claim follows.

This establishes the induction step and the lemma is proved.[]

Since the number of levels is $\log k$ and no hooking tree has more than n vertices, we have $n \geq 2^{T-\log k}$, where T is the number of time units used by the algorithm. Thus $T=O(\log n)$. We use a processor for each edge at each level. Thus this gives an $O(\log n)$ time algorithm on a CRCW PRAM with $O(m\log k\log n)$ work.

A methodological remark. The $O(\log^2 n)$ connectivity algorithm of [HiChSa] consisted of $\log n$ rounds. Each round included one step of hooking and $\log n$ steps of shortcutting. The contribution of [ShVi] was in designing the hookings so that each hooking is performed as as soon as it becomes possible. To sum extent, this can be viewed as saving a $\log n$ factor in the algorithm of [HiChSa] by pipelining the hookings. Here we showed how to proceed in each connectivity computation of Fig. 3.1 as soon as such computation becomes possible. Thereby, we saved a $\log n$ factor. So, what we have here is one "layer" of pipelining with respect to the algorithm in [ShVi] and two "layers" with respect to [HiChSa].

4. APPLICATION TO TRICONNECTIVITY

4.1. Review of Previous Work

We first review the triconnectivity algorithm of [MiRa2].

An *ear decomposition* [Lo,Wh] $D = [P_0, \cdots, P_{r-1}]$ of an undirected graph $G = (V,E)$ is a partition of E into an ordered collection of edge disjoint simple paths $P_0, \cdots, P_{r-1}$ such that P_0 is a simple cycle and each endpoint of $P_i, i = 1, \cdots, r-1$ is contained in some $P_j, j < i$. The P_i's are called the *ears* of D. D is an *open ear decomposition* if none of the $P_i, i = 1, \cdots, r-1$ is a simple cycle. A *trivial ear* is an ear consisting of a single edge. A graph has an open ear decomposition if and only if it is biconnected [Wh].

Let $G = (V,E)$ be a graph and let P be a simple path in G. If each bridge of P in G contains exactly one vertex not on P, and there is a bridge B of P with the endpoints of P as attachments, then we call G the *star graph of P and denote it by $G(P)$*. Let $G(P)$ be a star graph, and let $S_1, \cdots, S_k$ be some of the stars in $G(P)$. The operation of *coalescing* the stars $S_i, i = 1, \cdots, k$ removes these stars and replaces them by a new star S whose attachments are the union of the attachments of $S_1, \cdots, S_k$.

Let G be a biconnected graph with an open ear decomposition $D = [P_0, \cdots, P_{r-1}]$. Let the bridges of P_i in G that contain vertices on ears numbered lower than i be $B_{r_1}, \cdots, B_{r_n}$. We shall call these the *anchor bridges of P_i*. The *ear graph of P_i*, denoted by $G_i(P_i)$ is the graph obtained from the bridge graph of P_i by a) coalescing all stars corresponding to anchor bridges; and b) removing multiple copies of two-attachment bridges with the endpoints of the ear as attachments. We will call the star obtained by coalescing all anchor bridges, the *anchoring star* of $G_i(P_i)$. For any two vertices x,y on P_i, we denote by $V_i(x,y)$, the internal vertices of $P_i(x,y)$; we denote by $V_i[x,y]$, the vertices in $P_i[x,y] - \{x,y\}$ together with the vertices in anchor bridges. For a star graph $G(P)$, the set $V(x,y)$ represents the vertices in $P(x,y) - \{x,y\}$, and the set $V[x,y]$ represents the vertices in $P[x,y] - \{x,y\}$.

We now extend the notion of *interlacing*, which we introduced in section 2 for chords, to stars. Two stars S_j and S_k in a star graph $G(P)$, where P is a simple path, *interlace* (see also [Ev, p. 149]) if one of the following two hold: 1) there exist four distinct vertices a,b,c,d in increasing order in P such that a and c belong to $S_j(S_k)$ and b and d belong to $S_k(S_j)$; or 2) there are three distinct vertices on P that belong to both S_j and S_k.

Given a star graph $G(P)$, the *coalesced graph* G' of G is the graph obtained from G by coalescing all pairs of stars that interlace.

Let $G = (V,E)$ be a biconnected graph. A *(nontrivial) separating pair* in G is a pair of vertices u,v in V whose removal decomposes G into two or more connected components. A *trivial separating pair* is a pair of vertices u,v with (u,v), an edge (note that a pair of vertices can be both a trivial and a nontrivial separating pair). A *candidate pair* is a trivial or nontrivial separating pair; a *candidate set* is a set of vertices such that each pair in the set is a candidate pair. It is established in [MiRa2] that the following algorithm determines triconnectivity and finds all candidate sets in a biconnected undirected graph $G = (V,E)$.

Algorithm 3: Triconnectivity Algorithm

Input: Undirected biconnected graph $G = (V,E)$.

1) Find an open ear decomposition $D = [P_0, \cdots, P_{r-1}]$ for G.

2) *For $j = r-1, r-2, \cdots, 0$ pardo:*

A)   Construct the ear graph G_j.

B)   Coalesce all interlacing stars on $G_j(P_j)$ to form the coalesced graph G'_j. Find a planar embedding of G'_j, with P_j on the external face, and identify each set of vertices on P_j on a common inner face in this embedding as a candidate set.

C)   If $j > 0$ let the endpoints of P_j be u and v. If the anchor bridge of P_j has an internal attachment on P_j, then delete the candidate set $\{u,v\}$.

D)   Delete any doubleton candidate set that contains the endpoints of an edge in P_j.

Step 1 has an $O(\log n)$ time parallel algorithm using a linear number of processors [MaScVi,MiRa]. Steps 2C and 2D are straightforward to implement. [MiRa2] give an $O(\log^2 n)$ time parallel algorithms on a CRCW PRAM for steps 2A and 2B that performs $O(m\log^2 n)$ work. In the next subsection we present an $O(\log n)$ time parallel algorithm for step 2A that performs $O(m\log^2 n)$ work, and in subsection 4.3 we present an $O(\log n)$ parallel algorithm for step 2B that performs almost linear work.

4.2. Forming the Ear Graphs

Let $G=(V,E)$ be a biconnected graph with an open ear decomposition $D=[P_0, \cdots, P_{r-1}]$. Let $|V|=n$ and $|E|=m$. We give an $O(\log n)$ time efficient parallel algorithm for finding the ear graphs based on Algorithm 2 in section 3. We need to address two technical details that arise in the definition of ear graphs, which are not handled in Algorithm 2.

The first detail is that each bridge of ear P_i needs to know if it contains a vertex on an ear numbered lower than i. This is because we need to construct the ear graph, in which all such bridges are coalesced. Each bridge reflects a connected component in the graph $H_{i,i}$, using the notations of section 3. In level α of this computation the vertices that participate are taken from ears whose numbers are in the range $j(k/2^\alpha) \cdots (j+1)(k/2^\alpha)-1$ for some integer j. For each connected component which is updated in level α (i.e. it includes vertices that belong to these ears) we record this fact, unless a previous level showed that the component contains a vertex belonging to an ear whose number is lower than this range. That is, each time a component is transferred to the next level we update the lowest range of vertices it includes. Modifying Algorithm 2 to keep track of this information is straightforward.

The second detail is that Algorithm 2 assumes that we find the bridges of a *disjoint* collection of subgraphs of G; without this assumption, the linear bound on the number of edges present at any given step is no longer valid. For the ear graphs, our collection of subgraphs is the set of ears in D. These form a disjoint collection of subgraphs with the possible exception that pairs of ears may share an endpoint. However, we do not need to construct the bridge graph of each ear, but only the ear graph, in which all anchor bridges have been coalesced.

In order to take care of this second detail we use a variant of $H_{i,j}$ of Algorithm 2, which we call $G_{i,j}$. Given $G=(V,E)$ with an open ear decomposition $D=[P_0, \cdots, P_{r-1}]$ as above, we define $G_{i,j}$, the (i,j) *ear graph of G, for $i<j$*, as follows: Let Q be the set of paths $P_i, \cdots, P_j$, and let U be the set of vertices in Q that are contained in ears numbered lower than i (these are some of the endpoints of $P_i, \cdots, P_j$). Let $R_1, \cdots, R_l$ be the bridges of Q that contain a vertex on an ear numbered lower than i, and let $S_1, \cdots, S_k$ be the bridges of S whose attachments are all in U. Let G_B be the bridge graph of Q. Then $G_{i,j}$ is the graph G_B with the bridges $R_1, \cdots, R_l$ coalesced, and with the bridges $S_1, \cdots, S_t$ replaced by a new bridge with attachments to all vertices in U. Note that $G_{i,i}$ is simply the ear graph G_i. It is not difficult to see that the total size of all of the $G_{i,j}$ at a given level is linear in the size of G.

While implementing Algorithm 2 for finding the bridges of the ears, we will form $G_{i,j}$ instead of $H_{i,j}$ at each level; thus we will discard any connected component that is fully formed at a level, whose only attachments to the deleted ears are to those endpoints of these ears that lie on lower numbered ears. However, we will coalesce all bridges of an ear P_i that contain a vertex in an ear numbered lower than i only at the end of the algorithm. We claim that, with these modifications, the algorithm correctly constructs the ear graph of each ear.

Lemma 4 Let C be a bridge of $\bigcup\limits_{k=i}^{j} P_k$, whose only attachments are to the endpoints of ears $P_i, P_{i+1}, \cdots, P_j$ that lie on ears numbered lower than i. Then
a) if C is part of a bridge B of some $G_k(P_k), k=i, \cdots, j$ with an internal attachment, then B contains a vertex on an ear numbered lowered than k;
b) the attachments to each such P_k that are contributed by C are only the endpoints of P_k, i.e., internal attachments to P_k can be contributed only by vertices in B that are outside C.
Proof Straightforward.[]

Note that in the definition of ear graph of P_i, we delete all multiple two-attachment bridges with the endpoints of the ear as attachments. Any connected component formed while implementing Algorithm 2, whose only attachments to the deleted ears are to those endpoints of these ears that lie on lower numbered ears will finally contribute attachments only to the endpoints of some ears by Lemma 4. If this connected component finally results in a bridge of P_i whose only attachments are the endpoints of that ear, then deleting it earlier does not alter the final result. If this connected component becomes part of a larger bridge of P_i which has some internal attachment, then the attachments to the endpoints of the ear can be deduced without explicitly keeping this connected component since the corresponding bridge will contain contain a vertex from an ear numbered lower than P_i by Lemma 4. Finally, at the end of the algorithm we coalesce all bridges with a vertex on a lower numbered ear, and thus we obtain the ear graph of each ear at the end of the algorithm.

With these modifications we can run Algorithm 2 using the ears as the collection of subgraphs, and thus obtain the ear graphs in $O(\log n)$ time performing $O(m\log^2 n)$ work on a CRCW PRAM.

4.3. Forming the Coalesced Graph

We will use Algorithm 1 to perform step 2B of Algorithm 3. In step 2B we are given a simple path (an ear) together with a collection of stars, and we need to coalesce all stars that interlace. Since Algorithm 1 applies only to interlacing chords, we first reduce the computation in step 2C to that of coalescing interlacing chords.

Let $G(P)$ be a star graph with $P=<1,2,\cdots,n>$ and containing stars $S_1,\cdots,S_k$. We will decompose each S_i into a collection of chords such that each pair has an interlacing parity. Thus their coalesced graph will give back S_i.

We first split each vertex v on P into two adjacent vertices $v<v'$ on P and put in an edge between vertices 1 and n' so that P becomes the simple cycle $C=<1,1',2,2',\cdots,n,n'>$.

If S_i has only two attachments u and v, $u<v$, then we replace S_i by the chord (u',v).

If S_i has r attachments $v_1<\cdots<v_r,r\geq3$, then we replace it by the chord (v'_1,v'_2) and the set of chords $(v_2,v_i),i=3,\cdots,r$. Since every chord in this set interlaces with chord (v'_1,v'_2), every pair has an interlacing parity. Further the size of this set is linear in the number of attachments of S_i.

Let us call the resulting graph H. H consists of the simple cycle C together with a collection of chords on C. The following two lemmas are straightforward.

Lemma 5 Let star S_j be decomposed into chords $c_1,\cdots,c_r$ by the above method. Then every pair of chords $c_i,c_j,1\leq i<j\leq r$ has an interlacing parity.

Lemma 6 Let star S_i be decomposed into $c_1,\cdots,c_r$ and let star S_j be decomposed into $d_1,\cdots,d_s$. Then S_i and S_j interlace on P if and only if there exist i and j such that c_i and d_j interlace on C.

We now run Algorithm 1 on H. We then form a star on P for each connected component in H' with attachments to the endpoints of each chord in that component, where v and v' are both identified as v on P, for each v. This gives us the coalesced graph of $G(P)$. Since the reduction from G to H can be done easily in $O(\log n)$ time with a linear amount of work, the entire coalescing algorithm can be performed with the same processor and time bounds as Algorithm 1.

5. OTHER APPLICATIONS

5.1. Planarity of Graphs with a Known Hamiltonian Cycle

We can use Algorithm 1 to develop a simple, almost optimal, $O(\log n)$ time parallel algorithm to find a planar embedding of a graph G with a known Hamiltonian cycle, if such a planar embedding exists, or to report that no such planar embedding exists.

Let the Hamiltonian cycle in G be $C=<1,2,\cdots,n,1>$. Then each edge of G, not on C, appears as a chord on C, and G is planar if and only if no pair of chords have both odd and even interlacing parities. To determine a planar embedding of G, we run Algorithm 1 on it. Each chord gets either a color 0 or a color 1. We place chords with color 0 within the cycle C and chords with color 1 outside C. We now check if this is a valid planar embedding, i.e., we run step 1 of Algorithm 1 separately on the set of chords with color 0, and the set of chords with color 1: G is planar if and only if no interlacing is detected, i.e., if each of the derived graphs on the chords has no edge.

5.2. Testing Two-Stack Sortable Permutations

There is a straightforward reduction from this problem to the problem of testing if an associated graph with a known Hamiltonian cycle is planar [RoTa]. This reduction can be done optimally in parallel in $O(\log n)$ time, thus giving an almost optimal, $O(\log n)$ time algorithm for the original problem.

5.3. Recognizing Gauss Codes

There is a reduction from this problem to the problem of determining if a graph with a known Hamiltonian cycle is planar [RoTa]. The only nontrivial part of the reduction is the construction of two special types of Euler tours in a graph in which every vertex has degree 4. It is easy to use the Euler tour algorithm of [AtVi] to construct these special Euler tours; the resulting complexity for recognizing Gauss codes is the same as that of Algorithm 1.

REFERENCES

[AlSc] N. Alon, B. Schieber, "Optimal preprocessing for answering on-line product queries," tech. report, Tel Aviv University, May 1987.

[AtVi] M. J. Attalah, U. Vishkin, "Finding Euler tours in parallel," *Jour. Comput. and System Sciences,* vol. 29, pp. 330-337, 1984.

[CoVi] R. Cole, U. Vishkin, "Approximate and exact parallel scheduling with applications to list, tree and graph problems," *Proc. 27th Ann. Symp. on Foundations of Comp. Sci.,* Toronto, Canada, Oct. 1986.

[Ev] S. Even, *Graph Algorithms,* Computer Science Press, Rockville, MD, 1979.

[EvIt] S. Even, A. Itai, "Queues, stacks, and graphs," *Theory of Machines and Computations,* Z. Kohavi and A. Paz, eds., Academic Press, New York, NY, 1971, pp. 71-86.

[Ga] C. F. Gauss, "Werke," Teubner, Leipzig, 1900, pp. 272 and 282-286.

[HiChSa] D. S. Hirschberg, A. K. Chandra, D. V. Sarwate, "Computing connected components on parallel computers," *CACM,* vol. 22, 1979, pp. 461-464.

[LoMa] L. Lovasz, M. L. Marx, "A forbidden substructure characterization of Gauss codes," *Bull. Amer. Math. Soc.,* vol. 82, 1976, pp. 121-122.

[MaScVi] Y. Maon, B. Schieber, U. Vishkin, "Parallel ear decomposition search (EDS) and st-numbering in graphs," *Theoretical Computer Science,* vol. 47, 1986, pp. 277-298.

[MiRa] G. L. Miller, V. Ramachandran, "Efficient parallel ear decomposition with applications," unpublished manuscript, MSRI, Berkeley, CA, January 1986.

[MiRa2] G. L. Miller, V. Ramachandran, "A new graph triconnectivity algorithm and its parallelization," *Proc. ACM Symp. on Theory of Computing,* New York, NY, May 1987.

[Ro] P. Rosenstiehl, "Solution algebrique du probleme de Gauss sur la permutation des points d'intersection d'une ou plusiers courbes fermees du plan," *C. R. Acad. Sci. Paris,* vol. 283, no. 8, 1976.

[RoTa] P. Rosenstiehl, R. E. Tarjan, "Gauss codes, planar Hamiltonian graphs, and stack-sortable permutations," *J. Algorithms,* Vol. 5, no. 3, 1984, pp. 375-390.

[ShVi] Y. Shiloach, U. Vishkin, "An $O(\log n)$ parallel connectivity algorithm," *J. Algorithms,* vol. 3, 1982, pp. 57-63.

[Wh] H. Whitney, "Non-separable and planar graphs," *Trans. Amer. Math. Soc.,* vol. 34, 1932, pp. 339-362.

Subtree Isomorphism is in Random NC

Phillip B. Gibbons
University of California, Berkeley

Richard M. Karp
University of California, Berkeley

Gary L. Miller
University of Southern California

Danny Soroker
University of California, Berkeley

Abstract

Given two trees, a guest tree G and a host tree H, the subtree isomorphism problem is to determine whether there is a subgraph of H that is isomorphic to G. We present a randomized parallel algorithm for finding such an isomorphism, if it exists. The algorithm runs in time $O(\log^3 n)$ on a CREW PRAM, where n is the number of nodes in H. Randomization is used (solely) to solve each of a series of bipartite matching problems during the course of the algorithm. We demonstrate the close connection between the two problems by presenting a log space reduction from bipartite perfect matching to subtree isomorphism. Finally, we present some techniques to reduce the number of processors used by the algorithm.

1. Introduction

A <u>subtree</u> of a tree T is any subgraph of T that is a tree. Given two (unrooted) trees, a guest tree G and a host tree H, the <u>subtree isomorphism</u> problem is to determine whether there is a subtree of H that is isomorphic to G. There is an $O(n^{2.5})$ sequential algorithm for this problem due to Matula[5], where n is the number of nodes in H. In this paper, we present an $O(\log^3 n)$ time randomized algorithm for a CREW PRAM that exhibits the mapping between the trees, if such a mapping exists. We assume the word size of the PRAM is $c \log n$ for some constant c. With a few techniques to reduce the processor count, the algorithm uses $< \sqrt{n} \cdot P(n)$ processors, where $P(n)$ is the number of processors needed for *one* bipartite matching problem on n nodes using the fastest algorithm for bipartite matching to date[7], for a total of $o(n^{4.9})$ processors. More precisely, let $M(n)$ be the number of bit operations required by a CREW PRAM to multiply two $n \times n$ boolean matrices in $O(\log n)$ time. Then our algorithm uses $n^{2.5} M(n)/\log^3 n$ processors.

Our algorithm is based on Matula's sequential algorithm. The main obstacle to developing a fast parallel algorithm from this sequential algorithm is that its running time is proportional to the height of the tree. But by adapting the dynamic tree contraction algorithm of Miller and Reif[6], we show that subtree isomorphism is in random-NC (RNC). Dynamic tree contraction is one of two classic methods for achieving NC and RNC algorithms for problems involving potentially unbalanced trees; the other is recursively finding a vertex "$1/3 - 2/3$" separator for the tree[2]. In a complementary effort, Lingas and Karpinski[4] independently developed an RNC^3 algorithm for subtree isomorphism based on the latter method. In both algorithms, the randomization is used to perform the bipartite matching problems: if a (deterministic) NC algorithm is found for bipartite matching, then subtree isomorphism will also be in NC.

As in Matula's algorithm, we recast subtree isomorphism as a problem on limbs of G and H. A limb of a tree T is a subgraph of T rooted at a vertex u consisting of an edge $\{u, v\}$ of T, together with the connected component of $T - \{\{u, v\}\}$ which contains v. Let $T\langle u, v \rangle$ denote the limb of T defined by the root vertex u and the edge $\{u, v\}$. Any edge $\{v, w\} \in T$, $w \neq u$, will determine a limb $T\langle v, w \rangle$, a subgraph of $T\langle u, v \rangle$, which we call a child limb of $T\langle u, v \rangle$. If $\{t_1, t_2\}$ is the only edge incident to t_1 in T, then the limb $T\langle t_1, t_2 \rangle$ contains all of T, and is denoted a root limb, and $T\langle t_2, t_1 \rangle$ is denoted a leaf limb. We say $S\langle s_1, s_2 \rangle$ is limb imbeddable in $T\langle t_1, t_2 \rangle$ if and only if $S\langle s_1, s_2 \rangle$ is isomorphic to a subtree of $T\langle t_1, t_2 \rangle$ such that s_1 is mapped to t_1 and s_2 is mapped to t_2. The height of a limb $T\langle t_i, t_j \rangle$ denotes the maximum distance of any vertex of $T\langle t_i, t_j \rangle$ from the root vertex t_i. A unary chain in a tree T is a maximal sequence of edges $\{t_0, t_1\}, \{t_1, t_2\}, \ldots, \{t_{k-1}, t_k\}$, where $T\langle t_i, t_{i+1} \rangle$ is the only child limb of $T\langle t_{i-1}, t_i \rangle$, for $1 \leq i \leq k$, and $T\langle t_k, t_{k+1} \rangle$ is not a leaf limb.

Given two trees G and H, one can test whether there is subtree of H that is isomorphic to G as follows. First choose a root limb $G\langle g_1, g_2 \rangle$ of G. G can be imbedded in H if and only if $G\langle g_1, g_2 \rangle$ can be limb imbedded in some limb of H (typically not a root limb). To determine whether $G\langle g_i, g_j \rangle$ can be limb imbedded in $H\langle h_k, h_l \rangle$, one can apply the following theorem due to Matula:

Theorem 1 *[5]: Let $G\langle g_i, g_j \rangle$ be a limb of a tree G and $H\langle h_l, h_m \rangle$ be a limb of a tree H. Let A be a bipartite graph, where the boys are the child limbs of $G\langle g_i, g_j \rangle$ and the girls are the child limbs of $H\langle h_l, h_m \rangle$. There is an edge between boy $G\langle g_j, g_k \rangle$ and girl $H\langle h_m, h_n \rangle$ if and only if $G\langle g_j, g_k \rangle$ is limb imbeddable in $H\langle h_m, h_n \rangle$. Then there exists a matching in A that matches all the boys if and only if $G\langle g_i, g_j \rangle$ is limb imbeddable in $H\langle h_l, h_m \rangle$.*

In this way, one can solve an instance of the subtree isomorphism problem by starting at limbs of height 1 and progressing up to the root limb $G\langle g_1, g_2 \rangle$, at each height solving a series of bipartite matching problems in order to determine which limb imbeddings are possible.

In order to achieve polylogarithmic running time we will apply dynamic tree contraction[6] to the guest tree. At each phase of the contraction process the following two operations are applied to the current tree $\hat{G}$: the rake operation, which deletes the edges incident with the leaves of $\hat{G}$, and the compress operation, which pairs up edges along unary chains in $\hat{G}$ and replaces each pair by a single edge. As the contraction process proceeds, information about the possible limb imbeddings of G in H is accrued. In particular, for each limb in $\hat{G}$ with exactly one child limb, we compute and maintain a set of conditionals defining its limb imbeddability in H as a function of the as-yet-unknown limb imbeddability of its one child limb. As part of a compress operation, such conditionals for a limb can be readily composed with the conditionals for its child. By the time $\hat{G}$ is a single edge, sufficient information is available to determine whether the root limb of G is limb imbeddable in any limb of H. If so, an explicit isomorphism of G to a subtree of H is constructed during a second iterative process which expands G back to its original size by reversing the contraction process.

To reduce the number of processors, we will apply two bipartite matching algorithms: one for the *decision* problem (determine if a perfect matching exists) while contracting the tree, and one for the *search* problem (find the edges in a perfect matching) while expanding the tree. Further processor savings are achieved by solving groups of related matching problems at once. Finally, we show that bipartite perfect matching and subtree isomorphism are mutually NC reducible by presenting a log space reduction from the former to the latter.

Miller and Reif[6] use dynamic tree contraction to develop NC algorithms for the related problems of tree isomorphism, canonical labels for trees, and canonical labels for all subtrees. The latter problem assigns labels to all nodes in a rooted tree such that two nodes u and v have the same label if and only if the *maximal* subtree rooted at u is isomorphic to the *maximal* subtree rooted at v. This differs from the subtree isomorphism problem, in which the subtrees of H are not necessarily maximal.

2. The subtree isomorphism algorithm

Let n_G be the number of nodes in G and $n = n_H$ be the number of nodes in H. Let G be rooted at a root limb $G\langle g_1, g_2 \rangle$. This determines a set of $n_G - 1$ limbs associated with the rooted G. As the algorithm proceeds, it contracts G using suitably defined rake and compress operations. Throughout the algorithm, we will associate the limb $\hat{G}\langle g_i, g_j \rangle$ with its second vertex g_j, and name the limb $\hat{G}\langle \star, g_j \rangle$ to reflect the fact that the parent node of g_j may change as a result of compress operations. The host tree H is not rooted and thus has $2(n_H - 1)$ limbs to consider. As the algorithm does not alter H, we will name its limbs with two vertices, e.g. $H\langle h_k, h_l \rangle$.

As the contraction process proceeds, the algorithm maintains a correspondence between the limbs that remain in the contracted tree $\hat{G}$ and the limbs of the original rooted tree G. The limb $G\langle g_p, g_q \rangle$ of G corresponding to the limb $\hat{G}\langle \star, g_j \rangle$ of $\hat{G}$ is called the <u>original limb</u> of $\hat{G}\langle \star, g_j \rangle$. At the beginning of the process, each limb of G is its own original limb. The assignment of original limbs to current limbs changes whenever a compress operation takes place. Consider the case where edges $\{g_i, g_j\}$ and $\{g_j, g_k\}$ on a unary chain are compressed into a single edge $\{g_i, g_k\}$. Suppose that, before the compress operation, the original limb of $\hat{G}\langle \star, g_j \rangle$ is $G\langle g_p, g_q \rangle$. Then, after the compress operation, the limb $\hat{G}\langle \star, g_j \rangle$ is no longer present, and the original limb of $\hat{G}\langle \star, g_k \rangle$ becomes $G\langle g_p, g_q \rangle$. Based on this correspondence, we say a limb $\hat{G}\langle \star, g_j \rangle$ in $\hat{G}$ is limb imbeddable in $H\langle h_k, h_l \rangle$ if and only if the original limb of $\hat{G}\langle \star, g_j \rangle$ is limb imbeddable in $H\langle h_k, h_l \rangle$.

At the conclusion of each contract phase, the following data will have been computed:

- for each leaf limb $\hat{G}\langle \star, g_j \rangle$ currently in $\hat{G}$ or removed at an earlier rake step, a <u>value</u> giving the set of H limbs in which $\hat{G}\langle \star, g_j \rangle$ can be limb imbedded;

- for each limb $\hat{G}\langle \star, g_i \rangle$ with exactly one child limb, a <u>partial value</u> consisting of a collection of conditional sets.

The <u>conditional sets</u> are defined as follows: let $\hat{G}\langle \star, g_j \rangle$ be the child limb of $\hat{G}\langle \star, g_i \rangle$. Then the conditional set of the pair $\hat{G}\langle \star, g_i \rangle$, $H\langle h_k, h_l \rangle$ contains the limb $H\langle h_m, h_n \rangle$ if and only if determining that $\hat{G}\langle \star, g_j \rangle$ is limb imbeddable in $H\langle h_m, h_n \rangle$ would yield the fact that $\hat{G}\langle \star, g_i \rangle$ is limb imbeddable in $H\langle h_k, h_l \rangle$. A short hand notation for this situation is "$H\langle h_k, h_l \rangle$ if $H\langle h_m, h_n \rangle$". The partial value of a limb with exactly one child limb is the collection of its $2(n_H - 1)$ conditional sets.

2.1. Pseudo-code for the algorithm

Our algorithm uses the following data structures. Let **Imbed**$[,]$ be an $(n_G - 1) \times (2n_H - 2)$ matrix, with one entry for each pair $\hat{G}\langle \star, g_i \rangle, H\langle h_j, h_k \rangle$. During the course of the algorithm, **Imbed**$[g_i, \langle h_j, h_k \rangle]$ will be set to 1 if and only if $\hat{G}\langle \star, g_i \rangle$ is found to be limb imbeddable in $H\langle h_j, h_k \rangle$. If it is set to 1, then $\hat{G}\langle \star, g_i \rangle$ must be a leaf. Thus the original limb of $\hat{G}\langle \star, g_i \rangle$ will no longer change, and **Imbed**$[g_i, \langle h_j, h_k \rangle]$ will be valid for the rest of the algorithm. Let **Conditionals**$[,,]$ be an auxiliary $(n_G - 1) \times (2n_H - 2) \times (2n_H - 2)$ matrix, used for storing the conditional sets of all pairs of limbs. **Conditionals**$[g_i, \langle h_j, h_k \rangle, \langle h_l, h_m \rangle]$ will be set to 1 if and only if $H\langle h_l, h_m \rangle$ is in the conditional set of the pair $\hat{G}\langle \star, g_i \rangle, H\langle h_j, h_k \rangle$.

Algorithm C below gives a pseudo-code description of our subtree isomorphism algorithm.

Theorem 2 *Given two trees G and H, algorithm C determines if there is a subtree of H isomorphic to G.*

Proof: We omit the proof. It will appear in the full paper. $\Box$

2.2. Analysis of the algorithm

The resource bounds for our algorithm are bounded by the time and processor count needed for steps C9 and C13. We will use the Mulmuley, Vazirani, Vazirani randomized algorithm[7] for constructing a perfect matching in a bipartite graph. This algorithm takes $O(\log^2 n)$ time and $n^2 M(n) \log \log n / \log n$ processors, i.e. $o(n^{4.4})$ processors, where n is the number of nodes in each half of the bipartite graph.

For step C9, i.e. steps C19–C22, for each g_i, $O(n_H)$ bipartite matching problems of size $\leq n_H$ are solved in parallel. For each g_i, this requires $O(\log^2 n_H)$ time and $n_H^3 M(n_H) \log \log n_H / \log n_H$ processors. Similarly, for step C13, i.e. steps C23–C27, for each g_i, $O(n_H^2)$ bipartite matching problems of size $\leq n_H$ are solved. This requires $O(\log^2 n_H)$ time and $n_H^4 M(n_H) \log \log n_H / \log n_H$ processors for each g_i.

Algorithm C: Given a tree H on n_H vertices $\{h_1, \ldots, h_{n_H}\}$ and a tree G on n_G vertices $\{g_1, \ldots, g_{n_G}\}$, this algorithm determines if there is a subtree of H isomorphic to G.

1. Select a root limb $G\langle g_1, g_2 \rangle$ of G, and for all g_i except g_1,
    let $g_i.parent$ be the parent of g_i in the resulting rooted tree $\hat{G}$.

2. Initialize the **Imbed** and **Conditionals** matrices to all zeroes. Then for each
    leaf limb $\hat{G}\langle \star, g_i \rangle$, set **Imbed**$[g_i, \langle h_j, h_k \rangle]$ to 1 for all H limbs $H\langle h_j, h_k \rangle$.
    For each limb $\hat{G}\langle \star, g_i \rangle$ with exactly one child limb: for all H limbs $H\langle h_j, h_k \rangle$,
    set **Conditionals**$[g_i, \langle h_j, h_k \rangle, \langle h_k, h_l \rangle]$ to 1 for all its child limbs $H\langle h_k, h_l \rangle$.

3. WHILE there exists > 1 edges in $\hat{G}$ DO:
4.   IN PARALLEL for each limb $\hat{G}\langle \star, g_i \rangle$ in $\hat{G}$ DO:

        /* rake all leaves, update their parents accordingly */
5.   IF leaf limb
6.   mark g_i as deleted from $\hat{G}$

7.   ELSE IF all its child limbs are leaf limbs
8.   IF > 1 child limbs
9.   Find_Imbeddings_For_Limb(g_i)

            ELSE   /* exactly 1 child limb */
10.   determine the set of H limbs in which $\hat{G}\langle \star, g_i \rangle$ is limb imbeddable
            from its conditionals and the H limbs in which its remaining
            child limb $\hat{G}\langle \star, g_j \rangle$ is now known to be limb imbeddable,
            e.g. if **Conditionals**$[g_i, \langle h_k, h_l \rangle, \langle h_m, h_n \rangle] = 1$ and
            Imbed$[g_j, \langle h_m, h_n \rangle] = 1$, set **Imbed**$[g_i, \langle h_k, h_l \rangle]$ to 1.

11.   ELSE IF some of its child limbs are leaf limbs and some are not
12.   IF exactly 1 nonleaf child limb $\hat{G}\langle \star, g_j \rangle$
13.   Find_Conditional_Imbeddings_For_Limb(g_i, g_j)

        /* compress all unary chains */
        ELSE   /* none of its child limbs are leaf limbs */
14.   IF exactly 1 child limb AND edge $\{g_i.parent, g_i\}$
            is of even parity on its unary chain
15.   compose the conditionals associated with g_i and $g_i.parent$,
            e.g. "$H\langle h_i, h_j \rangle$ if $H\langle h_k, h_l \rangle$" in parent and "$H\langle h_k, h_l \rangle$
            if $H\langle h_m, h_n \rangle$" in g_i results in "$H\langle h_i, h_j \rangle$ if $H\langle h_m, h_n \rangle$",
            marking **Conditionals** accordingly.

16.   mark $g_i.parent$ as deleted from $\hat{G}$
17. $g_i.parent \leftarrow (g_i.parent).parent$

18. Let $\hat{G}\langle \star, g_i \rangle$ be the remaining limb in $\hat{G}$. If there is an $H\langle h_k, h_l \rangle$ such that
    Imbed$[g_i, \langle h_k, h_l \rangle] = 1$, then there is a subtree of H isomorphic to G.

48

PROCEDURE Find Imbeddings For Limb(g'):

19. IN PARALLEL for each limb $H\langle h, h'\rangle$ of H DO:

20. IF $H\langle h, h'\rangle$ has at least as many child limbs as $G\langle g'.\text{parent}, g'\rangle$

21. Set up a bipartite matching problem P where the boys are the child (leaf) limbs $\hat{G}\langle \star, g'_1\rangle, \ldots, \hat{G}\langle \star, g'_k\rangle$ of $\hat{G}\langle \star, g'\rangle$ currently in $\hat{G}$ or removed by an earlier rake operation, and the girls are the child limbs $H\langle h', h'_1\rangle, \ldots, H\langle h', h'_l\rangle$ of $H\langle h, h'\rangle$. There is an edge between boy $\hat{G}\langle \star, g'_i\rangle$ and girl $H\langle h', h'_j\rangle$ if and only if $\hat{G}\langle \star, g'_i\rangle$ can be limb imbedded in $H\langle h', h'_j\rangle$, i.e. **Imbed**$[g'_i, \langle h', h'_j\rangle] = 1$. Also, add dummy limbs (with edges to all the girls) to the set of boys, to make the number of boys equal the number of girls.

22. Find a perfect matching in P. If one exists, set **Imbed**$[g', \langle h, h'\rangle]$ to 1.

PROCEDURE Find Conditional Imbeddings For Limb(g', g'_x):

23. IN PARALLEL for each limb $H\langle h, h'\rangle$ of H DO:

24. IF $H\langle h, h'\rangle$ has at least as many child limbs as $G\langle g'.\text{parent}, g'\rangle$

25. IN PARALLEL for all child limbs $H\langle h', h'_j\rangle$ of $H\langle h, h'\rangle$ DO:

26. Set up a bipartite matching problem P' as in C21, except exclude child limbs $\hat{G}\langle \star, g'_x\rangle$ and $H\langle h', h'_j\rangle$ from the matching problem.

27. Find a perfect matching in P'. If one exists, set **Conditionals**$[g', \langle h, h'\rangle, \langle h', h'_j\rangle]$ to 1.

Lemma 1 *Let G be a tree rooted at a root limb. Let $G\langle g_i, g_j\rangle$ be a limb in G that has c child limbs. Let the contraction process (as defined in Algorithm C) be applied to G until the tree is contracted to 1 edge. Then (1) there are no bipartite matching problems solved for node g_j if $c \leq 1$, (2) there is exactly one phase in which there are bipartite matching problems solved for node g_j if $c > 1$, and (3) prior to the matching problems during this phase, $G\langle g_i, g_j\rangle$ will be the original limb of $\hat{G}\langle \star, g_j\rangle$.*

Proof: The main observation behind the proof is that we perform a *lazy* rake operation, one that solves matching problems at a node g_j only when the rake operation will result in $\hat{G}\langle \star, g_j\rangle$ having 0 or 1 child limbs. □

$O(\log n_G)$ iterations of the WHILE loop suffice to contract G to 1 edge[6], and so steps C3–C17 will take $O(\log n_G \log^2 n_H)$ time. By lemma 1, step C9 or step C13 will be executed at most once for each g_i. It follows that algorithm C runs in $O(\log n_G \log^2 n_H)$ time on a CREW PRAM with $o(n_G n_H^{6.4})$ processors. In the next section, we show how the processor count can be significantly reduced.

In order to exhibit an explicit isomorphism of G to a subtree of H, we make the following additions to algorithm C. While contracting the tree, count the number of contract phases applied so far, in order to save the "time" each node was deleted. When a node is deleted as a result of a rake operation, also save the name of its parent; for a compress operation, save the name of its child. Save all perfect matchings constructed.

After G has been contracted, we reconstruct G by an expansion process which reverses the contraction process, with each expand phase splicing back into $\hat{G}$ all nodes and edges deleted at the corresponding contract phase. At the conclusion of each expand phase, we will have computed the home limb for each limb in the current $\hat{G}$. The <u>home limb</u> of a limb $\hat{G}\langle\star, g_i\rangle$ is the limb $H\langle h_k, h_l\rangle$ such that the isomorphism being constructed maps the (current) original limb of $\hat{G}\langle\star, g_i\rangle$ to $H\langle h_k, h_l\rangle$. Because they are associated with the original limbs of the current $\hat{G}$ limbs, these home limbs typically will be scattered throughout H prior to the final expand phase. During the expansion process, new home limbs are computed based on the matchings performed during the contraction process and the home limbs of existing limbs in $\hat{G}$.

Clearly the time and processor count for exhibiting the imbedding is bounded by the time and processor count for algorithm C.

3. Processor efficiency

We have recast the subtree isomorphism problem as a problem on limbs, as in Matula's algorithm, in order to save having to try out all possible roots for the trees. In this section, we show how to reduce further the number of processors needed. First, use a *decision* algorithm for steps C22 and C27. In addition, while contracting the tree, save the "time" the matching problems were solved for each node. Then, construct the necessary matchings while expanding the tree. We expand the tree as described in section 2, with the following modification. At the beginning of each expand phase, if (decision) bipartite matching problems were solved for node g_i at the corresponding contract phase, construct the matching and save the results. There are two cases to consider. (1) If step C9 was performed, then the home limb $H\langle h_j, h_k\rangle$ of $\hat{G}\langle\star, g_i\rangle$ is known, so it suffices to solve only one (search) bipartite matching problem for g_i: the matching problem for the pair $\hat{G}\langle\star, g_i\rangle$, $H\langle h_j, h_k\rangle$. (2) If step C13 was performed, then both the home limb $H\langle h_k, h_l\rangle$ of $\hat{G}\langle\star, g_i\rangle$ and the home limb $H\langle h_l, h_m\rangle$ of the remaining child limb $\hat{G}\langle\star, g_j\rangle$ of $\hat{G}\langle\star, g_i\rangle$ are known, so it suffices to solve only one (search) bipartite matching problem for the pair $\hat{G}\langle\star, g_i\rangle$, $H\langle h_k, h_l\rangle$ where $\hat{G}\langle\star, g_j\rangle$ and $H\langle h_l, h_m\rangle$ are excluded from the matching problem.

Lemma 2 *During the expansion process, there is at most one bipartite matching problem solved for each node in the rooted H.*

Proof: By lemma 1, there will be at most one bipartite matching problem solved for each node g_j during the expansion process, and this matching problem determines the home limb of $G\langle g_i, g_j\rangle$. $\square$

The running time for expanding the tree is $O(\log n_G \log^2 n_H)$, using the Mulmuley, Vazirani, Vazirani algorithm for constructing perfect matchings. Define the work of an algorithm to be the sum over all processors p_i of the number of PRAM instructions/operations executed by p_i during the course of the algorithm. Clearly the work to expand the tree is dominated by the work to construct the matchings. Let d_j be the degree of node h_j. By lemma 2, the work is $\leq \sum_{j=1}^{n_H} d_j^2 M(d_j) \log d_j \log \log d_j$, i.e. $\in O(n_H^2 M(n_H) \log n_H \log \log n_H)$

We will now analyze the complexity of contracting the tree using a *decision* algorithm for bipartite matching instead of a *search* algorithm. Recall that a bipartite graph has a perfect matching if and only if a certain symbolic matrix is nonsingular[3]. Based on this fact, Borodin, von zur Gathen, Hopcroft[1] developed a randomized algorithm for deciding if a bipartite graph has a perfect matching that runs in $O(\log^2 n)$ time on a CREW PRAM. An improved version of their algorithm computes determinants over Z_p, the integers modulo some suitable prime p of size $O(n^4)$[9]. This can be done with $O(\sqrt{n} M(n))$ work on a CREW PRAM, using the Preparata and Sarwate algorithm[8] for computing the adjoint and the determinant of a matrix, since all operations involve $O(\log n)$-bit numbers.

While contracting the tree, we can save processors by solving groups of related matching problems at once as follows. The a_{ij} entry of the adjoint of a matrix A contains the determinant of the minor A_{ji} (the $(j,i)^{\text{th}}$ *cofactor*). Thus by testing whether a cofactor is 0, we can determine if a perfect matching exists when the parent limb and any one H limb are left out of the matching problem. From Rabin and Vazirani[9], it follows that this holds even when the adjoint is computed over Z_p. Thus, in order to find in which parent limbs $H\langle h_j, h_k \rangle$ adjacent to h_k the limb $\hat{G}\langle \star, g_i \rangle$ can be limb imbedded, we solve one bipartite matching problem where the boys are the child limbs of $\hat{G}\langle \star, g_i \rangle$(currently in $\hat{G}$ or removed by an earlier rake operation), and the girls are the child limbs $H\langle h_k, h_j \rangle$ adjacent to h_k. Add dummy boys to match the number of girls as in step C21, except that one of these dummy boys is designated to correspond to $\hat{G}\langle \star, g_i \rangle$. From the matrix adjoint computed, test the cofactors of the dummy row corresponding to the parent $\hat{G}\langle \star, g_i \rangle$. $\hat{G}\langle \star, g_i \rangle$ is limb imbeddable in $H\langle h_j, h_k \rangle$ if and only if the cofactor in row $\hat{G}\langle \star, g_i \rangle$ and column $H\langle h_j, h_k \rangle$ is $\neq 0$. In this way, for each node g_i, $\leq n_H$ bipartite matching problems (one per each h_j) are solved, each with $\leq d_j = \deg(h_j)$ children. The work to solve these matching problems is $\leq n_G \sum_{i=1}^{n_H} \sqrt{d_i} M(d_i)$, i.e. $\in O(n_G \sqrt{n_H} M(n_H))$ This technique can also be applied to conditional matching problems, where the parent is left out of the graph and the dummy row corresponds to the remaining child (or vice-versa). This results in $\leq 2n_H - 2$ bipartite matching problems solved for each node g_i, each with $\leq n_H$ children.

Let algorithm C' be the improved version of algorithm C, which uses the above steps to save processors and to exhibit the mapping.

Theorem 3 *Given a tree H on n_H vertices and a tree G on n_G vertices, algorithm C' determines, with probability $\geq 1/2$, if there is a subtree of H isomorphic to G, and exhibits the mapping. It runs in time $t \in O(\log n_G \log^2 n_H)$ on a CREW PRAM with $(n_G + n_H^5 \log n_H \log \log n_H) n_H^{1.5} M(n_H)/t$ processors, i.e. $O(\log^3 n)$ time with $n^{2.5} M(n)/\log^3 n$ processors.*

4. Reducing bipartite matching to subtree isomorphism

In this section we show that bipartite perfect matching is log space reducible to subtree isomorphism. (Lingas and Karpinski[4] independently discovered an NC[1] reduction of bipartite perfect matching to subtree isomorphism).

Let $A = (X, Y, E)$ be a bipartite graph, where $X = \{x_1, x_2, \ldots, x_n\}$ and $Y = \{y_1, y_2, \ldots, y_n\}$. We will construct trees T_x, T_y corresponding to the vertex sets X and Y, such that every imbedding of T_x in T_y yields, in a natural way, a perfect matching in A. It is convenient to view T_x and T_y as rooted at R_x and R_y respectively. This creates no obstacle since our construction forces R_x to be mapped to R_y in any imbedding. The structure of the trees is as follows:

T_x : R_x has $n + 2$ children - $X_1, X_2, \ldots, X_n, V_1, V_2$. X_i corresponds to vertex x_i in A. V_1 and V_2 have no children. For $1 \leq i \leq n$, X_i is the parent of i children, X_{ij}, each of which is a root of a path of length $n - i + 1$.

T_y : R_y has $n + 2$ children - $Y_1, Y_2, \ldots, Y_n, U_1, U_2$. Y_i corresponds to vertex y_i in A. V_1 and V_2 have no children. For $1 \leq i \leq n$, Y_i is the parent of n children, Y_{ij}, where Y_{ij} is the root of a path of length $n - j + 1$ if $\{y_i, x_j\} \in E$ and length $n - j$ otherwise.

Note that this reduction can clearly be performed in log space.

Theorem 4 *T_x is imbeddable in T_y if and only if A has a perfect matching.*

It follows that the problem of *deciding* if a bipartite graph has a perfect matching is log space reducible to the problem of deciding if a tree is isomorphic to a subtree of another tree, and the problem of *constructing* a perfect matching in a bipartite graph is log space reducible to the problem of constructing an imbedding of a tree into another tree. From the reduction, we observe that the number of imbeddings of T_x in T_y is $2n!(n-1)!(n-2)! \cdots 2!$ times the number of perfect matchings of A. Thus the problem of determining the number of imbeddings of a tree in another tree is $\#P$-complete.

5. Remarks

Suppose we use a search algorithm for bipartite matching while contracting the guest tree. Then we can extend our Monte Carlo algorithm to a Las Vegas algorithm as follows. Given a matching M, we can determine whether M is maximum as follows. By directing all matched edges from the boys to the girls, directing all unmatched edges from the girls to the boys, and applying a shortest path calculation in this directed graph, we can find an augmenting path or prove that none exists. The resulting algorithm runs in expected time $O(\log n_G \log^2 n_H)$ with $n_G n_H^2 M(n_H) \log \log n_H / (\log n_G \log n_H)$ processors.

The case where G (or H) has bounded maximum degree d can be done *deterministically* in time $O(d \log^2 n_H \log n_G)$ on a CREW PRAM. Simply solve each bipartite matching problem in our algorithm using d applications of the above method for finding an augmenting path in parallel.

With appropriate implementation, our algorithm is in RNC^3. To see this, observe (1) without the matchings, our algorithm runs in $O(\log^2 n)$ time, and can be implemented on an NC circuit of depth $O(\log^3 n)$, and (2) the bipartite matching algorithms used are in RNC^2.

Acknowledgement

The first author was supported by the International Computer Science Institute (ICSI), Berkeley, CA and by an IBM Doctoral Fellowship. The second author was supported in part by ICSI and NSF Grant #DCR-8411954. The third author was supported in part by NSF Grant #DCR-8514961 and by the Mathematical Sciences Research Institute (MSRI), Berkeley, CA. The fourth author was supported by ICSI and by Defense Advanced Research Projects Agency (DoD) Arpa Order #4871, monitored by Space and Naval Warfare Systems Command under Contract #N00039-84-C-0089. He is currently at IBM Almaden Research.

References

[1] A. Borodin, J. von zur Gathen, and J. Hopcroft. Fast parallel matrix and GCD computations. In *Proc. of the Symp. on Foundations of Computer Science (FOCS)*, Oct. 1982.

[2] R. Brent. The parallel evaluation of general arithmetic expressions. *JACM*, 21:201–208, 1974.

[3] J. Edmonds. Systems of distinct representatives and linear algebra. *J. Res. Nat. Bureau of Standards*, 71B:241–245, 1967.

[4] A. Lingas and M. Karpinski. Subtree isomorphism and bipartite perfect matching are mutually NC reducible. 1987. submitted for publication.

[5] D. W. Matula. Subtree isomorphism in $O(n^{5/2})$. *Annals of Discrete Mathematics*, 2:91–106, 1978.

[6] G. L. Miller and J. H. Reif. Parallel tree contraction and its applications. In *Proc. of the Symp. on Foundations of Computer Science (FOCS)*, Oct. 1985.

[7] K. Mulmuley, U. Vazirani, and V. Vazirani. Matching is as easy as matrix inversion. *Combinatorica*, 7(1):105–113, 1987.

[8] F. Preparata and D. Sarwate. An improved parallel processor bound in fast matrix inversion. *Information Processing Letters*, 7(3):148–150, 1978.

[9] M. O. Rabin and V. V. Vazirani. *Maximum Matchings in General Graphs through Randomization*. Technical Report TR-15-84, Aiken Computation Laboratory, Harvard University, Oct. 1984.

All Graphs Have Cycle Separators

and

Planar Directed Depth-First Search is in DNC

Ming-Yang Kao*

Department of Computer Science

Indiana University, Bloomington, IN 47405

Kao@Indiana.Edu

(Preliminary Version)

Abstract

All graphs have cycle separators. (A single vertex is regarded as a trivial cycle.) In sequential computation, a cycle separator can be found in $O(n + e)$ time for any undirected graph of n vertices and e edges; in $O((n + e) \log n)$ time for any directed graph. In parallel computation, it is in deterministic NC to convert any depth-first search forest of any graph into a cycle separator. Moreover, finding a cycle separator for any *planar directed* graph is in *deterministic* NC; consequently, finding a depth-first search forest in any planar directed graph is in deterministic NC, too.

1 Introduction

A *separator* of a vertex-weighted digraph G is a set S of vertices or edges such that $G - S$ has no strongly connected component heavier than a constant fraction of the total weight of G. This definition also applies to undirected graphs if each undirected edge is replaced by a pair of directed edges [LT79] [Lei80] [Mil84]. A *cycle separator* is a vertex-simple directed cycle such that its vertices form a separator. (A single vertex is regarded as a trivial cycle.) Finding graph separators is a very important part of the divide-and-conquer strategies for a wide range of graph theoretical problems and applications. Many applications require *small* separators [LRT79] [Lei80] [Val81] [JV83] [FJ86] [PR87]. Others can use separators of *any size*; for example, Smith [Smi86] shows that cycle separators of any size can be used to parallelize depth-first search on *planar undirected* graphs. In fact, this paper concentrates on the relationship between depth-first search and cycle separators of any size for *general directed* graphs.

Miller [Mil84] shows that *planar undirected* graphs have *small* cycle separators; these graphs are the only non-trivial class previously known to have cycle separators. In contrast, this paper shows that all *general directed* graphs have cycle separators of *any size*. Not all graphs, however, can have small cycle separators; complete graphs are obvious examples. The proof of the universal existence of cycle separators is based on depth-first search; consequently, the proof has many complexity implications on finding depth-first search forests and finding cycle separators for digraphs. These

---

*supported in part by a 1987 Summer Faculty Fellowship from Indiana University at Bloomington.

implications and other contributions are briefly presented now; in the following let n be the number of vertices in any given graph, and let e be the number of edges.

1. Sequential Complexity

   - For *planar undirected* graphs, Miller [Mil84] gives a linear-time algorithm to find *small* cycle separators.
   - This paper shows that cycle separators can be found in linear time for *general undirected* graphs, and in $O((n + e) \cdot \log n)$ time for *general directed* graphs.

2. Parallel Complexity

   (a) Between Depth-First Search Forests and Cycle Separators

   This paper shows that for *general directed* graphs, it is in *deterministic* NC to convert any depth-first search forest into a cycle separator.

   (b) Undirected Graphs

   - For *planar* undirected graphs, Gazit and Miller [GM87] present an efficient RNC algorithm for finding cycle separators of *small size*; Goldberg, Plotkin, and Shannon [GPS87] also show that cycle separators of *any size* can be found *deterministically* in $O(\log n)$ time and a linear number of processors on a CRCW PRAM.
   - For *general* undirected graphs, Reif shows that *ordered* depth-first search is log-space complete in P [Rei85]. Aggarwal and Anderson [AA87] show that *un-ordered* depth-first search is in RNC; this paper shows that immediately from the general conversion technique, finding a cycle separator is also in RNC.

   (c) Directed Graphs

   - For *acyclic* digraphs, Ghosh and Bhattacharjee [GB84] give a deterministic NC algorithm for lexicographically-first depth-first search.
   - For *planar* digraphs, this paper shows that finding cycle separators is in *deterministic* NC; consequently, depth-first search is also in *deterministic* NC, following a deterministic NC technique by the author, Aggarwal, and Anderson [AAK88] for using cycle separators to conduct depth-first search in general directed graphs.
   - For *general* digraphs, Schevon and Vitter [SV85] give a deterministic NC algorithm to test whether any given spanning forest is a DFS forest. The algorithm also produces a DFS order if the forest is a DFS forest.
   - Also for *general* digraphs, the author jointly with Aggarwal and Anderson has recently discovered an RNC DFS algorithm [AAK88].

This paper is organized as follows. Section 2 gives the DFS-based proof for the universal existence of cycle separators; this section also discusses the immediate algorithmic implications of the proof in sequential and parallel computation. Section 3 describes the DNC algorithm for finding a cycle separator in any planar digraph; this cycle separator algorithm yields a DNC algorithm for DFS on planar digraphs. Section 4 gives the conclusions.

2 The Existence Proof of Cycle Separators

2.1 The Proof

Throughout this paper a weighted graph always refers to one with non-negative vertex weights; to avoid triviality, at least one vertex weight is greater than zero. Let G be such a weighted digraph.

Let $W(v)$ denote the weight of any vertex v; let $W(K)$ denote the total weight of any vertex set K. K is *overweight* if $W(K) > W(G)/2$.

Definition 1 *(Generic Separators) Let G be any weighted digraph. A separator S of G is a vertex set of G such that $G - S$ has no overweight strongly connected component.*

A *cycle* (or *path*) separator is a vertex-simple directed cycle (or path, respectively) such that its vertices together form a separator. This section offers two separator theorems: *all weighted digraphs have path separators and cycle separators*. The theorems can be immediately applied to any weighted *undirected* graph by substituting each undirected edge with a pair of directed edges; in such a substitution, any directed cycle separator obtained in the theorems cannot degenerate into an undirected edge because it is one of the following three cases: the cycle of the empty set, the cycle of a single vertex, or a cycle of at least three vertices.

In the following, depth-first search is used to identify a path separator for any given weighted digraph; then the path separator is converted into a cycle separator.

Theorem 2 *(Path Separator) Every weighted digraph has a path separator.*

Proof. Let G be any given weighted digraph with n vertices. Assume w.l.o.g. that G is strongly connected. Otherwise replace G with its maximum-weight strongly connected component G' because every path separator for G' is also one for G.

Construct a path separator for G as follows. Apply depth-first search (DFS) to G from an arbitrary vertex r and list the vertices in the DFS last-visit order $x_1, ..., x_n$. Let f be the smallest index such that $1 \leq f \leq n$ and $W(x_1) + \cdots + W(x_f) \geq W(G)/2$. Then $W(x_1) + \cdots + W(x_{f-1}) < W(G)/2$ and $W(x_{f+1}) + \cdots + W(x_n) \leq W(G)/2$. Now let P be the vertex-simple directed path in the DFS tree from the root r to x_f.

Prove that P is a path separator of G as follows. Let $A = \{x_1, ..., x_{f-1}\}$ and $B = G - (A \cup P)$. Arrange the DFS tree in such a way that for all last-visit numbers i and j, if $i > j$, the vertex x_i is either above or to the right of the vertex x_j in the tree. In such an arrangement, the vertex x_f is above or to the right of every vertex in A because the index f is greater than the last-visit numbers of all vertices in A. Moreover, every vertex in B is to the right of x_f because the last-visit numbers of vertices in B are all greater than f and because P consists of x_f and all its ancestors. The above two observations imply that every vertex in B is to the right of every vertex in A. Because all cross edges in DFS are directed edges from the right to the left, there are no undirected edges between A and B, and no directed edges from A to B. Therefore, any strongly connected component of $G - P$ is contained either entirely in A or entirely in B. Consequently, the total weight of any strongly connected component of $G - P$ is no more than $W(A)$ or $W(B)$, which both are in turn no more than $W(G)/2$ because $A = \{x_1, ..., x_{f-1}\}$ and $B \subseteq \{x_{f+1}, ..., x_n\}$. ∎

Theorem 3 *(Cycle Separator) Every weighted digraph has a cycle separator.*

Proof. Let G be any given weighted digraph; let $P = u_1, ..., u_r$ be a path separator of G obtained from the Path Separator Theorem. Convert P to a cycle separator as follows. Let t be the largest index such that $1 \leq t \leq r$ and $G - \{u_1, ..., u_{t-1}\}$ has an overweight strongly connected component X; then the set X must contain u_t because the path $u_1, ..., u_t$ is still a separator. If such an index t does not exist, then the empty set is a trivial cycle separator; otherwise continue the conversion and let $P' = u_1, ..., u_t$. Let s be the smallest index such that $1 \leq s \leq t$ and $G - \{u_{s+1}, ..., u_t\}$ has an overweight strongly connected component Y; then the set Y must contain u_s because the path $u_s, ..., u_t$ is still a separator. To finish the conversion, there are two cases based on whether or not $s = t$. If $s = t$, then the vertex u_s is a trivial cycle separator. If $s \neq t$, then $u_s \notin X$ and $u_t \notin Y$. Furthermore, because X and Y are overweight and strongly connected, $X \cap Y$ must have a vertex

w such that there is a vertex-simple directed path Q from u_t through X to w and then from w through Y to u_s. Finally, because $X \cup Y$ does not contain any of $u_{s+1}, ..., u_{t-1}$, the path Q and the path $u_s, ..., u_t$ form a vertex-simple directed cycle of at least three vertices; this cycle is a separator because the path $u_s, ..., u_t$ is a separator. ■

2.2   The Immediate Algorithmic Implications

In the following, n is the number of vertices in any given graph, and e is the number of edges.

1. **Compute Path and Cycle Separators Sequentially**

   The proof of the Path Separator Theorem yields a simple linear-time algorithm for finding a path separator of any weighted digraph. The proof of the Cycle Separator Theorem can be easily translated into algorithms for converting a path separator into a cycle separator. In general, such a conversion takes $O((n+e) \cdot \log n)$ time if bisection is used to find the indices t and s; in the case of undirected graphs, the indices t and s can easily be found in linear time. Therefore, it takes $O(n+e)$ time to find a cycle separator for any weighted undirected graph, and $O((n+e) \cdot \log n)$ time for any weighted digraph.

2. **Convert DFS Forests into Path and Cycle Separators in Parallel**

   For any given weighted digraph G and any given depth-first search forest T of G, let MT be the subtree of T in a maximum weight strongly connected component of G. Because computing the strongly connected components is in NC, obtaining MT is also in NC. To convert T into a path separator, the proof of the Path Separator Theorem indicates that in MT, at least one tree path is a path separator; therefore a path separator can be found by simultaneously testing all tree paths from the root to the leaves. Furthermore, if the DFS last-visit order of MT is known, the proof explicitly gives a path separator by identifying the vertex x_f, which can be found by a standard prefix computation algorithm in $O(\log n)$ time and $O(n/\log n)$ processors on an EREW PRAM [KRS85].

   To convert any path separator into a cycle separator, the proof of the Cycle Separator Theorem can be implemented as follows. The index t can be found by bisecting and testing the path separator; after t is obtained, the index s can be found in the same way. The bisection process requires $O(\log n)$ tests of whether a path is a separator. To save time over the processor complexity, the index t can be found by simultaneously trying all possibilities; after t is obtained, the index s is treated in the same way.

3   An NC Depth-First Search on Planar Digraphs

Theorem 4 *Finding a cycle separator for any weighted planar digraph is in deterministic NC. Consequently, building a DFS forest for any planar digraph is also in deterministic NC, following a deterministic NC technique by the author, Aggarwal, and Anderson [AAK88] for using cycle separators to conduct depth-first search in general directed graphs.*

The theorem is proven by the following description of a cycle separator algorithm for any weighted planar digraph. The complexity of the cycle separator algorithm is $O(\log^5 n)$ time and $O(n^4)$ processors on an EREW PRAM for any graph of n vertices; the complexity for the corresponding DFS algorithm is $O(\log^7 n)$ time and $O(n^4)$ processors. Such complexity is beyond practical application and is not worth the space for a tedious straightforward analysis; thus the following discussion concentrates on showing that the cycle separator algorithm is in deterministic NC, and omits the analysis for its complexity. For ease of understanding, the cycle separator algorithm and its subroutines are presented in a top-to-bottom hierarchical way. Subsection 3.1 gives an overview of the cycle separator algorithm; the subsequent subsections discuss in detail the subroutines used in the algorithm.

To shorten terminologies, all cycles and paths in the following discussion refer to vertex-simple directed ones unless explicitly stated otherwise.

3.1   An Overview of the Cycle Separator Algorithm

For any given weighted planar digraph G of n vertices, a cycle separator of G can be found as follows. A vertex set is called a *block* if it is strongly connected in G. A *path-block separator* is a disjoint pair of a path and a block whose vertices form a separator. G has an obvious path-block separator, namely, the empty path and its maximum-weight strongly connected component. The cycle-separator algorithm starts with this trivial separator and iteratively cuts by half the number of vertices in the block while building up the path. After $O(\log n)$ cuts, the block is left with at most one vertex. The cycle separator algorithm finishes by merging this one-vertex block and the final path into a cycle separator, using the same idea as in the proof of the Cycle Separator Theorem.

The block-cutting procedure is described as follows. Let (P_{in}, B_{in}) be any path-block separator; the goal is to find another path-block separator (P_{out}, B_{out}) such that $\mid B_{out} \mid \leq \mid B_{in} \mid /2$, where $\mid \cdot \mid$ denotes the number of vertices in a vertex set.

Step 1. To break B_{in} into smaller pieces, first use the NC algorithm by Lovász [Lov85] to compute an ear decomposition $E_1, ..., E_k$ for B_{in}. Then find the *smallest* index f such that there are at least $\mid B_{in} \mid /2$ vertices in $E_1 \cup \cdots \cup E_f$. Let $B_{big} = E_1 \cup \cdots \cup E_f$; also let $B_{small} = E_1 \cup \cdots \cup E_{f-1}$. From the property of the ear decompositions, B_{big} and B_{small} are blocks; from the choice of the index f, $\mid B_{big} \mid \geq \mid B_{in} \mid /2$ but $\mid B_{small} \mid \leq \mid B_{in} \mid /2$. Now compute the strongly connected components $SC_1, ..., SC_s$ of $B_{in} - B_{big}$; the SC_i's are blocks such that $\mid SC_i \mid \leq \mid B_{in} \mid /2$. The structure of B_{in} can be simplified by treating B_{big} and the SC_i's as individual vertices; let $\overline{G}$ and $\overline{B_{in}}$ be the condensed versions of G and B_{in} respectively. Then in $\overline{B_{in}}$, all non-trivial cycles go through the same vertex B_{big}; a *daisy* will refer to any strongly connected planar digraph with such a centralization property. A *path-daisy separator* is one consisting of a disjoint pair of a path and a daisy; the result of this step is the path-daisy separator $(P_{in}, \overline{B_{in}})$ for $\overline{G}$.

Step 2. (Subsections 3.4 and 3.5.) This step starts with $(P_{in}, \overline{B_{in}})$, and operates in $\overline{G}$. The daisy $\overline{B_{in}}$ can be thought of as a collection of cycles centered at the vertex B_{big}; the goal is to iteratively cut by half the number of cycles in $\overline{B_{in}}$ while building up P_{in}. However, there may be an exponential number of cycles in $\overline{B_{in}}$. For this concern, a novel approach is invented to cut $\overline{B_{in}}$ by cutting the so-called petals; in a planar digraph, a *petal bounded by an edge* e is a cycle of a maximal interior among the cycles going through e. In any *daisy*, there are at most two petals *bounded by* each edge; therefore after $O(\log n)$ cuts, $\overline{B_{in}}$ is left with at most one cycle. Then this one-cycle daisy and the final path in the separator are merged into a cycle separator C of $\overline{G}$, using the same idea as in the proof of the Cycle Separator Theorem; the result of this step is the cycle separator C of $\overline{G}$.

Setp 3. (Subsections 3.2 and 3.3.) This step starts with C and operates in G. A *pseudo-cycle* of G will refer to a cycle in any condensed version of G; C is a pseudo-cycle in G. A pseudo-cycle separator of G is one consisting of a pseudo-cycle; C is a pseudo-cycle separator of G. A pseudo-cycle of G is in fact a cycle of G attached with disjoint blocks; C may have more than one block. Again a novel approach is used to iteratively cut by half the number of blocks in C while keeping C a separator; after $O(\log n)$ cuts, the result of this step is a pseudo-cycle separator of at most one block.

Step 4. The one-block pseudo-cycle separator can be decomposed into a path and a block; the remaining block is either B_{big} or some SC_i. In the former case, B_{big} can be further decomposed into the ear E_j and the block B_{small}. Therefore the separator consists of two or three paths and a block at most half the size of B_{in}. To finish the block-cutting procedure, merge the paths in the decomposition into a single path, using the same idea as in the proof of the Cycle Separator Theorem; the resulting path-block separator is of the required block size.

3.2   Basics of Pseudo-Cycle Separators

- Let $G = (V, E)$. A *block* B of G is a non-empty vertex set of G such that the digraph $(B, E \cap (B \times B))$ is strongly connected. If $B_1, ..., B_k$ are pairwise disjoint blocks of G, then the *condensed graph* $\overline{G}$ of G at $B_1, ..., B_k$ is a weighted *planar* digraph defined in the following expected way: (1) Each B_i is treated as a vertex in $\overline{G}$; each vertex in G but not in any B_i remains a vertex in $\overline{G}$. (2) For any vertices x and y in $\overline{G}$, there is a directed edge from x to y in $\overline{G}$ if and only if there is a directed edge from x to y in G. (3) For any vertex x in $\overline{G}$, the weight of x in $\overline{G}$ is the total weight of x in G.

- A pseudo-cycle of G is a cycle of the condensed graph of G at some pairwise disjoint blocks. A *pseudo-cycle separator* of G is a pseudo-cycle whose vertices form a separator of G.

- To shorten statements, Algorithm *Join-Paths-and-Cycles* will refer to a class of similar procedures that merge a *constant* number of paths, cycles and vertices into a path or a cycle, using the same idea as in the proof of the Cycle Separator Theorem. Examples of such processes can be found in the overview of the Cycle Separator Algorithm in Subsection 3.1.

- The next algorithm and its slight variations are a key subroutine for processing path-daisy separators and pseudo-cycle separators.

Algorithm 5 (*Find-a-Maximal-Set-of-Paths*)

Input: F, $x_1, ..., x_s$, and $y_1, ..., y_t$, where (1) F is any planar digraph such that its boundary is a vertex-simple cycle but not necessarily a directed cycle, and (2) $x_1, ..., x_s$ and $y_1, ..., y_t$ are the vertices on the boundary and in the clockwise order.

Output: a maximal set of pairwise vertex-disjoint paths from x_i's to y_j's through only interior vertices of F.

Ideas. The following discussion yields an NC algorithm to find an output. Let $\Pi_1 = \{(i, j) \mid$ there is a path from x_i to y_j through interior vertices of F.$\}$; let $\Pi_2 = \{(i, j) \mid$ the index j is the smallest for i such that $(i, j) \in \Pi_1$.$\}$; further let $\Pi_3 = \{(i, j) \mid$ the index i is the smallest for j such that $(i, j) \in \Pi_2$.$\}$. Now for each $(i, j) \in \Pi_3$, let $P(i, j)$ be an *arbitrary* path from x_i to y_j through interior vertices of F; let MP be the set of these $P(i, j)$'s. It is straightforward to show that MP is a valid output; furthermore, MP can be computed in parallel as follows. Let m be the number of vertices in F. To obtain the three Π_k's, first compute an $m \times m$ boolean matrix TC such that its $(p, q)^{th}$ entry is 1 if and only if F has a path from the pth vertex to the qth vertex through interior vertices. TC can be computed by slightly modifying the standard doubling-up process for transitive closure; with TC obtained, the Π_k's can be easily computed one by one. To compute an arbitrary $P(i, j)$ for each $(i, j) \in \Pi_3$, first compute from TC the vertex set $V(i, j) = \{x_i, y_j\} \cup \{v \mid v$ is a vertex reachable *both* from x_i *and* to y_j through interior vertices in F.$\}$. These $V(i, j)$'s are pairwise disjoint because of the minimality of the indices i and j; with the disjoint property, the $P(i, j)$'s can be simultaneously computed.

3.3 Trim Blocks off a Pseudo-Cycle Separator

The next algorithm is a detailed description for Step 3 in the overview of the Cycle Separator Algorithm in Subsection 3.1.

Algorithm 6 (*Trim-Blocks-off-a-Pseudo-Cycle-Separator*)
Input: G and a pseudo-cycle separator C_{in} containing k blocks $B_1, ..., B_k$.
Output: a pseudo-cycle separator C_{out} such that C_{out} contains at most one block, and this block is some original B_i.

Method. The idea is to use the circular arrangement of the blocks and the planarity of the given graph; it is similar to the petal-cutting process described in Subsubsection 3.5.2. The details are given in the full paper.

3.4 Basics of Path-Daisy Separators

- A *daisy* is a strongly connected planar digraph such that all non-trivial cycles go through the same vertex u. (A trivial cycle is a single vertex or the empty set.) The vertex u is called a *center* of the daisy.

- Let J be any daisy; let e be any edge of J. A *petal bounded by e* in J is a cycle C going through e such that the cycle has a *maximal* interior among all cycles going through e; in other words, for any cycle C' going through e, if the interior of C is a subset of that of C', then $C = C'$. The following discussion shows that for any daisy, there are at most two petals bounded by each edge, and that finding all these petals is in deterministic NC. Let w be a center of J; let $e = s \rightarrow t$. Also let T (or S) be the set of edges on paths *from t to w* (or respectively, *from w to s*). Because J is a daisy centered at w, the boundary of T (or S) *consists* of one or two paths from t to the center (or respectively, from the center to s). Furthermore, the following three items form a cycle: (1) the edge e, (2) either of the two boundary paths of T, and (3) either of the two boundary paths of S. There are at most four such combinations; these combinations include the two petals bounded by e. Finally, the above discussion has a straightforward NC implementation by using the planarity algorithm by Klein and Reif [KR86] and a standard transitive closure algorithm.

- A *simple daisy* is one such that the petals bounded by its boundary edges are the only non-trivial cycles in the daisy. A simple daisy has no edges and no vertices inside any petal; moreover, the petals have pairwise disjoint interiors and can be arranged in a circular order around the center of the simple daisy. These properties can be used to visualize a simple daisy; they are also essential for the petal-cutting process in the Subsubsection 3.5.2.

- Let $L = (V, E)$ be a weighted planar digraph; let P be a path of L; let $H = (V', E')$ be a daisy (or simple daisy) such that $V' \subseteq V$ and $E' \subseteq E$. The pair (P, H) is called a *path-daisy separator* (or respectively, *path-simple-daisy separator*) of L if P and H are vertex-disjoint and their vertices together form a separator of L. The edge set of H is not necessarily equal to $E \cap (V' \times V')$; in fact, edges of H are often removed to simplify its structure without turning (P, H) into a non-separator.

3.5 Cut Petals off a Path-Daisy Separator

The next algorithm is a detailed description for Step 2 of the overview of the Cycle Separator Algorithm in Subsection 3.1.

Algorithm 7 (*Convert-a-Path-Daisy-Separator-to-a-Cycle-Separator*)
 Input: $\overline{G}$ *and a path-daisy separator* (P_{in}, D_{in}) *of* $\overline{G}$.
 Output: *a cycle separator of* $\overline{G}$.

Overview. There are three steps.
 Step 1. (Subsubsection 3.5.1.) Preprocess the path-daisy separator (P_{in}, D_{in}) into a path-*simple*-daisy separator (P_{mid}, SD_{mid}) of $\overline{G}$.
 Step 2. (Subsubsection 3.5.2.) Convert (P_{mid}, SD_{mid}) to another path-simple-daisy separator (P_{final}, SD_{final}) such that SD_{final} has at most one petal, or in other words SD_{final} is a cycle.
 Step 3. Use Algorithm Join-Paths-and-Daisies in Subsection 3.2 to merge the path P_{final} and the cycle SD_{final} into a cycle separator of $\overline{G}$.

3.5.1 Convert a Path-Daisy Separator to a Path-Simple-Daisy Separator

The next algorithm is a detailed description for Step 1 of Algorithm 7.

Algorithm 8 (*Convert-a-Path-Daisy-Separator-to-a-Path-Simple-Daisy-Separator*)
 Input: *a weighted digraph* $\overline{G}$ *and a path-daisy separator* (P_{in}, H_{in}).
 Output: *a path-simple-daisy separator* (P_{out}, H_{out}) *and the petals of* H_{out}.

Method. First find all petals of H_{in}; the planar embedding for H_{in} is part of that for $\overline{G}$. Then there are two cases based on whether any petal has an overweight interior. Case I: no petal has an overweight interior; Case II: at least one petal has an overweight interior. These two cases are discussed separately as follows.
 Case I. No petal has an overweight interior. Let L be the boundary of H_{in}; a *boundary sink* (or *source*) is a vertex on L with zero out-degree (or in-degree, respectively) on L. Because H_{in} is a daisy, for any boundary sink (source), H_{in} has a path from the sink to the center of H_{in} (or respectively, from the center to the source) such that the path and L are disjoint except at its end(s); call such a path a *petal path*. Compute in parallel a set of petal paths, one for each source and one for each sink; because H_{in} is a daisy, the petals in any such set are pairwise disjoint except that they all share the center. These chosen petals divide the interior of H_{in} into disjoint regions; each region is bounded by the petal paths of an adjacent source-sink pair; so the boundary of each region is a cycle. Delete from H_{in} all edges and all vertices within each region; then P_{in} and the remainder of H_{in} still form a path-daisy separator of $\overline{G}$ because every cycle is contained in a petal and because no petal of H_{in} has an overweight interior. After the deletion, the remainder of H_{in} is a simple daisy and the boundaries of the disjoint regions are its new petals.
 Case II. At least one petal has an overweight interior. To reduce this case to Case I, H_{in} can be shrunk by the following deletion analysis. Take a pair of petals CA and CB with overweight interiors such that CA and CB run in the same direction, say, clockwise. Because both CA and CB have overweight interiors, the following three regions cannot be overweight: (1) the region outside both CA and CB, (2) the region inside CA but outside CB, and (3) the region outside CA but inside CB. In other words, only the region inside both CA and CB may be overweight; call this region the *central region*. Based on this possibility, there are two cases. Case A: The central region is not overweight. Then the two cycles already form a separator; keep these two cycles. Case B: The central region is overweight. Then the boundary of this central region is a cycle CC because H_{in} is a daisy and CA and CB run in the same direction; replace CA and CB by CC.
 Now pair up the petals with overweight interiors such that each pair runs in the same direction; perform the above analysis in parallel for all pairs. The outcome is either at least one pair is in Case A or all pairs are in Case B. In the former outcome, save only one pair; in the latter outcome, there remain one half as many overweight cycles as before the analysis. By repeating the analysis

$O(\log n)$ times, at most two cycles left; the following three cases and their symmetries are the only possibilities: (1) one clockwise cycle from Case B, (2) two clockwise cycles from Case A, or (3) one clockwise cycle and one counter-clockwise cycle both from Case B. The three cases are discussed separately as follows.

Case II-(1). Let R be the remaining clockwise cycle; let H_{mid} be the H_{in} without the region outside R; then (P_{in}, H_{mid}) is still a separator. Because of the above shrinking process, the interior of R is the intersection of all petals of H_{in} with overweight interiors; this fact can be used to show that no petal of H_{mid} except R has an overweight interior. Therefore, a segment CH can be chopped off R such that the remainder of H_{mid} is still a daisy; however, because the loss of CH destroys R, H_{mid} now has no petal of an overweight interior. To reduce the case back to Case I, use Algorithm Join-Paths-and-Cycles in Subsection 3.2 to merge CH and P_{in} into a path.

Case II-(2). Let $R1$ and $R2$ be the two remaining cycles; then $R1$ and the part of $R2$ within $R1$ form a separator. The remainder of $R2$ are actually chords on $R1$; to get a simpler separator, start with $R1$ and add the chords one by one. Adding a chord in effect chops off a region and forms a smaller cycle; in other words, the chords induce a decreasing chain of cycles. In this chain some pair of adjacent cycles still form a separator; finding such a pair is in NC. Moreover, an adjacent pair is just a cycle and a chord; so use Algorithm Join-Paths-and-Cycles in Subsection 3.2 to merge the adjacent pair into a cycle separator. This is a better output than requested for the output.

Case II-(3). Let CR and CL be the remaining cycles. Delete from H_{in} everything outside CR; then the remainder and P_{in} still form a path-daisy separator. Moreover, the remainder of CL consists of chords on CR; these chords divide the interior of CR into disjoint regions. Because the regions are disjoint, either exactly one region is overweight or no region is overweight. In the former case, delete everything in all the regions except the overweight region; the case is reduced to Case II-(1). In the latter case, delete everything in all the regions; the case is reduced to one very close to Case II-(2).

3.5.2 Convert a Path-Simple-Daisy Separator to a Path-Cycle Separator

The next algorithm is a detailed description for Step 2 of Algorithm 7.

Algorithm 9 (*Convert-a-Path-Simple-Daisy-Separator-to-a-Path-Cycle-Separator*)
Input: *a weighted digraph* $\overline{G}$, *a path-simple-daisy separator* (P_{in}, H_{in}) *of* $\overline{G}$, *and the petals* $E_1, ..., E_k$ *of* H_{in}.
Output: *a path-simple-daisy separator* (P_{out}, H_{out}) *such that* H_{out} *has at most one petal, or in other words* H_{out} *is a cycle of* $\overline{G}$.

Method. The idea is to repeatedly cut by half the number of petals in H_{in} while building up P_{in}; it takes $O(\log n)$ such cuts to reduce to one the number of petals. The petal cutting process is described as follows.

Step 1. W.l.o.g. let $E_1, ..., E_k$ be listed in the same circular order as their edges appear on the boundary of H_{in}. Call $E_1, ..., E_{\lceil k/2 \rceil}$ the *upper petals*; call $E_{\lceil k/2 \rceil +1}, ..., E_k$ the *lower petals*. Let U (or L) be the set of vertices on the boundary of H_{in} and simultaneously on the upper (or lower, respectively) petals. U and L contain one or two common vertices; delete them from U and L for simplicity.

Step 2. Use Algorithm Find-a-Maximal-Set-of-Paths in Subsection 3.2 to compute a maximal set of pairwise disjoint paths from L to U such that the internal vertices of these paths are outside the boundary of H_{in} and not on P_{in}; call such paths the *division paths*. For each division path, there is a cycle consisting of the path and parts of a lower petal and an upper one; the cycle goes through the center of H_{in}; call such cycles the *division cycles*. Because H_{in} is simple, the division cycle is unique, with one minor exception, for any division path. The exception happens when a division path ends at the shared vertex of two adjacent petals; in such a case, use some simple resolution rule such that the interiors of any two division cycles either are disjoint or contain one another.

Step 3. These division cycles form two groups, counter-clockwise and clockwise. If two cycles are in the opposite directions, their interiors are disjoint; for the cycles in the same group, their interiors form an increasing chain of the set inclusion. There are several cases based on whether any division cycle has an overweight interior. All of them are essentially the same; so the following discussion focuses on the case that in the clockwise group at least one cycle has an overweight interior and at least one cycle does not have an overweight interior. From the increasing chain property, there is an adjacent pair of clockwise division cycles $C1$ and $C2$ such that the larger $C1$ has an overweight interior and the smaller $C2$ does not; finding such an adjacent pair is in NC.

Step 4. Now discard petals as follows. Call the region between $C1$ and $C2$ the *central region*. Because $C1$ does not have an overweight *exterior* and because $C2$ does not have an overweight *interior*, the following three items form a separator of $\overline{G}$: (1) P_{in}, (2) C_1 and C_2, and (3) the petals in the central region. The third item can be further discarded as follows. From the maximality of the division path set, there are no directed paths from L to U through the central region; this implies that any strongly connected component of the central region cannot have vertices from L and U at the same time. Therefore, w.l.o.g. the vertices of L in the the central region can be deleted without turning the 3-item separator into a non-separator; at this point, no lower petals survive the cut even though $C1$ and $C2$ contain segments of them. The remainder of the three-item separator has another decomposition: (1) the simple daisy consisting of the remaining petals, (2) two paths which are part of $C1$ and $C2$, and (3) the path P_{in}. The first item is a simple daisy of at most $\lceil k/2 \rceil$ petals; the other two items can be merged into a path by Algorithm Join-Paths-and-Cycles in Subsection 3.2. The resulting path-simple-daisy separator is a desired output.

4 Conclusions

This paper has given very simple proofs to two rather unexpected theorems, namely, the Path Separator Theorem and the Cycle Separator Theorem. The proofs are based on depth-first search, and have many algorithmic implications. In sequential computation, the proofs have been translated into very efficient algorithms for computing path and cycle separators of any graphs. In parallel computation, the proofs have yielded deterministic NC algorithms for converting any depth-first search forests of any graphs to path and cycle separators.

The paper has also presented a deterministic NC algorithm for computing a cycle separator for any weighted planar digraph; consequently, finding a depth-first search forest in any planar digraph is also in deterministic NC. The cycle separator algorithm runs in $O(\log^5 n)$ time and $O(n^4)$ processors on an EREW PRAM; the corresponding depth-first search algorithm runs in $O(\log^7 n)$ time and $O(n^4)$ processors. Such complexity is beyond practical application; it remains an open question to find algorithms with much better complexity.

References

[AA87]   Alok Aggarwal and Richard J. Anderson. A random NC algorithm for depth first search. In *ACM Symposium on Theory of Computing*, pages 325–334, 1987.

[AAK88] Alok Aggarwal, Richard J. Anderson, and Ming Y. Kao. Parallel depth-first search in directed graphs. Manuscript, February 1988.

[FJ86]   Greg N. Frederickson and Ravi Janardan. Separator-based strategies for efficient message routing. In *IEEE Symposium on Foundations of Computer Science*, pages 428–437, 1986.

[GB84]   Ratan K. Ghosh and G. P. Bhattacharjee. A parallel search algorithm for directed acyclic graphs. *BIT*, 24:134–150, 1984.

[GM87]   Hillel Gazit and Gary L. Miller. A parallel algorithm for finding a separator in planar graphs. In *IEEE Symposium on Foundations of Computer Science*, pages 238–248, 1987.

[GPS87] Andrew V. Goldberg, Serge A. Plotkin, and Gregory E. Shannon. Parallel symmetry-breaking in sparse graphs. In *ACM Symposium on Theory of Computing*, pages 315–324, 1987.

[JV83]   Donald B. Johnson and Shankar M. Venkatesan. Partition on planar flow networks. In *IEEE Symposium on Foundations of Computer Science*, pages 259–264, 1983.

[KR86]   Philip N. Klein and John H. Reif. An efficient parallel algorithm for planarity. In *IEEE Symposium on Foundations of Computer Science*, pages 465–477, 1986.

[KRS85] Clyde P. Kruskal, Larry Rudolph, and Marc Snir. The power of parallel prefix. *IEEE Transactions on Computers*, c-34(10):965–968, October 1985.

[Lei80] Charles E. Leiserson. Area-efficient graph layouts (for VLSI). In *IEEE Symposium on Foundations of Computer Science*, pages 270–281, 1980.

[Lov85] L. Lovász. Computing ears and branchings. In *IEEE Symposium on Foundations of Computer Science*, pages 464–467, 1985.

[LRT79] R.J. Lipton, D.J. Rose, and R.E. Tarjan. Generalized nested dissection. *SIAM Journal of Numerical Analysis*, 16:346–358, 1979.

[LT79]   R.J. Lipton and R.E. Tarjan. A separator theorem for planar graphs. *SIAM Journal of Applied Mathematics*, 36:177–189, 1979.

[Mil84] Gary L. Miller. Finding small simple cycle separators for 2-connected planar graphs. In *ACM Symposium on Theory of Computing*, pages 376–382, 1984.

[PR87]   Victor Pan and John Reif. *Fast and Efficient Solution of Path Algebra Problems*. Technical Report 3, Computer Science Department, State University of New York at Albany, 1987.

[Rei85] John H. Reif. Depth-first search is inherently sequential. *Information Processing Letters*, 20:229–234, June 1985.

[Smi86] Justin R. Smith. Parallel algorithms for depth first searchs I. planar graphs. *SIAM Journal of Computing*, 15(3):814–830, August 1986.

[SV85]   Catherine A. Schevon and Jeffrey Scott Vitter. *A Parallel Algorithm for Recognizing Unordered Depth-First Search*. Technical Report 21, Department of Computer Science, Brown University, 1985.

[Val81] L.G. Valiant. Universality considerations in VLSI circuits. *IEEE Transactions on Computers*, 30(2):135–140, February 1981.

On Some Languages in NC[1] [*]

(Extended Abtract)

Oscar H. Ibarra, Tao Jiang, Bala Ravikumar[†]   *Jik H. Chang*

Department of Computer Science
University of Minnesota
Minneapolis, Minnesota 55455

Department of Computer Science
Sogang University
Seoul, South Korea

1. Introduction

Identifying classes of problems that admit efficient parallel algorithms is of great theoretical as well as practical interest. A significant step towards this direction was made by Pippenger [PIPP79] who introduced the class $NC = \bigcup_{k \geq 1} NC^k$, and by Ruzzo [RUZZ81] who characterized NC in terms of the computations of an alternating Turing machine (ATM) [CHAN81]. Here, NC^k stands for the class of languages accepted by uniform boolean circuits of polynomial size and depth $O(\log^k n)$. There is a strong circumstantial evidence that NC is the class of problems that can be solved efficiently using a parallel computer with a feasible number of processors. In the past few years, some important computational problems have been shown to be in NC. For an excellent summary of these results, refer to [COOK85]. In this paper, we show that some important language recognition problems admit very efficient parallel solutions. We feel that the techniques we use to prove these results may be useful in the further study of NC, particularly, classes such as NC^1 and NC^2.

In the definition of NC, it is conventional to enforce some notion of uniformity (such as log-space uniformity) in order to guarantee that the circuit designs can be carried out using reasonable amounts of resources [RUZZ81]. It appears that the language classes obtained by using various notions of uniformity are all different. Although there is no proof of this fact, [ALLE86] has shown strong evidence for this. Here, we have chosen the strongest notion of uniformity, viz., U_E-uniformity [RUZZ81]. Since the results we present are upper bounds, U_E-uniformity gives the strongest results. Since NC^1 with respect to U_E-uniformity is the class of languages accepted by an indexing ATM in $O(\log n)$ time, our approach would be to use an ATM as the model of computation. For related results on NC^1, the reader is referred to [BARR86] and [BUSS87].

Let $\mathbf{N}$ be the set of natural numbers and k be a positive integer. A set $Q \subseteq \mathbf{N}^k$ is a *semilinear set* (see [HARR78]) if there exist positive integers m, $r_1,...,r_m$ and vectors v_j^i in $\mathbf{N}^k$ for $1 \leq i \leq m$, $0 \leq j \leq r_i$ such that $Q = \bigcup_{i=1}^{m} \{v_0^i + \sum_{j=1}^{r_i} t_j v_j^i \mid t_j \text{ in } \mathbf{N}\}$. The language, L_Q, defined by Q is the set of binary encodings of the k-tuples in Q. More precisely, $L_Q = \{x_1 \# ... \# x_k \mid x_i \text{ in binary}, (x_1,...,x_k) \text{ in } Q\}$. We show

[*] This research was supported in part by NSF Grants MCS-8304756 and DCR-8604603.
[†] Current address: Dept. of Comp. Sci., Univ. of Rhode Island, RI 02881.

that L_Q can be accepted by an ATM in O(log n) time. Since the class of languages accepted by ATM's in O(log n) time is exactly NC^1 with respect to U_E - uniformity [RUZZ81], it follows that L_Q is in NC^1. Now it is known that languages accepted by ATM's in T(n) time are in DSPACE(T(n)) = the class of languages accepted by T(n) space-bounded DTM's [CHAN81]. Hence, L_Q is in DSPACE(log n). Earlier, it was shown in [ALT76] (see also [ALT79]) that the unary version of L_Q, i.e., the language $\{1^{x_1}\#...\#1^{x_k} \mid (x_1,...,x_k)$ in $Q\}$, is in DSPACE(log n).

It is well-known that context-free languages (CFL's) can be accepted by DTM's in $\log^2 n$ space [LEWI65], and it was shown in [RUZZ81] that CFL's are in NC^2. It is open if CFL's or even deterministic CFL's or linear CFL's are in NC^1. In fact, it is not known whether or not these languages can be accepted by DTM's in less than $\log^2 n$ space. However, there are special classes of CFL's that have been shown to be in DSPACE(log n), such as the one-sided Dyck languages on k letters (for any $k \geq 1$), bracketed CFL's, and structured CFL's (see, e.g., [SPRI72,HARR78]). We can show that these languages are in NC^1, thus improving the results of [SPRI72].

We also show that bounded CFL's are in NC^1. In fact, we prove something stronger: Every bounded language $L \subseteq w_1^*...w_k^*$ (each w_i a string) accepted by a one-way multihead nondeterministic pushdown automaton is in NC^1. A weaker result, which shows that every bounded CFL is in DSPACE(log n), was shown earlier in [SPRI72]. It is very unlikely that the boundedness condition on L can be removed, even for the case of L accepted by a one-way 2-head nondeterministic finite automaton. This is because, it is known that if every language accepted by a one-way 2-head nondeterministic finite automaton is accepted by a log n space-bounded DTM, then DSPACE(S(n)) = NSPACE(S(n)) for all $S(n) \geq \log n$ [SUDB75], where NSPACE(S(n)) is the class of languages accepted by S(n) space-bounded nondeterministic TM's (NTM's). The deterministic case is an interesting open question, i.e., whether or not every language accepted by a one-way k-head deterministic finite automaton can be accepted by an ATM in O(log n) time. Even the case k=2 is not known. We conjecture that the answer is negative, and we give some evidence to support this conjecture.

As mentioned above, it is not known whether linear CFL's (or equivalently, the languages accepted by nondeterministic PDA's whose pushdown stack makes exactly one reversal (or turn)) are in NC^1. In fact, it is not known whether or not the languages accepted by nondeterministic 1-reversal counter machines (i.e., nondeterministic finite automata augmented with one counter which makes at most 1 reversal) can be accepted by DTM's in less than $\log^2 n$. However, for the deterministic case, we can show that languages accepted by 1-reversal counter machines are in NC^1. This claim holds as well for deterministic reversal-bounded multicounter machines (i.e., each counter makes at most a fixed number of reversals) and for even more powerful machines. Let DFCM(k,r) be the class of one-way deterministic finite automata augmented with k counters, each counter making at most r reversals and SDFA(k) be the class of simple one-way deterministic k-head finite automata. Here simple means that only one head can read the input and the other k-1 heads *cannot* distinguish symbols in the input alphabet but these heads can detect whether they are on an input symbol or the right endmarker. (We assume that for all the machines we consider with one-way heads, the input tape contains a right endmarker.) The machines in $\bigcup_k$SDFA(k) are quite powerful. For example, the language $\{ x \mid x$ in $\{0,1\}^+$, number of 0's in x = number of 1's in x $\}$ can be accepted by a machine in SDFA(2). It is easy to show that machines in $\bigcup_{k,r}$DFCM(k,r) are strictly weaker than those in $\bigcup_k$SDFA(k), e.g., the language $\{ (0+1)^n 0^n \mid n \geq 1 \}$ can be accepted by a machine in SDFA(2), but not by any machine in $\bigcup_{k,r}$DFCM(k,r). We show that the languages accepted by machines in $\underset{k}{\bigcup}$SDFA(k) (and hence also the languages accepted by machines in $\underset{k,r}{\bigcup}$DFCM(k,r)) are in NC^1. This result cannot be obtained using a simple divide-and-conquer technique because the movements of the heads are independent. The nondeterministic versions of the above classes are denoted by NFCM(k,r) and SNFA(k), respectively. Consider the language $L_{U-KNAP} = \{$ $1^{n_1}\#1^{n_2}\# \ldots \#1^{n_t}\rlap{/}{c}\ 1^y \mid n_i$, y are nonnegative integers, there exist $x_i \in \{0,1\}$ such that $\sum_{i=1}^{t} x_i\,n_i = y\ \}$, which is the unary version of the well-known (NP-complete) knapsack problem. Clearly, L_{U-KNAP} can be accepted by a machine in NFCM(1,1), and $L_{U-KNAP} \in$ NSPACE(log n). It does not seem to be in DSPACE(log n) nor does it appear to be complete in NSPACE(log n) [COOK85]. So, even providing some evidence that $L_{U-KNAP} \notin$ DSPACE(log n) may be very difficult. Now L_{U-KNAP} cannot be accepted by any machine in SNFA(2). However, we çan show that if every language L accepted by a machine in SNFA(2) is in NC^1, then L_{U-KNAP} is in DSPACE(log n). In view of these observations, it appears that establishing the precise parallel complexity of the classes of languages accepted by machines in $\underset{k,r}{\bigcup}$NFCM(k,r) or $\underset{k}{\bigcup}$SNFA(k) may be formidable, even for the case of machines in NFCM(1,1) or SNFA(2).

It is well-known that NSPACE(S(n)) $\subseteq$ DSPACE(S^2(n)) for S(n) $\geq$ log n [SAVI70]. For S(n) = o(log n), the result is weaker: NSPACE(S(n)) $\subseteq$ DSPACE(S(n) log n) [MONI82] (see also [TOMP81]). However, it was shown in [LITO85] that languages L $\subseteq a_1^*...a_k^*$ (each a_i a symbol) accepted by NTM's in S(n)=o(log n) space are in DSPACE(S^2(n) + log n). We improve this: such languages can be accepted by ATM's in O(S^2(n) + log n) time, even if the a_i's are replaced by strings w_i's. Hence, when S(n) $\leq \log^{1/2}$n, these languages are in NC^1.

We also investigate the closure properties of the class NC^1. In particular, we show that NC^1 is closed under marked Kleene closure and inverse homomorphism.

Finally, we show that every language accepted by an alternating two-way finite automaton with a pebble is regular, generalizing a result in [LADN78].

We omit the proofs in this extended abstract. The proofs will be given in the full paper.

2. Semilinear Sets and Multihead PDA's

We shall show that for a semilinear set Q, the set L_Q of binary encodings of tuples in Q, can be accepted by an ATM in O(log n) time. Hence, L_Q is in NC^1.

An ATM [CHAN81, RUZZ81] is a generalization of an NTM whose state set is partitioned into "universal" and "existential" states. As with an NTM, we can view the computation of an ATM as a tree of configurations. A configuration is called universal (existential) if the state associated with the configuration is universal (existential). A computation tree of an ATM M on input w is a tree whose nodes are labeled by configurations of M on w, such that the root is the initial configuration and the children of any non-leaf node labeled by a universal (existential) configuration include all (one) of the immediate successors of that configuration. A computation tree is accepting if it is finite and all the leaves are accepting configurations. M accepts w if there is an accepting computation tree for M on input w. Note that NTM's are essentially ATM's with only existential states. An ATM has time complexity T(n) if for all accepted inputs of length n, there is an accepting computation tree of height at most T(n). To allow the ATM to operate in sublinear time, we modify the ATM to have "random-access" input. An indexing ATM [CHAN81, RUZZ81] has a distinguished halting state - the "read state", and a special worktape called "index tape". Whenever the ATM enters the read state with a guessed input symbol 'a' and a binary integer i written on the index tape, it accepts if and only if 'a' is the i-th input symbol. The class of languages accepted by ATM's in $O(\log n)$ time is exactly the class NC^1 with respect to U_E - uniformity [RUZZ81]. Hence, in all our constructions, we use the ATM.

Our first result is the following, which improves [ALT76].

Theorem 1. Let $Q \subseteq N^k$ be a semilinear set. Then L_Q can be accepted by an ATM in $O(\log n)$ time.

A one-way k-head NPDA (abbreviated k-NPDA) is a nondeterministic pushdown automaton with k independent one-way (read-only) input heads [HARR68]. The heads are "nonsensing" in that the heads cannot sense the presence of other heads on the same input position. As is well-known, 1-NPDA's accept exactly the CFL's [HOPC79].

Using Theorem 1, we can show

Corollary 1. Let L be a language accepted by a k-NPDA, and $L \subseteq w_1^* ... w_t^*$ for some strings $w_1,...,w_t$. Then L can be accepted by an ATM M in $O(\log n)$ time.

The proof of Corollary 1 uses the fact that the Parikh map, f(L), of L is semilinear. The assumption that the heads of the k-NPDA are nonsensing is necessary, since otherwise, f(L) may not be semilinear. For example, the language $\{0^{n^2} \mid n \geq 1\}$ can be accepted by a 2-head deterministic PDA if we allow the heads to sense each other. This language can also be accepted by a sensing one-way 3-head deterministic finite automaton (3-DFA). It is very unlikely that the boundedness condition in Corollary 1 can be removed, even for the case of L accepted by a one-way 2-head nondeterministic finite automaton (2-NFA). This is because it is known that if every language accepted by a 2-NFA is accepted by a $\log n$ space-bounded DTM, then $DSPACE(S(n)) = NSPACE(S(n))$ for all $S(n) \geq \log n$ [SUDB75]. The deterministic case is an interesting open question, i.e., whether or not every language accepted by a k-DFA (a k-head deterministic finite automaton) can be accepted by an ATM in $O(\log n)$ time. Even the case k=2 is not known. We conjecture that the answer is negative. In fact, we conjecture that if every language accepted by a 2-DFA can be accepted by an ATM in $O(\log n)$ time, then every language accepted by a 2-NFA can also be accepted by an ATM in $O(\log n)$ time. We can prove this latter claim for a restricted type of ATM, using the following result which is of independent interest.

Theorem 2. Let L be a language accepted by a k-NFA M. Then $L = h(L')$ for some *length-preserving* homomorphism h and language L' accepted by a k-DFA M'.

The above theorem also holds when k-NFA (k-DFA) is replaced by k-NPDA (k-DPDA). The converse of the theorem is not true since, e.g., the language $L = \{a^1 b^2 a^3 ... a^{2t-1} b^{2t} \mid t \geq 1\}$ can be accepted by a 2-DFA, but h(L), where $h(a) = h(b) = a$, cannot be accepted by any k-NFA since f(h(L)) is not semilinear.

Define a restricted (indexing) ATM as one in which in *every* accepting computation tree, an input position is accessed *exactly* once. Then, from Theorem 2, we have:

Corollary 2. If every language accepted by a 2-DFA can be accepted by a restricted ATM in $O(\log n)$ time, then every language accepted by a 2-NFA can also be accepted by a restricted ATM in $O(\log n)$ time.

It is easy to see that Corollary 2 is true under a weaker restriction on the ATM. The restriction is that in any computation tree (accepting or not) whose leaves are either accepting or query nodes, the following property holds: There cannot be two query nodes '$<a,i>$?' and '$<b,i>$?' with a $\neq$ b. (The query '$<a,i>$?' means 'Is the ith square symbol$=$a?'). It would be interesting to prove Corollary 2 for unrestricted ATM's.

3. Some Subclasses of CFL's in NC^1

It is known that CFL's are in NC^2 [RUZZ81]. It is an interesting open question whether CFL's are in NC^1. Here we show that some well-known subclasses are in NC^1.

A one-sided Dyck language on k letters (k $\geq$ 1) [HARR78] is a context-free language defined by the grammar $G_k = <\Sigma_k, \{S\}, P, S>$, where $\Sigma_k = \{[_1, [_2, \ldots, [_k,]_1,]_2, \ldots,]_k\}$ and P $= \{$ S $\to [_i$ S $]_i$, S $\to$ SS, S $\to [_i]_i \mid 1 \leq i \leq k \}$. We will show that the one-sided Dyck language $D_k = L(G_k)$ can be recognized by an ATM in $O(\log n)$ time. The proof of this result is based on a nice characterization of one-sided Dyck languages given in the next lemma. As far as we know, no characterization of Dyck languages on k letters has been given before for k $\geq$ 2. For k$=$1, there is a simple characterization [HOPC79] stated as follows: a string x is in D_1 if and only x has equal number of $[_1$'s and $]_1$'s, and in any prefix of x, the number of $[_1$'s is greater than or equal to the number of $]_1$'s. We call a string that satisfies this condition *balanced*.

For simplicity, we consider the language D_2. The generalization to arbitrary k can be carried out in an obvious manner.

Definition. For a string w, let $w_{i:j}$ denote the substring starting at the i-th position and ending at the j-th position. The single letter $w_{i:i}$ is denoted by w_i. Let h, h_1 and h_2 be homomorphisms defined as follows: $h([_1) = h([_2) = [$, $h(]_1) = h(]_2) =]$, $h_1([_1)=[_1$, $h_1([_2) = \epsilon$, $h_1(]_1)=]_1$, $h_1(]_2) = \epsilon$, and $h_2([_2)=[_2$, $h_2([_1) = \epsilon$, $h_2(]_2)=]_2$, $h_2(]_1) = \epsilon$. Let $w \in D_2$, $1 \leq i < j \leq |w|$ and let $w_i = [_1$ (or $[_2$). (i,j) is a *matched pair* if $h(w_{i:j})$ is balanced, i.e., $h(w_{i:j})$ has equal number of ['s and]'s, and in any prefix of $h(w_{i:j})$, the number of ['s is greater than or equal to the number of]'s.

Lemma 1. A string w is in D_2 if and only if the following conditions hold:

a) $(1,|w|)$ is a matched pair, and

b)   for all (i,j) such that (i,j) is a matched pair, the strings $h_1(w_{i:j})$ and $h_2(w_{i:j})$ are balanced.

Using Lemma 1, we can show

Theorem 3. The one-sided Dyck languages on k letters are in NC^1.

Let G $= <V,\Sigma,P,S>$ be an arbitrary context-free grammar (CFG). A structured CFG [SPRI72] G' induced by G is G' $= <V,\Sigma', P', S>$ where $\Sigma' = \Sigma \cup \{ [_A,]_A \mid A \in V \}$, and P' $= \{ A \to [_A \alpha]_A \mid A \to \alpha$ is in P $\}$. A structured CFL is a CFL accepted by a structured CFG. A bracketed CFL [HARR78] is the same as a structured CFL except that, in a bracketed CFL, the subscript of a parenthesis is a number indicating the production number (the rules of the original grammar are numbered) rather than the name of a nonterminal. It was shown in [SPRI72] that structured CFL's are in DSPACE(log n). We can show:

Theorem 4. The structured and bracketed CFL's are in NC^1.

4. Simple Multihead DFA's

As remarked earlier, proving the membership of even some special classes of CFL's such as deterministic CFL's or linear CFL's in NC^1 (for that matter, even in DSPACE(log n)) appears very difficult. Even the more restricted class of nondeterministic reversal-bounded counter machine languages is not known to be in DSPACE(log n). When restricted to deterministic case, we can show that these languages are in NC^1, even when the devices have multicounters. A closely related class is that of the languages accepted by simple multihead DFA's. A (one-way) multihead DFA in which only one head can read the input and the other heads (called counting heads) *cannot* distinguish symbols in the input alphabet but can detect whether they are on an input symbol or the right endmarker is called simple [IBAR76]. (We assume that for all the machines we consider with one-way heads, the input tape contains a right endmarker.) It can be shown that the latter class is larger than the former. (For example, the language $L = \{ (0+1)^n \, 0^n \mid n \geq 1 \}$ can be accepted by a simple 2-head DFA, but it cannot be accepted by any deterministic reversal-bounded multicounter machine.) We can show that the class of languages accepted by simple multihead DFA's is in NC^1. These results cannot be obtained using a simple divide-and-conquer technique because the movements of the heads (counters) are independent.

Definition. A k-SDFA is a 5-tuple $(Q, \Sigma, \delta, q_0, F)$, where Q is the finite set of states, Σ is the finite input alphabet with $\$ \notin \Sigma$, $q_0 \in Q$ is the initial state, $F \subseteq Q$ is the set of accepting states, and δ is the next move function, a mapping from $Q \times (\Sigma \cup \{\$\}) \times \{0,1\}^{k-1}$ to $Q \times \{0,1\}^k$.

We denote an *instantaneous description* (ID) of the k-SDFA M by $(a_1 a_2...a_n\$, q, s_0, s_1,...,s_{k-1})$, where $a_1 a_2...a_n\$$ is the input; q is the state of M; s_0 is the position of the read head R; $s_1, s_2,...,s_{k-1}$ are the positions of the counting heads $C_1, C_2,...,C_{k-1}$ respectively; and $1 \leq s_i \leq n+1$, $i=0,1,...,k-1$. We define a *move* of M as follows. Let $(a_1 a_2...a_n\$, q, s_0, s_1,...,s_{k-1})$ be an ID. Let

$$b_i = \begin{cases} 1 & \text{if } a_{s_i} = \$ \\ 0 & \text{otherwise} \end{cases}$$

for $i=1,2,...,k\text{-}1$. Here $a_{n+1}=\$$. Suppose $\delta(q, a_{s_0}, b_1,...,b_{k-1}) = (q', d_0, d_1,...,d_{k-1})$. Then the next ID of M will be $(a_1 a_2...a_n\$, q', s_0+d_0, s_1+d_1,...,s_{k-1}+d_{k-1})$ and write

$$(a_1 a_2...a_n\$, q, s_0, s_1,...,s_{k-1}) \vdash_M (a_1 a_2...a_n\$, q', s_0+d_0, s_1+d_1,...,s_{k-1}+d_{k-1}) .$$

If two ID's are related by $\vdash_M$, we say that the second one results from the first one by one move. If ID_2 results from ID_1 by t moves, then we write $ID_1 \vdash_M^t ID_2$ The subscript M could be omitted if there is no confusion. The language accepted by M, denoted by L(M), is $\{ w \mid w \in \Sigma^*,$ and there exists $t \geq 0$ such that $(w\$, q_0, 1, 1,...,1) \vdash^t (w\$, q_f, n+1, n+1,...,n+1)$ for some $q_f \in F$, where n is the length of w $\}$.

k-SDFA's (for $k \geq 2$) are more powerful than DFA's, e.g., it is easy to show that $\{ x \mid x$ in $\{0,1\}^+,$ the number of 0's in x = the number of 1's in x $\}$ can be accepted by a 2-SDFA.

Without loss of generality, we assume that all the k-SDFA's we are dealing with satisfy the condition that exactly one head moves at every step. Since each head can only move to the right, a k-SDFA accepts input $a_1 a_2...a_n\$$ if and only if it accepts this input in $k \cdot n$ moves. So the language accepted by a k-SDFA is L(M) = $\{ w \mid w \in \Sigma^*,$ and $(w\$, q_0, 1, 1,...,1) \vdash^{k \cdot n} (w\$, q_f, n+1, n+1,...,n+1)$ for some $q_f \in F$, where n=$|w|$ $\}$.

Now we show how a k-SDFA can be simulated by an ATM in $O(\log n)$ time. We will prove the theorem for 3-SDFA's. The generalization to arbitrary k-SDFA's is straightforward.

Let $M=(Q,\Sigma,\delta,q_0,F)$ be a 3-SDFA. Let R be the read head, C_1 and C_2 be the counting heads. Let $a_1a_2...a_n\$$ be an input. Define the following predicates:

$\text{Count1}(q_1,s_1,t,q_2,s_2) \quad <==> \quad 1\leq t\leq n-1 \quad \text{and} \quad (a_1...a_n\$,q_1,s_1,1,n+1) \vdash^{t+s_2-s_1}$

$(a_1...a_n\$,q_2,s_2,t+1,n+1)$

$\text{Count2}(q_1,s_1,t,q_2,s_2) \quad <==> \quad 1\leq t\leq n-1 \quad \text{and} \quad (a_1...a_n\$,q_1,s_1,n+1,1) \vdash^{t+s_2-s_1}$

$(a_1...a_n\$,q_2,s_2,n+1,t+1)$

$\text{Count12}(q_1,s_1,t_1,t_2,q_2,s_2) \quad <==> \quad 1\leq t_1\leq n-1, \; 1\leq t_2\leq n-1 \quad \text{and} \quad (a_1...a_n\$,q_1,s_1,1,1) \vdash^{t_1+t_2+s_2-s_1}$

$(a_1...a_n\$,q_2,s_2,t_1+1,t_2+1)$

When the counting head C_1 is not reading $\$$, its position is not important. So for each $1\leq t\leq n-1$,

$(a_1...a_n\$,q_1,s_1,1,n+1) \vdash^{t+s_2-s_1} (a_1...a_n\$,q_2,s_2,t+1,n+1)$ if and if only for all $1\leq r\leq n-t$,

$(a_1...a_n\$,q_1,s_1,r,n+1) \vdash^{t+s_2-s_1} (a_1...a_n\$,q_2,s_2,r+t,n+1)$.

Now it is clear that $\text{Count1}(q_1,s_1,t,q_2,s_2)$ is true if and only if when the read head R moves from square s_1 to s_2, M enters state q_2 from q_1 while counting head C_2 stays at the rightmost end of the input and C_1 makes t moves in $a_1...a_n$.

Similarly, $\text{Count2}(q_1,s_1,t,q_2,s_2)$ is true if and only if when R moves from square s_1 to s_2, M enters state q_2 from q_1 while C_1 stays at the rightmost end of the input and C_2 makes t moves in $a_1...a_n$. $\text{Count12}(q_1,s_1,t_1,t_2,q_2,s_2)$ is true if and only if when R moves from square s_1 to s_2, M enters state q_2 from q_1 while C_1 makes t_1 moves and C_2 makes t_2 moves in $a_1...a_n$.

It can be shown that the validity of the predicates Count1, Count2 and Count12 (with the given arguments) can be verified in $O(\log n)$ time. Using this fact, we can prove the following

Theorem 5. Every language accepted by a k-SDFA is in NC^1.

Definition. [IBAR78] A reversal-bounded deterministic multicounter machine is a DFA augmented with k counters which can make at most r reversals (for some fixed k and r). We denote such a device by (k,r)-DFCM. The nondeterministic analog of (k,r)-DFCM will be denoted by (k,r)-NFCM.

It is easy to show that any (k,r)-DCFM can be simulated by some k'-SDFA, where k' depends only on k and r. (As noted earlier, the converse is not true.) Hence, from Theorem 5, we have

Corollary 3. Let k and r be fixed positive integers. Then, every language accepted by a (k,r)-DFCM can be accepted by an ATM in $O(\log n)$ time.

It is not easy to extend Corollary 3 to the case of nondeterministic devices since the language $L_{U\text{-KNAP}}$ ("the unary knapsack problem" defined in the Introduction) can be accepted by a (1,1)-NFCM and hence such an extension would imply that $L_{U\text{-KNAP}}$ is in $DSPACE(\log n)$. The following shows that it is equally difficult to extend the result of Theorem 5 to k-SNFA, even for the case $k=2$, because such an extension would again imply that $L_{U\text{-KNAP}}$ is in $DSPACE(\log n)$:

Let $L_1 = \{\ 1^{n_1}\#1^{n_2}\#\ldots\#\ 1^{n_t}\not{c}\ 1^y\ |\ \sum_{i=1}^{t}(n_i+1) = 2y$ and there exist x_i in $\{0,1\}$ such that $\sum_{i=1}^{t}x_i(n_i+1) = y\}$. Then, the following can be shown:

i) $L_{U\text{-KNAP}}$ can be reduced to L_1 by a log-space TM transducer; thus, L_1 is in DSPACE(log n) implies that $L_{U\text{-KNAP}}$ is in DSPACE(log n).

ii) $L_1 = L_2 \cap L_3$ for some L_2 and L_3 such that L_2 can be accepted by a 2-SNFA and L_3 is in DSPACE(logn).

5. ATM Simulation of Nondeterministic Turing Machines Using Small Space

It is well-known that NSPACE(S(n)) $\subseteq$ DSPACE(S^2(n)) for S(n) $\geq$ log n [SAVI70]. For S(n) = o(log n), the result is weaker: NSPACE(S(n)) $\subseteq$ DSPACE(S(n) log n) [MONI82] (see also [TOMP81]). However, it was shown in [LITO85] that languages L $\subseteq$ a$_1$*...a$_k$* (each a$_i$ a symbol) accepted by NTM's in S(n)=o(log n) space are in DSPACE(S^2(n) + log n). We can improve this:

Theorem 6. If L $\subseteq$ w$_1$*...w$_k$* is accepted by an NTM M$_1$ in space O(S(n)), where S(n) = o(log n), then L can be accepted by an ATM M$_2$ in O(S^2(n)+log n) time.

This result improves Litow's result and complements the Monien-Sudborough-Tompa's result for the bounded case.

As an application of Theorem 6, we have

Corollary 4. If L $\subseteq$ w$_1$*...w$_k$* is accepted by an NTM in log$^{1/2}$n space, then L is in NC^1.

6. Closure Properties

In this section, we investigate the closure properties of NC^1. It is easy to show that NC^1 is closed under intersection, union, complementation, reversal and concatenation. Interestingly, we can show the following:

Theorem 7. NC^1 is closed under marked Kleene closure and inverse homomorphism. (If L is a language and # is a new symbol not used the language, then (L#)* is the marked Kleene closure of L.)

It is unlikely that NC^1 is closed under length-preserving homomorphism, since we can show that if NC^1 is closed under length-preserving homomorphism, then NP $\subseteq$ DSPACE(log n).

It would be interesting to know if NC^1 is closed under (unmarked) Kleene closure.

7. Alternating Finite Automata with a Pebble

Let M = $<$Q,U,Σ,δ,q$_0$,F$>$ be a 2-AFA (two-way alternating finite automaton) without pebble. Q is the set of states, U is the set of universal states, Σ is a finite alphabet (not containing $\not{c}$ and \$), q$_0$ is the start state, F is the set of accepting states, and δ is a mapping $\delta : Q \times \Sigma \cup \{\not{c},\$\} \rightarrow 2^{Q\times\{-1,0,1\}}$. ($2^S$ denotes the power set of S.) E=Q–U is the set of existential states.

We define a configuration of M as a triple $<$w,q,i$>$, where w is the input, q is the state, and i is the location of the head on the tape. (Note that the string w does not include the endmarkers.) 0 and n+1 denote the locations of the endmarkers $\not{c}$ and \$, respectively. A computation tree T(α) of M starting with a configuration α is a tree whose vertices are labeled by configurations of M, with the root having the configuration α, such that the children of any non-leaf node labeled by a universal (existential) configuration include all (one) of the immediate successors of that configuration. (A node is a leaf node if it has no children.) A computation tree is accepting if it is finite and all the leaves are accepting. A partial left (right) computation subtree $T_L(\alpha)$ $(T_R(\alpha))$ of a computation tree $T(\alpha)$ is a subtree of $T(\alpha)$ rooted at configuration $\alpha = <w,q,r>$, $(r>0)$ such that all the nodes in $T_L(\alpha)$ $(T_R(\alpha))$ have configurations of the form $<w,p,j>$ where $j \leq r$ $(j \geq r)$, (p arbitrary), the leaves have configurations of the form $<w,p,r>$ and the next move of M moves the head right (left). Note that $T_L(\alpha)$ $(T_R(\alpha))$ depends only on the prefix u (suffix v) of length r (n-r+1) of the input w. To emphasize this, we call the tree $T_L(\alpha)$ $(T_R(\alpha))$, a u-subtree (v-subtree).

We make the following assumption about M (without loss of generality): For any configuration α, M has a left (right) computation subtree rooted at α.

We define

$L(u,q) = \{Q' \mid Q' =$ the set of leaves of T for some left computation u-subtree T rooted at $\alpha = <u,q,|u|>\}$, and

$R(u,q) = \{Q' \mid Q' =$ the set of leaves of T for some right computation u-subtree T rooted at $\alpha = <u,q,1>\}$.

Let $q \in Q$ and $Q_1, Q_2, \ldots, Q_s \subseteq Q$. We define

$LL(q,Q_1,...,Q_s) = \{w \mid L(u,q) = \{Q_1,...,Q_s\}\}$ and $LR(q,Q_1,...,Q_s) = \{w \mid R(u,q) = \{Q_1,...,Q_s\}\}$.

Using the observation that for any q, $Q_1, \ldots, Q_s$, the language $L = LL(q,Q_1,...,Q_s)$ and $L' = LR(q,Q_1,...,Q_s)$ are regular, we can prove that

Theorem 8. Every language accepted by a 2-AFA with a pebble is regular.

8. Conclusion

In this paper, we showed how some classes of languages can be recognized in O(log n) time by ATM's, thus proving their membership in NC^1. These classes include: (i) the binary encodings of semilinear sets, (ii) bounded languages accepted by one-way multihead NPDA's, (iii) languages accepted by simple multihead DFA's, (iv) one-sided Dyck languages on k letters, and (v) bounded languages that can be accepted by NTM's operating in space $\log^{1/2} n$. We also showed that NC^1 is closed under marked Kleene closure and inverse homomorphism. We further gave evidence that the extension of result (ii) to unbounded languages and (iii) to nondeterministic devices may be very hard. The problems of whether any of the following languages are in NC^1 remain open: i) the family of languages accepted by k-DFA's, ii) the languages accepted by k-SNFA's, and iii) two-sided Dyck languages [HARR78]. It was recently shown in [BUSS87] that parentheses CFL's [LYNC77] are in NC^1.

We have also shown in Theorem 8 that any language accepted by a two-way AFA with a pebble is regular. This leads to a question about ATM's operating in space S(n)=o(log n). Does a pebble enhance the recognition power of such devices? This question was answered in the affirmative in [CHAN86] for DTM's and NTM's. We conjecture that the answer is 'yes' and suggest the language $L = \{ 0^i 1^j \mid j \geq i \}$ as a possible candidate. We can show that L can be accepted by a loglog n space-bounded ATM with a pebble. It remains to be shown that L cannot be accepted by a loglog n space-bounded ATM.

References

[ALLE86]   Allender, E., P-uniform circuit complexity, *Tech. Report, Department of Comp. Science, The Rutgers University*, DCS–TR–198.

[ALT76]   Alt, H., Eine untere Schranke fur den Platzbedarf bei der Analyse beschrankter kontextfreier Sprachen, Dissertation, Saarbrucken, (1976).

[ALT79]   Alt, H., Lower bounds on space complexity for context-free recognition, *Acta Informatica* 12, (1979), pp. 33-61.

[BARR86]   Barrington, D., Bounded-width polynomial-size branching programs recognize exactly those languages in NC^1, *Proc. 18th Annual Symp. on Theory of Computing* (1986), pp. 1–5

[BUSS87]   Buss, S., Boolean formula-value problem is in ALOGTIME, *Proc. of 19th Annual Symp. on Theory of Computing* (1987), pp. 123– 131.

[CHAN81]   Chandra, A., D. Kozen, and L. Stockmeyer, Alternation, *J. ACM* 28, 1 (1981), pp. 114-133.

[CHAN86]   Chang, J., O. Ibarra, M. Palis, and B. Ravikumar, On pebble automata, *TCS* 44 (1986), pp. 111-121.

[COOK85]   Cook, S., A taxonomy of problems with fast parallel algorithms, *Information and Control* 64 (1985), pp. 2-22.

[HARR68]   Harrison, M. and O. Ibarra, Multitape and multihead pushdown automata, *Information and Control* 13 (1968), pp. 433-470.

[HARR78]   Harrison, M., *Introduction to Formal Language Theory*, Addison-Wesley (1978).

[HOPC79]   Hopcroft, J. and J. Ullman, *Introduction to Automata Theory, Languages, and Computation*, Addison-Wesley (1979).

[IBAR76]   Ibarra, O. and C. Kim, A useful device for showing the solvability of some decision problems, *JCSS* 13 (1976), pp. 153-160.

[IBAR78]   Ibarra, O., Reversal-bounded multicounter machines and their decision problems, *JACM* 25 (1978), pp. 116-133.

[LADN78]   Ladner, R., R. Lipton, and L. Stockmeyer, Alternating pushdown and stack automata, *SICOMP* 13 (1984), pp. 135-155.

[LEWI65]   Lewis, P., J. Hartmanis, and R. Stearns, Memory bounds for the recognition of context-free and context-sensitive languages, *IEEE Conf. Record on Switching Circuit Theory and Logical Design*, pp. 191-202.

[LITO85]   Litow, B., On efficient deterministic simulation of Turing machine computations below logspace, *MST* 18 (1985), pp. 11-18.

[LYNC77]   Lynch, N., Log space recognition and translation of parenthesis languages, *JACM* 24 (1977), pp. 583-590.

[MONI82]   Monien, B. and I. Sudborough, On eliminating nondeterminism from Turing machines which use less than logarithm worktape space, *TCS* 21 (1982), pp. 237-253.

[PIPP79]   Pippenger, N., On simultaneous resource bounds, *Twentieth IEEE Foundations of Computer Science* (1979) pp. 307-310.

[RUZZ81]   Ruzzo, W., On uniform circuit complexity, *JCSS* 22 (1981), pp. 365-383.

[SAVI70]   Savitch, W., Relationships between nondeterministic and deterministic tape complexities, *JCSS* 4 (1970), pp. 177-192.

[SPRI72]   Springsteel, F. and R. Ritchie, Language recognition by marking automata, *Information and Control* 4 (1972), pp. 313-330.

[SUDB75]   Sudborough, I., Tape-bounded complexity classes and multihead finite automata, *JCSS* 10 (1975), pp. 62-76.

[TOMP81]   Tompa, M., An extension of Savitch's theorem to small space bounds, *IPL* 12, 2 (1981), pp. 106-108.

O(log(n)) Parallel Time Finite Field Inversion

Bruce E. Litow and George I. Davida
Electrical Engineering and Computer Science Department
University of Wisconsin-Milwaukee

Abstract

Let p be prime and assume that $GF(p^n)$ is given via an irreducible nth degree GF(p) polynomial. We exhibit a boolean circuit of size $n^{O(1)}$ and depth $O(\log(n))$ such that for any $x \in GF(p^n)$ the circuit produces x^{-1}. The circuit is based upon an interesting connection between finite field computations and the eigenvalues of certain matrices. The issue of circuit uniformity is also considered.

1 Introduction

Computation of matrix inverses and related operations by circuits of polylog depth and polynomial was established in [5]. These results were not explicitly formulated within the NC framework but subsequent work has shown that matrix inversion over any field can be done in NC^2. For example in [2] this is achieved with circuit size $O(n^{\alpha+1})$ where n^α is optimal boolean circuit size for $O(\log(n))$ depth matrix multiplication over commutative rings. In that paper it is observed that the original inversion method involved division of $O(n\log(n))$ bit numbers. Most of the more recent research in this area has sought more algebraic methods in order to avoid such large scale arithmetic and to extend parallelism to quite general algebraic structures. However,there are instances where large scale arithmetic offers advantages over more abstract approaches. Finite field inversion is one such case. In this short paper we prove

Theorem Finite field inversion can be done in $O(\log(n))$ parallel time.

Before sketching the idea behind the proof of the theorem we mention one point related to NC uniformity. Details on the NC model of parallel computing may be found in [12] or [4]. We will show that given an irreducible nth degree GF(p) polynomial,g(x),upon which is based a representation of $GF(p^n)$ that our inversion circuit can be constructed in DTIME($n^{O(1)}$). However,if we also require that g(x) be computed as part of the circuit,then we appear to require a PTIME monte carlo routine in generating the circuit. This matter is taken up in more detail in section 2.

It is straightforward to model finite field inversion as matrix inversion over GF(p). This results in $O(log^2(n))$ depth. In order to achieve $O(log(n))$ depth we will replace the GF(p) matrices with integer matrices which will then be diagonalized. The precomputation needed for diagonalization including explicit determination of approximations to eigenvalues and similarity matrix is in $DTIME(n^{O(1)})$.

The paper is in three sections. Section 2 outlines our algorithm. Section 3 verifies size and depth and takes up the uniformity issue. All material is numbered by section. We let $\diamond$ denote end of proof. The ideas in this paper grew out of but are distinct from ideas in [8] and [6].

2

In this section we work exclusively over GF(2). This is a matter of convenience and the extension to GF(p) is given in section 3. We represent $GF(2^n)$ via an irreducible nth degree polynomial, $g(x) = x^n + g_1 x^{n-1} + ... + g_n$, where of course $g_n = 1$. Details on finite field structure can be found in [3] or [9]. The column vector $< g >$ with transpose $< g >^T = < g_n, ..., g_1 >$ is identified with g(x). $GF(2^n)$, then is just the set of GF(2) polynomials of degree $< n$ under polynomial addition and multiplication taken mod g(x).

We identify field elements with order n GF(2) matrices. For any matrix, M, let M_{ij} denote the entry at row i and column j. We also let $M_{,j}$ denote column j and $M_{i,}$ denote row i. Let G denote the matrix associated with $< g >$.

$$1 < i \leq n, 1 \leq j < n, G_{ij} = \begin{cases} 0 & \text{if } i \neq j+1 \\ 1 & \text{if } i = j+1 \end{cases}$$

$$i = 1, 1 \leq j < n, G_{ij} = 0 \text{ and } G_{,n} = < g >$$

For every field element $< f >^T = < f_n, ..., f_1 >$ we associate the matrix $F = f_1 G^{n-1} + ... + f_n I$ where I is the order n identity matrix. If $A = a_1 G^{n-1} + ... + a_n I$ is a matrix representing a field element, then we let $< A >^T = < a_n, ..., a_1 >$. This representation makes certain important facts about the ground field polynomial arithmetic transparent. In particular g(x) is irreducible iff $I, G, G^2, ..., G^{n-1}$ are linearly independent over GF(2). The recent algorithm [11] makes this characterization especially useful.

We next define integer matrices $\hat{G}$ and $\hat{F}$. For the moment we work over GF(p).

$$\hat{G}_{,j} = \begin{cases} G_{,j} & \text{if } j < n \\ p - G_{,j} & \text{if } j = n \end{cases}$$

Now we put $\hat{F} = f_1 \hat{G}^{n-1} + ... + f_n I$. We assume real arithmetic in all expressions involving 'hat' matrices. We also write

$$DET(xI - \hat{F}) = x^n + c_1 x^{n-1} + ... + c_n \tag{1}$$

where the c_j are integers.

We proceed to study $\hat{G}$. Let $\alpha_1, ..., \alpha_n$ be the eigenvalues of $\hat{G}$ (which are distinct by the Frobenius theorem since $\hat{G}$ is a nonnegative, irreducible matrix). It is also straightforward to check that when viewed as an incidence matrix (basically replacing all nonzero entries with 1's) it represents a connected graph. By the Frobenius theorem its eigenvalues are all distinct.

Lemma 1 $\hat{G}$ is diagonalizable.

proof We will exhibit a similarity matrix $\hat{S}$ such that $\hat{D} = \hat{S}\hat{G}\hat{S}^{-1}$ where $\hat{D}$ is diagonal and $\hat{D}_{jj} = \alpha_j$.

$$\hat{S}_{jk} = \alpha_j^{-(n-k)} \tag{2}$$

By computation using the fact that $g(\alpha_j) = 0$ one verifies the claim. $\Diamond$

Lemma 2 The eigenvalues of $\hat{F}$ are given by $\lambda_1, ..., \lambda_n$ where $\lambda_j = f_1\alpha_j^{n-1} + ... + f_n = f(\alpha_j)$.

proof By lemma 1 $\hat{G}$ possesses an n-dimensional eigenbasis which is therefore also an eigenbasis for $\hat{F}$. The λ_j are clearly eigenvalues of $\hat{F}$ and since $\hat{G}$ and $\hat{F}$ share an eigenbasis they are the only eigenvalues. $\Diamond$

The next lemma is the starting point for the method in [5] but we apply it in a very different manner. Let us put $\prod_{j=1}^n (x - \lambda_j) = x^n + c_1x^{n-1} + ... + c_n$. We also define $r_j = c_j \bmod 2$.

Lemma 3 $F^{n-1} + r_1F^{n-2} + ... + r_{n-1}I = F^{-1}$.

proof Note that arithmetic in the above equation is over GF(2). Thus F^{-1} is the matrix form of the GF(2^n) inverse of f(x). Now $\hat{F}^n + c_1\hat{F}^{n-1} + ... + c_nI = 0$ by the Cayley-Hamilton theorem. Elementary properties of modular arithmetic and this fact imply that $F^n + r_1F^{n-1} + ... + r_nI = 0$. If $r_n = 1$, then we have the lemma by multiplying through by F^{-1}. Now there is a primitive GF(2) nth degree polynomial, h(x) with matrix H, such that for some positive integer k, $H^k = F = \hat{F} = \hat{H}^k \bmod 2$. Thus $DET(\hat{F}) = c_n = r_n = DET(\hat{H}) \bmod 2$. It is easy to check that $DET(\hat{H}) = \pm h_n = 1 \bmod 2$. $\Diamond$

The following lemma is really implicit in the matrix representation for GF(2^n), however it is sufficiently important to be stated explicitly.

Lemma 4 If A and B are matrices corresponding to GF(2^n) elements $< A >$ and $< B >$, then $< AB > = A< B >$.

proof It is a basic fact that $G< G > = < G^2 >$ thus $G^j < G^k > = < G^{j+k} >$.

A and B can be written as GF(2) polynomials in $I, G, ..., G^{n-1}$. The lemma follows from these facts and linearity. $\diamond$

In order to apply lemma 3 we need the r_j which we obtain from the c_j. Let β_j be a rational approximation to α_j.

$$\gamma_j = f_1 \beta_j^{n-1} + ... + f_n \tag{3}$$

We regard the γ_j as rational approximations to the λ_j.

$$\prod_{j=1}^{n}(x - \gamma_j) = x^n + d_1 x^{n-1} + ... + d_n \tag{4}$$

That is we use eq.(3) to define rational approximations, d_j, to the integers, c_j. We define a diagonal matrix $\hat{E}$ by $\hat{E}_{jj} = \gamma_j$. We also define a matrix $\hat{T}$ by taking $\hat{S}$ in eq.(2) and replacing the α_j by the β_j.

$$For\ 1 \le j \le n, |c_j - d_j| < 1/4 \tag{5}$$

For $1 \le i, j \le n$, and $1 \le k \le n$:

$$|\hat{F}_{ij}^k - (\hat{T}^{-1}\hat{E}^k\hat{T})_{ij}| < 1/4n \tag{6}$$

The next lemma is obvious.

Lemma 5 If eq.(5) holds, then $c_j = \text{FLOOR}(d_j + 1/4)$. If eq.(6) holds, then $\hat{F}_{ij}^k = \text{FLOOR}((\hat{T}^{-1}\hat{E}^k\hat{T})_{ij} + 1/4n)$. $\diamond$

Lemma 6 For $1 \le j \le n$, $|\alpha_j| < 2$ and $|\lambda_j| < 2^n$.

proof By computation the Euclidean matrix norm of $\hat{G}$ is bounded above by 2 which is therefore an upper bound on $|\alpha_j|$. The bound on $|\lambda_j|$ follows from that for $|\alpha_j|$ and lemma 2. $\diamond$

Put $\beta_j = \alpha_j + \delta_j$.

Lemma 7 If for $1 \le j \le n$, $|\delta_j| \epsilon\ 2^{-O(n^2)}$, then eq.(5) holds.

proof We sketch the straightforward but rather tedious proof. In eq.(5) repeatedly using the triangle inequality and the binomial expansion in eq.(3) for the powers $(\alpha_j - \delta_j)^k$ occurring there we get $|\lambda_j - \gamma_j| \epsilon\ 2^{-O(n^2)}$. Using eq.(4), writing $\gamma_j = \lambda_j + \epsilon_j$ and using the same method we obtain eq.(5).

In eq.(6) we can establish by straightforward but even more tedious arguments that the matrix norms of $\hat{S} - \hat{T}$ and $\hat{E} - \hat{S}\hat{F}\hat{S}^{-1}$ are of order $2^{-O(n^2)}$. From this one can get a bound on the error terms for the matrix entries obeying eq.(6). $\diamond$

METHOD

Initial Data: g(x)

Step 1 Compute the c_j and r_j

Step 2 Compute: $\hat{E}, \hat{E}^2, ..., \hat{E}^{n-1}$.

Step 3 Compute $\hat{K} = \hat{T}^{-1}(\hat{E}^{n-1} + r_1\hat{E}^{n-2} + ... + r_{n-1}I)\hat{T}$.

Step 4 Compute K where $K_{ij} = \text{FLOOR}(\hat{K}_{ij} + 1/4) \bmod 2$

Step 5 Compute $K^2 < F >$.

We conclude this section by showing that step 3 computes $< F^{-1} >$.

Lemma 8 $K^2 < F > = < F^{-1} >$

proof By lemma 4 it is enough to show that $K = F^{-1}$. By lemma 7 $|(\hat{T}^{-1}\hat{E}^j\hat{T})_{km} - \hat{F}^j_{km}| < 1/4n$. Thus $|\hat{K}_{km} - \hat{F}^{-1}_{km}| < 1/4$ which means that $\text{FLOOR}(\hat{K}_{km} + 1/4) = K_{km} = F^{-1}_{km}$. Therefore $K = F^{-1}$. $\Diamond$

3

We now prove the theorem stated in the introduction. First we make explicit all the precomputation i.e. circuit generation. Note that in the steps of the method in section 1 the only arithmetic operations needed are:

1. addition

2. multiplication

3. comparison

4. computing x mod 2

Comparison is required in steps 1 and 3. All numbers may be regarded as $O(n^2)$ bit numbers. By methods of [1] ,NC^1 circuits for (1) - (4) can be DTIME($n^{O(1)}$) computed. We point out two facts. By the method of [7] $O(\log(n)\cdot\log\log(n))$ depth and circuit size $O(n^3)$ can be achieved. We also note that (1) and (2) can be done in $O(1)$ parallel time with $O(n^3)$ size. We will elaborate on optimizing circuit size in a subsequent paper. Based on [10] the approximations β_j can be DTIME($n^{O(1)}$) computed. It is also evident that $\hat{T}$ and $\hat{T}^{-1}$ can be DTIME($n^{O(1)}$) computed.

We turn to the steps themselves. In step 1 the c_j can be computed in NC^1 given the γ_j and these can also be computed in NC^1 from the β_j. By our remarks on (1) - (4) we see that the r_j are also computable in NC^1. In step 2 since $\hat{E}$ is diagonal the required powers can all be NC^1 computed. Step 3 is clearly in NC^1. It is plain that steps 4 and 5 are in NC^1. Thus we have established the theorem. $\diamond$

We show that the method works for any prime p. The lemmas go through unchanged except for a small detail in lemma 3. In lemma 3 we will compute - $(r_n F^{-1})$ mod p. It is still true,however,that $r_n \neq 0$ mod p. Also note that for (4) above, x mod 2 is replaced by x mod p. This still leads to the same size and depth circuits as in the mod 2 case.

The circuit generation is PTIME uniform provided that g(x) is a given. The only precomputation that is not known to be in PTIME is the determination of g(x). If g(x) is given,then finite field inversion can be computed by PTIME uniform NC^1 circuit families. If we are required to produce g(x) i.e. if the initial data is just p and n,then we can proceed as follows. The density of irreducibles is roughly $O(1/n)$ e.g. [3]. Thus in producing $O(n^2)$ random nth degree GF(p) polynomials at least one irreducible polynomial occurs with probability $> 1\text{-}2^{-n}$. Actually the probability is somewhat better than that. Using [11],for example,each polynomial can be checked for irreducibility in PTIME. This gives a good,simple Monte Carlo approach to circuit generation but ordinary uniformity is lost.

References

1. P.Beame,S.Cook,H.Hoover,"Log Depth Circuits for Division and Related Problems",SIAM J. Comput.,vol.15,no.4,pp.994-1003,1986

2. S. Berkowitz,"On Computing the Determinant in Small Parallel Time using a Small Number of Processors",Inf.Proc.Lett.,vol.18,pp.147-150,1984

3. E. Berlekamp,Algebraic Coding Theory,McGraw-Hill,1968

4. S. Cook,"A Taxonomy of Problems with Fast Parallel Algorithms",Inf. and Control, vol.64,pp.2-22,1985

5. L. Csanky,"Fast Parallel Matrix Inversion Algoritms",SIAM J. Comput.,vol.5,pp.618-623, 1976

6. G.Davida,B.Litow,"Fast Parallel Inversion in Finite Fields",19th CISS at Johns Hopkins,pp.305-308,1985

7. G. Davida,B.Litow,"Fast Parallel Comparison and Division in Modular Representation 1987,in preparation

8. G. Davida,"Inverse of Elements of a galois Field",Electronics Lett.,vol.8,21,Oct.1972

9. N.Koblitz,p-Adic Analysis and Zeta Functions,GTM 58,SpringerVerlag,1977

10. B.Litow,"Complexity of Polynomial Root Approximation",manuscript,1987

11. K. Mulmuley,"A Fast Parallel Algorithm to Compute the Rank of a matrix Over an Arbitrary Field",Proc. 18th ACM STOC,pp.338-339,1986

12. W. Ruzzo,"On Uniform Circuit Complexity",JCSS,vol.22,pp.365-383,1981

Deterministic Parallel List Ranking

Richard J. Anderson[*]   Gary L. Miller[†]

Abstract

In this paper we describe a simple parallel algorithm for list ranking. The algorithm is deterministic and runs in $O(\log n)$ time on EREW P-RAM with $n/\log n$ processor. The algorithm matches the performance of the Cole–Vishkin [CV86a] algorithm but is simple and has reasonable constant factors.

1 Introduction

List ranking is a fundamental operation on lists. The problem is: given a linked list, compute the distance each cell is from the end of the list. The problem can be solved by a straightforward sequential algorithm that traverses the list. However, the problem is much more difficult to solve efficiently in parallel. The problem was first proposed by Wyllie [Wyl79] who gave an algorithm that ran in $O(\log n)$ time using n processors. A substantial amount of effort has been put in to finding a deterministic parallel algorithm that achieves the same time bound with $n/\log n$ processors. Such an algorithm would be optimal in the sense that it would have a time processor product equal to that of the sequential algorithm. The result of Wyllie has been gradually improved in a series of papers [Vis84] [WH86] [KRS85] [MR85] [CV85], until Cole and Vishkin [CV86a] succeeded in giving an optimal algorithm with $O(\log n)$ runtime on an EREW P-RAM. The draw back to their algorithm is that it is complicated and has very large constant factors. They rely on an expander graph construction to solve a scheduling problem that arises. In this paper we give an algorithm that runs in $O(\log n)$ time and uses $n/\log n$ processors. Our algorithm is much simpler, and does not rely on an expander graph construction.

There are two main reasons why the parallel list ranking problem has attracted so much attention. First of all, list ranking is a fundamental operation that has many applications. List ranking can be used to compute a prefix sum for any associative operation over a list; this can either be done by using the ranks to reorganize the list into an array and using an efficient algorithm for data independent prefix sum, or by

---

[*]Department of Computer Science, FR-35, University of Washington, Seattle, Washington 98195, USA. Supported by an NSF Presidential Young Investigator Award.

[†]Department of Computer Science, University of Southern California, Los Angeles, California 90089, USA. Supported by NSF Grant DCR-8514961

embedding the associative operation into the list ranking algorithm. Another application of list ranking is to perform traversal operations, such as preorder numbering, on trees [TV85]. The problem of expression evalution on trees can be reduced to list ranking [GMT86] which leads to optimal algorithms for evaluating certain types of expressions. The second reason for looking at parallel list ranking is that the problem has been a rich source for ideas about parallel algorithms in general. For example certain arbitration techniques have been developed for list ranking that have turned out to have much wider an application [GPS87]. Many ideas about methodologies for parallel algorithms and schedluling have come out of this work.

Our list ranking algorithm follows the same scheme as other efficient list ranking algorithms. We first describe the general scheme and then give the basic version of our algorithm. Although the first algorithm is correct, it does not achieve the $O(\log n)$ running time because the processor workload may be unbalanced. We show how to fix the problem by attending to certain details. The final subsection gives the analysis of the algorithm which establishes it as an optimal algorithm.

2   List ranking Algorithms

The standard parallel algorithm for list ranking is due to Wyllie and runs in $O(\log n)$ time using n processors. The algorithm uses a path-doubling strategy. Each cell v contains a pointer $D(v)$ and at each time step the assignmemt $D(v) := D(D(v))$ is made. The distance covered by a pointer doubles at each time step, so in $\log n$ steps, all pointers are at the end of the list. It is straight forward to embed the computation of the distance to the end of the list in this process. The drawback of this algorithm is that the *work* (time–processor product) performed is $O(n \log n)$ as opposed to $O(n)$ work for the natural sequential algorithm. The goal of subsequent work on the list ranking problem has been to reduce the processor requirement while still maintaining a run time of approximately $O(\log n)$.

The basic step of the list ranking algorithm can be viewed as splicing out an element from the list. When the pointer $D(v)$ is replaced by $D(D(v))$ the cell $D(v)$ is removed from the list starting at v. The source of the inefficiency in Wyllie's algorithm is that the same cell is spliced out of a number of different lists instead of being left alone once it is spliced out. The basic approach to get improved list ranking algorithms has been to splice out a large number of the list cells and then solve the list ranking problem on the smaller list. Most of the list ranking algorithms subsequent to Wyllie contain the following steps:

1. Splice out elements from the list so that $O(n/\log n)$ elements remain.

2. Solve the list ranking problem on the reduced list with Wyllie's algorithm.

3. Reconstruct the entire list by processing the elements that were spliced out.

Step 2 can be done in $O(\log n)$ time and $n/\log n$ processors since the list has been reduced in length. Step 3 can also be done in $O(\log n)$ time with $n/\log n$ processors.

The details of this step depend on the data structure used to represent the spliced out elements. However, standard parallel techniques suffice for this step. The key step in order to have an optimal parallel algorithm is step 1 which must also be solved in $O(\log n)$ time with $n/\log n$ processors. It is this step that has been gradually improved in the series of papers on list ranking and is the step that we address in our algorithm.

There are two major issues that arise in the algorithms to splice elements out of the list. The issues are identifying elements to remove and resolving contention. In order to perform the splice out phase efficiently, it is necessary to have fewer processors than list cells, so processors must identify cells to work on. As cells are removed, processors must be reallocated and find cells still in the list to work on. Naive strategies run into the problem that as elements get removed from the list, it becomes more difficult to find elements still in the list to work on. Various scheduling and reorganizing schemes have been developed to get around this difficulty. Note that to get an optimal $n/\log n$ processor algorithm, a constant fraction of the processors must succeed in finding an element to remove from the list at each time step. A feature of our algorithm is a rather simple scheduling strategy that allows the processors to keep busy with cells still in the list.

The most frequently used reorganization scheme is to move the remaining cells to the start of the array [Vis84] [WH86] [KRS85] [CV86b] [MR85]. The basic way to do this is to compute a prefix summation of the number of live cells in the array, and then move cells forward. The summation can become a bottleneck since it takes $O(\log n)$ time on an EREW P-RAM. The summation can be performed less than $O(\log n)$ time on the more powerful CRCW P-RAM [Rei85] [CV86b], so this step need be dominant. An alternate approach proposed by [MR85] is to compress with a probabilistic routine that does not give perfect compression, but does concentrate the live cells sufficiently so that they can be easily identified. A different approach to allocating the work is to have processors randomly probe into the array to find cells to work on [Vis84]. To construct their optimal deterministic list ranking algorithm Cole–Vishkin introduced a complicated rescheduling scheme based on expander graphs [CV86a]. All of these schemes can be viewed as dynamic rescheduling, since the processor assignments depend upon the data. In contrast to these algorithms, our algorithm relies upon a static scheduling of the list cells to the processors.

The second issue in splicing out elements is avoiding contention. Two adjacent list cells can not be spliced out at the same time. A natural strategy to resolve contention is with randomization [MR85] [Vis84]. When processors identify adjacent cells to remove, random bits are chosen and then a protocol is used that insures that adjacent cells are not selected. The deterministic case for resolving contention is more difficult. Cole–Vishkin and Han [CV85] [Han87] independently developed deterministic protocols based on the cells' addresses to resolve contention. The Cole-Vishkin method (deterministic coin-tossing) finds a subset of the list cells such that no adjacent cells are in the subset and every cell is within a distance of k of an element of the subset. The subset is referred to as a k-ruling set. They show that a 3-ruling set can be constructed in $O(\log^* n)$ time and a $\log^{(k)} n$-ruling set can be constructed in $O(k)$ time. Our algorithm will use their subroutine for constructing a $\log\log n$-ruling set.

3   Basic Algorithm

Our list ranking algorithm splices items out of the list until $n/\log n$ items remain. When the algorithm is reduced to $n/\log n$ items, Wyllies algorithm can be applied with one processor per list cell. The ranks of the items can then be computed by adding them back into the list in the reverse of the order that they were removed.

One way to view our algorithm is that the processors make probes into the list to find cells to work on. The probing is done in a manner that assures that the processors making probes find distinct cells that have not been removed. If a processor identifies a cell with neither of its neighbors identified by other processors, then the cell can be spliced out immediately. The difficult case is when the cell identified by a processor is adjacent to a cell identified by another processor. The processors cannot splice out their cells simultaneously. One of the key components of our algorithm is the manner in which contention is resolved.

We refer to a set of adjacent cells that are selected by processors as a *chain*. The first cell is the *ruler* and the remaining cells are *subjects*. It is easy to determine which cells are rulers and which cells are subjects in constant time. All of the work of removing the subject cells is assigned to the processors of the ruler. The processors of the subject cells are released and go on to work on other cells. A ruler removes its subjects by splicing them out one at a time. This removal takes place while other cells are probing the list and determining whether they are rulers or subjects.

The algorithm is divided into *stages*. During a stage, a processor may remove a cell from the list or find another cell to work on. A stage can be performed in constant time. Each ruler removes one of its subjects in a stage. If a ruler removes its last subject, it becomes *active* and will participate in the reallocation phase, otherwise it waits for the next stage to remvoe its next subject. After the rulers have removed subjects, all active processors pick cells and arbitrate to find rulers and subjects. At this time, the selected cells not adjacent to other selected cells are spliced out. The processors associated with subject cells are set to being active at the next stage so that they are reallocated.

The allocation of processors to list cells is done in a static manner. Initially, the memory is divided up into $n/\log n$ blocks of $\log n$ items each. Each processor is assigned one of these blocks which we refer to as its *queue*. The items in a queue are not necessarily adjacent in the list. A processor acts upon the items in its queue in order, starting with the first item. We refer to the number of items left in a queue as its *height*, and the item currently under consideration as its *head*. There are two ways that an item can cease to be the head of a queue, it can be spliced out, or it can become a subject. In either case, the processor takes the next item in the queue as its new head. If a queue becomes empty, then the processor ceases to work.

We give a more formal description of this algorithm by giving the code for a single stage. Each processor runs the following code. The queue for the processor is represented by an array, with *head* the location of the queue head. The list is assumed to be doubly linked. The cell p has pointers $p.next$ and $p.prev$ to its successor and predecessor respectively. A cell also has a status field, which can have the value *ruler, subject, active, inactive* or *removed*. Initially, all cells have *inactive* status, active status indicates that a processor is assigned to them.

The code breaks into the three indicated parts. The first part is where the rulers can remove cells, the second part is to identify new rulers and subjects, and the last part is to advance the queue where necessary.

```
Stage
  p := Q[head];
                    -- Rulers splice out cells
  if p.status = ruler then
    SpliceOut(p.next);
    if p.next.status ≠ subject then
      p.status := active;
    end
  end
                    -- Identify new rulers
  if p.status = active then
    if p.prev.status ≠ active ∧ p.next.status ≠ active then
      SpliceOut(p); p.status := removed;
    else if p.prev.status = active then
      p.status := ruler;
    else
      p.status := subject;
    end
  end
                    -- Advance queue heads
  if p.status = removed ∨ p.status = subject then
    head := head − 1;
    if head < 0 then
      stop
    else
      Q[head].status := active;
    end
  end
```

It must be shown that this algorithm correctly splices out the list elements and that $O(\log n)$ stages are sufficient to reduce the list to $n/\log n$ cells. The correctness of the algorithm follows from the facts that adjacent cells are never removed at the same time, and every cell is eventually looked at by a processor. The run time is more difficult to establish, (especially since, as currently stated, it is greater than $O(\log n)$.) The following arguement shows that as long as most of the processors have nonempty queues, then the algorithm is performing work at a reasonable rate. We do this by establishing a simple accounting scheme for the work done. There are two ways that a cell can be spliced out, it can either be removed as an isolated cell, or by becoming a subject and then removed by a ruler. When an isolated cell is spliced out, we assign the processor one unit, when a ruler splices out a cell we assign the processor one half unit and when we make a cell a subject we assign the processor one half unit. Thus, for each cell that is removed, one unit is assigned. At the start of a stage, each processor is associated with a cell that has status *ruler* or status *available*. If there are k *available* cells, at least $k/4$ units are assigned, since the number of subjects created is at least as large as the number of rulers created. If all processors were to remain busy, this argument would show that $4 \log n$ phases would suffice to remove all of the cells from the list. However, some processors could exhaust there queues, while others had made very little progress. One bad case that could occur is for a ruler to have a large number of subjects (as many as $n/\log n$) which would take a very long time to remove. To solve this problem, we alter the selection of rulers, so that a ruler never has too many subjects, and the cells that are made rulers tend to be lower down in their queues.

4 Load Balancing

In the previous section we saw that the problem with our proposed algorithm is that some queues could be delayed working on long chains, while others could finish quickly and become idle. The solution to this problem has two parts, we insure that the chains do not become too long and also insure that the shorter queues are assigned work as opposed to the longer queues. The major tool that we use for this is the ruling set algorithm of Cole and Vishkin [CV85].

Definition 4.1 *In a list L, a k-ruling set is a subset R of L such that there are no adjacent elements in R, and every element of L is within distance k of an element of R.*

Cole and Vishkin give an efficient parallel algorithm for constructing a ruling set. We use a version of their algorithm that finds a $\log \log n$ ruling set in a list of length n in constant time with one processor assigned to each element.

The process of identifying rulers consists of taking the sets of adjacent elements that are picked by processors at that phase. In the original algorithm, we just take the first cell in a chain as a ruler, and the remaining cells as subjects. To balance the work load, we break up the longer chains that we identify. There are a number of things that we can do while making chains and still have a stage run in constant time. If we break up a chain so that we have some non-adjacent pieces of length one, we can splice them out when they are encountered. We can also traverse the list backwards, instead of forwards to splice out elements. This allows us to have as rulers the cells of minimum height in a chain, and then remove cells working uphill. Finally, we can mark the first and last elements of a chain so that rulers know when to stop removing items. With the flexibility to perform these operations, it is easy to break the chains into increasing height, decreasing height, and constant height chains. The constant height chains can then broken up into subchains of length at most $\log \log n$ time with the ruling set algorithm. The code given above can be modified to perform these additional tasks.

The effect of the modification is to have the following two constraints on rulers and chains: a ruler never has more that $\log \log n$ subjects, and the height of a ruler is no greater than the height of its subjects. The performance analysis given in the next section shows that these conditions are sufficient to make this an $O(\log n)$ algorithm.

5 Performance Analysis

In this section we prove that the algorithm with the enhanced method of choosing rulers is an $O(\log n)$ algorithm. The proof relies on an accounting scheme that divides the work for removing an item amongst the various steps.

Each cell is assigned a weight related to its height. The i-th item from the top is assigned a weight of $(1 - \alpha)^i$ where $\alpha = \frac{1}{\log \log n}$. The accounting scheme is the same as the one sketched above: the removal of an isolated item is the full weight of the item, the removal of an item by a ruler is half the weight, the identification of a subject is half the weight, and the identification of a ruler is 0. We show that each stage reduces the total weight by a factor of at least $1 - \frac{\alpha}{4}$. This allows us to bound the number of phases that are required to reduce the number of items to $\frac{n}{\log n}$.

Claim 5.1 *A single phase of the algorithm reduces the weight by a factor of at least* $1 - \frac{\alpha}{4}$.

To facilitate the argument, we use the following bookkeeping trick: the weight of a queue is considered to be the weight of the remaining elements, plus the remaining weight of the subjects of the queue head. In Figure 1, queue 1 has weight $(1 - \alpha)^2 + (1 - \alpha)^3 + (1 - \alpha)^4 + \frac{1}{2}(1 - \alpha)^2 + \frac{1}{2}(1 - \alpha)^1$, with the first three terms coming from queue 1, the fourth term coming from queue 2 and the fifth term coming from queue 3.

We cover the three major cases separately. First we look at what happens when an isolated cell is removed. Suppose that we remove a cell of wieght $(1 - \alpha)^i$. The queue has weight at most $\sum_{i \leq j < \log n}(1 - \alpha)^j < \sum_{i \leq j}(1 - \alpha)^j = (1 - \alpha)^i \frac{1}{\alpha}$. Thus, the factor of reduction is

$$\frac{\sum_{i+1 \leq j < \log n}(1 - \alpha)^j}{\sum_{i \leq j < \log n}(1 - \alpha)^j} = 1 - \frac{(1 - \alpha)^i}{\sum_{i \leq j < \log n}(1 - \alpha)^j} < 1 - \alpha$$

The second case is when a vertex is identified as a subject. We account for the weight of all of the queues associated with the chain. The queues that have cells that become subjects lose one cell each and the queue that has the ruler picks up the cells. Each cell that becomes a subject has half of its weight removed by the accounting.

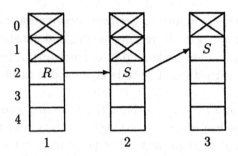

Figure 1: Figure 1. Computing the weight of a chain

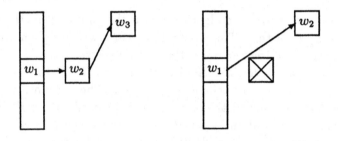

Figure 2: Figure 2. Rearranging weights

Suppose the weight of the ruler is $(1-\alpha)^{i_1}$, and the weight of the subjects are $(1-\alpha)^{i_j}$ for $2 \leq j \leq k$. Since the ruler is chosen to be the element of minimum height, we have $i_1 \leq i_j$. Let $Q_j = \sum_{i_j \leq l < \log n}(1-\alpha)^l$. The factor of reduction is:

$$1 - \frac{\frac{1}{2}\sum_{2 \leq j \leq k}(1-\alpha)^{i_j}}{\sum_{1 \leq j \leq k} Q_j} < 1 - \frac{\alpha \sum_{2 \leq j \leq k}(1-\alpha)^j}{2 \sum_{1 \leq j \leq k}(1-\alpha)^j} \leq 1 - \frac{\alpha}{4}.$$

The final case to consider is when a ruler removes a cell. In this case we have to account for the weight of the queue as well as the weight of the chain of subjects. We shall perform the accounting as if the heaviest element was removed (even though it is not necessarily the case). It can be simulated by rearranging the weights as indicated in the Figure 2.

Suppose the ruler has weight $(1-\alpha)^i$ and the subjects have weight $(1-\alpha)^{i_1}, \ldots, (1-\alpha)^{i_k}$, where $(1-\alpha)^{i_k}$ is the maximum weight and $k \leq \log \log n$. (Finally the value of α and the length's of the chains come into play!) The fraction of reduction is:

$$1 - \frac{\frac{1}{2}(1-\alpha)^{i_k}}{\sum_{i \leq j < \log n}(1-\alpha)^j + \frac{1}{2}\sum_{1 \leq j \leq k}(1-\alpha)^{i_j}} \leq 1 - \frac{(1-\alpha)^{i_k}}{\frac{2}{\alpha}(1-\alpha)^{i_k} + (1-\alpha)^{i_k} \log \log n} = 1 - \frac{\alpha}{3}.$$

We can now state the main theorem concerning the runtime of this algorithm.

Theorem 5.2 *The number of list cells remaining after $5 \log n$ stages is at most* $n / \log n$.

Proof: At the start of the algorithm, the total weight of all of the items is:

$$\frac{n}{\log n} \sum_{0 \le i < \log n} (1 - \alpha)^i < \frac{n}{\log n} \frac{1}{\alpha}.$$

Since the weight of the smallest item is $(1-\alpha)^{\log n - 1}$, if we show that the total weight is reduced to at most $\frac{n}{\log n}(1-\alpha)^{-\log n + 1}$ at time $t = 5 \log n$, we will have established that there are at most $n / \log n$ items. Since the total weight is reduced by a factor of at least $1 - \frac{\alpha}{4}$ each stage, the weight at time t is at most

$$\frac{n}{\log n} \frac{1}{\alpha} (1 - \frac{\alpha}{4})^t \le \frac{n}{\log n} \log \log n (1 - \frac{\alpha}{4})^{5 \log n} < \frac{n}{\log n}(1 - \alpha)^{\log n}.$$

■

6 Conclusions

We have shown that list ranking can be solved by a relatively simple optimal algorithm. It is important to pursue the idea further and develop truely practical list ranking algorithms. In practice, Wyllie's algorithm is still probably the best deterministic algorithm, although we believe that our algorithm is a contribution towards finding a better practical algorithm. In this paper we have considered the synchronous P-RAM model. An important area of research is to look at the list rank problem in other models such as asynchronous models [LG87].

References

[CV85]   R. Cole and U. Vishkin. Deterministic coin tossing and accelerating cascades: micro and macro techniques for designing parallel algorithms. In *Proceedings of the 18th ACM Symposium on Theory of Computation*, pages 206–219, 1985.

[CV86a] R. Cole and U. Vishkin. Approximate scheduling, exact scheduling, and applications to parallel algorithms. In *27th Symposium on Foundations of Computer Science*, pages 478–491, 1986. Part I to appear in SIAM Journal of Computing.

[CV86b] R. Cole and U. Vishkin. *Faster Optimal Parallel Prefix Sums and List Ranking*. Technical Report 56/86, Tel Aviv University, December 1986.

[GMT86] H. Gazit, G. L. Miller, and S. H. Teng. Optimal tree contraction in the EREW model. 1986. Extended abstract.

[GPS87]   A. V. Goldberg, S. A. Plotkin, and G. E. Shannon. Parallel symmetry-breaking in sparse graphs. In *Proceedings of the 19th ACM Symposium on Theory of Computation*, pages 315–324, 1987.

[Han87]   Y. Han. *Designing Fast and Efficient Parallel Algorithms*. PhD thesis, Duke University, 1987.

[KRS85]   C. Kruskal, L. Rudolf, and M. Snir. The power of parallel prefix computation. In *International Conference on Parallel Processing*, pages 180–185, 1985.

[LG87]   B. D. Lubachevsky and A. G. Greenberg. Simple, efficient asynchronous parallel prefix algorithms. In *International Conference on Parallel Processing*, pages 66–69, 1987.

[MR85]   G. L. Miller and J. H. Reif. Parallel tree contraction and its applications. In *26th Symposium on Foundations of Computer Science*, pages 478–489, 1985.

[Rei85]   J. H. Reif. An optimal parallel algorithm for integer sorting. In *26th Symposium on Foundations of Computer Science*, pages 496–503, 1985.

[TV85]   R. E. Tarjan and U. Vishkin. An efficient parallel biconnectivity algorithm. *SIAM Journal on Computing*, 14(4):862–874, 1985.

[Vis84]   U. Vishkin. Randomized speed-ups in parallel computation. In *Proceedings of the 16th ACM Symposium on Theory of Computation*, pages 230–239, 1984.

[WH86]   R. A. Wagner and Y. Han. Parallel algorithms for bucket sorting and the data dependent prefix problem. In *International Conference on Parallel Processing*, pages 924–930, 1986.

[Wyl79]   J. C. Wyllie. *The Complexity of Parallel Computation*. PhD thesis, Department of Computer Science, Cornell University, 1979.

Optimal parallel algorithms for expression tree evaluation and list ranking

Richard Cole‡ and Uzi Vishkin†

Courant Institute of Mathematical Sciences, New York University‡† and

Sackler Faculty of Exact Sciences, Tel Aviv University†

Abstract. Two related results are presented. The first is a simple $n/\log n$ processor, $O(\log n)$ time parallel algorithm for list ranking. The second is a general parallel algorithmic technique for computations on trees; it yields the first $n/\log n$ processor, $O(\log n)$ time deterministic parallel algorithm for expression tree evaluation, and solves many other tree problems within the same complexity bounds.

I. Introduction

Two variants of the PRAM model of parallel computation are used in this paper, namely the EREW and CRCW PRAMs. A PRAM employs p synchronous processors all having access to a common memory. An EREW PRAM does not allow concurrent access by more than one processor to the same common memory location, while a CRCW PRAM allows concurrent access for both reads and writes; in a concurrent write, among the processors attempting the concurrent write, an arbitrary one is successful. See [Vi-83] for a survey of PRAM results.

Let $Seq(n)$ be the fastest known worst-case running time of a sequential algorithm, where n is the length of the input for the problem at hand. Obviously, the best upper bound on the parallel time achievable using p processors, without improving the sequential result, is of the form $O(Seq(n)/p)$. A parallel algorithm with this running time is said to have *optimal speed-up* or more simply to be *optimal*.

This paper studies two related problems, namely list ranking and expression tree evaluation. In addition, the paper identifies an interesting hierarchical structure among general parallel techniques for trees. Thereby, we make a methodological contribution to the general topic of parallel tree algorithms. The two problems are defined as follows.

‡This research was supported in part by NSF grants DCR-84-01633 and CCR-8702271, ONR grant N00014-85-K-0046 and by an IBM faculty development award.

†This research was supported in part by NSF grants NSF-CCR-8615337 and NSF-DCR-8413359, ONR grant N00014-85-K-0046, by the Applied Mathematical Science subprogram of the office of Energy Research, U.S. Department of Energy under contract number DE-AC02-76ER03077, and by the Foundation for Research in Electronics, Computers and Communication, Administered by the Israeli Academy of Sciences and Humanities.

List ranking.

Input: n nodes, each occupying an entry in an array of size n. Each node has at most one successor in the array. A node having a successor has the array index of this successor. Consider a path that starts at any node and follows the successor relation. We assume that such a path never comprises a circuit. Each node v has a weight $w(v)$.

The problem: For each node compute the total weight of the nodes following it in its linked list.

Previous results: [CV-87] and [AM-88] gave logarithmic time parallel algorithms for this problem that use an optimal $n/\log n$ processors, in the EREW PRAM model. There, the list ranking problem is defined for inputs where only one node does not have a successor. Both algorithms, however, can be applied to the above extended list ranking algorithm within the same complexity bounds.

New result: A logarithmic time algorithm that uses an optimal $n/\log n$ processors in the CRCW model. The contribution here is that the algorithm is strikingly simple. (This result was first reported in [CV-87b].)

Expression tree evaluation.

Input. An expression tree: each internal node stores an operator $(+, -, \times, +)$ and each leaf stores an integer. The nodes are stored in an array, and each node has the array indices of its children and parent.

The problem. Compute the value of the expression represented by the expression tree.

Previous results. Miller and Reif [MR-85] gave a deterministic algorithm that runs on n processors in $O(\log n)$ time. They also gave a randomized algorithm that runs on $n/\log n$ processors in $O(\log n)$ time. We also mention the following related result. Concurrent with our work, [GR-86] gave an $O(\log n)$ time, $n/\log n$ processor deterministic parallel algorithm for the problem of expression evaluation for expressions provided the expressions are given in the form needed for the parsing algorithm of [BV-85].

New result. A deterministic algorithm that runs on $n/\log n$ processors in $O(\log n)$ time. (This result was first reported in [CV-86c].)

Postscript. Recently, Gazit, Miller and Teng obtained a deterministic algorithm with the same complexity bounds [GMT-87].

Remark. By definition, an expression tree is a binary tree. In the full paper we discuss applications of our ideas to general trees; we give parallel deterministic algorithms with the same complexity bounds for the minimum vertex cover problem and the dominating vertex set problem on trees. There, we require the children of each internal node to occupy a contiguous portion of the array and each internal node to have pointers to, or rather array indices of, its first and last child in the array.

We turn to the methodological contribution to tree algorithms. Previously, [MR-85] gave an elegant parallel algorithmic technique for a class of tree problems including expression tree evaluation. Their contribution was not only to provide a new technique but also to characterize the class of problems that it can solve. Our new technique benefits greatly from their characterization. We propose an alternative *accelerated centroid decomposition (ACD)* technique. Typically, the Miller/Reif technique provides a deterministic algorithm that runs in $O(\log n)$ time on n processors, and a randomized algorithm that runs in $O(\log n)$ time on $n/\log n$ processors; for all such problems, the

ACD technique provides a deterministic algorithm that runs in $O(\log n)$ time on $n/\log n$ processors. Thus the advantage of the ACD technique is that it provides optimal deterministic algorithms, where Miller and Reif had obtained optimal randomized algorithms (and non-optimal deterministic algorithms). In addition, we believe our technique provides simpler algorithms than the randomized approach.

Basically, both in [MR-85] and here, the approach is to reduce the size of the input problem by many simultaneous local operations on the input tree. This then leads to the obvious recursive algorithm. The major problem is to schedule these local operations. In [MR-85] the scheduling is done on the fly, being performed as the tree is being reduced. By contrast, in the ACD technique, the schedule is computed beforehand, and is then used to direct the tree reduction. This separation of concerns appears to yield more efficient algorithms.

Another advantage of the ACD technique is that it builds, in a structured fashion, on a known set of techniques for parallel algorithms on trees. More specifically, we identify the following problem and techniques: (1) list ranking, (2) the Euler tour technique, (3) the centroid decomposition (CD) technique, and (4) the accelerated centroid decomposition (ACD) technique. The first three are already known, the fourth is new. The list ranking problem is the most basic of these. The Euler tour technique needs a list ranking algorithm as a subroutine; its computational complexity is the same as that of the list ranking algorithm. Each of the CD and ACD techniques uses both the Euler tour technique and a list ranking algorithm as subroutines. Using any optimal logarithmic time list ranking algorithm the CD technique yields optimal $O(\log^2 n)$ time algorithms for many tree problems. The ACD technique is an enhancement of the CD technique, and on many problems it yields optimal $O(\log n)$ time algorithms rather than $O(\log^2 n)$ time algorithms; however, the CD technique is of wider applicability.

We describe the optimal CRCW list ranking algorithm in Section 2. We refer the reader to [TV-85] and [Vi-85] for the Euler tour technique. In Section 3, we review the CD technique (this technique is well known; it was used by Megiddo [Me-83] and was implicit in the work of Winograd [Wi-75] on expression evaluation). Also, in Section 3, to motivate the use of the CD technique, we show its application to the expression tree evaluation problem. Then, in Section 4, we describe the ACD technique.

2. Optimal List Ranking in $O(\log n)$ time

The list ranking algorithm given below is similar to the Basic List Ranking algorithm of Section 4 in [CV-86a]. There are two changes: we use both a new algorithm for finding a maximal independent set in a list and the new parallel prefix sums algorithm of [CV-86b]. Both algorithms are CRCW algorithms that use $n \log \log n/\log n$ processors and $O(\log n/\log \log n)$ time. The maximal independent set algorithm is described later in this Section; it is the main new component.

We recall Wyllie's list ranking algorithm [W-79]; it runs in time $O((n \log n)/p + \log n)$ using p processors.

The new list ranking algorithm has the following form.

Step 1

> **While** list has size $> n/\log n$ **do**
>> (i) Compute a maximal independent set of nodes.
>> (ii) Shortcut over the nodes in the maximal independent set.
>> (iii) Compress the remaining nodes to the front of the array and then reassign the processors evenly among these nodes.
>
> **end while**

Step 2. Apply Wyllie's list ranking algorithm to the reduced list.

Step 3. By backtracking through Step 1, compute ranks for all nodes.

Complexity. Given the above two changes, it is straightforward to show that the algorithm performs $O(n)$ operations in $O(\log n)$ time. It is then a simple application of Brent's Theorem [Br-74] to schedule $n/\log n$ processors so as to perform the whole algorithm in $O(\log n)$ time (see [CV-86c] for a more detailed description of this analysis).

We conclude

Theorem 2.1. There is a CRCW PRAM algorithm for list ranking that takes $O(\log n)$ time using $n/\log n$ processors.

It remains to give the new algorithm for computing a maximal independent set. The algorithm outlined in this section is based on a few changes to the optimal logarithmic time 2-ruling set (another name for maximal independent set) algorithm of Section 2.3 in [CV-86a], and on the more recent presentation by Goldberg, Plotkin and Shannon [GPS-87] of the deterministic coin tossing technique of [CV-86a]. The algorithm first computes a $5 \log n/\log\log n$-coloring of the input list and then extracts a maximal independent set using this coloring. The coloring is computed in $O(1)$ time using n processors (or more generally, in $O(n/p)$ time using p processors) by the algorithm of [GPS-87]. Let this coloring be stored in the array *COLOR*. Note that adjacent vertices have distinct colors. Below, we show how to use the *COLOR* numbers so as to produce a maximal independent set.

Step 1

for $i = 0$ to $5 \log n$ / $\log\log n - 1$ **do**
　for each vertex v for which $COLOR(v) = i$ **pardo**
　　if neither v nor its neighbors are in the independent set
　　　then add v to the independent set

At each of its $5 \log n$ / $\log\log n$ "rounds", Step 1 selects a set of non-adjacent vertices. When Step 1 is finished, any vertex that was not selected must have a selected vertex as a neighbor. Thus this algorithm selects a maximal independent set, as we wanted.

It remains to implement Step 1 in $O(\log n$ / $\log\log n)$ time using $n \log\log n$ / $\log n$ processors. We use three substeps.

Step 1.1 Sort the vertices by their *COLOR* number. The outcome of this sort is that each vertex v will be given a number $RANK(v)$, $1 \leq RANK(v) \leq n$. No two vertices will have the same *RANK*. This step simply needs a bucket sort of n numbers in the range $[0, 5 \log n$ / $\log\log n - 1]$. This is implemented essentially as in [CV-86a]; the only change is to use the $O(\log n/\log\log n)$ time prefix sum algorithm in order to speed up the bucket sort to $O(\log n/\log\log n)$ time.

Step 1.2 For each v, $RANK(v) := RANK(v) + i\,5n\mathrm{loglog}\,n\,/\,\log n$, where $i = COLOR(v)$.
Step 1.3 Implement the high-level description of Step 1 in $10\log n\,/\,\mathrm{loglog}\,n$ rounds (and not $5\log n\,/\,\mathrm{loglog}\,n$ rounds). In round j ($1 \le j \le 10\mathrm{log}\,n\,/\,\mathrm{loglog}\,n$), process all vertices v such that $(j-1)n\mathrm{loglog}\,n\,/\,\log n < RANK(v) \le jn\mathrm{loglog}\,n\,/\,\log n$. Note that we never simultaneously process two vertices whose $COLOR$ numbers are different.

We conclude

Theorem 2.2. A maximal independent set can be obtained in $O(\log n\,/\,\mathrm{loglog}\,n)$ time using $n\mathrm{loglog}\,n\,/\,\log n$ processors.

3. Centroid Decomposition

We begin by discussing some approaches to parallel algorithms for the tree evaluation problem. This will motivate the use of the centroid decomposition technique, which yields an $O(\log^2 n)$ time algorithm for the tree evaluation problem. The centroid decomposition uses both the Euler tour technique and an optimal parallel algorithm for list ranking. To keep the discussion simple, we consider expression trees in which all the operators are either $+$ or $\times$ (this restriction is removed in the full paper).

A sequential algorithm proceeds by repeatedly reducing the size of the problem, as follows. Select an internal node v, both of whose children are leaves. Evaluate the expression represented by the subtree rooted at v (either an addition or a multiplication), store the result at v, and make v a leaf (by removing its children). To make this into a parallel algorithm, we can allow the above reduction to be performed simultaneously at all such nodes v. Unfortunately, there are expressions on which this (parallel) algorithm uses $\Theta(n)$ time.

To produce a faster parallel algorithm we allow an internal node with only one child to store an expression of the following form: $ax + b$, where x is an indeterminate, representing the value of the unevaluated subtree, and a and b are integers. Now, we can simultaneously remove every leaf u, by the following PRUNE operation, applied to each leaf u. Let v be the parent of u.
Case (i). v has only one child. Let the expression stored at v be $ax + b$, and the value stored at u be c. Then place the value $ac + b$ in v, and remove u (thereby making v into a leaf).
Case(ii). The sibling of u is a leaf. We proceed as in the previous paragraph.
Case (iii). The sibling of u is not a leaf. Suppose that v stores the operator $+$ (resp. $\times$) and u holds the value c. Then place the expression $x + c$ (resp. cx) in v and remove u.

For the accelerated centroid decomposition technique, it will be convenient to allow more general expressions to be stored at internal nodes. Specifically, we allow an internal node to store an expression $axy + bx + cy + d$, where x and y are indeterminates representing, respectively, the values of the left and right subtrees, and a, b, c, d are integers. So, initially, an internal node stores one of the expressions xy, $x + y$, while at a node with a single child the expression stored will be of the form $bx + d$ (or $cy + d$). The PRUNE operation is performed (essentially) as before. In addition, we introduce the SHORTCUT operation (it is similar to the COMPRESS operation of [MR-85]). The SHORTCUT operation can be applied to any node v that has only one child u; the effect of the SHORTCUT is to remove v from the tree. The SHORTCUT operation proceeds as follows. Let w be the parent of v, and let $ax_1 + b$, $cx_2y + dx_2 + ey + f$ be the expressions stored at nodes v and u, respectively. Then, on removing node v (by making u the child of w), the new expression to be stored at u is $acx_2y + adx_2 + aey + (af + b)$. We have to avoid applying the SHORTCUT operation simultaneously at adjacent nodes. Also, we have to avoid simultaneously applying a SHORTCUT at node w and a PRUNE at node u, where w is the grandparent of u (since both operations affect node v, the parent of u). We note that an individual PRUNE or SHORTCUT operation requires $O(1)$ time on a single processor.

The key to obtaining a fast efficient parallel algorithm for expression tree evaluation is to schedule the PRUNE and SHORTCUT operations so that the time required to reduce the tree to a single node is small. The centroid decomposition technique yields a simple schedule; it provides an optimal $O(\log^2 n)$ time parallel algorithm.

Before describing the centroid decomposition, we need a few definitions. Let $T = (V, E)$ be a rooted tree, where r is its root, and let $SIZE(v)$ be the number of nodes in the subtree of T rooted at v. Let $n = |V|$. We define the *centroid level* of v, denoted $C\_LEVEL(v)$, to be $\lceil \log SIZE(v) \rceil$ (the base of all logarithms in this paper is two).

Observation. Each node v has at most one child u such that $C\_LEVEL(u) = C\_LEVEL(v)$.

Accordingly, we define the *centroid path* of v to be the longest directed path of nodes and tree edges, including v, where all the nodes on the path have the same centroid level as v. We note that there might be several disjoint centroid paths at the same centroid level. This partition of the nodes of T into centroid paths is called the *centroid decomposition* of T.

The algorithm for expression tree evaluation follows.

> **for** $i := 0$ **to** $\lceil \log n \rceil$ **do**
>
> **invariant:** each centroid path at level $< i$ has been removed, so each node on a centroid path at level i has at most one child.
>
> > **for** each centroid path at level i **pardo**
> >
> > > by repeated application of SHORTCUTs and PRUNEs remove the centroid path

We have yet to explain how a centroid path is removed. So consider a level i centroid path of length m. We proceed in $\lceil \log m \rceil$ steps; at each step the length of the path is roughly halved by SHORTCUT operations at alternate nodes (so after $\lceil \log m \rceil$ steps the path is reduced to a single leaf node); one PRUNE operation completes the path removal. More precisely, on the jth step, SHORTCUT at those nodes originally in positions (ranks) $k \, 2^j + 2^{j-1}$, $k = 0, 1, 2, \cdots$, where the first node is the node nearest the root. Prior to the above expression evaluation, we determine these ranks using a list ranking algorithm.

We summarize. The expression tree evaluation divides into two stages: a Scheduling Stage and an Evaluation Stage. In the Scheduling Stage, we form the centroid decomposition and rank the nodes on each centroid path. This yields the schedule for performing the SHORTCUT and PRUNE operations. The Scheduling Stage requires

$O(\log^2 n)$ time on $n/\log^2 n$ processors, as we explain below. The Evaluation Stage performs the PRUNE and SHORTCUT operations using the schedule described above. As we show below, it too takes $O(\log^2 n)$ time using $n/\log^2 n$ processors. So the whole computation can be performed in $O(\log^2 n)$ time on $n/\log^2 n$ processors.

Recall that the *centroid path* of v is the longest directed path of nodes and tree edges, including v, where all the nodes on the path have the same centroid level as v. Thus to compute the centroid paths for T is suffices to compute the *SIZE* of each node. Since the nodes of the tree are provided in an array, the array SIZE can be computed using the Euler tour technique, in time $O(\log n)$ using $O(n/\log n)$ processors.

To create the centroid paths, we add a pointer from each node to its centroid child, the child whose centroid level is the same as its own, if any. We note that the nodes are still stored in an array. So we can apply the list ranking algorithm to obtain the rank of each node in its centroid path; this also takes $O(\log n)$ time using $n/\log n$ processors. Thus the Scheduling Stage can be performed in $O(\log n)$ time using $n/\log n$ processors, or more generally, in time $O(n/p)$ using p processors, for any $p \le n/\log n$.

The Evaluation Stage consists of applying SHORTCUT or PRUNE operations to each node of the tree. Since each such application needs $O(1)$ operations, these applications require a total of $O(n)$ operations. Had we had a processor standing by each node, it would have been trivial to perform the Evaluation Stage in $O(\log^2 n)$ time. In order to see that the Evaluation Stage can be performed in $O(\log^2 n)$ time using only $n/\log^2 n$ processors we refer the reader to the comment following the Evaluation Stage for the accelerated centroid decomposition method in Section 4.

Note that nodes in centroid paths at level i are not being considered until the treatment of all centroid paths at lower levels is finished. By contrast, the new method (which is given in the next section) works on nodes in centroid paths at level i *in parallel* to working on nodes in centroid paths at lower levels. This motivates our naming the new method the *accelerated centroid decomposition (ACD)* method.

4. The accelerated centroid decomposition method

The new method has three stages: the Preparatory Stage, the Scheduling Stage and the Evaluation Stage.

The Preparatory Stage. The input for this stage is a rooted tree where each node has a linked list of its children (incoming edges) and its parent (outgoing edge). We compute the following for each node v.

(i) *SIZE* (v).

(ii) *TAIL* (v) and *HEAD* (v). *TAIL* (v) (resp. *HEAD* (v)) is a bit vector that records whether v is the tail (resp. head) of its centroid path. The tail (resp. head) of a centroid path is the vertex on the path furthest from (resp. nearest to) the root.

Let $v_1, v_2, \cdots, v_{\alpha+1}$ be a centroid path at level $i + 1$, where v_1 is the head of the path and $v_{\alpha+1}$ is the tail of the path. Let $SIZE_j$ be $SIZE(v_j) - SIZE(v_{j+1})$, $1 \le j \le \alpha$. In words, $SIZE_j$ is the size of the tree rooted at v_j excluding the tree rooted at v_{j+1}. Observe that $\sum_{j=1}^{\alpha} SIZE_j = SIZE(v_1) - SIZE(v_{\alpha+1})$. Thus $\sum_{j=1}^{\alpha} SIZE_j < 2^i$. We also compute:

(iii) All α prefix sums $A_k = \sum\limits_{j=1}^{k} SIZE_j$, for $1 \le k \le \alpha$, and we define $A_0 = 0$.

The Evaluation Stage. We assume that the scheduling stage provides a schedule for the PRUNE and SHORTCUT operations that lasts $O(\log n)$ time. We also assume that each node knows when it is scheduled for removal. We implement this schedule on $n/\log n$ processors in $O(\log n)$ time, as follows.

We sort the nodes according to their schedules. Lemma 3.4 in [R-85] gives a logarithmic time optimal parallel *deterministic* algorithm for bucket sorting n positive numbers which are $\le \log n$. It is easy to extend this to positive numbers which are $O(\log n)$ without changing the time and processor complexities. We use this algorithm to sort the schedules. The assignment of any number of $p \le n/\log n$ processors to nodes is easy. We implement each time unit of the schedule in turn: At each time step of the implementation we perform the next p jobs from the current time unit of our schedule. If there are less than p jobs remaining we perform all of them; this will leave some of the processors idle for this time step of the implementation (see [CV-86a] for details of a similar assignment of processors to jobs). Nonetheless, it is easily seen that this assignment of $n/\log n$ processors to nodes uses a total of $O(\log n)$ time.

Comment. Given n positive numbers which are $O(\log^2 n)$, the same bucket sort algorithm takes $O(\log^2 n)$ time using $n/\log^2 n$ processors. For the Evaluation Stage of the Centroid Decomposition method (Section 3) we perform the assignment of processors to jobs in exactly the same way as here, using this sorting algorithm. So the Centroid Decomposition method will take $O(\log^2 n)$ time using $n/\log^2 n$ processors.

The Scheduling Stage. It computes the schedule for PRUNE and SHORTCUT operations in the Evaluation Stage. The schedule comprises $2\lceil \log n \rceil + 1$ steps. The following Inductive Goal describes the reduced tree, obtained during the Evaluation Stage, following the performance of the first $2i - 1$ steps, and the first $2i$ steps of the schedule:

The Inductive Goal.

A. The reduced tree, following the performance of the first $2i - 1$ steps of the schedule, satisfies the following: If the centroid level of node v is $< i$, then v has been removed.

B. The reduced tree, following the performance of the first $2i$ steps of the schedule, satisfies the following: If the centroid level of node v is i, v is present if and only if it was the tail of its centroid path originally; in addition, if v is present it is a leaf.

For $i = 0$ the Inductive Goal is trivially achieved. Next, we assume the Inductive Goal for i and show how to satisfy it for $i + 1$ within 2 time steps past time $2i$, by defining an appropriate schedule. First, we show how to satisfy A of the Inductive Goal for $i + 1$. At step $2i + 1 = 2(i + 1) - 1$ we PRUNE the level i leaves (these are the nodes that were originally tails of the level i paths). Next, we show how to satisfy B of the Inductive Goal for $i + 1$.

Consider the tree rooted at the head of a centroid path at level $i + 1$, excluding the subtree rooted at the tail. Each node on the centroid path (excluding the tail) has a *centroid child* and perhaps another, non-centroid, child. We would like to schedule the SHORTCUT operations for the nodes on the centroid path. However, complications arise as different nodes on the centroid path become ready for SHORTCUT at different times. We focus on a single centroid path at level $i + 1$: $v_1, v_2, \cdots, v_{\alpha+1}$. Let u_j be the current non centroid child of v_j, for $1 \le j \le \alpha$, if any. Let $SIZE_j$ and their prefix sums A_j, for $1 \le j \le \alpha$, be as above. Our algorithm uses:

1. The inductive assumption that if the tree rooted at any non-centroid child has x nodes then this tree has been removed after $2\lceil \log x \rceil + 1$ steps.

2. The fact that for each centroid path at level $i+1$, $\sum_{j=1}^{\alpha} SIZE_j < 2^i$.

3. The data which were computed in the Preparatory Stage.

By the inductive hypothesis, node u_j, $1 \le j \le \alpha$, was removed after $t_j = 2\lceil \log SIZE(u_j) \rceil + 1$ steps. Thus it is safe to SHORTCUT node v_j after time t_j. Recall that we must avoid performing simultaneous SHORTCUTs at adjacent nodes.

We explain when to schedule the SHORTCUTs for nodes on the level $i+1$ centroid path, other than the tail. Suppose $\alpha \ge 1$ (otherwise the inductive hypothesis already holds). We will ensure that by the end of step $2i+2 = 2(i+1)$ of the schedule the nodes $v_1, \cdots, v_\alpha$ have all been SHORTCUT. For each j, $1 \le j \le \alpha$, consider the binary representations of A_{j-1} and A_j. Let K_j be the index of the most significant bit in which these two binary representations differ. (*Example.* Let the representations of A_{j-1} be 010101 and A_j be 011011. Then K_j is 4.)

Remark 4.1.1. To ensure that K_j is well defined, we concatenate zeros to the left of the binary representation of A_{j-1}. A_0 is simply $0...0$.

Remark 4.1.2. Bit K_j is zero in (the representation of) A_{j-1} and one in A_j (since A_j is larger than A_{j-1}).

The schedule follows: SHORTCUT at node v_j, $1 \le j \le \alpha$, at time $2K_j + 2$.

In order to establish the validity of our solution we show:

(i) All SHORTCUTs and PRUNEs are legal (Validity Claims 1-3).

(ii) All the SHORTCUTs on centroid paths at level $i+1$ occur by time $2i+2$ (Validity Claim 4). Since every non-tail node is scheduled for a SHORTCUT, we conclude that B of the inductive hypothesis holds for $i+1$.

Validity Claim 1. We never SHORTCUT at a node before it is ready.

Proof: Observe that $SIZE_j = A_j - A_{j-1} \le 2^{K_j} - 1$. If v_j has a non-centroid child u_j, the subtree rooted at u_j originally had size $SIZE_j - 1$; by the inductive hypothesis, this subtree was removed by time $2K_j + 1$. $\square$

Validity Claim 2. Two simultaneous SHORTCUTs are never performed at adjacent nodes of the path. (That is, it is impossible that the two nodes are adjacent at the time at which the SHORTCUTs are performed. Earlier we called such SHORTCUTs illegal.)

To show this, we need the following observation:

Observation. For every $1 \le j < l \le \alpha$ with $K_j = K_l$, there exists $j < \beta < l$, for which $K_\beta > K_j$. In other words, given two nodes v_j, v_l at which simultaneous SHORTCUTs are performed we are guaranteed to have node v_β between them at the time of these SHORTCUTs, and therefore the SHORTCUTs are legal, so Validity Claim 2 follows.

Proof of Observation. Remark 4.1.2 implies that bit K_j at A_j is 1, and bit $K_j (= K_l)$ at A_{l-1} is 0. Therefore, for some $j < x < l$, bit K_j was 1 at A_{x-1} and 0 at A_x. Apply Remark 4.1.2 again to conclude $K_x > K_j$, which proves the Observation. $\square$

Validity Claim 3. SHORTCUT and PRUNE operations are never performed simultaneously.

Validity Claim 4. All SHORTCUTs on a centroid path at level $i+1$ occur by time $2i+2$.

Proof. Recall that for $1 \leq j \leq \alpha$, $A_j = \sum_{k=1}^{j} SIZE_k < 2^i$, and thus $K_j \leq i$. $\square$

Complexity Analysis. Omitted.

We conclude:

Theorem 4.1. There is an EREW algorithm for expression tree evaluation that takes $O(\log n)$ time using $n/\log n$ processors.

Acknowledgement. Helpful comments by Zvi Kedem, Yael Maon and Baruch Schieber are gratefully acknowledged. A careful reading of the paper by Ofer Zajicek is also appreciated.

References

[AM-88] R.J. Anderson and G.L. Miller, Optimal parallel algorithms for list ranking, this proceedings.

[B-74] R.P. Brent, "The parallel evaluation of general arithmetic expressions," *J. ACM* 21,2 (1974), 201-206.

[BV-85] I. Bar-On and U. Vishkin, Optimal parallel generation of a computation tree form, ACM Trans. on Prog. Lang. and Sys. 7,2 (1985), 348-357.

[CV-86a] R. Cole and U. Vishkin, Deterministic coin tossing with applications to optimal parallel list ranking, Information and Control 70 (1986), 32-53.

[CV-86b] R. Cole and U. Vishkin, Approximate and exact parallel scheduling with applications to list, tree and graph problems, Proc. 27th Symp. on Foundations of Computer Science, 1986, 478-491.

[CV-86c] R. Cole and U. Vishkin, The accelerated centroid decomposition technique for optimal parallel tree evaluation in logarithmic time, Computer Science Department Technical Report #242, Courant Institute, 1986.

[CV-87] R. Cole and U. Vishkin, "Approximate parallel scheduling. Part I: The basic technique with applications to optimal parallel list ranking in logarithmic time", to appear, *SIAM Journal on Computing*.

[GPS-87] A. Goldberg, S. Plotkin, and G. Shannon, "Parallel symmetry-breaking in sparse graphs", *Nineteenth Annual ACM Symp. on Theory of Computing*, 315-224.

[GMT-87] H. Gazit, G. Miller and S. H. Teng, Optimal tree contraction in EREW model, manuscript, Computer Science Department, University of Southern California.

[GR-86] A. Gibbons and W. Rytter, An optimal parallel algorithm for dynamic tree expression evaluation and its applications, Research Report 77, Dept. of Computer Science, Univ. of Warwick, Coventry, CV4 7AL, England, 1986.

[H-86] X. He, The general tree algebraic computations and its applications in parallel algorithms design, preprint, 1986, Dept. of Computer and Information Science, Ohio State University, Columbus, OH 43210.

[Me-83] N. Megiddo, Applying parallel computation algorithms in the design of serial algorithms, JACM 30(1983), 852-865.

[MR-85] G.L. Miller and J.H. Reif, Parallel tree contraction and its applications, Proc. 26th Symp. on Foundations of Computer Science, 1985, 478-489.

[R-85] J.H. Reif, An optimal parallel algorithm for integer sorting, Proc. 26th Symp. on Foundations of Computer Science, 1985, 496-503, to appear *SIAM J. Comput.*

[TV-85] R.E. Tarjan and U. Vishkin, An efficient parallel biconnectivity algorithm, SIAM J. Comput. 14,4 (1985), 862-874.

[Vi-83] U. Vishkin, Synchronous parallel computation - a survey, TR 71, Dept. of Computer Science, Courant Institute, New York University, 1983.

[Vi-85] U. Vishkin, On efficient parallel strong orientation, Information Processing Letters 20 (1985), 235-240.

[Wi-75] S. Winograd, On the evaluation of certain arithmetic expressions, JACM 22(1975), 477-492.

[W-79] J.C. Wyllie, "The Complexity of Parallel Computation", Ph.D. thesis, TR 79-387, Dept. of Computer Science, Cornell Univ., Ithaca, NY, 1979.

Optimal Parallel Evaluation of Tree-Structured Computations by Raking (Extended Abstract)

S. Rao Kosaraju*   *Arthur L. Delcher*

Department of Computer Science
Johns Hopkins University
Baltimore, Maryland 21218

Abstract. *We show that any arithmetic expression of size n can be evaluated on an EREW PRAM with $O(n/\log n)$ processors in $O(\log n)$ steps. A major contribution is the simplicity of the algorithm. In contrast with existing algorithms which require independent* RAKE *and* COMPRESS *operations, our algorithm combines the* RAKE *and* COMPRESS *into one simple operation. In fact, our algorithm can be viewed as avoiding* COMPRESS*es entirely and simply performing* RAKE*s. The algorithm can be modified easily to evaluate every subexpression of the original arithmetic expression. We also show how it can be applied to the following problem: Given a positive constant λ and a tree with weighted nodes, partition the tree into minimal components subject to the constraint that the sum of the node weights in each component is at least λ.*

1. Introduction

Arithmetic expression evaluation has been established as an important paradigm for solving several problems. Miller & Reif [6] and Rytter [8] independently gave exclusive-read exclusive-write (EREW) parallel random access machine (PRAM) algorithms for evaluating any arithmetic expression of size n, which run in $O(\log n)$ time using $O(n)$ processors. Subsequently Cole & Vishkin [4] and Gibbons & Rytter [5] independently developed $O(\log n)$-time, $O(n/\log n)$-processor EREW PRAM algorithms. In the next section we give an extremely simple $O(\log n)$-time, $O(n/\log n)$-processor EREW PRAM algorithm for the same problem. We view our algorithm as a dramatically simplified version of the algorithm in [5]. We also show how the algorithm can be modified to yield the values of all the subexpressions in $O(\log n)$ time using an $O(n/\log n)$-processor EREW PRAM.

In section 3, we apply the technique to obtain an $O(n/\log n)$-processor, $O(\log n)$-time EREW PRAM algorithm for the following problem: Given a positive constant λ and a tree T with non-negative weights assigned to its nodes, delete a maximal set of edges from T, such that the total weight of the nodes in each resulting component is at least λ.

2. Main Algorithm

For simplicity of presentation we assume that the expression to be evaluated consists of just constants and the operations of addition and multiplication. The algorithm can be modified easily to accommodate subtraction and division, and other non-arithmetic

---
*Supported by the National Science Foundation under Grant DCR-8506361

operations. We can even drop the commutativity requirement for the $+$ and $\times$ operations. Again for simplicity, we assume that this expression is given in the form of a binary tree in which each leaf has an associated constant value, and each internal node has an associated operator, $+$ or $\times$. Each internal node in this tree has exactly 2 children—a property of binary trees we shall refer to as *properness*. For any node u in this tree, let $Par(u)$, $Gpar(u)$, $Sib(u)$, $Left(u)$ and $Right(u)$ denote its parent, grandparent, sibling, left child and right child respectively. We call u a *left node* (resp. *right node*) if u is the left child (resp. right child) of $Par(u)$. We measure the size of the expression as the number of its leaves.

Our goal is to compute the value of the entire expression, *i.e.*, the value of the root. To motivate the reader, we first give an informal description of our algorithm.

A. Informal Description

We evaluate the expression by reducing the size of its tree until only the root is left. The operation we employ to remove nodes from the tree is called **RAKE** because it is applied only to leaf nodes. ([7] describes a similar operation, called *shunt*.) When we **RAKE** a leaf w, we remove w and $Par(w)$ from the tree, and make $Sib(w)$ the child of $Gpar(w)$ in the position vacated by $Par(w)$. Notice that this operation preserves properness of binary trees. To prevent modifying the value of the root as nodes are removed from the tree, we adjust the information stored at $Sib(w)$.

Suppose that at some stage the leaves of the tree in left-to-right order are $w_1, w_2, \cdots, w_m$. In the next phase we will **RAKE**, in parallel, all the odd-subscripted leaves $w_1, w_3, w_5, \cdots$ by an $O(m)$-processor EREW PRAM in $O(1)$ steps. The resulting proper binary tree will have $m/2$ leaves. Thus starting with an n-leaf proper binary tree, we need to perform $\log_2 n$ phases before the tree is reduced to a single node. Note that the total number of steps performed by the complete algorithm is $O(n)$. Hence, by Brent's Theorem, we have an $O(n/\log n)$-processor EREW PRAM algorithm to evaluate the given tree in $O(\log n)$ steps. By performing the phases in reverse, we can compute the value of each subexpression within the same processor and time bounds.

We now make the above description more precise.

B. Basic Algorithm

At any intermediate stage we maintain a proper binary tree with a constant c_w at each leaf w, and an operator, $\bigcirc_u$ ($+$ or $\times$), at each internal node u. Every node u stores pointers to its parent, left child and right child, as well as a bit, $Side(u)$, indicating whether it is the left or right child of its parent. Each node u also stores a pair of numbers (a_u, b_u) which represent the linear expression $a_u X + b_u$ where X is an indeterminate. For simplicity, we shall say that u stores the linear expression $a_u X + b_u$.

We now recursively define the *value at a node* in the tree. The *value at a leaf* w is its associated constant c_w. For any internal node u with operator $\bigcirc_u$, if the values at its left and right children are Val_L and Val_R, and the expressions at the left and right children are $a_L X + b_L$ and $a_R X + b_R$, respectively, then the *value at* u is defined to be the result of the expression

$$(a_L \, Val_L + b_L) \bigcirc_u (a_R \, Val_R + b_R) \ .$$

Note that the value at node u does not involve the expression $a_u X + b_u$; this expression indicates the *contribution* of the value at u to the value at $Par(u)$. Further note that the linear expression at the root is insignificant.

At each phase of the algorithm, we ensure that the value at the root of the tree defined in this way equals the value of the original arithmetic expression. To make this true initially, we begin with the expression X, *i.e.*, $a_u = 1$ and $b_u = 0$, stored at every node u in the tree.

The **RAKE** operation on any leaf w achieves the following:

1) Disconnects w and $Par(w)$ from the tree by connecting $Sib(w)$ as the child of $Gpar(w)$ on the side where $Par(w)$ was.

2) Sets $(a_{Sib(w)}, b_{Sib(w)})$ to the coefficients that result from simplifying either

$$a_{Par(w)} \cdot ((a_w c_w + b_w) \circ_{Par(w)} (a_{Sib(w)} \cdot X + b_{Sib(w)})) + b_{Par(w)}$$

if w is a left node, or

$$a_{Par(w)} \cdot ((a_{Sib(w)} \cdot X + b_{Sib(w)}) \circ_{Par(w)} (a_w c_w + b_w)) + b_{Par(w)}$$

if w is a right node.

The figure below shows the effect of this **RAKE** operation applied to a left or a right leaf w, where we denote $Par(w)$, $Gpar(w)$ and $Sib(w)$ by p, g and s, respectively.

a) RAKEing a Left Leaf w

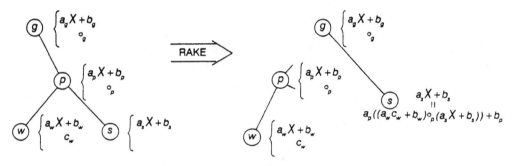

b) RAKEing a Right Leaf w

Note that s inherits the same side of g that p had. The side of g is not affected at all. We further note that:

- Each **RAKE** decreases the number of leaves in the tree by 1, while preserving the properness of the binary tree.

- RAKEing w does not change any value stored in either w or $Par(w)$.

- RAKEing w does not change the value at $GPar(w)$, because the new expression stored in $Sib(w)$ combines the former contributions of $Par(w)$ and $Sib(w)$ to $GPar(w)$. Therefore, each **RAKE** leaves invariant the value at each remaining node in the tree. In particular, the value at the root remains unchanged.

We can now describe the global behaviour of the algorithm. We assume that the leaves (or a pointer to them) are stored in consecutive locations of an array. This can be achieved for any reasonable input representation by an $O(\log n)$-time, $O(n/\log n)$-processor EREW PRAM algorithm [1] [2] [3] [4]. In order to simplify the assignment of processors to leaves and to prevent removing the root from the tree, we do not allow the left-most and right-most leaves in the original expression to be RAKEd. Their siblings, however, will be RAKEd and will modify these nodes' expressions during the course of the algorithm. The details follow:

ALGORITHM Basic Evaluation *(one phase)*:

At any stage, let the remaining leaves (not counting the left-most and right-most in the original expression) be $w_1, w_2, \cdots, w_m$ stored in left-to-right order in m consecutive locations of an array. Let k and j be the largest odd and even subscripts, respectively. Then apply the following 3 steps in synchronized succession.

> *STEP 1:* In parallel, **RAKE** each of the odd-subscripted leaves
> $w_1, w_3, w_5, \cdots, w_k$ which is a left leaf.

> *STEP 2:* In parallel, **RAKE** the rest of $w_1, w_3, w_5, \cdots, w_k$, namely, those that are right leaves.

> *STEP 3:* Collect all the remaining leaves $w_2, w_4, \cdots, w_j$ into locations
> $1, 2, \cdots, j/2$ of the array, maintaining the left-to-right order.

Since each **RAKE** removes 1 leaf, STEPS 1 and 2 together decrease the number of leaves by $\lceil m/2 \rceil$. Hence after STEP 3 there will be $\lfloor m/2 \rfloor$ leaves remaining in the tree.

We now establish that each of STEPS 1-3 can be performed by an $\lceil m/2 \rceil$-processor EREW PRAM in $O(1)$ steps. We assign processor P_i to leaf w_{2i-1} throughout.

We first show that **RAKE** operations in STEPS 1 and 2 can be accomplished without read or write conflicts. The **RAKE** operation for leaf w uses the following values:

- all of the information in nodes w and $Par(w)$,

- the child pointer in $Gpar(w)$ on the side of $Par(w)$,

- the parent pointer and expression values in $Sib(w)$.

We think of these values as *local* to leaf w. We will prevent read/write conflicts in STEPS 1 and 2 of the algorithm if we ensure no overlap between local values of any two leaves being RAKEd simultaneously. Clearly the leaf nodes themselves are entirely distinct, so the only potential overlap involves adjacency of the parents of the leaves. Such overlapping occurs iff two parents are either identical or adjacent in the tree, *i.e.*, one is the parent of the other. The following result shows that this cannot happen in STEP 1 of the algorithm.

Lemma: If w_1 and w_2 are two distinct non-consecutive (in left-to-right order) left leaves in binary tree T, then $Par(w_1)$ and $Par(w_2)$ are neither identical nor adjacent in T.

Proof: Since T is binary and w_1 and w_2 are distinct left leaves, $Par(w_1)$ and $Par(w_2)$ are distinct also. If $Par(w_1)$ and $Par(w_2)$ were adjacent in T with, say, $Par(w_1)$ the parent of $Par(w_2)$, then since w_1 and w_2 are both left leaves, w_1 would be the immediate predecessor of w_2 in the left-to-right order of leaves, contradicting our assumption that w_1 and w_2 are non-consecutive. $\square$

We note that it is possible for two non-consecutive left leaves to have the same grandparent node. These leaves' parent nodes are distinct, however, and so are on opposite sides of the grandparent. Thus the RAKE procedures for such leaves reference different child pointers in the grandparent node. Since STEP 1 removes no even-numbered leaves, the remaining odd-numbered leaves to be removed in STEP 2 are still non-consecutive, and thus a symmetric argument applies to them.

In STEP 3 every w_{2i} moves to location i in the array. This is easily done in parallel by an $\lfloor m/2 \rfloor$-processor EREW PRAM in $O(1)$ steps.

Since RAKE operations do not change the expression value at the root of the tree, and since the algorithm consists of nothing but non-overlapping RAKE operations, we have shown that the algorithm correctly computes the value of the original expression on an EREW PRAM. Thus, if we start with an n-leaf expression tree and exclude the left- and right-most leaves L and R, repeating BASIC EVALUATION $\lfloor \log_2(n-2) \rfloor + 1$ times will reduce the original tree to just its root and the two excluded leaves. In the final phase we compute the value of this 3-node tree, namely $(a_L \cdot c_L + b_L) \bigcirc_{Root} (a_R \cdot c_R + b_R)$. Thus, the algorithm consists of at most $\lceil \log_2 n \rceil + 1$ phases, where phase i can be performed by a $\lceil n/2^i \rceil$-processor EREW PRAM in $O(1)$ steps. Recalling the following result:

Brent's Theorem: Any EREW PRAM algorithm that runs in α phases, where phase i requires p_i processors and runs in $O(1)$ steps, can be implemented on a p-processor EREW PRAM to achieve a speed of $O(\Sigma p_i/p + \alpha)$.

we have now proven:

Theorem: Any arithmetic expression of size n can be evaluated by a p-processor EREW PRAM in $O(n/p + \log n)$ steps.

C. Simplified Algorithm

It is worth noting that STEP 3 in BASIC EVALUATION can be avoided completely, if we observe that no new leaves are created during any RAKE operation. It is easy to verify that if the leaves to be RAKEd are initially in locations $1, 2, \cdots, n-2$ of an array, we can RAKE leaves $1, 3, 5, \cdots$ in the first phase, $2, 6, 10, \cdots$ in the second, etc. More precisely, we have:

ALGORITHM Simplified Evaluation:

At phase i, where i goes from 0 to $\lceil \log_2 (n-2) \rceil$

> *STEP 1:* In parallel RAKE each of the leaves in locations $2^i, 3 \cdot 2^i, 5 \cdot 2^i, 7 \cdot 2^i, \cdots$ that is a left leaf.

> *STEP 2:* In parallel RAKE each of the leaves in locations $2^i, 3 \cdot 2^i, 5 \cdot 2^i, 7 \cdot 2^i, \cdots$ that is a right leaf.

D. Evaluating Subexpressions

It is easy to modify the algorithm to yield the value of each subexpression of the given expression. These subexpression values are precisely the values at each node u of the tree, as defined earlier in subsection B. We have already seen that at any point in the algorithm, when a RAKE removes a leaf w and its parent $Par(w)$, the value at $Gpar(w)$ does not change. Note, however, that when w is RAKEd, the value at $Par(w)$ has not yet been evaluated. This value cannot be calculated until after the value of $Sib(w)$ gets calculated, which in general will occur at a later phase in our algorithm.

Therefore, we calculate the subexpression values in two stages. The first stage is just SIMPLIFIED EVALUATION, but with a small modification to the RAKE procedure to store information needed in the second stage. The second stage essentially consists of running the first stage in reverse. As we do this we compute values for the parents of the nodes RAKEd in the corresponding phase of the forward stage. By the end of the second stage, every node will have been assigned a value corresponding to the subexpression for which it was the root in the original tree.

The procedure relies upon the following result:

Lemma: At every stage of the evaluation, the value at each node in the current tree is the same as the value of the same node in the original tree.

As the algorithm progresses from phase to phase, however, the linear expressions at the nodes will be modified.

We add to each internal node u in the tree a field $Val(u)$ in which the final value of its corresponding subexpression can be stored. For leaf nodes w, we identify $Val(w)$ with c_w. To each leaf node w we add a pair of numbers $(\hat{a}_w, \hat{b}_w)$ in which we will save the values of $(a_{Sib(w)}, b_{Sib(w)})$ just before w is RAKEd. It is important to save these values since the value at $Par(w)$ depends on them, and the RAKE operation modifies $(a_{Sib(w)}, b_{Sib(w)})$. This extra step in the RAKE procedure is the only difference between the first stage and SIMPLIFIED EVALUATION as described earlier.

We now describe the second, or reverse, stage of the algorithm. As a result of the first stage, we start with a 3-node tree with the correct value $Val(u)$ stored in each node. As the reverse stage proceeds, leaf nodes and their parents will be reconnected to the tree (UNRAKEd) in exactly the reverse order from that in which they were removed (RAKEd) in the first stage. When a leaf w is UNRAKEd, both w and $Par(w)$ store exactly the same pointers and expressions they had at the time they were RAKEd, since the RAKE operation does not affect these. In particular, we can find $Gpar(w)$ and $Sib(w)$ using the pointers in $Par(w)$.

The UNRAKE operation applied to a leaf node w is given below:

1) Reconnect w and $Par(w)$ to the tree between $Gpar(w)$ and $Sib(w)$ exactly as they were before the RAKE operation was applied to w in the first stage.

2) Copy $(\hat{a}_w, \hat{b}_w)$ to $(a_{Sib(w)}, b_{Sib(w)})$. This restores the expression stored at $Sib(w)$ to exactly what it was just before w was RAKEd.

3) Set $Val(Par(w))$ either to

$$(a_w \cdot Val(w) + b_w) \circ_{Par(w)} (a_{Sib(w)} \cdot Val(Sib(w)) + b_{Sib(w)})$$

if w is a left node, or

$$(a_{Sib(w)} \cdot Val(Sib(w)) + b_{Sib(w)}) \circ_{Par(w)} (a_w \cdot Val(w) + b_w)$$

if w is a right node. If $Val(Sib(w))$ is correct, this is exactly how we defined the value at $Par(w)$.

The earlier figures, read from right to left, illustrate the action of this UNRAKE operation.

The entire reverse stage of the algorithm can now be defined as follows:

ALGORITHM Reverse Evaluation:

At phase i, where i goes from $\lceil \log_2(n-2) \rceil$ down to 0

> *STEP 1:* In parallel UNRAKE each of the leaves in locations $2^i, 3 \cdot 2^i, 5 \cdot 2^i, 7 \cdot 2^i, \cdots$ that is a right leaf.
>
> *STEP 2:* In parallel UNRAKE each of the leaves in locations $2^i, 3 \cdot 2^i, 5 \cdot 2^i, 7 \cdot 2^i, \cdots$ that is a left leaf.

We now argue for the correctness of the algorithm.

- The locality of operations in the reverse stage is identical to that of the forward stage. In particular all operations can be performed on an EREW PRAM.

- The reverse stage of the algorithm performs UNRAKEs in exactly the same phases as RAKEs were performed in the forward stage, only in reverse order. Since each RAKE operation preserved the value at each node in the tree, and each UNRAKE simply undoes the effect of the corresponding RAKE, it follows that at each phase of the reverse stage the value at each node is invariant.

- At the start of the reverse stage, the correct value at each of the three nodes is contained in $Val(u)$. If the correct value is in $Val(Sib(w))$ when w is UNRAKEd, then the correct value is stored in $Val(Par(w))$. Therefore, it follows by induction that the reverse stage stores the correct value in $Val(Par(w))$ for all such nodes, namely every internal node in the tree.

Since the computations performed in the second stage exactly mirror those performed in the first stage of the algorithm, it clearly has the same processor and time bounds on an EREW PRAM. Thus we have the following:

Theorem: All subexpressions of an arithmetic expression of size n can be evaluated by a p-processor EREW PRAM in $O(n/p + \log n)$ steps.

It is worth emphasizing how simple the entire algorithm is. It requires only constant storage in each processor, and the only requirements are that $+$ and $\times$ be commutative and that $\times$ distributes over $+$.

3. Partitioning a Weighted Tree

We now apply the technique of our algorithm to the following problem: Given a positive constant λ and a tree T with a non-negative weight α_u at each node u, remove a maximal set of edges from T, such that the total weight of the nodes in each resulting component is at least λ. Note that such a decomposition is not unique. This problem was mentioned in [4] as one to which the tree evaluation technique did not seem to apply.

A. Proper Binary Tree Case

For simplicity, let us initially assume that T is a proper binary tree. We start with the algorithm of the previous section used to compute the sum of the weights in each subtree of T. The forward stage of this algorithm needs no modification. In the reverse stage, however, we modify the UNRAKE operation, so that when each leaf w is UNRAKEd, we can decide which of the two edges from $Par(w)$ to its children can be removed from T.

We now describe some of the details. To modify the previous algorithm to compute the sum of the weights of the nodes in each subtree of the original tree T, simply redefine the value at each internal node u to be its weight α_u plus the values of its left and right children. Because only addition is being performed, the expression stored at each node u is now a single constant b_u. Note that at each phase in this algorithm, both forward and reverse stages, the value of b_u is precisely the sum of the weights of all nodes in the original tree that were *between* u and its current parent $Par(u)$. By *between*, we mean all nodes on the path from u to the current $Par(u)$ and all their descendants that are not in u's subtree.

The forward stage of this algorithm is left completely unchanged. It is only during the reverse stage that we decide which edges to remove. At each phase i of the reverse stage, we have a proper binary tree T_i consisting of a subset of nodes from the original tree T. Each edge (u,v) in T_i corresponds to a path of subtrees in the original tree T, namely the nodes that were *between* u and v in the sense of the preceding paragraph. At each phase we shall select edges for removal. The edges selected will be child edges relative to the original tree T, of nodes that are in the current tree T_i. Thus, the removed edges induce a decomposition of the current tree T_i into components, although the actual edges removed are in the original tree T.

Let us define an augmented weight W_u for each node u in T_i to be $\alpha_u + b_u$, which is the weight of u plus the total weight of all nodes in the original tree T between u and its current parent in T_i. The property we maintain throughout the reverse stage of the algorithm is the following:

At each phase i of the reverse stage of the algorithm, the edges selected for removal induce a maximal decomposition of T_i into components of size at least λ, relative to the augmented weight function W.

Note that after the final phase, when T_i is the original tree T, the augmented weight function W is the same as the weight of each node in the original tree, so that the above property guarantees a correct decomposition of T.

It is easy to select edges to remove from the 3-node tree that begins the reverse stage, in such a way that the above property is true initially. To maintain it throughout the reverse stage of the algorithm, in each node u of the current tree T_i we store some additional values. These values reflect the total weight of nodes that are in the current component of u, i.e., the connected component of u as determined by the edges removed up till the current phase i. In particular we maintain separate information about three portions of u's current component: the total weight in each of the two subtrees of u, and the total weight of nodes in neither subtree of u. Because of space limitations we omit the details of the definitions of these values, the selection of edges for deletion, the updating of the values in each node, and the avoidance of read/write conflicts during the **UNRAKE** operations.

B. Non-Binary Tree Case

We now very briefly indicate how to extend the above idea to non-binary trees. Given any such tree T, we can transform it into a proper binary tree T' by adding a dummy child of weight 0 to each degree-1 node, and adding $d-2$ extra 0-weight nodes in a straight-line under each node of degree $d > 2$, attaching its extra children to them. We would like to be able to apply our algorithm directly to T' to obtain a maximal decomposition, but such a decomposition might include edges between the dummy nodes. Such edges in general have no corresponding edges in T that can be removed to yield the desired partition. Therefore, we extend our algorithm to the more general problem of computing a maximal decomposition of a proper binary tree in which some edges have been designated as *unremovable*. In our particular case, the unremovable edges are precisely those we added in converting the original tree T to a proper binary tree T'. The details of this construction will be provided in the final version of the paper.

Acknowledgements: We sincerely thank Richard Beigel and Dwight Wilson for their many constructive comments.

References

[1] Anderson, R. J. and G. L. Miller, Optimal Parallel Algorithms for List Ranking, Extended Abstract, 1986.

[2] Bar-On, I. and U. Vishkin, Optimal Parallel Generation of a Computation Tree Form, *ACM Trans. Prog. Lang. and Sys.* **7**, pp. 348-357, 1985.

[3] Cole, R. and U. Vishkin, Deterministic Coin Tossing with Applications to Optimal Parallel List Ranking, *Information and Control* **70**, pp. 32-53, 1986.

[4]   Cole, R. and U. Vishkin, The Accelerated Centroid Decomposition Technique for Optimal Parallel Tree Evaluation in Logarithmic Time, TR, Courant Institute, June, 1987.

[5]   Gibbons, A. and W. Rytter, An Optimal Parallel Algorithm for Dynamic Expression Evaluation and Its Applications, RR 77, Dept. of Computer Sci., Univ. of Warwick, April, 1986.

[6]   Miller, G. L. and J. H. Reif, Parallel Tree Contraction and Its Applications, *Proc. 26th Symp. Found. Comp. Sci*, pp. 478-489, 1985.

[7]   Miller, G. L., V. Ramachandran, and E. Kaltofen, Efficient Parallel Evaluation of Straight-Line Code and Arithmetic Circuits, *Proc. Aegean Workshop on Computing*, pp. 236-251, July, 1986.

[8]   Rytter, A., Remarks on Pebble Games on Graphs, *Combinatorial Analysis and Its Application (ed. M. Syslo)*, 1985.

ON FINDING LOWEST COMMON ANCESTORS:

SIMPLIFICATION AND PARALLELIZATION

(Extended Summary)

Baruch Schieber [†][1,3]   *Uzi Vishkin* [†][1,2]

[†][1]Department of Computer Science
School of Mathematical Sciences
Sackler Faculty of Exact Sciences
Tel Aviv University
Tel Aviv, Israel 69978

[†][2]Department of Computer Science
Courant Institute of Mathematical Sciences
New York University
251 Mercer St., New York, NY 10012

ABSTRACT

We consider the following problem. Suppose a rooted tree T is available for preprocessing. Answer on-line queries requesting the lowest common ancestor for any pair of vertices in T. We present a linear time and space preprocessing algorithm which enables us to answer each query in $O(1)$ time, as in Harel and Tarjan [HT-84]. Our algorithm has the advantage of being simple and easily parallelizable. The resulting parallel preprocessing algorithm runs in logarithmic time using an optimal number of processors on an EREW PRAM. Each query is then answered in $O(1)$ time using a single processor.

1. Introduction

We consider the following problem. Given a rooted tree $T(V,E)$ for preprocessing, answer on-line LCA queries of the form, "Which vertex is the Lowest Common Ancestor (LCA) of x and y?" for any pair of vertices x,y in T. (Let us denote such a query $LCA(x,y)$.) We present a preprocessing

1. The research of both authors was supported by the Applied Mathematical Sciences subprogram of the Office of Energy Research, U.S. Department of Energy under contract number DE-AC02-76ER03077.
2. The research of this author was supported by NSF grant NSF-CCR-8615337, ONR grant N00014-85-K-0046 and the Foundation for Research in Electronics, Computers and Communication, administered by the Israeli Academy of Sciences and Humanities.
3. Current address: I.B.M. - Research Division, T.J. Watson Research Center, P.O. Box 218, Yorktown Heights, NY 10598.

algorithm which runs in linear time and linear space on the serial RAM model. (For the definition of a RAM model see, e.g., [AHU-74].) Given this preprocessing we show how to process each such LCA query in constant time.

We consider also parallelization of our algorithm. The model of parallel computation used is the exclusive-read exclusive-write (EREW) parallel random access machine (PRAM). A PRAM employs p synchronous processors all having access to a common memory. An EREW PRAM does not allow simultaneous access by more than one processor to the same memory location for either read or write purposes. See [Vi-83] for a survey of results concerning PRAMs.

Let $Seq(n)$ be the fastest known worst-case running time of a sequential algorithm, where n is the length of the input for the problem at hand. A parallel algorithm that runs in $O(Seq(n)/p)$ time using p processors is said to have *optimal speed-up* or, more simply, to be *optimal*. A primary goal in parallel computation is to design optimal algorithms that also run as fast as possible.

Our preprocessing algorithm is easily parallelized to obtain an optimal parallel preprocessing algorithm which runs in $O(\log n)$ time using $n/\log n$ processors on an EREW PRAM, where n is the number of vertices in T. Parallelizing the query processing is straightforward provided read conflicts are allowed: k queries can be processed in $O(1)$ time using k processors.

In their extensive paper [HT-84], Harel and Tarjan gave a serial algorithm for the same problem. The performance of their algorithm is the same as ours. However, our algorithm has two advantages: (1) It is considerably simpler in both the preprocessing stage and the query processing. (2) It leads to a simple parallel algorithm. Below, we discuss similarities and differences with respect to [HT-84]. *Similarities*: Both algorithms use two basic observations. (1) It is possible to answer LCA queries in simple paths in constant time. (2) It is possible to answer LCA queries in complete binary trees in constant time. Both algorithms pack information regarding several vertices into a single $O(\log n)$ bits number. *Differences*: The subtler part of both algorithms is to show how to use the above two observations for answering an LCA query. In this part, our approach is completely different. In the preprocessing stage we compute a mapping from the vertices of the input tree T to the vertices of a complete binary B. The mapping has two properties: (i) all the vertices of T mapped into the same vertex in B form a path, and (ii) for each vertex v in T, the descendants of v are mapped into descendants of the image of v in B. This mapping together with some additional information enable us to answer an LCA query in constant time. In

[HT-84], on the other hand, the vertices of the input tree T are mapped to the vertices of an arbitrary tree of logarithmic height, called the compressed tree. The preprocessing consists of a quite involved manipulation of this compressed tree. This manipulation includes partitioning the compressed tree into three plies and preprocessing each ply separately and also embedding the compressed tree in a complete binary tree.

Our parallel algorithm improves on the following results. Tsin [Ts-86] gave two parallel algorithms for the LCA problem. In his first algorithm both the preprocessing stage and the query processing take logarithmic time with a linear number of processors. In his second algorithm the preprocessing stage takes $O(\log n)$ time using n^2 processors and processing a query takes $O(1)$ time using a single processor. Vishkin [Vi-85] includes a parallel algorithm for the LCA problem. The processing of an LCA query takes logarithmic time (as in the first algorithm of Tsin). The preprocessing stage takes $O(\log n)$ time using $n/\log n$ processors (as in the present paper).

Observe that using our parallel preprocessing algorithm we can process k off-line LCA queries in $O(\log n)$ time using $(n+k)/\log n$ processors provided read conflicts are allowed. This affects the performance of parallel algorithms for three problems: (1) Given an undirected graph orient its edges so that the resulting digraph is strongly connected (if such orientation is possible) [Vi-85]. (2) Computing an open ear decomposition and st-numbering of a biconnected graph [MSV-86]. Using the new parallel connectivity and list ranking algorithms of [CV-86a] it has become possible to solve each of these problems in logarithmic time using an optimal number of processors only when $m \geq n \log n$, where n is the number of vertices and m is the number of edges in the input graph. Our off-line LCA computation enables extending the range of optimal speed-up logarithmic time parallel algorithms for these problems to sparser graphs, where $m \geq n \log^* n$ as in the above connectivity algorithm. (3) Approximate string matching [LV-86]. The new parallel suffix tree construction of [LSV-87] together with the present parallel LCA computation lead to a considerable simplification of the parallel algorithm of [LV-86]. This simplification has already been described in [AILSV-88].

The paper is organized as follows. Section 2 gives a high-level description of the algorithm. Section 3 describes the preprocessing stage. In Section 4 we show how to process LCA queries in T using the outcome of the preprocessing stage. Section 5 presents parallelization of our preprocessing stage. Because of space constraints some proofs and implementation details are omitted from the extended summary. The reader is referred to the full paper [SV-87].

2. High-level description

The whole algorithm is based on the following two observations (made also in [HT-84]): (1) Had our input tree been a simple path, it would have been possible to preprocess it (by way of computing the distance of each vertex from the root, as explained below) and later answer each LCA query in constant time. (2) Had our input tree been a complete binary tree, it would have been possible to preprocess it (by way of computing its inorder numbering, as explained below) and later answer each LCA query in constant time.

The preprocessing stage assigns a number $INLABEL(v)$ to each vertex v in T. Motivated by observation (1), these numbers satisfy the following *Path Partition property*: The $INLABEL$ numbers partition the tree T into paths, called $INLABEL$ paths. Each $INLABEL$ path consists of the vertices which have the same $INLABEL$ number.

Let B be the smallest complete binary tree having at least n vertices. Our description identifies each vertex in B by its inorder number. Motivated by observation (2), the $INLABEL$ numbers satisfy also the following *Descendence Preservation property*: The $INLABEL$ numbers map each vertex v in T into the vertex $INLABEL(v)$ in B, such that the descendants of v are mapped into descendants of $INLABEL(v)$ in B (v is considered both a descendant and an ancestor of itself).

Consider a vertex v in T. By the Descendance Preservation property all the ancestors of v are mapped into ancestors of $INLABEL(v)$. This implies that there are $O(\log n)$ distinct numbers among the $INLABEL$ numbers of all the ancestors of v. Later, we show how to record all these $INLABEL$ numbers using a single string of $O(\log n)$ bits. In the preprocessing stage we compute this string, for each vertex v in T, into $ASCENDANT(v)$.

In the preprocessing stage we also compute the table $HEAD$. It contains the highest vertex in every $INLABEL$ path.

Section 4 describes how to process a query $LCA(x,y)$ for any pair of vertices x,y in T. The processing breaks into two cases. The simpler case is where x and y belong to the same $INLABEL$ path. In the preprocessing stage we compute for each vertex v in T its distance from the root into $LEVEL(v)$. So, $LCA(x,y)$ is simply the vertex among x and y which is closer to the root. The more complicated case is where $INLABEL(x) \neq INLABEL(y)$. We proceed in four steps. In the first step, we find the LCA of $INLABEL(x)$ and $INLABEL(y)$ in the complete binary tree B, denoted by b. Let $z = LCA(x,y)$ in T. In the second step, we find $INLABEL(z)$. $INLABEL(z)$ is the lowest ancestor of b in B which is the $INLABEL$ number of a common ancestor of x and y in T. For this, we use information provided by $ASCENDANT(x)$ and $ASCENDANT(y)$. In the third and fourth steps we find z in the $INLABEL$ path defined by $INLABEL(z)$. In the third step, we find the lowest ancestor of x, denoted $\hat{x}$, and the lowest ancestor of y, denoted $\hat{y}$, in the path defined by $INLABEL(z)$ in T. This is done in an indirect fashion. Consider the path in B from $INLABEL(z)$ to $INLABEL(x)$. We derive from $ASCENDANT(x)$ the first $IBLABEL$ number (i.e., vertex of B) of an ancestor of x in this path. Table $HEAD$ gives the highest ancestor of x in T having this $INLABEL$ number. Finally, $\hat{x}$ is the father in T of this ancestor. We find $\hat{y}$ similarly. In the fourth step, we find z which is simply the vertex among $\hat{x}$ and $\hat{y}$ which is closer to the root.

3. The Preprocessing Stage

The outcome of the preprocessing stage consists of labels which are assigned to the vertices of T and a look-up table, called $HEAD$. The label of each vertex v in T consists of three numbers: $INLABEL(v)$, $ASCENDANT(v)$ and $LEVEL(v)$.

We start with computing $INLABEL(v)$, for each vertex v in T. This is done in two steps. After a discussion of these two steps we show how to implement them.

Let $PREORDER(v)$ be the serial number of v in preorder traversal of T and $SIZE(v)$ be the number of vertices in the subtree rooted at v. Definition of preorder traversal can be found, e.g., in [AHU-74], pp. 54-55.

Step 1. Compute $PREORDER(v)$ and $SIZE(v)$.

Note that the $PREORDER$ numbers of the vertices in the subtree rooted at v range between $PREORDER(v)$ and $PREORDER(v) + SIZE(v) - 1$, therefore, the closed interval $[PREORDER(v), PREORDER(v) + SIZE(v) - 1]$ is called *the interval of v.*

In Step 2 we consider the binary representation of the (integer) numbers in the interval of v. We remark that throughout this paper we alternately refer to numbers and to their binary representations. No confusion will arise.

Step 2. Find the (integer) number which has the maximal number of rightmost "0" bits in the interval of v. This number is assigned to *INLABEL* (v).

For an example of computations described in this section see Fig. 3.1.

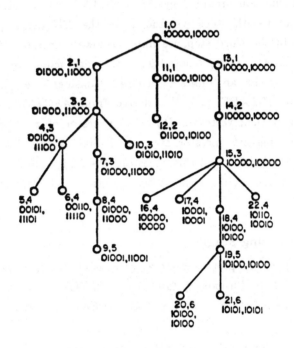

Fig. 3.1

Example. A tree with four numbers: *PREORDER* , *LEVEL* , *INLABEL* and *ASCENDANT* at each vertex. (The last two are given in binary representation.)

Discussion. We show that the *INLABEL* numbers satisfy the two properties defined in the high-level description of the previous section.

Lemma 1. The *INLABEL* numbers satisfy the *Path Partition property.*

Proof: Omitted. See the full paper [SV-87].

Lemma 2. The *INLABEL* numbers satisfy the *Descendence Preservation property.*

Proof: Let d be any descendant of v in T. We show that *INLABEL* (d) is a descendant of *INLABEL* (v) in the complete binary tree B. (Recall that our description identifies each vertex in B by its inorder number.) Thus, proving the Lemma. Consider any two vertices b and c in B. We, first, give a necessary and sufficient condition for c to be a descendant of b in B and then show that *INLABEL* (d) and *INLABEL* (v) satisfy this condition. Let $l = \lfloor \log n \rfloor$† and i be the number of rightmost "0" bits in b. That is, b consists of $l-i$ leftmost bits followed by a single "1" and i "0"s.

Claim: A vertex c is a descendant of b if and only if (1) the $l-i$ leftmost bits of c are the same as the $l-i$ leftmost bits of b, and (2) the number of rightmost "0" bits in c is at most i.

For an example of a complete binary tree and its inorder numbering see Fig. 3.2.

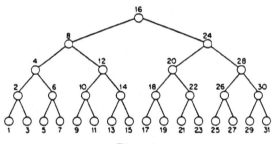

Fig. 3.2

Example. Inorder numbering of the complete binary tree with 31 vertices.

We return to the proof of Lemma 2. Let i be the number of rightmost "0" bits in *INLABEL* (v). Since *INLABEL* (d) belongs to the interval of v and *INLABEL* (v) has the maximal number of rightmost "0" bits in this interval, the number of rightmost "0" bits in *INLABEL* (d) is at most i. The $l-i$ leftmost bits are the same for all numbers in the interval. In particular, the $l-i$ leftmost bits in *INLABEL* (d) are the same as the $l-i$ leftmost bits in *INLABEL* (v). This implies that *INLABEL* (d) is the descendant of *INLABEL* (v) in B. Lemma 2 follows. □

Implementation: Step (1) is implemented in linear time and linear space, using preorder traversal of T. Given *PREORDER* (v) and *SIZE* (v), for each vertex v in T, Step (2) is implemented in constant time per vertex in two substeps.

---

† The base of all logarithms in this paper is two.

2.1. Compute $\lfloor \log[(PREORDER(v)-1) \textbf{ xor } (PREORDER(v)+SIZE(v)-1)] \rfloor$ into i. Let us explain this. The bitwise logical exclusive OR (denoted **xor**) of $PREORDER(v) - 1$ and $PREORDER(v) + SIZE(v) - 1$ assigns "1" to each bit in which $PREORDER(v) - 1$ and $PREORDER(v) + SIZE(v) - 1$ differ. The floor of the (base two) logarithm gives the index of the leftmost bit of difference (counting from the rightmost bit whose index is 0). Note that the bit indexed i must be "0" in $PREORDER(v) - 1$ and "1" in $PREORDER(v) + SIZE(v) - 1$, since the second number is larger.

Step 2.2 shows how to "compose" $INLABEL(v)$. For this, we need two observations: (1) The $l-i+1$ leftmost bits of $INLABEL(v)$ are the same as the $l-i+1$ leftmost bits in $PREORDER(v) + SIZE(v) - 1$. (2) The i other bits in $INLABEL(v)$ are "0"s.

2.2. Compute $2^i \left\lfloor \dfrac{PREORDER(v)+SIZE(v)-1}{2^i} \right\rfloor$ into $INLABEL(v)$. This assigns the $l-i+1$ leftmost bits in $PREORDER(v) + SIZE(v) - 1$ to the $l-i+1$ leftmost bits in $INLABEL(v)$ and "0"s to the other bits of $INLABEL(v)$.

We proceed to the computation of the $ASCENDANT$ numbers. The general idea is that for each vertex v, the single number $ASCENDANT(v)$ will record the $INLABEL$ numbers of *all* the ancestors of v in T. We observe that, from the viewpoint of vertex v the $INLABEL$ number of each of its ancestors can be fully specified by the index of its rightmost "1". This is, since the bits which are to the left of this "1" are the same as their respective bits in $INLABEL(v)$. Like the $INLABEL$ numbers, $ASCENDANT(v)$ is also an $(l+1)$-bit number. Denote the binary representation of $ASCENDANT(v)$ by the binary sequence $A_l(v),...,A_0(v)$. We set $A_i(v) = 1$ only if i is the index of a rightmost "1" in the $INLABEL$ number of an ancestor of v in T. To compute the $ASCENDANT$ numbers, we scan the vertices of T from its root r down to its leaves (use, for instance, Breadth-First Search). We start with $ASCENDANT(r) = 2^l$. Consider an internal vertex v in T and let $F(v)$ be the father of v in T. If $INLABEL(v) = INLABEL(F(v))$ then we assign $ASCENDANT(F(v))$ to $ASCENDANT(v)$, otherwise, we assign $ASCENDANT(F(v)) + 2^i$ to $ASCENDANT(v)$, where i is the index of the rightmost "1" in $INLABEL(v)$. It can be easily verified that i is given by $\log(INLABEL(v) - [INLABEL(v) \textbf{ and } (INLABEL(v)-1)])$, where **and** denotes bitwise logical AND.

Recall that $LEVEL(v)$, for each vertex v in T, is the distance, counting edges, of the path from v to the root r. Computation of the $LEVEL$ numbers is straightforward and can be done using, e.g., Breadth-First Search.

Recall that Fig. 3.1 gives an example of the labels.

We conclude by describing how to compute the table $HEAD$. $HEAD(k)$ contains the vertex which is closest to the root in the path consisting of all vertices whose $INLABEL$ number is k. $HEAD(k)$ is sometimes called the *head* of the $INLABEL$ path k. Computation of the table $HEAD$ is trivial. For each vertex v, such that $INLABEL(v) \neq INLABEL(F(v))$ we assign v to $HEAD(INLABEL(v))$. This, again, takes linear time and linear space.

4. Processing LCA queries

In this section we show how to answer LCA queries using the outcome of the preprocessing stage.

Consider a query $LCA(x,y)$, for any pair of vertices x,y in T. (To illustrate the presentation the reader is referred to Fig. 3.1.) There are two cases:
(Case A) $INLABEL(x) = INLABEL(y)$. It must be that x and y are in the same $INLABEL$ path. We conclude that $LCA(x,y)$ is x if $LEVEL(x) \leq LEVEL(y)$ and y otherwise.
(Case B) $INLABEL(x) \neq INLABEL(y)$. Let z be $LCA(x,y)$. We find z in four steps:
Step 1. Find b, the LCA of $INLABEL(x)$ and $INLABEL(y)$ in the complete binary tree B, as follows. Let i be the index of the rightmost "1" in b. Since b is a common ancestor of $INLABEL(x)$ and $INLABEL(y)$ in B, i must satisfy the following two conditions: (1) the $l-i$ leftmost bits in $INLABEL(x)$ and in $INLABEL(y)$ are the same as these bits in b, and (2) the index of the rightmost "1" in $INLABEL(x)$ and in $INLABEL(y)$ is at most i. Since b is the *lowest* common ancestor of $INLABEL(x)$ and $INLABEL(y)$ in B, i is the *minimum* index satisfying both conditions. We distinguish three cases. *Case (1)*. $INLABEL(x)$ is an ancestor of $INLABEL(y)$. Let i_1 be the index of the rightmost "1" in $INLABEL(x)$. Note that in this case the $l-i_1$ leftmost bits in $INLABEL(x)$ and in $INLABEL(y)$ are the same and that the index of the rightmost "1" in $INLABEL(y) < i_1$. Hence, i equals i_1. *Case (2)*. $INLABEL(y)$ is an ancestor of $INLABEL(x)$. Similar to Case (1), i is the index of the rightmost "1" in $INLABEL(y)$. *Case (3)*. Not cases (1) and (2). In this case i is the *minimum* index such that the $l-i$ leftmost bits in $INLABEL(x)$ and

INLABEL (y) are the same.

We can deal with all these three cases at once by simply taking i to be the maximum among: the index of the leftmost bit in which INLABEL (x) and INLABEL (y) differ, the index of the rightmost "1" in INLABEL (x) and the index of the rightmost "1" in in INLABEL (y). b consists of the $l-i$ leftmost bits in INLABEL (x) (or INLABEL (y)) followed by a single "1" and i "0"s.

In Step 2 we find INLABEL (z) (where z is LCA (x,y)). The Descendence Preservation property of the INLABEL numbers implies that INLABEL (z) is a common ancestor of INLABEL (x) and INLABEL (y). Notice that INLABEL (z) is not necessarily b, the *lowest* common ancestor of INLABEL (x) and INLABEL (y). This is, since the vertices in T mapped into b are not necessarily ancestors of x or y. However, it is not difficult to see that INLABEL (z) is the lowest ancestor of b in B which is the INLABEL number of an ancestor of both x and y in T.

Step 2. Find INLABEL (z). For this we find the index of the rightmost "1" in INLABEL (z), denoted by j. Since z is a common ancestor of x and y in T, $A_j(x) = 1$ and $A_j(y) = 1$. Since INLABEL (z) is the *lowest* ancestor of b which is a common ancestor of x and y, the index j must be the index of the *rightmost* "1" in $A_l(x),...,A_i(x)$ and $A_l(y),...,A_i(y)$. INLABEL (z) consists of the $l-j$ leftmost bits of INLABEL (x) (or INLABEL (y)) followed by a single "1" and j "0"s.

In the next steps we find z, the lowest vertex in the path defined by INLABEL (z) which is a common ancestor of x and y in T. For this we find $\hat{x}$, the lowest ancestor of x in the path defined by INLABEL (z) and $\hat{y}$, the lowest ancestor of y in this same path. z is the highest vertex among these two vertices.

Step 3. Find $\hat{x}$ and $\hat{y}$. We show how to find $\hat{x}$. $\hat{y}$ is found similarly. If INLABEL $(x) =$ INLABEL (z) then $\hat{x} = x$ and nothing has to be done. Suppose INLABEL $(x) \neq$ INLABEL (z). We set the following intermediate goal, as the main step towards finding $\hat{x}$: Find the son of $\hat{x}$ which is also an ancestor of x. Denote the vertex that we search by w and let k be the index of the rightmost "1" in INLABEL (w). It is not difficult to verify that k is the index of the leftmost "1" in $A_{j-1}(x),...,A_0(x)$. So, we find k. Clearly, INLABEL (w) consists of the $l-k$ leftmost bits of INLABEL (x) followed by a single "1" and k "0"s. Observe that w is the head of its INLABEL path (since the INLABEL number of its father $\hat{x}$ is different from INLABEL (w)). Therefore, w is

$HEAD\ (INLABEL\ (w\ ))$ and our intermediate goal is achieved. Finally, $\hat{x}$ is the father of w in T.

Step 4. $LCA\ (x,y)$ is $\hat{x}$ if $LEVEL\ (\hat{x}\ ) \leq LEVEL\ (\hat{y}\ )$ and $\hat{y}$ otherwise.

The additional implementation details required for the above query processing are given in the full paper [SV-87].

5. The Parallel Preprocessing Algorithm

In this section we describe the parallel version of our preprocessing stage. It runs in $O\ (\log n\ )$ time using $n\ /\log n$ processors. We make the following assumption regarding the representation of the input tree T. Its $n-1$ edges are given in an array, where the incoming edges of each vertex are grouped successively. By our definition of the tree T, its edges are directed towards the root.

Computing the labels in parallel. To compute the labels of the vertices in T we apply the Euler tour technique for computing tree functions, which was given in [TV-85] and [Vi-85]. We will implement it, however, using the $O\ (\log n\ )$ time optimal parallel list ranking algorithm of [CV-86a].

Below, we first recollect the construction required for the Euler tour technique. We then show how to use it for computing the labels. The only reason which forced us to present anew the Euler tour technique is that the computation of the $ASCENDANT$ numbers has not appeared elsewhere.

Step 1. For each edge $(v \rightarrow u\ )$ in T we add its anti-parallel edge $(u \rightarrow v\ )$. Let H denote the new graph.

Since the in-degree and out-degree of each vertex in H are the same, H has an Euler path that starts and ends in r. Step 2 computes this path into the vector of pointers D, where for each edge e of H, $D\ (e\ )$ will have the successor edge of e in the Euler path.

Step 2. For each vertex v of H we do the following. (Let the outgoing edges of v be $(v \rightarrow u_0),...,(v \rightarrow u_{d-1}).)$ $D\ (u_i \rightarrow v\ ) := (v \rightarrow u_{(i+1)\ \text{mod}\ d}\ )$, for $i = 0,...,d-1$. Now D has an Euler circuit. The "correction" $D\ (u_{d-1} \rightarrow r\ ) := end-of-list$ (where the out-degree of r is d) gives an Euler path which starts and ends in r.

We show how to use the Euler path in order to find $PREORDER\ (v\ )$, $PREORDER\ (v\ ) + SIZE\ (v\ ) - 1$ and $LEVEL\ (v\ )$ for each vertex v in T.

Step 3. We assign two weights: $W_1(e\ )$ and $W_2(e\ )$ to each edge e in the Euler path as follows. (1) $W_1(e\ ) = 1$ if e is directed from r (that is, if e is not a tree edge) and $W_1(e) = 0$ otherwise. (2) $W_2(e) = 1$ if e is directed from r and $W_2(e) = -1$ otherwise.

Step 4. We apply twice an optimal logarithmic time parallel list ranking algorithm to find for each e in H its (weighted) distance from the *start* of the Euler path: The first application is relative to the weights W_1 and the result is stored in $DISTANCE_1(e)$. The second application is relative to the weights W_2 and the result is stored in $DISTANCE_2(e)$. Consider a vertex $v \neq r$ and let u be its father in T. $PREORDER(v)$ is $DISTANCE_1(u \to v) + 1$, $PREORDER(v) + SIZE(v) - 1$ is $DISTANCE_1(v \to u) + 1$, and $LEVEL(v)$ is $DISTANCE_2(u \to v)$. (These claims can be readily verified by the reader.)

Step 5. Given $PREORDER(v)$ and $PREORDER(v) + SIZE(v) - 1$ for each vertex v in T we compute $INLABEL(v)$ in constant time using n processors as in the serial algorithm.

Next, we show how to use the Euler path in order to find $ASCENDANT(v)$ for each vertex v in T.

Step 6. We assign a (new) weight $W(e)$ to each edge e in the Euler path as follows. For each vertex $v \neq r$ we do the following. Let u be the father of v in T and let i be the index of the rightmost "1" in $INLABEL(v)$. If $INLABEL(v) \neq INLABEL(u)$, we assign $W(u \to v) = 2^i$ and $W(v \to u) = -2^i$. The weight of all other edges is set to zero.

Step 7. We apply again a parallel list ranking algorithm to find for each e in H its (weighted) distance from the start of the Euler path. Consider a vertex $v \neq r$ and let u be its father in T. $ASCENDANT(v)$ is the distance of the edge $(u \to v)$ plus 2^l. Clearly, $ASCENDANT(r) = 2^l$.

We note that, given the labels, the table $HEAD$ can be computed in constant time using n processors.

Complexity. Each of steps 4 and 7 needs $n/\log n$ processors and $O(\log n)$ time. Each of steps 1,2,3,5,6 and the computation of $HEAD$ needs n processors and $O(1)$ time and can be readily simulated by $n/\log n$ processors in $O(\log n)$ time. Thus, the parallel preprocessing stage can be done in a total of $O(\log n)$ time using $n/\log n$ processors.

Acknowledgements. We are grateful to Noga Alon and Yael Maon for stimulating discussions.

REFERENCES

[AHU-74]   A.V. Aho, J.E. Hopcroft and J.D. Ullman, *The Design and Analysis of Computer Algorithms*, Addison-Wesley, Reading, MA, 1974.

[AILSV-88]   A. Apostolico, C. Iliopoulos, G.M. Landau, B. Schieber and U. Vishkin, "Parallel construction of a suffix tree with applications", to appear in *Algorithmica* special issue on parallel and distributed computing.

[CV-86a]   R. Cole and U. Vishkin, "Approximate and exact parallel scheduling with applications to list, tree and graph problems", *Proc. 27th Annual Symp. on Foundations of Computer Science*, (1986), pp. 478-491.

[CV-86b]   R. Cole and U. Vishkin, "Faster optimal parallel prefix sums and list ranking", TR 56/86, the Moise and Frida Eskenasy Institute of Computer Science, Tel Aviv University (1986).

[HT-84]   D. Harel and R.E. Tarjan, "Fast algorithms for finding nearest common ancestors", *SIAM J. Comput.*, 13 (1984), pp. 338-355.

[LV-86]   G.M. Landau and U. Vishkin, "Introducing efficient parallelism into approximate string matching", *Proc. 18th ACM Symposium on Theory of Computing*, 1986, pp. 220-230.

[LSV-87]   G.M. Landau, B. Schieber and U. Vishkin, "Parallel construction of a suffix tree", *Proc. 14th Int. Colloq. on Automata Lang. and Prog.*, Lecture Notes in Computer Science 267, Springer-Verlag (1987), pp. 314-325.

[MSV-86]   Y. Maon, B. Schieber and U. Vishkin, "Parallel ear decomposition search (EDS) and st-numbering in graphs", *Theoretical Computer Science*, 47 (1986), pp. 277-298.

[SV-87]   B. Schieber and U. Vishkin, "On finding lowest common ancestors: simplification and parallelization", to appear in *SIAM J. Comput.*, also TR 63/87, the Moise and Frida Eskenasy Institute of Computer Science, Tel Aviv University (1987).

[Ts-86]   Y.H. Tsin, "Finding lowest common ancestors in parallel", *IEEE Tran. Comput.* C-35 (1986), pp. 764-769.

[TV-85]   R.E. Tarjan and U. Vishkin, "An efficient parallel biconnectivity algorithm", *SIAM J. Comput.* 14 (1985), pp. 862-874.

[Vi-83]   U. Vishkin, "Synchronous parallel computation - a survey", TR-71, Dept. of Computer Science, Courant Institute, NYU, (1983).

[Vi-85]   U. Vishkin, "On efficient parallel strong orientation", *Information Proc. Letters* 20 (1985), pp. 235-240.

A Scheduling Problem Arising From Loop Parallelization on MIMD Machines

Richard J. Anderson*   Ashfaq A. Munshi†   Barbara Simons‡

Abstract

Automatic parallelization of code written in a sequential language such as FOR-TRAN is of great importance. One natural approach to loop parallelization is to assign separate invocations of a loop to different processors. However, it is often necessary to delay the starting times of loops to avoid violating data dependences. In this paper we study a scheduling problem, called the Delay Problem, that models an aspect of the loop parallelization problem. We show that the Delay Problem is NP-Complete when the precedence constraints are a set of arbitrary trees. Our major result of the paper is a polynomial time algorithm for the case where the precedence constraints are a forest of in-trees or a forest of out-trees. This covers an important practical case for the Delay Problem.

1   Introduction

While much research and development is being devoted to designing parallel machines and parallel algorithms, there is a large amount of code that has been written for sequential machines in languages such as FORTRAN. It is expensive and time-consuming to rewrite this code so that it will run efficiently on parallel machines. To further complicate the problem, people are frequently afraid to rewrite the code, because they do not know precisely what it does or what side-effects it might have. Consequently, there has been considerable work, especially at the University of Illinois [KKLW80] and Rice [AK87], on the problem of generating parallel code from sequential code. Much of this work has focused on detecting parallelism in loops, especially FORTRAN DO loops. One of the methods, known as DOACROSS [Cyt84], assigns a processor to execute the loop for each value of the loop iteration variable. If a loop is to be executed N times, and if there are at least N processors, then one processor is assigned to each iteration of the loop. Ideally, all N processors start at the same time and consequently finish at approximately the same time. Because of data dependences within the loop, however, this idealized version of DOACROSS usually is not possible. Data dependences may force certain processors to wait for results that must be computed by other processors before they can continue their local computations. In other words, processors may be forced to delay executing their local task due to data dependences.

More precisely, let K be the iteration variable of the loop and let $s_1, \ldots, s_n$ be the statements of the loop in the order which they occur. Suppose that there is a dependence

---

*Department of Computer Science, University of Washington. This work supported by NSF Presidential Young Investigator Award CCR-8657562.

†Ridge Computers, Santa Clara, CA. This work done while the author was at IBM Almaden Research Center.

‡IBM Almaden Research Center.

between statements s_i and s_j such that statement s_j accesses a memory location that is also accessed by statement s_i. Then we say that s_j *depends* on s_i. If $i > j$, then processor K must wait until processor $K - 1$ has computed s_i before K can compute s_j. If $i < j$, then no delay is necessary since by the time processor K reaches s_i, processor $K - 1$ will have already computed s_j. Finally, if $i = j$, then the delay must be at least 1, since processor K must wait for s_i to be executed by processor $K - 1$.

We assume that all statements require unit processing time. This assumption closely models both machine level language and many forms of intermediate language that are generated by compilers. In this case, the length of a data dependence is $|j - i|$, and it is said to be *backward* if $j - i < 0$. A simple extension of the above discussion shows that the amount of delay is precisely the length of the longest backward dependence [Cyt84]. If there are no backward dependences, the delay is zero. A loop is "fully parallel" if it has zero delay and "fully sequential" if the delay is equal to the number of statements in the loop plus one.

A natural question to ask is how to minimize the delay associated with a loop by rearranging the code to make the loop more parallel. This question is of considerable practical importance, since its solution would allow improve our ability to run existing codes on parallel machines.

2 Loop parallelization

The following code fragment illustrates the two types of data dependences that arise between statements. Within a loop, statement s_1 must be executed before statement s_2 since s_2 uses a value updated by s_1. We refer to this as a *loop independent* dependence. The other dependence is between statements in separate iterations of the loop. Statement s_3 in iteration $K - 1$ must be completed before statement s_1 in iteration K is executed. This is a *loop carried* dependence, and is the type of prime concern to us. [1]

```
     DO 10 K = 1, N
s₁      A(K) = B(K − 1)
s₂      D(K) = A(K)
s₃      B(K) = C(K)
10 CONTINUE
```

As the code fragment is written, it is necessary to have a delay of two time units between invocations of the loop to avoid violating the dependence. If statements 2 and 3 are interchanged, then the delay necessary is only one unit. The basic problem that we are interested in is how to reorganize statements within a loop, subject to the loop independent dependences, so as to minimize the delay.

In practice, the situation can be more complicated since loops may be nested, or loop carried dependences might arise because of dependences on data computed several iterations prior to the current iteration. These complications can be eliminated by relying on techniques such as loop unwinding. For a further discussion see [Cyt84] and [Mun86].

Given a loop L, the *dependence graph*, $G_L(N, A)$, of L is a directed graph where the nodes of G_L represent statements (or jobs) and the arcs represent dependences. In particular

---

[1]There is another type of dependence, called a *control* dependence, that is derived from an IF statement. Since control dependences can be treated as loop independent data dependences in the problem formulation, we shall assume that all dependences are data dependences. For more information on control dependences, see [AKPW83]

$(s_j, s_i) \in G_L(A)$ iff s_i depends on s_j. The delay optimization problem can now be phrased succinctly: Given a loop L, find a permutation π of the dependence graph $G_L(N, A)$ such that the semantics of the loop are preserved and $\max_{(s_i,s_j) \in G_L(A)}\{\pi(s_i) - \pi(s_j)\}$ is minimized over all backward arcs.

As was mentioned above, two statements participating in a loop carried dependence can be permuted while preserving the semantics. Therefore, a first approximation to the delay problem is to consider a loop that contains only loop carried dependences, i.e. every permutation of the statements preserves semantics. Unfortunately, even this simplified problem is NP-hard [Cyt84]; the reduction is a straightforward reduction from bandwidth minimization [GJ79]. In general, it is not possible to make arbitrary permutations of the statements within the loop, since loop independent data dependences force a partial order on the statements of the loop. Note that if the dependence graph does not have any circuits, then a topological sort of the graph results in no backward arcs which in turn implies no delay.

Since the general problem is NP-hard, we must rely on heuristics which will hopefully give good results. The basic approach that we have adopted is to deal with the two types of constraints separately. We first use the loop independent constraints to impose some restrictions upon the problem, and then consider the loop carried dependences. The main results of this paper are on the complexity of the subproblem of dealing with the loop carried dependences.

3 Delay problem

Our approach to the general scheduling problem is to restrict the permutations of the statements that we consider. We can then look for an optimal schedule within the class of permutations that we allow. We divide time into a set of disjoint intervals. The jobs are assigned to the intervals, so that each interval receives an independent set of jobs. The assignment is made so that any permutation of the jobs with a given interval is a valid schedule. The question is then to minimize the delay with the jobs assigned to the intervals.

The division of the jobs into intervals can be accomplished in many different ways. A natural approach that has been used in implementing this algorithm [SM87] is to divide the precedence graph of loop independent dependences into levels. The result of this is that we have a partition of the statements $\Sigma = \{\Sigma_1, \ldots, \Sigma_k\}$ such that the statements within Σ_i are independent, and if a statement in Σ_j depends upon a statement in Σ_i when $i > j$ (i.e. a 'backwards' edge), then that dependence is loop carried. The jobs in Σ_1 will be assigned to time units 1 to $|\Sigma_1|$, the jobs in Σ_2 to time units $1 + |\Sigma_1|$ to $|\Sigma_1| + |\Sigma_2|$ and so on. Note that any ordering of the jobs is possible within Σ_i.

Now consider a new directed acyclic graph which has a node corresponding to each statement in the loop and which has an edge (u, v) if and only if $u \in \Sigma_i, v \in \Sigma_j$ and $i > j$. The problem is to schedule the statements to minimize the distance covered by any of the precedence edges in this new graph. The following is a formal description of the Delay Problem. An instance of the delay scheduling problem consists of a set of jobs, a set of intervals, a precedence order on the jobs, an assignment of the jobs to the intervals, and a bound on the delay. The problem is to find an assignment of the jobs to time slots that has a delay no greater than the given bound. The following descriptions of the components of the problem and the desired schedule introduces notation that will be used in the next section.

1. The set of jobs is $\mathcal{J} = \{j_1, \ldots, j_n\}$. Each job requires one unit of time for its execution.

2. The set of intervals is $\mathcal{I} = \{I_1, \ldots, I_m\}$. The intervals are assumed to be disjoint. For intervals I', I'' we write $I' < I''$ if all elements of I' are less than elements of I''.

3. The precedence order is a directed acyclic graph on the set of jobs $\mathcal{J}$. For jobs j', j'', we write $j' \prec j''$ if j' must come before j'' because of the precedence order.

4. The jobs are assigned to intervals by a function $I : \mathcal{J} \to \mathcal{I}$. The interval associated with j is $I(j)$. The assignment to intervals respects the precedence order, so if $j' \prec j''$, then $I(j') < I(j'')$.

5. The bound on the delay is denoted by K.

A schedule is a mapping $s : \mathcal{J} \to Z$, where each job is mapped to a distinct time. The mapping obeys the precedence order, so if $j' \prec j''$, then $s(j') < s(j'')$, and the interval assignment, so $s(j) \in I(j)$. To satisfy the delay bound, the distance spanned by a precedence edge is at most K, so if $j' \to j''$ is an edge in the precedence graph, then $s(j'') - s(j') \leq K$.

4 Out-tree algorithm

In this section we give an algorithm for the scheduling problem for the case that the precedence graph is a forest of out-trees. Since the type of precedence graph that tends to arise for the delay problem is a tree with a few additional edges, this could be used as the basis of a heuristic for the delay problem. It also includes as a special case graphs that consist of a set of independent chains. These were used in the heuristic of Munshi and Simons [SM87]. The case of a forest of in-trees can be handled by running our algorithm in reverse.

Given an instance of the scheduling problem where the precedence graph is a forest of out-trees and a delay bound K, our algorithm constructs a valid schedule if one exists. The algorithm is a relatively simple greedy algorithm, which makes a pass through the intervals one at a time and fills them with jobs. The key to the algorithm is to define an order on the jobs within an interval which determines how they will be scheduled.

A chain is an ordered set of jobs $j_1 \cdots j_r$ where $j_i \to j_{i+1}$ is an edge in the precedence graph for $1 \leq i \leq r - 1$. The chains are ordered lexicographically based on the intervals associated with the jobs, i.e., $j_1 \cdots j_r \ll k_1 \cdots k_q$ if for $i < l$, $I(j_i) = I(k_i)$ and $I(j_l) < I(k_l)$ or $r < q$ and for $i \leq r$, $I(j_i) = I(k_i)$. We can use this ordering on chains to define an ordering on jobs assigned to the same interval. Suppose that j and k are assigned to the same interval and $jj_1 \cdots j_r$ and $kk_1 \cdots k_q$ are the maximal chains (w.r.t. $\ll$) starting at j and k. Then $j \ll k$ iff $jj_1 \cdots j_r \ll kk_1 \cdots k_q$.

We can now give the algorithm to schedule the jobs. The algorithm is a greedy algorithm which schedules the intervals in order. Once a job is given a time slot it will not have its slot changed because of the placement of a job in a later interval.

A job is first assigned to the latest feasible slot, where the feasibility is determined by the slot of the job's predecessor. In the case where several jobs are assigned the same slot, the maximal job (w.r.t. $\ll$), is assigned to the slot and the other jobs are moved to earlier slots.

Suppose the jobs $j_1, \ldots, j_k$ are assigned to the interval $I = [a, b]$. Let $j'_1, \ldots, j'_k$ be the predecessors of $j_1, \ldots, j_k$. For each i, $a \leq i \leq b$, we maintain a set of jobs S_i which have i as their latest feasible slot. The algorithm for scheduling the interval is:

for each j_i
$\quad x := \min(b, s(j'_i) + K);$
$\quad S_x := S_x \cup \{j_i\};$
for $l := b$ downto a
$\quad$ if $S_l \neq \emptyset$ then
$\quad\quad$ Let j_i be the maximal job in S_l;
$\quad\quad s(j_i) := l;$
$\quad\quad S_{l-1} := S_{l-1} \cup S_l - \{j_i\};$

If the algorithm fails to schedule any interval, it reports failure, otherwise it has constructed a valid schedule.

4.1 Correctness Proof

Now we must establish that if there is a valid schedule, then the greedy algorithm constructs one. To understand the intuition behind the algorithm, consider the case of scheduling a single chain. To schedule the chain, we place the first job at the right endpoint of the appropriate interval, and then work down the chain, scheduling jobs as late as possible subject to the constraints that the jobs fall within intervals and are not separated by more than K. If we were to do this simultaneously for a number of chains, we would then have to make the decision as to which chains to slide to the left to get down to one job per time slot. The key to the proof is to show that the decision on which job to shift can be made based on the ordering of chains that we gave.

The formal proof is by induction on the number of intervals. To make the induction go through, we generalize the problem slightly by allowing the jobs without predecessors to be given deadlines before which they must be scheduled.

The following technical lemma is the basis of the proof. It allows us to reorder the jobs in a manner consistent with the greedy algorithm.

Lemma 4.1 *Suppose that I_1, the first interval, is scheduled $Axz_1 \cdots z_s yB$ with y at time t, $x \gg y$, x has deadline at least t and for $1 \leq i \leq s$, $z_i \gg x$ and z_i has deadline earlier than t, and suppose that the remainder of the intervals are packed by the greedy algorithm. There exists a feasible schedule with I_1 having the form $Ayz_1 \cdots z_s xB$.*

Proof:

Let $X = xx_1x_2 \cdots$ and $Y = yy_1y_2 \cdots$ be the maximal chains starting from x and y respectively. Look at the first point where Y has separation less than K, i.e., let j be the minimum integer such that $s(y_{j+1}) - s(y_j) < K$. Our claim is that when this happens, the jobs of Y start coming before the jobs of X, so $s(y_{j+1}) < s(x_{j+1})$.

First we need to establish that for $i \leq j$, all jobs scheduled between x_i and y_i have priority greater than x_i. We show this by induction on i. We know that the jobs scheduled between x and y have priority greater than x. Suppose that $z'_1, \ldots, z'_p$ are scheduled between x_i and y_i and have greater priority than x_i. Let $z''_1, \ldots, z''_p$ be the maximal descendents of $z'_1, \ldots, z'_p$. Since all of the z'''_q's must have priority greater than x_{i+1}, they must be scheduled after x_{i+1}; since Y has separation precisely K up to and including I_{y_j}, they must be scheduled before y_{i+1}. It follows that time slots strictly between $s(x_i) + K$ and $s(y_i) + K$ fill up with items of priority greater than x_{i+1}. Since x_{i+1} is eligible to be scheduled at time $s(x_i) + K$, any job which forces x_{i+1} to be scheduled earlier has higher priority.

We can now show that $s(y_{j+1}) < s(x_{j+1})$. Suppose $I(x_j)$ is scheduled $Ax_jz'_1 \cdots z'_l y_j B$. By assumption, $s(y_{j+1}) < s(y_j) + K$. Let $z''_1, \ldots, z''_l$ be the maximal descendents of $z'_1, \ldots, z'_l$.

They all have greater priority than y_{j+1} and will consequently force y_{j+1} to be scheduled earlier. These l items mean the latest y_{j+1} could be scheduled is at $s(x_j) + K$, but then x_{j+1} is also available, forcing y_{j+1} earlier than x_{j+1}.

The chains $X = xx_1x_2 \cdots$ and $Y = yy_1y_2 \cdots$ are scheduled:

$$x \quad y \qquad x_1 \quad y_1 \qquad x_2 \quad y_2 \quad \cdots \quad x_j \quad y_j \qquad y_{j+1} \quad x_{j+1} \quad \cdots$$

For $i \leq j$, x_i and y_i belong to the same interval. We can now interchange the pairs x_i and y_i for $i \leq j$ without violating any of the delay constraints. This allows us to schedule the first interval $Ayz_1 \cdots z_sxB$ and still have a feasible schedule for the remaining intervals. ∎

Theorem 4.2 *The greedy algorithm finds a feasible solution to the scheduling problem if one exists.*

Proof: The proof is by induction on the number of intervals. The induction hypothesis is: For $\leq n$ intervals, the greedy algorithm finds a feasible schedule if one exists.

The base case of $n = 1$ is trivial.

Suppose that the induction hypothesis holds for $\leq n$ intervals. Let $\langle \mathcal{J}, \mathcal{I} \rangle$ be an instance of the scheduling problem with $n + 1$ intervals. Suppose that there exists a feasible schedule for this problem.

We can apply the lemma to rearrange the first interval so that it corresponds to greedy algorithm. We can assume that the intervals $I_2, \ldots$ are scheduled according to the greedy algorithm. As long as we have jobs that can be switched, we apply the lemma. When the lemma no longer applies, the first interval is scheduled in the manner that it would be scheduled by the greedy algorithm. ∎

4.2   Run time analysis

The run time of the algorithm to test if there is a schedule with delay K is $O(n \log n)$. The basis for an $O(n \log n)$ implementation is to do the sorting of the jobs in an efficient manner. The naive approach to the sorting takes $O(n)$ time per comparison. However, if the jobs are assigned to intervals, and the intervals are sorted one at a time starting with the last interval, then the cost per comparison can be reduced to $O(1)$. If we use binary search to find the minimum delay, it takes us $\log K = O(\log n)$ steps if the scheduling problem has been derived from the compiler problem. This gives us a $O(n \log^2 n)$ time algorithm for the delay problem.

If the scheduling problem is not tied to the compiler problem, then K could be arbitrarily large. (For example, some neighboring intervals could be very far apart). In this case we first compute K_0, a lower bound for K as follows. Let I_{max} be the interval of greatest length that has assigned to it some node that has both a predecessor and a successor. (Note that this implies that I_{max} is neither the first nor the last interval). Let $K_{I_{max}}$ be half the difference between the left endpoint of the interval immediately following I_{max} and the right endpoint of the interval immediately preceding I_{max}. Let K_d be the maximum distance separating two intervals containing nodes that are endpoints of an edge. Set K_0 to be the larger of K_d and $K_{I_{max}}$; this is clearly a lower bound on the value of K. By beginning the binary search with K_0, the number of iterations required will be $O(\log n)$, as above.

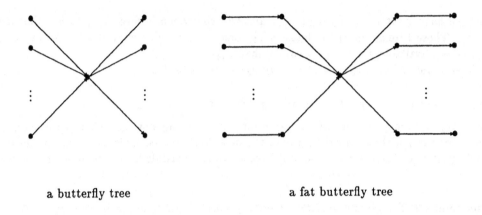

a butterfly tree   a fat butterfly tree

There is some opportunity for improvement in the run time of the algorithm. It appears possible to use bucket sort to reduce the basic algorithm to an $O(n \log^{1/2} n)$ algorithm.

5   NP-Completeness results

In this section we analyze the delay problem on a more general family of precedence graphs than we considered in the previous section. We prove that the delay problem is NP-complete for forests of directed trees[2]. Directed trees occur naturally in the loop scheduling context, so this not a contrived family of precedence constraints. The proof of the theorem is a reduction from 3-PARTITION, which is defined as follows.

An instance of the 3-PARTITION problem consists of a set A containing $3m$ elements, with item i having weight a_i, $1 \leq i \leq 3m$, and a positive integer bound B. In addition, $B/4 < a_i < B/2$ for all $i \in A$ and the sum of the weights is mB. Is there a partition of A into m sets of triplets such that the sum of the weights in each set equals B?

The idea is to have some of the trees correspond to the items so that the assignment of their nodes to two large intervals is equivalent to a division into sets. The remaining trees are used to induce a partition of the two large intervals.

Theorem 5.1 *The problem of determining whether or not there exists a schedule of delay less than K is NP-complete when the precedence graph is a forest of directed trees.*

Membership in NP is obvious since we can easily verify whether or not a proposed schedule satisfies the problem constraints.

Given an instance of the three partition problem, we now construct an instance of the scheduling problem when the DAG is a set of arbitrary trees. The DAG consists of a forest of trees that we refer to as *butterfly trees* and *fat butterfly trees*. There are $3m$ butterfly trees, each of which corresponds to an element of A, and $m-1$ fat butterfly trees, which partition the two largest intervals into subintervals of length B. There are $2m+1$ intervals, all but two of which are used to force the partitioning of the two largest intervals. Only the three largest intervals contain elements from every tree; every other intervals is assigned to a unique tree. Finally, the value of K which we test for feasibility is $(2m-1)B$.

---

[2]We consider a directed graph to be a tree if its underlying undirected graph is a tree. A directed tree may have nodes with out-degree greater than one and nodes with in-degree greater than one.

For each $i \in A$ we define the corresponding butterfly tree (bt_i) as follows. (See Figure 1 for an example of a butterfly tree and a fat butterfly tree). The tree has three layers, the first of which consists of a_i nodes, each of which is a source (no incoming edges) and has only a single outgoing edge. All outgoing edges from the first layer are incoming edges to a single node that is the second layer. This node in turn has outgoing edges to all the nodes in the third layer, of which there are a_i.

A fat butterfly tree *(fbt)* is similar to a butterfly tree except that it has two additional layers of nodes, one before the first layer of the butterfly tree and one after the last layer of the butterfly tree. With the exception of the middle layer, which consists of a single node as in the butterfly tree, all the layers contain B nodes. Each node in the first layer of the *fbt* is connected to a unique node in the second layer. The second layer is connected to the middle layer node as in the butterfly tree, and the middle layer node is connected to the fourth layer also as in the butterfly tree. Each node in the fourth layer is connected to a unique node in the fifth.

The intervals are constructed as follows. There are $m-1$ intervals of length B, $I_1, I_2, \ldots, I_{m-}$ with start times of $0, 2B, 4B, \ldots, (2m-4)B$, respectively. $I_m = [(2m-2)B, (4m-3)B-1]$, $I_{m+1} = [(4m-3)B, (6m-5)B]$, and $I_{m+2} = [(6m-5)B+1, (8m-6)B]$. Note that both I_m and I_{m+2} have length $(2m-1)B$, and I_{m+1} has length $(2m-2)B+1$. I_{m+2} is followed by $m-1$ intervals of length B, with start times of $(8m-5)B+1, (8m-3)B+1, \ldots, (10m-9)B+1$, respectively.

The nodes in the first layer of fbt_i, $1 \leq i \leq m-1$, are all assigned to I_i, the nodes in the second, third, and fourth layers are assigned to I_m, I_{m+1}, and I_{m+2}, respectively. The nodes in the fifth layer are all assigned to $I_{(m+2)+i}$. The nodes in the first, second, and third layers of the butterfly trees are all assigned to I_m, I_{m+1}, and I_{m+2}, respectively.

We claim that there is a schedule for $K = (2m-1)B$ if and only if the corresponding 3-PARTITION problem has a solution. The proof is based on the following two lemmas.

Lemma 5.2 *The start times for each layer of the nodes in the fat butterfly trees is predetermined for any schedule in which $K = (2m-1)B$.*

Proof:

The lemma clearly holds for the first and fifth layers of the fat butterfly trees. To see that it holds for the other layers, observe that if the nodes in the second layer are scheduled as late as possible relative to the start times of the nodes in the first layer, and if the nodes in the fourth layer are scheduled as early as possible relative to the nodes in the fifth layer, then there is a unique integer value that is reachable both from the earliest scheduled node of the second layer and the latest scheduled node of the fourth layer.

For example, the latest time that the first node from the second layer of fbt_1 can be scheduled is $(2m-1)B$; the earliest time that the last node from the fourth layer of fbt_1 can be scheduled is $(6m-3)B$ (since the last node from the fifth layer of fbt_1 must start at time $(8m-4)B$). Consequently, the only time at which the node in the third layer of fbt_1 can be scheduled is $(4m-2)B$. ∎

If all of the elements of the fat butterfly trees are scheduled as described in the above lemma, intervals I_m and I_{m+2} are partitioned into blocks of length B, with blocks of available time alternating with blocks of nodes from the fat butterfly trees. We assume without loss of generality that the nodes from each butterfly tree are scheduled consecutively in I_m and I_{m+2}.

Lemma 5.3 *Assume that the butterfly trees are scheduled in the same order in I_m and I_{m+2}. If the nodes from bt_i for some i that are scheduled in I_m are separated by B nodes from a fat butterfly tree, then there is some edge in bt_i having the middle layer node as an endpoint that has length greater than K.*

Proof:

A straightforward computation shows that there is a pair of nodes of bt_i, one in I_m and one in I_{m+2}, such that the distance between them is at least $(4m - 2)B + 2$. Therefore, some edge must have length at least $(2m - 1)B + 1 > K$. ∎

All that remains to prove is that the assumption that the butterfly trees are scheduled in the same order in I_m and I_{m+2} is not overly restrictive. It is easy to show that if there is a schedule for the tree scheduling problem with delay no greater than K, then by a simple rearrangement argument we can construct another schedule satisfying the assumption without increasing the value of the delay. This observation together with the above lemmas prove the theorem. ∎

This NP-completeness result leaves a very narrow gap between the types of graphs that we can solve in polynomial time and the types of graphs where the problem is NP-complete. The current boundary problem is for a mixed forest. A mixed forest is a directed acyclic graph whose components are in-trees or out-trees [May81].

6 Conclusions

In summary, it has been shown that important special cases of the Delay Problem can be solved optimally in polynomial time. We have also given an NP-Completeness proof for a slightly more general case of the Delay Problem, showing that the problem becomes hard for some natural precedence graphs. The Delay Problem is important for the automatic parallelization. The paper [SM87] reports empirical results on using algorithms for the Delay Problem in code parallelization.

7 Acknowledgement

We would like to thank Dave Alpern, Ron Cytron, Jeanne Ferrante, and Larry Stockmeyer for several helpful discussions.

References

[AK87]   R. Allen and K. Kennedy. Automatic translation of fortran programs to vector form. *ACM Transactions on Programming Languages and Systems*, 491–542, October 1987.

[AKPW83] J. R. Allen, Ken Kennedy, Carrie Porterfield, and Joe Warren. Conversion of control dependences to data dependences. In *ACM Priciples of Programming Languages*, pages 177–189, 1983.

[Cyt84]   R. G. Cytron. *Compile-time Scheduling and Optimization for Asynchronous Machines*. PhD thesis, University of Illinois at Urbana-Champagne, 1984.

[GJ79]   M. R. Garey and D. S. Johnson. *Computers and Intractability, A Guide to the Theory of NP-Completeness.* W. H. Freeman and Company, San Francisco, 1979.

[KKLW80] D. J. Kuck, R. H. Kuhn, B. Leasure, and M. Wolfe. The structure of an advanced vectorizer for pipelined processors. In *Proceedings of the IEEE 4th International COMPSAC*, pages 207–218, IEEE, 1980.

[May81]   E. W. Mayr. *Well Structured Parallel Programs are Not Easier to Schedule.* Technical Report, Department of Computer Science, Stanford University, September 1981.

[Mun86]   A. A. Munshi. Compiling sequential codes for parallel machines. 1986. Unpublished manuscript.

[SM87]   B. Simons and A. A. Munshi. *Scheduling Loops on Processors: Algorithms and Complexity.* Technical Report RJ 5546 (56570), IBM, 1987.

SCHEDULING DAGS TO MINIMIZE TIME AND COMMUNICATION

Foto Afrati[1], Christos H. Papadimitriou[2], and George Papageorgiou[1]

ABSTRACT: *We study the complexity of a generalization of the unit-execution-time multiprocessor scheduling problem under precedence constraints, in which the number of communication arcs is also minimized. Most versions of the problem are shown NP-complete, and two polynomial algorithms are presented for specialized cases.*

1. INTRODUCTION

Scheduling partially ordered tasks with equal execution times on identical processors to minimize the latest finishing time is one of the best-studied problems in Computer Science (see [Co] for an early review of related results). When the partial order is a tree, [AH] showed that the highest-level-first strategy is optimal. For general graphs, the case in which two processors are available can be solved in polynomial time [CG]; the case in which the number of processors is part of the input is NP-complete; and the cases with 3, 4, etc. processors remains open.

This problem has attracted so much attention mainly because of its obvious relevance to multiprocessor systems. However, in such systems finishing time is only one of the performance criteria, the other being the *communication* required among the processors [PU]. We measure communication as the number of arcs of the precedence relation, the two end points of which are executed by different processors. For example, the tree shown in Figure 1 can be executed by two processors in 3 time units and 2 units of communication (or, of course, in 5 time units and 0 units of communication by a single processor). Can it be executed in both three time units *and* a single unit of communication? The answer is "no."

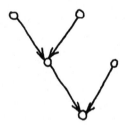

Figure 1: An example.

It turns out that the two-parameter version is a very interesting new twist to this classical problem. In this paper we study the complexity of the following problem: Given a dag (sometimes a tree), a number m of processors (fixed, part of the input, or infinite), and two positive integers t and c, can we schedule the dag on m processors within time t and communication c? Notice

---

[1] National Technical University of Athens.
[2] University of California at San Diego.

that the case of infinite processors is trivial when communication is not taken into account, as the optimum time is precisely the depth of the dag. It turns out that most of these versions are NP-complete. In particular, for general dags, when the number of processors is fixed (even 2) the problem is NP-complete, and likewise for infinite processors. For trees, the case of infinite processors, or a number of processors that is part of the input, are both NP-complete. Finally, scheduling a tree on two processors (in fact, on any fixed number of processors) is a very interesting problem left open here.

We also present algorithms for two somewhat interesting special cases of the problem. The first is that of *layered dags*, that is, dags for which the number of nodes in any maximal path is the same, and equal to t, the time bound. Using matching techniques, we can solve this problem in time $O(n^{2.5})$. Finally, when the amount of communication c is fixed, we show that the problem for general dags is polynomial. Notice that the original scheduling problem (no communication) is NP-complete even when $t = 3$ [PS], so the two parameters t and c appear to behave very differently in this respect.

2. NP-COMPLETENESS RESULTS

In the *two-parameter scheduling problem* we are given a dag $G = (V, E)$ and three integers m, t, and c. We are asked to find a schedule S, that is, a one-to-one (not necessarily onto) mapping from the nodes of V to $\{1, 2, \ldots, t\} \times \{1, 2, \ldots, m\}$ such that: (1) if $(u, u') \in E$ and $s(u) = (\tau, \mu)$, $s(u') = (\tau', \mu')$, then $\tau < \tau'$ (that is, S respects precedences), and (2) the set of arcs $(u, u') \in E$ such that $s(u) = (\tau, \mu)$, $s(u') = (\tau', \mu')$, and $\mu \neq \mu'$ has cardinality at most c. We shall call the same problem without restriction (2) the *original* scheduling problem.

The problem for general dags is trivially NP-complete, since the NP-complete original scheduling problem (with the communication bound c sufficiently large, say $|E|$) is a special case. We now show that it remains NP-complete even if G is a tree (recall that the original problem is polynomial for trees [AH]).

Theorem 1. The two-parameter scheduling problem for trees is (strongly) NP-complete.

Sketch: We reduce the 3-PARTITION problem [GJ] to it. We are given a set of $3m$ integers $a_1, \ldots, a_{3m}$, all between $\frac{B}{4}$ and $\frac{B}{2}$, and we are asked whether they can be divided into m triples, all adding up to B. Now, for each a_i construct a *tassel* of size a_i (that is, an in-tree with a_i leaves and depth one, see Figure 2), and then connect the roots of all tassels to a common root. We argue that the resulting tree can be scheduled on m processors in time $B + 4$ and communication $3m - 3$ if and only if the 3-PARTITION instance has a solution. $\square$

The problem, for general dags, remains NP-complete even when the number of processors is two (compare with [CG]):

Theorem 2. The two-parameter scheduling problem with two processors is NP-complete.

Sketch: The reduction is from ONE-IN-THREE 3-SAT, in which we are given clauses with three literals each, and asked whether there is an assignment such that exactly one literal in each clause is made true. We employ two "gadgets," shown in Figure 3(a) and 3(b) (M is a sufficiently large number). There is one copy of the dag in Figure 3(a) for each variable, and one copy of the dag in Figure 3(b) for each clause. We think that the three middle nodes of the clause dag (Figure 3(b)) correspond to the three literals of that clause. We connect all these dags in series, identifying the sink of one with the source of the next, with the dags corresponding to varables first. Then we draw an arc from all M nodes in the left (respectively, right) path of a variable dag (Figure 3(a)) to the middle nodes of clause dags that correspond to the positive (respectively, negative) occurrences of the variable. It can now be shown that we can schedule the dag on two processors

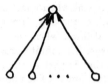

Figure 2: A tassel.

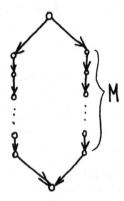

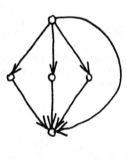

Figure 3: The two gadgets.

within time $n(M + 1) + 3m + 1$ and communication $2n + 3m$ (where n is the number of variables and m the number of clauses) if and only if there is a satisfying truth assignment. $\square$

Whether the above result holds when G is a tree is a question that we have been unable to settle (as with any other fixed number of processors). When the number of processors is infinite, the problem is NP-complete even for trees. The reduction is from EXACT COVER BY 3-SETS, and is omitted.

Theorem 3. The two-parameter scheduling problem for trees and an infinite number of processors is NP-complete. $\square$

3. TWO ALGORITHMS

Let us call a dag *layered* if all maximal paths have the same length, called the *depth* of the dag.

Theorem 4. We can solve the two-parameter scheduling problem with infinite processors for layered dags with depth is equal to t in $O(|V|^{2.5})$ time.

Sketch: Minimizing the communication can be shown to be equivalent to determining maximum bipartite matchings between subsequent layers. $\square$

Finally, when the communication bound c is fixed, the problem can also be solved in polynomial time:

Theorem 5. The two-parameter scheduling problem can be solved in time $O(|V|^{4c+2} \log n)$.

Sketch: By exhaustively examining all c-tuples of arcs, all assignments of processors to the weakly connected components formed by deleting each c-tuple, and all possible times for executing the tails of communication arcs, the problem reduces to single-processor scheduling with release times and deadlines [La], solvable in $O(n \log n)$ time. The total time required is $O(|E|^c \min(m, c + 1)^{c+1} t^c |V| \log |V|)$, dominated by the stated bound (naturally, we can assume that $t < |V|$). $\Box$

Note that, in contrast, fixing the other parameter, t, to any constant greater than 2 does not affect the NP-completeness of the problem (in fact, of the original problem) [PS].

4. RELATED WORK

Our definition of communication (adapted from [PU]) is slightly inaccurate, in that we may overestimate communication by counting twice the communication corresponding to two arcs emanating from the same node v and leading to two different nodes, if these two nodes are executed by the same processor (but different from the processor that executed v). This is of no major consequence for dags with bounded outdegree (such as those studied in [PU], or those used in our NP-completeness constructions). By using variants of our methods as well as more sophisticated techniques, [Pr] has established that our NP-completeness results can actually be extended to the more accurate variant of the problem, in which such arcs are counted only once.

There are two even more accurate versions of the problem. First, the one in which we wish to minimize the weighted sum $t + \tau c$, for some constant τ denoting the number of processor cycles it takes to send a message from a processor to another. This problem is also NP-complete [Pe], essentially under the same assumptions. Even closer to the desired performance measure, consider the scheduling problem in which (1) tasks may be executed more than once in the processors (another inaccuracy of the current formulation), and (2) the endpoints of an arc executed in different processors must be at least τ time units apart (if the tail has many instantiations, we take the least restrictive one, that is, the latest or the one executed on the same processor). We wish to minimize finish time. This problem is shown NP-complete in [PY]. In that paper we also give a very simple linear-time heuristic that approximates the optimum within a ratio of 2. We argue that this can be a useful tool for analysing parallel algorithms in a manner that is independent of the underlying architecture.

REFERENCES

[AH] D. Adolphson and T. C. Hu "Optimal Linear Ordering", *SIAM J. Appl. Math.*, 25, pp. 403-423, (1973).

[CG] E. G. Coffman, R. L. Graham "Optimal Scheduling for Two Processor Systems", *Acta Informatica*, 1, pp. 200-213, (1972).

[Co] E. G. Coffman, ed. *Computer and Jobshop Scheduling Theory*, Wiley, 1978.

[GJ] M. R. Garey, D. S. Johnson *Computers and Intractability: A Guide to the Theory of NP-completeness*, Freeman, 1979.

[La] E. L. Lawler "Optimal Sequencing of a Single Machine Subject to Precedence Constraints", *Management Science*, 19, pp. 544-546, (1973).

[PS] C. H. Papadimitriou, K. Steiglitz *Combinatorial Optimization: Algorithms and Complexity*, Prentice-Hall, 1983.

[PU] C. H. Papadimitriou, J. D. Ullman "A Communication-Time Trade-off", *Proc. 1985 STOC Conference*; also, *SIAM J. Comp.*, 1987.

[PY] C. H. Papadimitriou, M. Yannakakis "Towards an Architecture-Independent Analysis of Parallel Algorithms", *Proceedings 1988 STOC*, to appear.

[Pe] E. Petrohilos, Diploma Thesis, National Technical University of Athens, 1986 (in Greek).

[Pr] M. Prastein, manuscript, University of Illinios at Urbana-Champaign, 1987.

Computing a Perfect Matching in a Line Graph

Joseph Naor

Department of Computer Science
University of Southern California
Los Angeles, CA, 90089-0782

Abstract. *We show in this paper that a perfect matching of a line graph can be computed in NC. A necessary and sufficient condition for a line graph to have a perfect matching is an even number of vertices. To compute the perfect matching, we use a technique of dividing the graph into kingdoms. This result is equivalent to partitioning the edge set of a graph into edge disjoint paths of even length.*

1 Introduction

Determining whether a graph has a perfect matching and computing it is one of the important open questions in parallel computation. A random solution was proposed in [KUW] [MVV], yet the problem still awaits a determinstic solution. A partial solution was given in the case that the permanent of the adjacency matrix is bounded by a polynomial [GK].

Since the general problem seems hard, we restrict ourselves to classes of graphs where the solution might be easier. A necessary and sufficient condition for a line graph to have a perfect matching is an even number of vertices [CPS], [S]. Our main result in this paper is that a perfect matching of a line graph can be computed in NC. We use a similar technique of partitioning the graph into kingdoms that was used in [KN]. Recently, this result has been improved by [CN] who showed that a perfect matching in a claw free graph can be computed with a linear number of processors in $O(\log^2 n)$ time.

The perfect matching of a line graph is equivalent to a partition of the edges of the original graph into edge disjoint paths of even length. Our result implies that computing this partition is in NC.

2   Terminology and theoretical background

Let $G = (V, E)$ denote an undirected graph. The *line graph* $L(G)$ of G is defined as follows: there is a 1-1 correspondence between the edge set of G and the vertex set of $L(G)$; vertices v, w in $L(G)$ are adjacent if their corresponding edges in G are adjacent. The number of vertices in G will be denoted by either $|V|$ or $|G|$.

A *path* is a sequence of vertices $v_1, v_2, \ldots v_r$ such that for every i, v_i and v_{i+1} are adjacent and, for every i and j, $v_i \neq v_j$. A *matching* in a graph is a set of independent edges, i.e., edges that are not adjacent. A matching is called *perfect* if it contains all the vertices in the graph and *maximal* if it is not contained in any other matching. Given a matching M, a vertex v is called *exposed* if none of the edges adjacent to v are contained in M. An *augmenting path* with respect to M is a path connecting two exposed vertices v and w through matched vertices. In matching theory, an augmenting path is an alternating path of edges from M and $E - M$.

A *kingdom* k is a connected set of vertices. The *border* of a kingdom is the set of vertices which have neighbors outside the kingdom. *Internal* vertices are those not on the border. In a *well connected* kingdom we require: (i) interior vertices induce a connected subgraph; (ii) every border vertex is at distance at most two from an internal vertex. The *area* of a kingdom is defined as its number of vertices. Two disjoint kingdoms are *neighbors* if two vertices in the respective borders are at distance at most two from each other. Any subset U of the vertices induces a subgraph G_U in the natural way. For a vertex v, define *bfs(v)* as a list of the vertices of the graph ordered according to their distance from v, where vertices which have the same distance are ordered according to their names.

The following is the main theorem used in this paper, proven in [CPS] and [S].

Theorem 2.1 *A necessary and sufficient condition for a line graph G to have a perfect matching is that $|G|$ is even.*

In matching theory, an alternating path is computed between two exposed vertices in order to augment a given matching. The following theorem shows that in a line graph, **any path** connecting two exposed vertices can be used to augment a given matching.

Theorem 2.2 *Any path connecting two exposed vertices v and w in a line graph G is an augmenting path with respect to a matching M.*

Proof: Let $v = v_0, v_1, \ldots, v_m = w$ be a path p connecting v and w and assume w.l.o.g that all the vertices $v_1, \ldots, v_{m-1}$ are matched in M. We prove the theorem by induction on m. If $m = 1$, the theorem is obviously true. Assume the theorem holds for $l < m$. We have two cases:

- v_1 is matched to v_2. Match v_0 to v_1 and by the induction hypothesis, $v_2 \ldots, v_l$ is an augmenting path.

- v_1 is matched to $z \notin p$. G is a line graph and thus, either $(z, v_0) \in E$ or $(z, v_2) \in E$. In the first case, match z with v_0, and $v_1, \ldots, v_m$ is an augmenting path by the induction hypothesis. In the second case, match v_1 with v_0, and $z, v_2, \ldots, v_m$ is an augmenting path by the induction hypothesis.

∎

This proof implies a procedure to "move" an exposed vertex along a designated path untill it "meets" another exposed vertex.

3 Motivation

The crucial observation that enables us to compute in parallel a perfect matching in a line graph is that any path is an augmenting path, as opposed to general graphs, where only alternating paths serve as augmenting paths. In the next section we show how to implement in parallel the movement of an exposed vertex along a given path.

The main problem is how to move many exposed vertices simultaneously in the graph. The paths along which exposed vertices travel may overlap or, may even be in distance one from each other. To understand the latter case, consider the following scenario: $u_0, u_1, \ldots$ and $v_0, v_1, \ldots$ are two augmenting paths where u_0 and v_0 are exposed; $(u_0, v_1), (u_1, v_1), (u_1, v_2) \in E$ and u_1 and v_1 are matched together. If the two vertices travel simultaneously, both u_0 and v_0 will match themselves with v_1, leading to a contradiction. To overcome this problem, we use a similar technique to the one employed in the NC implementation of Brooks' theorem [KN]. We partition the graph into disjoint kingdoms such that each exposed vertex has its own kingdom, and can move independently in that kingdom.

Each kingdom is well connected; moreover, kingdoms are pairwise disjoint. Since our goal is to match the exposed vertices in pairs, we would like each kingdom to have a neighbor. We cannot always acomplish this, but we can guarantee that, except possibly for a kingdom with area larger than $|G|/2$, every other kingdom has a neighbor. A graph which reflects the dependencies between the neighboring kingdoms is constructed and, according to it, we decide upon the movements of the exposed vertices. Either a kingdom is a host (i.e., at least some other exposed vertex visits it) or its exposed vertex is a visitor (i.e., travels to another kingdom).

Once the exposed vertices are gathered in their destinated kingdoms, we continue the process recursively and independently in each of the kingdoms. If there are just a few exposed vertices in a kingdom, they travel sequentially (one by one) to their destination. The key observation is that the area of a visited kingdom is always smaller than $|G|/2$, which implies that gathering the exposed vertices in pairs terminates after $O(\log n)$ phases. After each gathering, we get rid of at least half of the exposed vertices.

4 The Algorithm

We present the algorithm in a top to bottom fashion. We start by presenting the parallel implementation of the movement of one exposed vertex v_0 along a path $p = v_0, v_1, \ldots, v_k$ where $v_1, v_2, \ldots v_{k-1}$ are matched in M and v_k is exposed. Let $\tilde{M}$ denote the new matching constructed after v_0 has traveled. We denote by m_i the exposed vertex generated after the i-th stage of the proof of theorem 2.2, and let $f, \tilde{f} : V \to V$ be functions that return for a vertex v its mate in M and $\tilde{M}$ respectively. If a vertex v_i is matched to a vertex $z \notin p$, then either $(v_{i-1}, z) \in E$, or $(v_{i+1}, z) \in E$. The reason is that G is a line graph.

Travel(G, v_0, p):
An exposed vertex v_0 travels along a path $p = v_1, v_2, \ldots, v_k$ in the graph G. In $\tilde{M}$, $v_0, v_1, \ldots, v_k$ are matched.
$\tilde{f} \leftarrow f$
$m_0 \leftarrow v_0$
for all $i \geq 1$ **do** *in parallel*:
   if $f(v_i) = v_{i-1}$ **then**
      idle { we "jump" from step $i - 1$ to step $i + 1$.}
   else if $f(v_i) = v_{i+1}$ **then**
      $m_i \leftarrow v_{i+1}$
   { otherwise, v_i is matched to a vertex $z \notin p$ }
   else if $(v_{i-1}, z) \in E$ **then**
      $m_i \leftarrow v_i$
   else if $(v_{i+1}, z) \in E$ **then**
      $m_i \leftarrow z$
end *parallel* **do**
$\tilde{f}(v_0) \leftarrow v_1 \cup f(v_1) - m_1$
$\tilde{f}(v_k) \leftarrow m_k$
for all $i \geq 1$ *s.t. the i-th stage was non idle* **do**
   $\tilde{f}([v_i \cup f(v_i)] - m_i) \leftarrow m_{i-1}$
end Travel

4.1 General outline

We are now ready to present our algorithm. A basic procedure which appears in our subalgorithms is computing the shortest path between two vertices. This can be done by raising the adjacency matrix to the n-th power and reconstructing the path from it.

Perfect Matching: Compute a perfect matching in the line graph G
  Initialize: Compute a maximal matching in G.
  { Let $Ev = \{v_1, v_2, \ldots, v_k\}$ be the set of exposed vertices. }
  { Ev constitutes an independent set.}
  while $k > 2$ **do**
    reduce(P,G) { number of exposed vertices is decreased by half }
    travel(G,v,p)
{ only two exposed vertices v and w remain, and
  let p be a path connecting them.}
end

We next present algorithm *reduce*. We have to justify our claim that exposed vertices in different kingdoms can proceed independently without interfering with each other. The paths from the exposed vertices to the borders of the kingdoms are only through internal vertices because the kingdoms are well connected. Thus, we can call procedure *travel* independently in different kingdoms. We partition the nodes of G into two subsets B and I (border and interior). Initialy, B is empty and $I = V$. Shortest paths will be computed relative to G_I.

Our goal is that each exposed vertex will be adjacent to another exposed vertex. If there are just a few exposed vertices (say 10) in the graph, they travel to their destination sequentially, i.e., one after the other. Otherwise, we construct the kingdoms around each exposed vertex, and then decide which exposed vertices travel and which kingdoms become hosts. By traveling to a kingdom we mean crossing the border and being contained in it. We apply the algorithm recursively in each nonempty kingdom.

reduce$(Ev, G = (B \cup I, E))$:
Decreases the number of exposed vertices Ev in G by half. At any point of time, Ev is the set of exposed vertices and $|Ev|$ is their number.

  { suppose $Ev = \{v_1, \ldots, v_r\}$, note that $r < |G|$ }.
  if $r < 10$ **then**
    for $i = 1, \ldots, r/2$ **do** { sequentially }
      { let p_i be a path from v_i to v_{2i} }
      travel(G, p_i, v_i)
      v_i and v_{2i} are matched.

else

form-kingdoms(G,Ev)
{ r well-connected kingdoms $k_1, \ldots, k_r$ are constructed
such that v_i is in k_i and every k_i (except, possibly,
for one with area larger than $|G|/2$) has a neighbor.
v_i is called the king of k_i }
if *there exists a kingdom k_i without a neighbor* **then**
{ there exists at most one such kingdom }
    v_i finds a path l to v_1
    { w. l. o. g. $v_i \neq v_1$ }
    travel (G, v_i, l)
    v_i *and v_1 are matched*
{ each kingdom has a neighbor }
form a graph $U'=(V',E')$ with
    $V' = \{(k_i, k_j)|k_i$ and k_j are neighbors }
    { both (k_i, k_j) and $(k_j, k_i) \in V'$. }
    $E' = \{[(k_i, k_j), (k_i, k_l)]\} \cup \{[(k_i, k_j), (k_j, k_l)]\}$
{ a king is called a host if its kingdom is a host. Intuitively,
vertices in V' represent possible motions of kings and edges
represent restrictions on the motions. In particular, a king cannot
go to two places and a king cannot be both a host and a visitor }
find a maximal independent set S in U'. { [L] or [GS] }
for each $(k_i, k_j) \in S$ **do**
    v_i travels to k_j { i.e., to the closest internal vertex of k_j }
{ let $W \in P$ be kings that are neither a host nor a visitor }
for each *king $v \in W$* **do**
    choose k_j such that k_j is a neighbor of the kingdom of v
    { v_j is a visitor, i.e., $(k_j, k_l) \in S$ for some l }
    if *no other $q \in W$ chose k_j* **then**
        { v_j is empty }
        v *travels to k_j*
        v *travels to k_l*
    else v *travels to k_j*
        { k_j has at least two visitors }
{ each non-empty kingdom has at least two exposed vertices. }
for each *non-empty k_i with exposed vertices* { $v_{i_1}, \ldots, v_{i_l}$ } **do**
    { $I_i = k_i \setminus border(k_i); B_i = border(k_i);$
    E_i are the edges of the induced graph $G_{I_i \cup B_i}$ }
    $reduce(\{v_{i_1}, \ldots, v_{i_l}\}, G_i = (B_i \cup I_i, E_i))$

end

It was proved in [KN] that an exposed vertex whose kingdom is not a host can
find a host kingdom in a radius of one or two kingdoms.

4.2 Crossing the border

We have to explain what we mean by the sentence v_i *travels to* k_j in the algorithm *reduce*. A path in k_i is computed from v_i to the closest vertex u (in k_i) to the kingdom k_j. By our definitions, u is matched to a vertex u'. Vertices u and u' are declared as exposed and v_i travels to u'. Thus, u becomes exposed and v_i and u' are matched in procedure *travel* which can be called independently in each kingdom. Let w be the closest vertex of k_j to u. The shortest path from u to w is of length five at most and v_i's aim is to travel along that path to w's mate in the matching.

All the paths that cross a border move simultaneously, yet sequentially, one vertex at a time step. After each step we compute a maximal matching on the exposed vertices that are adjacent so that the exposed vertices remain an independent set. There are two cases that should be noted:

- At most two exposed vertices will try to move into the same vertex at the same time. In that case however, only one of them (arbitrarily chosen) will do it. At the end of the step, the two exposed vertices will be adjacent and can be matched.

- Suppose that two exposed vertices v and w want to move into vertices l and m where l and m are matched. In this case, however, there is a perfect matching in the induced subgraph of v, w, l and m.

4.3 The kingdoms

We now present the algorithm *form-kingdoms*. Initially, kingdoms are the exposed vertices themselves. In each stage, every active kingdom finds out the distance to the closest neighbor and annexes the vertices within half of that distance. This is done until either a kingdom finds a neighbor or until it grows beyond $|G|/2$. In this case, it gives up vertices so that its area is $|G|/2 + 1$ and then becomes inactive. We note that the kingdoms, although disjoint, do not constitute a partition of the graph.

Form kingdoms (G, Ev):
Let $Ev = \{v_1, ..., v_r\}$
r well-connected kingdoms $k_1, \ldots k_r$ are constructed such that v_i is in k_i and every v_i (except, possibly, for one with area larger than $|G/2)$ has a neighbor.

> **for all** $v_i \in Ev$ **do**
>   *compute bfs(v_i).*
>   *active(i)=true*
> $k_i = \{v_i\}$

```
while active(i) do
    let r(i)= distance to k_i's closest neighbor
    if r(i) ≤ 2 then
        { a border was found }
        active(i)=false
    else
        k_i = k_i∪ next ⌊r(i)/2⌋ layers in bfs(v_i)
        if |k_i| > n/2 then
            drop vertices from the last layers of bfs(v_i)
            so that |k_i| = |G|/2 + 1
            active(i) = false
end
```

It is clear that *form-kingdoms* takes time $T_k = O(\log^2 n)$. This is because after each $while-loop$, the distance between a kingdom and its closest neighbor is reduced by at least half; thus, the algorithm has at most $\log n$ phases and each one takes $O(\log n)$ time. We have to show that the algorithm constructs the desired kingdoms. Each kingdom is well connected because it grows according to a breadth first search tree rooted at its exposed vertex. We point out the following observations:

(i) At every stage, kingdoms are disjoint.

(ii) A kingdom becomes inactive in any of the following cases:

   (a) It encounters a common border with another kingdom.

   (b) It grows beyond $|G|/2$. (note that only one such kingdom can exist).

The conditions of the construction are now satisfied. Note that the only kingdom that might be isolated is that with area larger than $|G| \setminus 2$.

Theorem 4.1 *A perfect matching of a line graph can be computed in NC.*

Proof: Follows from the above discussion. ∎

5 Discussion

In this paper we gave the combinatorial arguments for finding a perfect matching of a line graph. The number of processors used is clearly dominated by the number of processors needed to compute the partition into kingdoms. To compute that we need a breadth first search tree rooted at the king of every kingdom. Minimizing the number of processors in the computation of a breadth first search tree is an outstanding open question in parallel computation.

It is well known that constructing a perfect matching in a general graph is NC-equivalent to constructing a perfect matching in a cubic graph, i.e., 3-regular graph.

Finding a perfect matching in a cubic graph can be formulated as a path partition problem using a theorem proved originaly by Petersen [P].

Theorem 5.1 *A perfect matching in a cubic graph is equivalent to partitioning the edge set into edge disjoint paths of odd length such that every path is the start (end) of exactly one path.*

Acknowledgement

I would like to thank Professor Gary Miller for simplifying the presentation of the algorithm.

References

[CN] Marek Chrobak and Joseph Naor, "Perfect Matchings in Claw Free Graphs", *In preparation.*

[CPS] G. Chartrand, A. D. Polimeni and M. James Stewart, "The Existance of 1-Factors in Line Graphs, Squares and Total Graphs", *Indag. Math.* 35, pp. 228-232 (1973).

[GM] D. Y. Grigoriev and M. Karpinski, "The Matching Problem for Bipartite Graphs with Polynomially Bounded Permanents is in *NC*", *28th Annual IEEE Symposium on Foundations of Computer Science*, Los Angeles, CA, 1987.

[GS] M. Goldberg and T. Spencer, "A New Parallel Algorithm for the Maximal Independent Set Problem", *28th Annual IEEE Symposium on Foundations of Computer Science*, Los Angeles, CA, 1987.

[KUW] R. M. Karp, E. Upfal and A. Wigderson, "Constructing a Perfect Matching is in Random *NC*", *Combinatorica*, Vol. 6, 1, pp. 35-48, (1986).

[KN] Mauricio Karchmer and Joseph Naor, "A Fast Parallel Algorithm for Coloring a Graph with Δ Colors", *Journal of Algorithms*, Vol. 9, 1, pp. 83-91, (1988).

[L]   M. Luby, "A simple parallel algorithm for the maximal independent set", *Siam Journal on Computing*, Vol. 15, 4, (1986).

[MVV] K. Mulmuley, U. V. Vazirani and V. V. Vazirani, " Matching is as easy as Matrix Inversion", *Combinatorica*, Vol. 7, 1, pp. 105-113, (1987).

[P]   J. Petersen, "Die Theorie der Regulären Graphs", *Acta Math.* Vol. 15 , pp. 193-220 (1891).

[S]   D. P. Sumner, "On Tutte's Factorization Theorem", *Proceedings of the Capital Conference on Graph Theory and Combinatorics,* George Washington University, Washington D.C., 1973, pp. 350-355. Lecture Notes in Math., Vol. 406, Springer, Berlin, 1974.

Separation Pair Detection[1]

Donald Fussell   *Ramakrishna Thurimella*

Department of Computer Sciences
The University of Texas at Austin
Austin, TX 78712.

Abstract *A separation pair of a biconnected graph is a pair of vertices whose removal disconnects the graph. The central part of any algorithm that finds triconnected components is an algorithm for separation pairs. Recently Miller and Ramachandran have given a parallel algorithm that runs in $O(\log^2 n)$ time using $O(m)$ processors. We present a new algorithm for finding all separation pairs of a biconnected graph that runs in $O(\log n)$ time using $O(m)$ processors. A direct consequence is a test for triconnectivity of a graph within the same resource bounds.*

1 Introduction

Triconnected components are useful in testing a graph for planarity, finding how reliable a computer network is against node failures, etc. The central part of any triconnected component algorithm is an algorithm to find separation pairs. A separation pair of a biconnected graph is a pair of vertices whose removal disconnects the graph.

There are potentially $O(n^2)$ pairs that can qualify to be separation pairs. Consider a simple cycle of n vertices. It is biconnected and every non-neighboring pair of nodes is a separation pair. Therefore any parallel algorithm that is required to list all the pairs must have a processor-time product of $\Omega(n^2)$. But such a list is not necessary when finding triconnected components. The output of our algorithm is a set of paths, where a pair of non-neighboring vertices of G is a separation pair if both vertices appear on a path. Notice that it is possible to obtain a time complexity of $O(\log n)$ with $O(mn)$ processors for this algorithm by applying the parallel biconnectivity algorithm of Tarjan and Vishkin [TV 84] n times. In comparison, our approach uses only $O(m)$ processors while still retaining the same time bound.

The paper is organized as follows. The next section contains definitions and some preliminary results. Section 3 contains two transformations that preserve separation pairs. The first transformation results in what we call a *reach* graph of ears and the second planarizes the reach graph. The last section establishes the complexity bounds of the algorithm.

A major part of the work reported here was developed prior to the announcement by Miller and Ramachandran [MR 87] of their parallel algorithm for triconnected components. The reach graphs described in section 3.1 are similar to the bridge graphs of [MR 87], whereas the graphs resulting from planarizing the reach graphs and the star graphs of [MR 87] are the same.

The techniques we use differ significantly from the divide-and-conquer approach used by Miller and Ramachandran. The novelty in our approach comes from the extensive use of a method we refer to as "local replacement" to obtain an efficient parallel reduction of this problem to another which is solvable with known efficient techniques. The method of local replacement involves replacing each vertex v of the original graph by a sparse graph G_v. The sparse graph is such that $|V(G_v)| =$ degree(v). The technique of local replacement is of independent interest, and we have found it to be of use in other parallel graph algorithms as well [FT 87].

---

[1] This work was supported in part by the ONR under Contract N00014-86-K-0597.

2   Preliminaries

Let $V(G)$ and $E(G)$ stand for the vertex set and the edge set of a given graph G respectively. The subgraph G' *induced* by a subset $V'(G)$ of $V(G)$ is a graph whose vertex set is $V'(G)$ and the edge set $E(G')$ is a subset of $E(G)$ such that the edges in $E(G')$ are those with both end nodes in $V'(G)$. A connected graph G is *r-connected* if at least r vertices must be removed to disconnect the graph. A *biconnected graph (or a block)* is 2-connected. A *ear decomposition starting with a node* P_0 of an undirected graph G is $G = P_0 \cup P_1 \cup ... \cup P_k$ where P_{i+1} is a path whose end nodes belong to $P_0 \cup P_1 \cup ... \cup P_i$ but its internal nodes do not. In addition, $E(G) = E(P_0) \cup E(P_1) \cup ... \cup E(P_k)$ and $V(G) = V(P_0) \cup V(P_1) \cup ... \cup V(P_k)$. Each of these paths P_i is an *ear*. If the two end nodes of a path P_i do not coincide then P_i is an *open ear*, otherwise the ear is *closed*. In an *open ear decomposition* every ear P_i, $1 < i \leq k$, is open. An ear is a *short ear* if it consists of a single edge, otherwise it is *long* . For any $i \leq k$, the subgraph induced by the ears $E_0, ..., E_i$ of an open ear decomposition is biconnected for a given biconnected graph. A pair of vertices $\{x, y\}$ of a biconnected graph is a *separation pair* if the number of components of the subgraph induced by $V(G) - \{x, y\}$ is more than one. A vertex is *internal* to an ear if it is not one of the end nodes of that ear. Notice that, except for the root, each v is internal to exactly one ear, call it $ear(v)$.

Remark : It is assumed that $r \notin \{x, y\}$ throughout this paper. The pairs in which one of the vertices is r can be detected as a special case by finding the articulation points of the graph induced by $V(G) - \{r\}$ within the resource bounds claimed.

Lemma 1 *If $\{x, y\}$ is a separation pair of a graph G then there exists a long ear E_l in any open ear decomposition of G such that $\{x, y\}$ is a pair of nonconsecutive nodes on E_l.*

Proof : Let C be a connected component induced by $V(G) - \{x, y\}$ such that $r \notin V(C)$. Then E_l is the minimum of $\{ear(v) \mid v \in C\}$. See Fig. 1.   □

The following definition labels each vertex v depending on the position of v on $ear(v)$.

Starting with one of the end nodes $p \in \{x, y\}$ of E_l, define $pos(p, E_l)$ to be zero. For the rest of $V(E_l)$, $pos(v, E_l)$ is the distance from p to v on E_l. The value of $pos(x, E_l)$, for $x \notin V(E_l)$ is undefined. If v is an internal node of E_l then it is denoted by $pos(v)$; since the second argument is unique it can be omitted.

Given an internal node v of E_l, define

$$reach(v) = \begin{cases} [-m, m] & \text{If } \exists \text{ a path } P : v \to r \text{ and } V(P) \cap V(E_l) = \{v\} \\ [left_v, right_v] & \text{Otherwise} \end{cases}$$

where $left_v$ ($right_v$) is minimum (resp. maximum) over the set

$$\{pos(v', E_l) \mid \exists \text{ path } P : v \to v' \text{ and } V(P) \cap V(E_l) = \{v, v'\})\}$$

and m is any integer larger than the number of vertices in the graph. We will assume $m = 2n$ throughout. Refer to the first (second) component of the interval as $reach(v).1$ ($reach(v).2$).

Theorem 1 *$\{x, y\}$ is a separation pair iff $\exists$ a long ear E_l containing x and y such that x and y are not consecutive nodes on E_l and all nodes v between x and y on E_l satisfy $reach(v) \subseteq [pos(x, E_l), pos(y, E_l)]$ assuming $pos(x, E_l) < pos(y, E_l)$.*

Proof : **(only if)** Let E_l be as defined in in lemma 1. Assume, for contradiction, that the forward implication is not true. Then, there exists a vertex v such that either $(pos(x, E_l) > reach(v).1)$ or $(pos(y, E_l) < reach(v).2)$. In either case, we will exhibit a path from v to r that does not use x or y, thus implying $v \in V(C_1)$. As we assumed $v \in V(C_i)$, for some $C_i \neq C_1$, we would have a contradiction.

Assume $pos(x, E_l) > reach(v).1$. If $pos(x, E_l) = 0$ then $reach(v).1$ has to be $-2n$ as it is the only number less than 0 in our numbering. That is, $reach(v) = [-2n, 2n]$ and there is a path from v to r that does not use any vertices of E_l, including x and y. Hence $pos(x, E_l)$ cannot be zero. Assume $0 \leq reach(v).1 < pos(x, E_l)$. This implies there is a path P from v to v_1, where $pos(v_1, E_l) < pos(x, E_l)$. Let v_2 be the vertex with $pos(v_2, E_l) = 0$. From v_1 extend P first by taking edges of E_l to vertex v_2. To extend P further to r, consider the block B induced by the ears $P_0, ..., ear(v_2)$. B cannot contain x but can contain y. But since there are two vertex-disjoint paths from v_2 to r in B it is always possible to extend P to r without having to use y. Therefore there is a path from v to r that does not use either x or y.

The argument is similar when $pos(y, E_l) < reach(v).2$.

(if) Since x and y are not neighbors on E_l, there exists at least one vertex between x and y on E_l. If the segment of the path E_l that is between x and y does not get disconnected from the component containing r, then there is a path from the root to a vertex v of the segment that avoids both x and y. That means either $pos(x, E_l) > reach(v).1$ or $pos(y, E_l) < reach(v).2$, a contradiction. □

3 An Algorithm for Separation pairs

The point of characterizing separation pairs in terms of the *reach* labeling is to be able to generate separation pairs efficiently. Let G_r be the graph obtained by adding to two edges $(v, reach(v).1)$ and $(v, reach(v).2)$ to $E(G)$ for each vertex $v \in V(G)$, for $reach(v) \neq [-2n, 2n]$, and by deleting all short ears from $E(G)$. In the case when $reach(v) = [-2n, 2n]$ add single edge (v, r). Define G_r to be the *reach graph of G*. It follows from Theorem 1 that the set of separation pairs of G_r and G is identical since the reach labeling of their respective vertices is the same.

In Section 3.1 we show that an approximation of the *reach* labeling can be computed efficiently using the parallel biconnectivity algorithm of Tarjan and Vishkin [TV 84]. Theorem 3 shows that the suggested approximation, in fact, suffices for extracting separation pairs.

Even though G_r is structurally simpler than G, a further simplification appears to be necessary to generate separation pairs rapidly in parallel.

Define R_i to be the *reach graph of E_i* as follows. The graph R_i is the same as E_i with the following additions. Create a new vertex r_i and add an edge (v, r_i) if $reach(v) = [-2n, 2n]$ or v is an end node of E_i; otherwise add two edges $(v, reach(v).1)$ and $(v, reach(v).2)$ to $E(R_i)$. Section 3.2 includes the definition of the planar versions P_i of R_i and a procedure that finds the P_i by a reduction to the connected component algorithm of Shiloach and Vishkin [SV 81]. See Fig 3. A local modification is suggested at the end of Section 3.2 that yields a set of paths where each nonconsecutive pair of vertices on each path is a separation pair.

3.1 Compute the *reach* Labeling

This consists of two stages : first, we define an approximation $reach'$ and show how to compute it efficiently; next, we make local modifications of the original graph so as to offset the error introduced by the approximation. The graph G' that results after the local modifications has the following desirable property: corresponding to each separation pair $\{x, y\}$ and the E_l (as defined in Theorem 1), the $reach'$ labeling of E_l in G' is the $reach$ labeling of E_l of the original graph G.

Define $reach'(v)$ the same way as $reach(v)$ but with the weaker restriction on the nodes P can use: P should not use the internal nodes of $ear(v)$ but is allowed to use the end nodes of $ear(v)$. Notice that if each vertex of G is an end node of at most one ear then $reach$ and $reach'$ are identical. Also, for each v, $reach(v) \subseteq reach'(v)$ and if $reach(v) \neq reach'(v)$ then $reach'(v) = [-2n, 2n]$. Next we show how to compute the $reach'$ labeling efficiently.

Consider the auxiliary multigraph G_e constructed from the given graph G as follows. Let E_v be a long ear $v_0, v_1, \ldots, v_k$ of length greater than two, i.e. $k \geq 3$. Now, G_e is obtained from G by contracting all such ears E_v by merging the internal vertices $v_1, v_2, ..., v_{k-1}$. For all v, $v \in V(G_e)$, label v with the ear number it represents. See Fig 2. Let $B_1, B_2, ..., B_l$ be the blocks of G_e. Assume $r \in V(B_1)$. For $1 \leq i \leq l$, let v_{a_i} be the node with minimum node label in $V(B_i)$. (Recall that the node labels in G_e correspond to the ear labels of long ears of G.) Define $B_2', ..., B_l'$ to be the subgraphs of $B_2, ..., B_l$ respectively, where for $2 \leq i \leq l$, B_i' is the subgraph of B_i induced by the vertex set $V(B_i) - \{v_{a_i}\}$. It is easily verified that the $V(B_i')$s define a partition over $V(G_e)$.

Consider a labeling α of the B_i's. This labeling will help us find the $reach'$ values of a node v locally, that is by looking at the neighbors of v. For all $v \in V(B_i')$, $\alpha(v) = \langle v_{a_i}, x, y \rangle$ where x (y) is minimum (resp. maximum) over the set

$$\{pos(s) \mid (s, t) \in E(G), s \text{ and } t \text{ correspond to } v_{a_i} \text{ and some vertex of } B_i' \text{ resp.}\}$$

Extend the α labeling from the vertices of G_e to the vertices $V(G) - \{r\}$ as follows. Each v is an internal node of exactly one long ear E_x and the node corresponding to E_x belongs to exactly one B_i'.[2] For $v \in V(G)$, $\alpha(v)$ is the α label of the node in G_e that corresponds to the $ear(v)$. From the definition of an ear it follows that for all $v \in V(G)$, $\alpha(v).1 < ear(v)$.

It is easy to show that the α labeling satisfies the following property.

Lemma 2 If $(v, w) \in E(G)$ then $\alpha(w).1 \leq ear(v)$.

Let an edge (v, w) be an *escape-edges(v)* if $ear(v) \neq ear(w)$ and w is not adjacent to v on $ear(v)$. The following theorem characterizes $reach'(v)$ in terms of the labels of the nodes that are adjacent to v in G. See Fig 4.

Theorem 2 For a vertex v, $v \neq r$,

    (i) $reach'(v) = [-2n, 2n]$ iff $\exists(v, w) \in escape\text{-}edges(v)$ and $\alpha(w).1 < ear(v)$.

    (ii) if $reach'(v) \neq [-2n, 2n]$ then $reach'(v).1$ ($reach'(v).2$) is the min (resp. max) over the union of $\{pos(v)\}$, $\{pos(w) \mid \exists(v, w) \in E(G) - E(ear(v))$ and $ear(v) = ear(w)\}$, and $\{\alpha(w).2$ (resp. $\alpha(w).3$) $\mid (v, w)$ is an escape-edge(v) $\}$

Proof : (i) (only if) $reach'(v) = [-2n, 2n]$ implies that there is a path P from v to r in G that avoids all the internal nodes of $ear(v)$. Let the first two vertices of P be v and w. From the definition of $reach'(v)$, $ear(v) \neq ear(w)$. If $ear(w) < ear(v)$, then from Lemma 2 we can conclude that $\alpha(w).1 < ear(w) < ear(v)$.

Assume $ear(w) > ear(v)$. Corresponding to each long ear E_x of G there is a node in G_e (call it e_x). We will show that e_v and e_w belong to the same block B in G_e where B is attached to an articulation point e_y, where $ear(y) < ear(v) < ear(w)$. Since $\alpha(e_w).1$ would be e_y, it would follow that $\alpha(w).1 < ear(v)$.

Consider two paths P_1 and P_2 in G_e from e_v to e_0. Let P_1 be $e_v e_{v_1} e_{v_2} ... e_0$, where $e_v > e_{v_1} > e_{v_2} ... > e_0$. Such a path is guaranteed to exist given the definition of ear decomposition. Let P_2 be $e_v e_w ... e_0$, the path in G_e that corresponds to the path P. Let e_s be the first vertex, after e_v, common to P_1 and P_2, in the order they occur on the paths. Since node labels on P_1 decrease monotonically and as $e_v \neq e_s$, we conclude that $e_s < e_v < e_w$. Let B be the block that contains e_s, e_v and e_w. Then for all $e_l \in V(B)$, $\alpha(e_l).1 \leq e_s$. So $\alpha(e_w).1 \leq e_s < e_v$. Therefore $\alpha(w).1 < ear(v)$.

(i) (if) Notice that e_v and e_w belong to the same block B in G_e because of the edge (v, w) and $\alpha(w).1$ is the articulation point in G_e to which B is attached. Therefore there are two vertex-disjoint paths from e_w to $\alpha(w).1$. Hence there exists a path from e_w to $\alpha(w).1$ that does not contain e_v.

---

[2] This statement would be incorrect if B_i' is replaced by B_i because if the node corresponding to E_x were an articulation point in G_e then it would be split and therefore present in more than one component.

From $\alpha(w).1$ take a path with monotonically decreasing node labels to e_0. So, we can always a find a path in G_e, from e_w to e_0 that avoids e_v. Hence, by taking a corresponding path in G, the root r can be reached without using any internal nodes of $ear(v)$.

(ii) From (i) we know that $reach'(v) \neq [-2n, 2n]$ implies that for all $(v, w) \in escape\text{-}edges(v)$, $\alpha(w).1 \geq ear(v)$. But from lemma 2 we know that $\alpha(w).1 \leq ear(v)$. Therefore, for all such w, $\alpha(w).1' = ear(v)$. This means that all the paths starting with an escape-edge lead back to an internal node of $ear(v)$. Rest of the proof directly follows from the definition of $reach'(v)$, $\alpha(v).2$ and $\alpha(v).3$. □

Since we do not how to find the *reach* labeling, we need a theorem similar to theorem 1 with *reach* replaced by *reach'*. To achieve this goal, we propose the following local modifications to G resulting in G'.

Define two ears to be *parallel ears* if they have the same pair of end nodes. Let a *group of parallel ears* be parallel ears that are in the same connected component when the two end vertices are removed. See Fig. 5. In the following, assume E_x consists of $v, v_1, ..., v_k$, and w, in that order.

Algorithm *Build G'*

1. Identify each ear as either a single ear or a parallel ear. Also mark the ear with the minimum ear label in a group of parallel ears.

2. Partition the edge set into a set of paths induced by ears by "splitting" at the end nodes. Rename each $v \in V(E_x)$ by v_{E_x}.

3. "Glue" these paths, to get G', as described below.

   (a) Let E_1 be the first long ear. Pick one of the two end edges of E_1 arbitrarily and give it a direction. Orient rest of E_1 in the opposite direction.

   (b) Give directions to the remaining ears of G so that the resulting digraph G_d is acyclic.

   (c) Remove the last edge in each directed ear in G_d to obtain a directed spanning tree T_d.

   (d) If E_x is a single ear or a parallel ear with minimum ear label in that group then do the following.

   Let (v_k, w) be the edge that was removed in step 3c for the ear E_x. If $lca(v_k, w) = v$ in T_d then let E_y be the ear number of the first edge in the tree path from v to w. Add an edge between v_{E_x} and v_{E_y}. If $lca(v_k, w) \neq v$ in T_d then add an edge between v_{E_x} and v_{E_y}, where v is an internal node of E_y in G.

   (e) If E_x is a parallel ear and E_x is not the minimum ear label in that group then do the following. Connect the end node v_{E_x} to v_{E_w} where E_w comes after E_x in the descending order of ear labels in that group.

   (f) Reverse the direction of edges of G_d and repeat steps 3c and 3d.

Fig 7 illustrates the local modifications suggested by the above algorithm. Before we prove that *reach'* labeling of G' is sufficient to find separation pairs, we need the following definitions.

If a vertex v is not an end node of any ear in G then it retains the same vertex label in G' and we say v of G and v of G' *correspond* to each other. Otherwise vertex v of G and v_{E_i} of G' *correspond* to each other if the latter is obtained from the former by splitting in step 1. We also say a path (ear) of G *corresponds* to a path (ear) of G' if their vertices correspond to each other. Let E_x *dominate* E_y if no vertex of E_y belongs to the component containing the root r in $V(G) - V(E_x)$. See Fig 6.

The following lemmas are straightforward.

Lemma 3 *Let E_x dominate $E_{x_1}, E_{x_2}, ..., E_{x_k}$, for $x < x_1 < ... < x_k$. Let v be the end node of E_x, where the edge labeled E_x that is incident on v is in T_d. The subtree of T_d rooted at v contains all the vertices of the ears $E_{x_1}, E_{x_2}, ..., E_{x_k}$.*

Lemma 4 *If E_x dominates E_y and if both have the same end-vertex v then both ears are either outgoing or incoming at v in G_d.*

Assume for E_x and E_y, $E_x < E_y$, and that if they are parallel then they are in the same group. Let v be an end node of E_x in G. Let T'_d be the spanning tree of G' corresponding to T_d in which E_x is attached by a vertex corresponding to v.

Lemma 5 *If E_x dominates E_y then the subtree of T'_d rooted at v_{E_x} contains all the vertices of E_y.*

Let the end nodes of E_y (E_x) be v and w_1 (resp. w_2). Assume $E_x < E_y$, and that T_d is the spanning tree in which both E_x and E_y are attached to v.

Lemma 6 *If the fundamental cycle created by including the end edge of E_y that is incident at w_1 contains any edges of E_x then for every vertex $w \neq v$ on the cycle we have $ear(w) \geq E_x$.*

Remark : Notice that if $w_1 = w_2$, that is E_x and E_y are parallel, then the lemma is vacuously true since the fundamental cycle never contains edges of two parallel ears.

Let $\{x, y\}$ be a separation pair. From Theorem 1, we know that there exists a long ear E_l containing two nonconsecutive nodes x and y with $pos(x, E_l) < pos(y, E_l)$.

Theorem 3 *$\{x, y\}$ is a separation pair in G iff for all nodes v that are between x_{E_l} and y_{E_l} on E_l in G' we have $reach'(v) \subseteq [pos(x_{E_l}, E_l), pos(y_{E_l}, E_l)]$.*

Proof : The reverse implication is trivial. The following argument establishes the forward implication. Let E (E') be the set of ears both of whose end nodes are in $C \cup \{x, y\}$ $(C' \cup \{x_{E_l}, y_{E_l}\})$ where C (C') is the component containing the segment of E_l that is between x and y in $V(G) - \{x, y\}$ (resp. between x_{E_l} and y_{E_l} in $V(G') - \{x_{E_l}, y_{E_l}\}$). It is easy to see that $E \subseteq E'$. We will show that $E' \subseteq E$. Let C'_E be the part of the component corresponding to the ears in E. Let $E_n \in E' - E$.

We will first show that E_l is the minimum of E'. Notice that there cannot be an edge from a vertex that does not correspond to either x or y to a vertex of E_n as that would imply the corresponding edge is in $V(G) - \{x, y\}$ and $E_n \in E$. Therefore it is sufficient to prove that, for each split vertex x_{E_n} of x that if x_{E_n} is in C' then $E_n > E_l$. If there were a vertex x_{E_n} with $E_n < E_l$ then consider the ear label E_q of the edge (x_{E_n}, x_{E_q}) that connects x_{E_n} to one of the vertices of C'_E. Notice that this edge has to be either in the spanning tree T'_d or the tree obtained by reversing the directions. Assume, wlog., that it is in T'_d. That means the subtree of T'_d rooted at x_{E_l} does not contain the ear E_q, contradicting Lemma 5.

Assume E_n greater than the maximum of E and is connected directly to the component C'_E. But if the only edges of attachment of E_n are two edges connected to x_{E_l} and y_{E_l} then E_n would get disconnected when these two vertices are removed. Let x_{E_m} be a vertex of attachment of E_n to C'_E, for some $E_l < E_m < E_n$. If E_m and E_n are parallel then they are in the same group and therefore there is a path in G from an internal node of E_m to E_n that does not use either x or y. Hence $E_n \in E$. If E_m and E_n are not parallel, then by Lemma 6 there is a path in G from an internal node of E_m which uses vertices whose ear label is greater than or equal to E_m. But as $E_n > E_m$ neither x nor y can occur on this path, hence $E_n \in E$. Therefore such an E_n cannot exist. $\square$

3.2 Planar Versions of Reach Graphs

The reach graphs R_i for the ears E_i are defined at the beginning of Section 3. Refer to the edges in $E(R_i) - E(E_i)$ as *arcs*. Define an equivalence relation on $V(E_i)$ as follows. Two vertices x and y are related if there is an arc between them or there exists a pair of arcs (x, u), (y, v) with $pos(x) <$

$pos(y) < pos(u) < pos(v)$. Extend the relation to include r_i as follows. Define x to be related r_i if there is edge $(x, r_i) \in E(R_i)$ or there is an arc (x, y) and (u, r_i) such that $pos(x) < pos(u) < pos(y)$. Now consider the partition $p_1, p_2, ..., p_k$ induced by this relation. The planar version P_i of R_i is obtained as shown below. The vertex set $V(P_i)$ is the union of $V(E_i)$, $\{r_i\}$ and $\{v_{p_i} \mid p_i$ is a partition that does not include $r_i\}$. The edge set $E(P_i)$ is the union of $E(E_i)$, $\{(x, v_{p_i}) \mid x$ belongs to the partition $p_i\}$ and $\{(x, r_i) \mid x$ and r_i are in the same partition $\}$.

The following algorithm builds P_i using the connected component algorithm [SV 81]. See Fig 3.

Algorithm *Build P_i*

1. For each arc (x, y), identify two arcs (a, b) and (c, d) where $(a, b) = \max \{ pos(f) \mid (e, f)$ is an arc with $pos(x) < pos(e) < pos(y)\}$. Define (c, d) similarly with max replaced by min.

2. Replace each arc (x, y) with two edges $(x, v_{<x,y>})$, $(v_{<x,y>}, y)$. Refer to $v_{<x,y>}$ as an *arc vertex*.

3. If (a, b) is identified with (x, y) in the first step then add an edge $(v_{<x,y>}, v_{<a,b>})$.

4. Assume p_i is a connected component in the subgraph induced by the arc vertices after the Step 3. For all i, merge all the arc vertices of p_i into a single vertex v_{p_i} using the connected component algorithm.

Refer to the vertices resulting from the Step 4 of the previous algorithm also as *arc vertices*. For each arc vertex v_{p_i}, define $left(v_{p_i})$ $(right(v_{p_i}))$ to be the *pos* value of the vertex on E_i that is connected to v_{p_i} and that has a minimum (resp. maximum) *pos* value. Call the edges connecting v_{p_i} to its *left* and *right* vertices in P_i, *extreme edges*. It is easy to see that each vertex of E_i can have at most one non-extreme edge incident on it in P_i.

Lemma 7 *The above algorithm produces the desired equivalence classes.*

Proof : In other words, it is required to show that it suffices to identify the arcs (a, b) and (c, d) for each (x, y), where these edges are as defined in Step 1 of the algorithm. Define the *right-max(x, y)* to be the arc (a, b). Similarly the *left-max(x, y)* to be the arc (c, d). Consider the arcs (x, y) and (u, v) where $(u, v) \neq (a, b)$, and $pos(a) < pos(u) < pos(b) < pos(v)$. Let left-max$(u, v) = (e, f)$ and right-max$(x, y) = (g, h)$. Assume $pos(e) < pos(f)$ and $pos(g) < pos(h)$. If e, f, g and h are in the same partition then a, b, u and v also belong to the same partition. Otherwise we have two arcs $(e, f), (g, h)$ where $(pos(h) - pos(e)) > (pos(v) - pos(x))$. Repeat the argument if e, f, g and h are not in the same partition. Every application of the argument results in a new pair of arcs for which the distance between the vertices at the extreme is strictly greater than the corresponding distance for the old pair of arcs. As the length of E_i is finite we conclude that a, b, u and v are in the same partition. $\square$

Given the structure of R_i and the above theorem it is easy to see that each P_i is planar. Assume each P_i is embedded as shown in Fig. 3. We can conclude that

Theorem 4 *The following statements are equivalent:*

1. $\{x, y\}$ *is a separation pair in G.*

2. $\{x, y\}$ *is a separation pair in P_i.*

3. $\{x, y\}$ *lie on a bounded face in P_i.*

The following algorithm shows how to obtain separation pairs from the P_i. Basically, starting with E_i each node is expanded linearly by considering the edges incident on the corresponding vertex in P_i. This expanded path is then broken off at places where the vertices from the opposite sides cannot form a separation pair.

Algorithm *Generate-separation-pairs*

For a vertex $v \in V(E_i)$, let $x_1, x_2, ..., x_k$ be v's neighbors in P_i. Assume $x_1, x_k \in V(E_i)$. Assume $left(x_i)$, for $1 < i < k$, is in sorted order. In addition, for all j, if $left(x_j) = v$ then assume the relative ordering of x_j's is with respect to their $right(x_j)$ in decreasing order. Assume $Pred(v)$ and $Succ(v)$ refer to the vertex before and the vertex after v on the line segment under consideration.

1. If there is a non-extreme edge (x, x_j) incident on x in P_i, then replace x in E_i by the chains $v_{x_2} - v_{x_3} - ... - v_{x_j}$ and $v'_{x_j} - v_{x_{j+1}} - ... - v_{x_{k-1}}$. Otherwise replace x in E_i by $v_{x_2} - v_{x_3} - ... - v_{x_j} - v_{x_c} - v_{x_{j+1}} \cdots v_{x_{k-1}}$ where, for all $2 \le i \le j$, $left(x_i) < pos(v)$ and, for all $j < i < k$, $left(x_i) = pos(v)$.

2. For an arc vertex x, let p and q be vertices such that $left(x) = pos(p)$ and $right(x) = pos(q)$. For each arc vertex $x \ne r_i$, add an edge $(Pred(p_x), Succ(q_x))$ and destroy the edges $(Pred(p_x), p_x)$ and $(q_x, Succ(q_x))$.

3. For the arc vertex r_i, destroy the edges as in the previous step but add an edge only if there are no non-extreme edges incident on r_i. Discard all the resulting single-vertex-paths.

Fig 8 illustrates the local modifications suggested by the above algorithm. It is easy to see that two vertices x, y are on a path p produced by the above algorithm if and only if x, y is a separation pair in P_i.

4 Complexity on a CRCW Pram

Maon, Vishkin and Schieber [MSV 86] describe how to find open ear decomposition in at most $O(m \log n)$ operations. Let T be the spanning tree that was used to find open ear decomposition in the above algorithm. The *pos* labeling can be computed using the Euler-tour technique of Tarjan and Vishkin [TV 84] on T by doubling in $O(\log n)$ time with $O(m)$ processors.

Consider the complexity of *reach'* labeling.

The auxiliary multigraph G_e can be constructed as follows. Assume that the processor assigned to the first edge of each ear is responsible for creating the edge between the appropriate vertices of G_e. Recall that in the case of short ears there is only one edge which is also the first edge. Each processor P_i assigned to the first edge of an ear E_i finds the end nodes u and v of E_i. The processor P_i also finds the ear numbers E_j, E_k for which the pair u, v are interval nodes and the values $pos(u, E_j)$ and $pos(v, E_k)$. Assume the corresponding nodes in G_e are e_j, e_k. The processor P_i creates the edge (e_j, e_k) and labels the edge with a 2-tuple $< pos(u, E_j), pos(v, E_k) >$.

The α labeling of the vertices of G_e consists of the following steps. Find the blocks of G_e by the algorithm given by Tarjan and Vishkin [TV 84]. Treat each block B_i separately and construct a spanning tree T_{b_i} in each block using the modification of the connected component algorithm of Shiloach and Vishkin [SV 81]. The articulation point v_{a_i} that represents the minimum ear label in B_i can be found by using the Euler-tour technique. The values x and y can be found by examining the 2-tuple labels of the edges incident on v_{a_i}. These values are broadcast to all the vertices in the block B_i, again, using the Euler-tour technique .

Extending the α labeling to the vertices of G is straightforward. Let P_{e_i} be the processor assigned to the vertex $e_i \in V(G_e)$. As mentioned in the construction of G_e, P_{e_i} can be the same processor that was assigned to the first edge of E_i in G. Now P_{e_i} can broadcast the value of $\alpha(e_i)$ to all the internal vertices of E_i using the Euler-tour technique on the spanning tree T of G.

Theorem 2 gives a relation between $reach'(v)$ and $\alpha(v)$. Therefore $reach'(v)$ can be found by finding the minimum (or maximum as appropriate) of the components of α values of the edges incident on v. This can be performed using the doubling technique by the processors assigned to the edges incident on v.

From the explanation given above, it is easy to see that the $reach'$ labeling can be done in $O(\log n)$ time using $O(m)$ processors. The following argument shows that G' can be constructed within the same resource bounds.

To identify parallel ears do the following. Let $(u, v_1), (u, v_2), ..., (u, v_k)$ be the edges incident on a vertex v and let $E_{x_1}, E_{x_2}, ..., E_{x_k}$ be their ear labels respectively. Let the end nodes of $E_{x_1}, E_{x_2}, ..., E_{x_k}$ be $(u, w_1), (u, w_2), ..., (u, w_k)$ respectively. Now assume the edge list at v is in the increasing order of the pos values of $w_1, w_2, ..., w_k$. This can be achieved by any of the parallel sorting algorithms [AKS 83]. Consider an auxiliary graph G_p where the vertices correspond to the long ears of G. Now the processor P_i assigned to the first edge of each ear E_i examines the neighboring edge with ear label E_j to see if E_i and E_j have the same end nodes. If so, P_i adds an edge between the corresponding vertices in G_p. Now using the connected component algorithm of Shiloach and Vishkin on G_p, we can identify all groups of parallel ears.

Splitting and renaming can be achieved, for example, by making the node labels a 2-tuple: the first component representing the vertex label, and the second representing the property by which the graph is being split. In our case the second component is the ear label of the edge that is incident on that vertex. The edge list for the split version of the graph can be constructed by using a parallel sorting algorithm as explained in [TV 84].

In general the above approach can be adopted for the construction of any locally modified graph of a given graph.

The digraph G_d can be constructed efficiently as shown in [MSV 86]. The same paper also contains a method to compute lca values of non-tree edges in $O(\log n)$ parallel time.

The only nontrivial step in the construction of the P_i is the identification of the arcs (a, b) and (c, d) for each arc (x, y). It is easily seen that (a, b) and (c, d) can be found by doubling for each (x, y) in $O(\log n)$ time using at most $O(m)$ processors.

Finally the implementation of the algorithm that generates separation pairs involves building a locally modified graph. This can be done within the claimed resource bounds as explained before.

References

[AKS 83] M. Ajtai, J. Komlos and E. Szemeredi, "An $O(n \log n)$ sorting network," *Combinatorica 3:1*, 1983, pp. 1-19.

[FT 87] D. Fussell and R. Thurimella, " Finding a sparse graph that preserves biconnectivity," *manuscript*.

[MR 87] G.L. Miller and V. Ramachandran, "A new triconnectivity algorithm and its applications," *Proc. 19th annual STOC* , NY, May 1987, pp. 335-344.

[MSV 86] Y. Maon, B. Schieber and U. Vishkin, "Parallel ear decomposition search (EDS) and ST-numbering in graphs," *VLSI Algorithms and Architectures*, Lecture Notes in Computer Science Vol. 227, 1986, pp. 34-45.

[SV 81] Y. Shiloach and U. Vishkin, "An $O(\log n)$ parallel connectivity algorithm," *J. Algorithms 2*, (1981), pp. 57-63.

[TV 84] R. E. Tarjan and U. Vishkin, "An efficient parallel biconnectivity algorithm," *SIAM J. Computing*, 14, (1984), pp. 862-874.

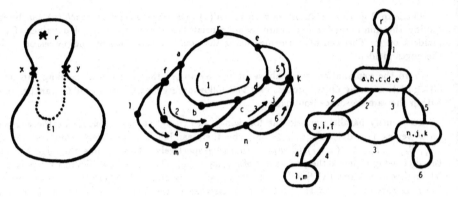

Fig. 1. E_1 contains x,y 　　　Fig. 2. A Graph G and its auxiliary graph G_e

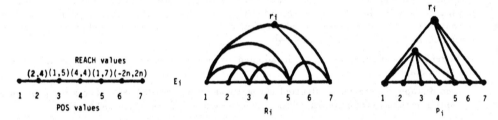

Fig. 3. The reach graph R_i of E_i and its planar version P_i

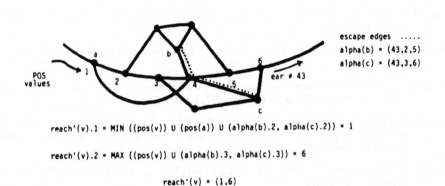

escape edges .....
alpha(b) = (43,2,5)
alpha(c) = (43,3,6)

reach'(v).1 = MIN ((pos(v)) U (pos(a)) U (alpha(b).2, alpha(c).2)) = 1

reach'(v).2 = MAX ((pos(v)) U (alpha(b).3, alpha(c).3)) = 6

reach'(v) = (1,6)

Fig. 4. Computing reach'(v)

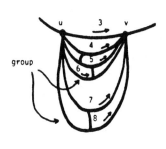

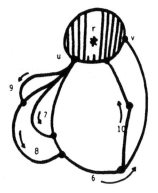

Fig. 5. Parallel Ears and a Group of Parallel Ears

Fig.6. Ear #6 dominates 7,8 and 9

group

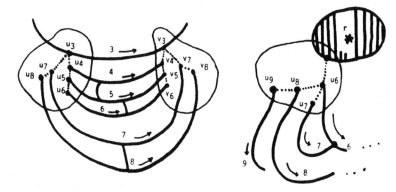

Fig. 7. The graphs of figures 5 & 6 after local modifications

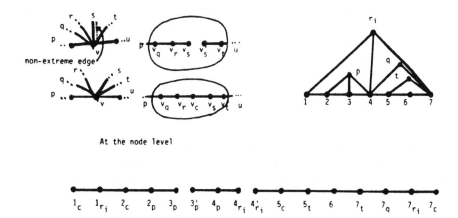

non-extreme edge

At the node level

Fig. 8. After Step 1 of Algorithm Generate-Separation-Pairs

GRAPH EMBEDDINGS 1988:
Recent Breakthroughs, New Directions

Arnold L. Rosenberg

Department of Computer and Information Science
University of Massachusetts
Amherst, MA 01003

Abstract. The past few years have seen a number of results on graph embeddings that deserve to be called *breakthroughs* because of their settling hard open problems and/or their raising important issues that promise to alter the direction of subsequent research on graph embeddings. We describe and discuss several such results.

1. THE FORMAL SETTING

1.1. Background

Except where otherwise noted, we operate with the following basic notion of graph embedding. Let G and H be simple undirected graphs. An *embedding* of G in H is a one-to-one association of the vertices of G with the vertices of H, together with a specification of paths in H connecting the images of the endpoints of each edge of G. The *dilation* of the embedding is the maximum length of any of these G-edge-routing paths. The *expansion* of the embedding is the ratio $|H|/|G|$ of the number of vertices in H to the number of vertices in G. (We need to consider both dilation and expansion since, somewhat counterintuitively, one can sometimes decrease dilation by embedding G in a larger relative of H [HMR].) The *congestion* of the embedding is the maximum number of edges of G that are routed through a single edge (or vertex) of H.

Sections 2-4 of our tour of the frontiers of graph-embedding theory deal with graph embeddings used to model *logical* or *inherent* phenomena: issues like incompatibilities in structure or intersimulatability of graph families. Sections 5 and 6 deal with *physical* phenomena: tolerance to faults and the physical mapping problem.

1.2. Graphs of Interest

The *d-dimensional Hypercube* $Q(d)$ is the graph whose vertex-set is the set $\{0, 1\}^m$ of length-m binary strings and whose edges connect just those pairs of vertices/strings that differ in precisely one bit-position.

The *m-level Butterfly graph* $B(m)$ has vertex-set

$$V_m = \{0, 1, \cdots, m-1\} \times \{0, 1\}^m.$$

The edges of $B(m)$ form *butterflies* between consecutive levels of vertices, with wraparound in the sense that level 0 is identified with level m. Each butterfly connects vertices

$$\langle \ell, \ \beta_0\beta_1 \cdots \beta_{\ell-1}0\beta_{\ell+1} \cdots \beta_{m-1} \rangle \text{ and } \langle \ell, \ \beta_0\beta_1 \cdots \beta_{\ell-1}1\beta_{\ell+1} \cdots \beta_{m-1} \rangle$$

$(0 \leq \ell < m;$ each $\beta_i \in \{0, 1\})$ with vertices

$$\langle \ell + 1(\text{mod } m), \ \beta_0\beta_1 \cdots \beta_{\ell-1}0\beta_{\ell+1} \cdots \beta_{m-1} \rangle$$

and

$$\langle \ell + 1(\text{mod } m), \ \beta_0\beta_1 \cdots \beta_{\ell-1}1\beta_{\ell+1} \cdots \beta_{m-1} \rangle$$

The *height-h complete binary tree* $T(h)$ is the graph whose vertices comprise all binary strings of length at most h and whose edges connect each vertex x of length less than h with vertices $x0$ and $x1$. The ℓth *level* of $T(h)$ $(0 \leq \ell \leq h)$ consists of all vertices/strings of length ℓ.

The *height-h X-tree* $X(h)$ is obtained from the complete binary tree $T(h)$ by adding edges that connect the vertices at each level of $T(h)$ in a path (with the vertices in lexicographic order).

The $s \times s$ *mesh* $M(s)$ is the graph whose vertex-set is

$$\{1, 2, \cdots, s\} \times \{1, 2, \cdots, s\}$$

and whose edges connect vertices $\langle a, b \rangle$ and $\langle c, d \rangle$ just when $|a - c| + |b - d| = 1$.

2. EMBEDDING GRAPHS IN BUTTERFLIES

2.1. Background

It has been known for many years (cf. [PV]) that a large class of algorithms run as fast on the 4-valent Butterfly network $B(m)$ as on the m-valent Hypercube $Q(m)$. Indeed, this observation motivates much of the interest in butterfly-like parallel architectures. Although it was widely believed that the Hypercube was strictly more powerful than the Butterfly – because of its interconnection structure rather than just its larger valence – no one had been able to prove that this was the case. Sandeep Bhatt, Fan Chung, Jia-Wei Hong, Tom Leighton, and I [BCHLR] have recently found such a proof. The following parameters of a graph are central to our result.

- A *1/3-2/3 (vertex-)separator* of G is a set of vertices whose removal partitions G into subgraphs, each having $\geq |G|/3$ vertices; we denote by $\Sigma(G)$ the size of the smallest 1/3-2/3 vertex-separator of G.

- When G is planar and we are given a witnessing planar embedding ϵ, we denote by $\Phi_\epsilon(G)$ the number of vertices in G's largest interior face in the embedding. When ϵ is clear from context, we omit the subscript.

2.2. A Nontrivial Lower Bound

Theorem 1 [BCHLR] *Any embedding of a nontree planar graph G in a Butterfly graph has dilation* $\Omega\left(\frac{\log \Sigma(G)}{\Phi(G)}\right)$. *This bound cannot be improved in general.*

Theorem 1 holds, in fact, for a wide variety of butterfly-like graphs. Two instantiations of this result are particularly interesting.

Corollary 1 *Any embedding of the height-h X-tree $X(h)$ in a Butterfly graph must have dilation* $\Omega(\log h) = \Omega(\log \log |X(h)|)$.

Corollary 2 *Any embedding of the $s \times s$ mesh $M(s)$ in a Butterfly graph must have dilation* $\Omega(\log s) = \Omega(\log |M(s)|)$.

Techniques in [BCLR] and [Gr] show that both of these graphs can be embedded very efficiently – simultaneous dilation $O(1)$ and expansion $O(1)$ – in the Hypercube. Thus, these graphs yield the desired examples.

2.3. Remaining Challenges

The bounds in Corollaries 1 and 2 are tight, to within constant factors [BCHLR], but the method of demonstration raises an important issue. The embeddings that witness the upper bounds embed the subject guest graphs in a complete binary tree, with the indicated dilations, and then embed the tree in the Butterfly, with simultaneous dilation $O(1)$ and expansion $O(1)$. The embeddings thus do not exploit the structure of the Butterfly and, as a consequence, have horrendous congestion. The moral of this story is:

> *Graph-embedding theory is going to have to start paying more attention to the issue of congestion.*

The second question raised Theorem 1 concerns extending the result to other networks. In just the same way that butterfly-like graphs can often simulate Hypercubes with no time loss, *coset graphs* of butterfly-like graphs – the Shuffle-Exchange graph, for instance – can often simulate butterfly-like graphs with little or no time loss [ABR]. Almost certainly, butterfly-like graphs are more powerful than such coset graphs in general; but, no proof to this effect has yet appeared. In its strongest form the problem is:

Problem 1 *Do there exist n-vertex graphs that are embeddable in Butterflies with dilation $O(1)$ but that require dilation $\Omega(\log n)$ when embedded in the Shuffle-Exchange graph?*

3. DYNAMIC GRAPH EMBEDDINGS

3.1. Background

In 1986, Sandeep Bhatt, Fan Chung, Tom Leighton, and I [BCLR] proved, via a very complicated embedding strategy, that every binary tree could be embedded efficiently – with simultaneous dilation $O(1)$ and expansion $O(1)$ – in the Hypercube. One major motivation for seeking this result was to be able to argue that Hypercube networks could efficiently execute any divide-and-conquer algorithm. The complication of our embedding was due to three stringent demands, that are inherent in our notion of graph embedding, but that one might argue are "overkill," given our motivation. We insisted

1. that we assign tasks to processors using *global* knowledge of how each execution of the algorithm unfolds

2. that the execution time of the algorithm, as measured by dilation, be $O(1)$

3. that the entire algorithm/tree reside in the network throughout execution, with each task occupying its own private processor.

3.2. A Dynamic Embedding

Sandeep Bhatt and Jin-Yi Cai [BC] have abandoned these demands and have discovered a very simple algorithm that dynamically maps a growing (and shrinking) binary tree onto the Hypercube, with small dilation; however, their mapping need not be an embedding: it may map several tree vertices onto the same Hypercube vertex, but only boundedly many.

Theorem 2 [BC] *Any n-vertex binary tree can be dynamically mapped into the Hypercube $Q(\lceil \log n \rceil)$, with dilation $O(\log \log n)$; there is a constant c such that, with probability $1 - n^{-c}$, only $O(1)$ tree vertices are mapped to any single Hypercube vertex. No randomized algorithm produces a mapping with smaller dilation.*

3.3. Remaining Challenges

Most obviously, one would like to generalize Theorem 2 to guest graphs other than trees and host graphs other than Hypercubes. It is not clear that the Bhatt-Cai strategy of mapping via random walks will admit either generalization.

I have recently begun to look at a genre of "embedding" that abandons only the third of the above demands. A similar but technically distinct relaxation appears in [Fe] and, with different motivation, in [Be, GH]. None of these approaches has yet produced any truly sophisticated embeddings; but the motivation is great, and I remain optimistic.

4. NEW RESULTS ON BOOK-EMBEDDINGS

4.1. Background

A *book* is a finite set of half-planes (the *pages*) that share a common boundary (the *spine*). One *embeds a graph in a book* by ordering the vertices of the graph along the spine and assigning the edges of the graph to pages in such a way that each edge lies on just one page, and edges that lie on the same page do not cross. One's goal is to embed one's graph in a book having as few pages as possible, and as small a cutwidth as possible on each page. Thus, we refer to the *pagenumber* of a graph G, i.e., the number of pages in the smallest book in which G can be embedded, and the *pagewidth* of G, i.e., the smallest possible maximum cutwidth on any page in a book in which G can be embedded.

4.2. Pagenumber and Graph Size

In an attempt to find a nontrivial predictor of the pagenumber of a graph, Fan Chung, Tom Leighton, and I [CLR] proved the following.

(a) *Every n-vertex d-valent graph has pagenumber $O(d\sqrt{n})$.*

(b) *For every $d > 2$ and all large n, there are n-vertex d-valent graphs whose pagenumber is at least*

$$\Omega\left(\frac{\sqrt{n}}{n^{1/d}\log^2 n}\right)$$

Except when d is in the neighborhood of $\log n$, these upper and lower bounds are quite far apart.

Seth Malitz [Ma] has narrowed the gap considerably; indeed, when the valence $d \geq \log n$, Malitz's bounds are tight to within constant factors.

Theorem 3 [Ma] **(a)** *Every e-edge graph has pagenumber $O(\sqrt{e})$.*
(b) *For every $d > 2$ and all large n, there are n-vertex d-regular graphs whose pagenumber is at least*

$$\Omega\left(\frac{\sqrt{dn}}{n^{1/d}}\right)$$

4.3. Pagenumber and Graph Structure

Also seeking a nontrivial predictor of the pagenumber of a graph, Mihalis Yannakakis [Ya] and Lenny Heath and Sorin Istrail [HI] have made a giant stride toward correlating the pagenumber of a graph with its genus:

[Ya] *Every planar graph has pagenumber at most 4. There exist planar graphs whose pagenumber is 4.*

[HI] *Every graph of genus $g \geq 1$ has pagenumber $O(g)$. For each such g, there exist graphs of genus g whose pagenumber is $\Omega(\sqrt{g})$.*

Seth Malitz [Ma] has built a sophisticated superstructure on top of the framework of [HI] to lower the upper bound in the latter result to within a constant factor of the lower bound.

Theorem 4 [Ma] *Every graph of genus $g \geq 1$ has pagenumber $O(\sqrt{g})$.*

4.4. Pagenumber vs. Pagewidth

From my vantage point, the greatest remaining challenge in the area of book-embeddings resides in the following pairs of facts, which suggest that it is generally impossible to minimize pagenumber and pagewidth simultaneously.

1. *Every d-valent n-vertex graph with pagenumber 1 is outerplanar* [BK]*, hence has cutwidth $O(d \log n)$.*

2. *There exist trivalent n-vertex outerplanar graphs every 1-page book-embedding of which has pagewidth $\Omega(n)$* [CLR]*.*

3. *Every d-valent n-vertex graph with pagenumber 2 is planar* [BK]*, hence has cutwidth $O(d\sqrt{n})$.*

4. *There exist 4-valent n-vertex planar graphs every 2-page book-embedding of which has pagewidth $\Omega(n)$* [CLR]*.*

Lenny Heath [He] has taken the sting out of the first pair of assertions, by showing that increasing pagenumber by 1 allows one to approach minimum cutwidth.

Theorem 5 [He] *Every d-valent n-vertex outerplanar graph admits a 2-page book-embedding with pagewidth* $O(d \log n)$.

The challenge alluded to above is to find an analog of Heath's result that takes the sting out of the second pair of facts also.

Problem 2 *Does there exist a constant c such that every d-valent n-vertex 2-page-embeddable graph admits a c-page book-embedding with pagewidth* $O(d\sqrt{n})$?

Indeed it is not even known that every 2-page-embeddable d-valent n-vertex graph admits a $(\log n)$-page book-embedding with pagewidth $O(d\sqrt{n})$.

If the open problem has an affirmative solution, then the next challenge will be to show that one can *always* find a close-to-minimum-cutwidth layout of a graph G via a book-embedding using only modestly more than pagenumber(G) pages.

5. SALVAGING HYPERCUBES

5.1. Background

The many techniques that have been proposed for rendering a parallel architecture tolerant to faults suffer from one of two deficiencies that render them inapplicable to "dense" interconnection networks, such as Butterflies and Hypercubes.

1. Some strategies merely maintain a degree of connectivity in the network. This is far too modest a goal, since parallel algorithms typically exploit the structure of the network, not just its degree of connectivity.

2. Strategies that preserve network structure typically have such high overhead that they are useful only for sparse networks, like meshes and trees.

To the extent that this assessment is valid, the only viable approach to fault tolerance for dense networks is to try to *salvage* the network, in the sense that [GE, LL] salvage meshes, i.e., to embed a copy of a smaller version of the network in the surviving portion of the current version.

5.2. Salvaging a Faulty Hypercube

Johan Hastad, Tom Leighton, and Mark Newman [HLN] have recently discovered a disarmingly simple technique for salvaging the Hypercube efficiently. We paraphrase only the simplest of their results, which differ in the assumed degree of debilitation of the Hypercube due to the faults.

Theorem 6 [HLN] *Let the vertices of $Q(d)$ be colored* red *and* green *at random, independently, with $p = Pr(\text{red}) < 1/2$. There is a deterministic algorithm that, with probability $1 - 2^{-c_p d}$, embeds $Q(d-1)$ in the* green *vertices of $Q(d)$ with dilation 3 and congestion $O(\log n)$, where c_p is a constant depending only on p.*

5.3. Remaining Challenges

Given the structural kinship of the Butterfly to the Hypercube, and of the Shuffle-Exchange graph to the Butterfly, one would hope that a strategy similar to that in [HLN] would yield the following.

Problem 3 *Can one find an analog of Theorem 6 that embeds $B(m-1)$ in $B(m)$ (or, the order-$(n-1)$ Shuffle-Exchange graph in the order-n Shuffle-Exchange graph) with dilation $O(\log \log n)$?*

The conjectured dilation in this problem is due to the 4-valence of the sparse graphs, in contrast to the $(\log n)$-valence of the Hypercube.

6. THE PHYSICAL MAPPING PROBLEM

6.1. Background

Even when techniques for tolerating faults in networks do not incur excessive overhead, they create a problem that has received little attention in the literature. The problem is exemplified by the following scenario. Say that we wish to realize a complete binary tree on our idealized architecture and that after fabrication and testing, we find that 510 of the processors of our physical architecture are free of faults. Since the largest complete binary tree we can realize on 510 PEs has only 255 nodes, we must decide which 255 fault-free processors to use. It is quite likely that our choice of processors will have an effect on the run-time efficiency of the tree; hence, we do not want to make the choice haphazardly. We need to find an efficient, efficiency-enhancing way to map the processors of our logical architecture on the surviving processors of our physical architecture.

6.2. An Instance of Physical Mapping

Motivated by the relationship between the book-embedding problem and the DIO-GENES methodology for designing fault-tolerant processor arrays [Ro], Lenny Heath, Bruce Smith, and I [HRS] studied the physical mapping problem for graphs that are embedded in books, considering both the average cost of an embedding and the maximum-wire-run cost *MCOST*. For the latter case, we established the following.

Say that we are given a book-embedding Λ of an m-vertex graph G, together with a linear sequence Π of "processors," $n \geq m$ of which are fault-free, hence available to "receive" the vertices of G. We wish to *assign* the vertices of Λ in an order-preserving manner to the fault-free "processors" in Π. The *MCOST* of an assignment is just the maximum distance in the sequence Π between the endpoints of any edge of G.

Theorem 7 [HRS] *Given a book-embedding Λ and a sequence of "processors" Π, there is an algorithm that finds an MCOST-optimal assignment of the vertices of Λ to the "processors" of Π, that operates in time*

$$O(m \cdot (n - m) \cdot \log(m \cdot (n - m)) \cdot \log M),$$

where M is the largest inter-"processor" distance.

When one allows "shortcuts" in the sequence Π, the general problem of finding *MCOST*-optimal assignments becomes *NP*-complete, although there are efficient approximation algorithms; cf. [HRS].

The obvious remaining challenge is to identify other genres of layout strategies where one can attack the physical mapping problems.

ACKNOWLEDGMENT. This work was supported in part by NSF Grant DCI-87-96236. It is a pleasure to acknowledge the contribution – to the field as well as to this paper – of all the authors whose work is cited here.

7. REFERENCES

[ABR] F. Annexstein, M. Baumslag, A.L. Rosenberg (1987): Group-action graphs and parallel architectures. Submitted for publication.

[Be] F. Berman (1983): Parallel computation with limited resources. *Johns Hopkins Conf. on Information Sciences and Systems.*

[BK] F. Bernhart and P.C. Kainen (1979): The book thickness of a graph. *J. Comb. Th. (B) 27*, 320-331.

[BC] S.N. Bhatt and J.-Y. Cai (1987): Take a walk, grow a tree. Typescript, Yale Univ.

[BCLR] S.N. Bhatt, F.R.K. Chung, F.T. Leighton, A.L. Rosenberg (1986): Optimal simulations of tree machines. *27th IEEE Symp. on Foundations of Computer Science,* 274-282.

[BCHLR] S.N. Bhatt, F.R.K. Chung, J.-W. Hong, F.T. Leighton, A.L. Rosenberg (1988): Optimal simulations by Butterfly networks. *20th ACM Symp. on Theory of Computing,* to appear.

[CLR] F.R.K. Chung, F.T. Leighton, A.L. Rosenberg (1987): Embedding graphs in books: A layout problem with applications to VLSI design. *SIAM J. Algebr. Discr. Meth.* *8*, 33-58.

[Fe] M.R. Fellows (1985): *Encoding graphs in graphs.* Ph.D. Dissertation, Univ. California at San Diego.

[FP] J. Friedman and N. Pippenger (1986): Expanding graphs contain all small trees. Typescript, IBM Almaden Research Center.

[Gr] D.S. Greenberg (1987): Minimum expansion embeddings of meshes in hypercubes. Tech. Rpt. DCS/RR-535, Yale Univ.

[GE] J.W. Greene and A. El Gamal (1984): Configuration of VLSI arrays in the presence of defects. *J. ACM 31*, 694-717.

[GH] A.K. Gupta and S.E. Hambrusch (1988): Embedding large tree machines into small ones. *MIT Conf. on Advanced Research in VLSI.*

[HLN] J. Hastad, F.T. Leighton, M. Newman (1987): Reconfiguring a hypercube in the presence of faults. *19th ACM Symp. on Theory of Computing.*

[He] L.S. Heath (1987): Embedding outerplanar graphs in small books. *SIAM J. Algebr. Discr. Meth. 8*, 198-218.

[HI] L.S. Heath and S. Istrail (1987): The pagenumber of genus g graphs is $O(g)$. *19th ACM Symp. on Theory of Computing.*

[HRS] L.S. Heath, A.L. Rosenberg, B.T. Smith (1988): The physical mapping problem for parallel architectures. *J. ACM,* to appear.

[HMR] J.-W. Hong, K. Mehlhorn, A.L. Rosenberg (1983): Cost tradeoffs in graph embeddings. *J. ACM 30*, 709-728.

[LL] F.T. Leighton and C.E. Leiserson (1985): Wafer-scale integration of systolic arrays. *IEEE Trans. Comp., C-34*, 448-461.

[Ma] S.M. Malitz (1988): Embedding graphs in small books. Manuscript, MIT.

[PV] F.P. Preparata and J.E. Vuillemin (1981): The cube-connected cycles: a versatile network for parallel computation. *C. ACM 24*, 300-309.

[Ro] A.L. Rosenberg (1983): The Diogenes approach to testable fault-tolerant arrays of processors. *IEEE Trans. Comp., C-32*, 902-910.

[Ya] M. Yannakakis (1986): Four pages are necessary and sufficient for planar graphs. *18th ACM Symp. on Theory of Computing,* 104-108.

Simulating Binary Trees on Hypercubes

by

Burkhard Monien[1]

Math. and Computer Science

University of Paderborn

4790 Paderborn, W. Germany

I. Hal Sudborough

Computer Science Program

University of Texas at Dallas

Richardson, TX 75083-0688 U.S.A.

Abstract

We describe how to embed an arbitrary binary tree with dilation 3 and O(1) expansion into a hypercube. (In fact, we show that all binary trees can be embedded into their optimal hypercube with dilation 3, provided that all binary trees with no more than B vertices, for some fixed number B, can be embedded with dilation 3.) We also show how to embed all binary trees into their optimal hypercube with dilation 5.

I. Introduction and Overview

Hypercube networks are popular parallel computer architectures. A hypercube of dimension n, denoted by $Q(n)$, is an undirected graph of 2^n vertices labelled by the integers between 0 and 2^n-1. There is an edge between two vertices when the binary representation of their labels differ in exactly one bit position. Their popularity is primarily due to the regularity of their structure, the relatively small number of connections at each processor, and the relatively small distance traversed in getting from one side of the network to the other. An important consideration in using these networks is their ability to execute programs written for simpler structures without lengthy communication delay.

This paper considers the simulation of binary trees. Such trees reflect common data structures and the type of program structure found in common divide-and-conquer recursive

---

[1] The work of this author was partly done during a visit to UTD and was supported by a grant from the German Research Association (DFG).

routines. We investigate embeddings of binary trees into hypercubes, i.e. one-to-one mappings of the vertices of the tree into the nodes of the hypercube. Given an embedding E, its *dilation* is the maximum distance in the hypercube between images of adjacent vertices of the tree. Our goal is to minimize the dilation, as the dilation corresponds to the number of clock cycles needed in the hypercube network to communicate between formerly adjacent processors in the tree. It is also important to minimize the size of the network. The *expansion* of such an embedding E is the ratio of the size of the hypercube network divided by the size of the tree embedded. Given a binary tree T with n vertices, we call a hypercube with the smallest number of points of any hypercube that T can be embedded into, namely $Q(\lceil \log_2 n \rceil)$, its *optimal hypercube* and the next largest hypercube is called *next-to-optimal*.

The ability of boolean hypercubes to simulate binary tree structures was previously studied in several papers [BCLR], [HL], [H], [MSUW], [N], and [BI]. These papers show that complete binary trees can be embedded into their optimal hypercubes with dilation 2 (in fact, with just one pair of adjacent vertices in the tree at distance 2 in the hypercube and all other pairs of adjacent vertices mapped to adjacent hypercube nodes) and into their next-to-optimal hypercubes with dilation 1. It was previously shown that complete binary trees are not subgraphs of their optimal hypercubes [BI] [H] [N], so dilation ≥ 2 is required. It is an open question whether all binary trees can be embedded into its optimal hypercube with dilation 2 or into its next-to-optimal hypercube with dilation 1. The ability of boolean hypercubes to simulate meshes of various dimensions has also been studied in several papers [BMS], [CC], [C], using binary reflected gray codes and refinements of techniques to *square up a grid* discussed in [AR] and [E].

A principal result in [BCLR] is that arbitrary binary trees can be embedded into hypercubes with constant expansion and dilation 10. We improve this result. We give an embedding with dilation 3 and O(1) expansion which maps an arbitrary binary tree into a hypercube. In fact, we show that any binary tree can be embedded into its optimal hypercube with dilation 3, provided that all binary trees with no more than B vertices can be so embedded, for some integer bound B. We also give a proof that all binary trees can be embedded into their optimal hypercubes with dilation 5.

As a lower bound it is only known that some binary trees (such as complete ones) are not subgraphs of their optimal hypercubes. The hypercube is a bipartite graph in which each of its two color classes contains exactly half of its vertices. Following [H] we call a bipartite graph *balanced* if it satisfies this property, i.e. divides into equal size color classes. A graph with 2^n vertices can only be a subgraph of $Q(n)$ if it is balanced and this implies that the complete binary tree of height h is not a subgraph of its optimal hypercube for h $\geq$ 2. As yet we don't know whether there exists a binary tree which is balanced and which is not a subgraph of its optimal hypercube. The situation is different for trees of degree larger than 3. There exists a balanced tree of degree 4 which *requires* dilation 2 for an embedding into its optimal hypercube [HL]. I. Havel has conjectured in [H] that every balanced binary tree is a subgraph of its optimal hypercube. This conjecture was proved to be true ([HL], [MSUW]) so far only for the class of trees with *search number* 2, i.e. the class of trees often called *caterpillars*.

II. Collision Trees

We begin by describing a mapping of complete binary trees into hypercubes so that (a) more than one hypercube location is assigned to each tree vertex (b) only hypercube locations at small distance from each other are assigned to adjacent nodes in the tree. The specific number of hypercube locations assigned to a vertex is represented by an integer weight at that vertex.

A *binary collision tree of height m* is a complete binary tree of height m with integer node weights. Our purpose in defining a binary collision tree T' is to consider it as a host for embedding a binary tree T. It is used therefore in a similar way to that of *thistle trees* in [BCLR]. We want to map T into T' with small dilation so that a node in T' with weight k is the image of at most k nodes of T. (Thus nodes in T *collide* into the same node of T'.) Our description of these results contains as a second step an embedding of T' into Q such that each node with weight k is assigned k distinct hypercube locations. Thus we have a two stage process: (1) embed a binary tree T into a binary collision tree T' of appropriate height, and (2) embed T' into its optimal hypercube Q.

We use standard notation for representing nodes in a binary tree. Each string $\gamma \in \{0,1\}^*$ represents a unique node. The empty string denotes the root and, if $\gamma \in \{0,1\}^*$ denotes a vertex x, then $\gamma 0$ denotes the left child of x and $\gamma 1$ denotes the right child of x.

Let us consider now the second phase of this process, namely embedding a collision tree T' of height m into its optimal hypercube. We consider first the collision tree T' where a node at level i receives weight m-i, call this a *type 1* collision tree. The root of T' has weight m, so we assign it m hypercube locations. They are the nodes in the set $\{1\alpha \,|\, \alpha \in 0^*10^* \text{ and } |\alpha| = m\}$. In general, the node γ, which has weight m$-|\gamma|$, is assigned the locations in the set $\{\gamma 1\alpha \,|\, \alpha \in 0^*10^* \text{ and } |\alpha| = m-|\gamma|\}$.

For every pair of nodes x and y in T', the locations assigned to x are distinct from the locations assigned to y. If x and y are at different levels in T', say at levels i < j, respectively, then every location assigned to x has a single occurrence of a 1 in its last m-i bits and every location assigned to y will have at least two occurrences of a 1 in its last m-i bits. If x and y are at the same level, then they are assigned nodes in $S(\gamma)$ and $S(\beta)$, for distinct strings γ and β of the same length. Hence the sets of strings must be disjoint.

Observe that the maximum distance in the hypercube between locations assigned to *cousins*, namely a pair of nodes with a common grandparent in T', is 4. Such nodes are assigned, for some $b_1 b_2 b_3 b_4 \in \{0,1\}$ and $\gamma \in \{0,1\}^*$, the sets $S(\gamma b_1 b_2) = \{\gamma b_1 b_2 1\alpha \mid \alpha \in 0^*10^*$ and $|\alpha| = m-|\gamma|-2\}$, and $S(\gamma b_3 b_4) = \{\gamma b_3 b_4 1\alpha \mid \alpha \in 0^*10^*$ and $|\alpha| = m-|\gamma|-2\}$. The distance in the hypercube between any two strings in these sets is at most 4.

In a similar manner one can verify that, for any node x in a collision tree T', the distance is at most 5 between locations assigned to x and any locations assigned to its grandchildren. That is, the maximum distance, for any $b_1, b_2 \in \{0,1\}$ and $\gamma \in \{0,1\}^*$, between positions in the set $S(\gamma b_1 b_2) = \{\gamma b_1 b_2 1\alpha \mid \alpha \in 0^*10^*$ and $|\alpha| = m-|\gamma|-2\}$, and positions in the set $S(\gamma)$ $= \{\gamma 1\alpha \mid \alpha \in 0^*10^*$ and $|\alpha| = m-|\gamma|\}$ is at most 5.

We consider a *type 2* collision tree $T^{(2)}$ in which each node at level i has weight $2(m-i)$. The root is assigned the nodes in the set $\{1\alpha b \mid \alpha \in 0^*10^*$ and $|\alpha| = m-1$ and $b \in \{0,1\}\}$. In general, a node $\gamma \in \{0,1\}^*$ is assigned positions in the set $\{\gamma 1\alpha b \mid \alpha \in 0^*10^*$ and $|\alpha| = m-|\gamma|$ and $b \in \{0,1\}\}$.

Using similar arguments as before, one can verify that no two nodes are assigned the same hypercube position and that the maximum distance in the hypercube (a) between positions assigned to a node and its grandchild is 6 and (b) between positions assigned to cousins is 5.

The distances indicated in the last paragraph are larger than desired, but we use special circumstances to decrease them. All edges in the tree T connect vertices that get mapped to nodes in the collision tree that are cousins or where one is a grandparent of the other. We can achieve dilation 3 by judiciously picking the hypercube positions at these nodes to be the images of neighbors in T. If x is mapped to the position $\gamma b_1 b_2 1\alpha b_3$, then the node mapped to its cousin is assigned the position $\gamma b_4 b_5 1\alpha b_6$, for some $b_1,...,b_6$, where b_3 may be unequal to b_6. It is easily seen that the distance between these positions is 3. If x is mapped to the position $\gamma 1\alpha b$, then the nodes mapped to its grandchildren are assigned the positions $\gamma b_1 b_2 1\beta b$ and $\gamma b_3 b_4 1\beta b$, respectively, where b_1, b_2, b_3, b_4, and β are such that $b_1 b_2 1\beta$ disagrees with 1α in at most 3 bits and $b_3 b_4 1\beta$ disagrees with 1α in at most 3 bits. Thus we can achieve a dilation 3 embedding of a binary tree T into a hypercube by simply embedding T into a collision tree $T^{(2)}$ so that adjacent vertices of T either (a) get mapped to cousins in $T^{(2)}$ or (b) get mapped to a node and its grandchild in $T^{(2)}$.

III. Partitioning a Binary Tree

We consider now methods for partitioning binary trees that form an important part of our embedding theorems. A set S of vertices in a tree T is *collinear* if there is a simple path (i.e. without repeated vertices) that includes all vertices in S. Let S_0 and S_1 be collinear sets of vertices in T. S_0 and S_1 are a *matched pair* if for each vertex in S_0 (S_1) there is exactly one vertex in S_1 (S_0) to which it is connected. S_0 and S_1 are a *nearly matched pair* if either it is a matched pair or there is at most one vertex $x \in S_i$ ($i=0$ or $i=1$), adjacent to exactly two vertices of degree 1 in S_{1-i}, say y and z, such that $S_i-\{x\}$, $S_{1-i}-\{y,z\}$ is a matched pair. A nearly matched pair of sets S_0 and S_1 are of *size k* if either S_0 or S_1 has k elements and the other has k or k-1 elements. The set of edges connecting vertices in S_0 with those in S_1 is called the *matching* and the set of vertices in S_0 and S_1 incident to edges in the matching are called the *matched vertices*.

Lemma 1. Let T be a n vertex tree with root r and let $\Delta \leq n/2$. Then there is a $k \leq \lceil \log_3(2\Delta+1) \rceil$ and a nearly matched pair of sets S_0 and S_1 of size k such that the deletion of all edges in the matching results in forests $F(S_0)$ and $F(S_1)$, containing S_0 and S_1, respectively, where $F(S_1)$ has Δ points, $F(S_0)$ has $n-\Delta$ points, and $F(S_0)$ contains r.

We observe that the partitioning of T described in Lemma 1 is accomplished by following a path away from the root always choosing to go in the direction of an appropriately large subtree. Along this path some subtrees are pruned off to be on one side of the partition and the remaining subtrees are left to be on the other side. The nearly matched pair of sets is simply the set of all pairs of vertices encountered, at least one of which lies on the path, which are on opposite sides of the partition, and which are connected by an edge in the original tree T. The partitioning, described in the proof of Lemma 1, but not included here due to space limitations, creates forests $F(S_0)$ and $F(S_1)$ with each tree in the forest contains one or more matched vertices. We describe in Lemmas 2 and 3 ways to partition trees with more than one *designated* vertex, where a vertex is designated if, for example, it is one of the matched vertices from a use of Lemma 1. (We will define other meanings for the term designated vertices later on.) Let T_i be a tree in either $F(S_0)$ or $F(S_1)$. T_i is *an interval* if it contains two designated vertices and *a segment* if it has more than two designated vertices. The following lemmas describe how to partition intervals and segments. Note that Lemma 1 already describes how to partition a tree with one designated vertex, i.e. one simply chooses the designated vertex to be the root.

Lemma 2. (*partition of an interval*) Let T be an interval with n vertices and two designated vertices v_0 and v_1 and let $\Delta \leq n/2$. Then there exists a $k \leq 1 + \lceil \log_3(2\Delta+1) \rceil$ and a nearly matched pair of sets S_0 and S_1 of size k such that the deletion of all edges in the matching results in forests $F(S_0)$ and $F(S_1)$, containing S_0 and S_1, respectively, where $F(S_1)$ has Δ points, $F(S_0)$ has $n-\Delta$ points, $F(S_0)$ contains v_0 and $F(S_1)$ contains v_1.

Lemma 3. (*partition of a segment*) Let T be a segment with n vertices and let A be the set of all designated vertices in T. Then there exist at most two pairs of nodes $x,y \in A^c$ and $a,b \in A$ such that the deletion of the edges $\{x,a\}$ and $\{y,b\}$ partitions T into forests F_1 and F_2 (each consisting of at most two segments) such that (a) F_1 and F_2 both have at most half of the designated vertices, (b) the nodes x and b belong to F_1 and the nodes y and a belong to F_2, and (c) there is a tree in the forest F_1, with no designated vertices, such that removing it from F_1 results in a forest $F_1{}'$ such that $F_1{}'$ and F_2 each have at most $\lceil n/2 \rceil$ vertices.

We note that the collinearity of the vertices in the nearly matched pair of sets S_0 and S_1, discussed in Lemmas 1 and 2, is needed in the proof of Lemma 3. It is also needed in our algorithm for embedding a binary tree into a collision tree in order to ensure that certain cases suffice. We now describe our use of Lemmas 1-3 to obtain a small dilation mapping of a binary tree into a collision tree T' of appropriate height.

V. Mapping a Binary Tree to a Collision Tree

Let T be a binary tree with n vertices. By Lemma 1 we partition T into two forests, say $F(0)$ and $F(1)$, each having at most n/2 vertices with at most $\lceil \log_3(2\Delta+1) \rceil$ pairs of matched vertices, with half of the matched vertices in $F(0)$ and half in $F(1)$. We assign $F(0)$ to the subtree of T' with root 0 and $F(1)$ to the subtree with root 1. Now we want to partition each of these forests into two smaller forests, each with at most n/4 vertices and no more than half the total number of matched nodes. To do this we view each forest as a segment and use Lemma 3 to obtain a partition. For instance, consider $F(0)$. By Lemma 3, $F(0)$ can be partitioned into forests F_1, F_2 (each consisting of at most two segments) such that (a) each of F_1 and F_2 have at most half of the matched vertices and (b) F_1 contains a tree T, with no matched vertices, such that removing it from F_1 results in a forest $F_1{}'$ such that both $F_1{}'$ and F_2 have at most $\lceil n/2 \rceil$ vertices. Partition T into two trees, say T_1 and T_2, using Lemma 1, so that two forests $F(00)$ and $F(01)$ can be created, consisting of (a) $F_1{}'$ and T_1 and (b) F_2 and T_2, respectively. Each has at most n/4 vertices and half of the matched vertices in $F(0)$. $F(00)$ and $F(01)$ are assigned to the subtrees of T' with the indicated root, i.e. $F(00)$ is assigned to the subtree with root 00. The vertices matched in the first partition are now mapped to locations at the roots of these subtrees.

In general, at any node in T' there are matched nodes from a partition made two levels higher, matched nodes from a partition made one level higher, and matched nodes from a partition made at the current level. To distinguish these matched nodes we refer to them as *grandparent-matched, parent-matched* or simply *matched* depending on how many levels up in T' the matching was made. A vertex in T that is actually mapped to a position at a node in T' is, in a similar manner as described above, *grandparent-fixed*, *parent-fixed*, or simply *fixed*. The basic idea in our mapping of vertices of T to positions in T' can be expressed as follows. Let $F(\gamma)$ be a forest assigned to the subtree of T' with root γ. Our objective is to map the grandparent-matched and the neighbors of grandparent-fixed nodes to γ and then partition the remaining unmapped vertices into forests $F(\gamma 0)$ and $F(\gamma 1)$ such that each of these forests has half the total number of remaining vertices and simultaneously half the parent- matched vertices, half the matched vertices, half the neighbors of parent-fixed vertices, and half the neighbors of fixed-vertices. The term *designated vertex* in Lemmas 1-3, we can now say, refers to these types of vertices: neighbors of parent-fixed, neighbors of fixed, parent-matched, and matched.

(a) map all grandparent-matched and all neighbors of grandparent-fixed nodes to γ,

(b) consider the subforest of trees in $F(\gamma)$ containing all parent-matched vertices (this will be the set of trees containing the matched vertices from one side of the partition done at the parent of γ) to be a single segment and partition it, using Lemma 3, into forests F_1, F_2 (each consisting of at most two segments) such that (a) each of F_1 and F_2 have at most half of the parent-matched vertices and (b) F_1 contains a tree T, with no parent-matched vertices, such that removing it from F_1 results in a forest F_1' such that both F_1' and F_2 have at most $\lceil n/2 \rceil$ vertices.

(c) place F_1' in $F(\gamma 0)$ and F_2 in $F(\gamma 1)$ and let T_* be a tree among all those remaining in the forest $F(\gamma)$, including the tree T removed from F_1 to form F_1', with the largest number of vertices,

(d) using Lemma 4, which will shortly be described, place all remaining trees in $F(\gamma)$, except for T_*, in either $F(\gamma 0)$ or $F(\gamma 1)$, so that the difference in the number of nodes in $F(\gamma 0)$ and $F(\gamma 1)$, respectively, is at most the number of vertices in T_* and each forest has half the number of parent-matched, neighbors of parent-fixed, and neighbors of fixed vertices, and

(e) using Lemma 1 or Lemma 2, partition T_* into two forests and then add them to $F(\gamma 0)$ and $F(\gamma 1)$, respectively, so that $|F(\gamma 0)|$ and $|F(\gamma 1)|$ differ by at most one (the nearly matched pair of sets S_0 and S_1 created by this partition is divided by placing S_0 and S_1 in $F(\gamma 0)$ and $F(\gamma 1)$, respectively).

Lemma 4. Let F be a n vertex forest in which some of the vertices are colored with one of at most k colors, for some $k \geq 2$. Let every tree of F contain exactly one colored vertex and let n_i be the number of vertices of color i, for each i ($1 \leq i \leq k$). Then there is a $k \leq \lceil \log_3(n+1) \rceil$ and a nearly matched pair of sets S_0 and S_1 of size k, such that $F(S_0)$ and $F(S_1)$ both have at most $\lceil n/2 \rceil$ vertices and each one contains, for each i ($1 \leq i \leq k$), at most $\lceil n_i/2 \rceil$ vertices of color i.

We will in fact need a stronger version of Lemma 4. The stronger version is that, for every i ($1 \leq i \leq k$), let there be some set P_i of forbidden pairs $\{u,v\}$ of vertices with color i. We may assume that every node appears in a forbidden pair at most once. The generalization of Lemma 4 asserts that, in addition to what is indicated above, for each forbidden pair $\{u,v\}$, the nodes u and v are placed in opposite forests, namely $F(S_0)$ and $F(S_1)$.

Lemma 4 is accomplished by a rather straightforward idea. Let F be an arbitrary forest containing trees in which some vertices are colored white and others are colored black. Let T_* be the largest tree in F. Consider the forest $F' = F - \{T_*\}$. Divide the forest F' into (a) a forest F_1 of all trees containing a black vertex and (b) a forest F_2 of all trees containing a white vertex. Then we partition the forests into $F(0)$ and $F(1)$ by taking two trees from F_1 (or two trees from F_2) and adding the larger one of the two to whichever of $F(0)$ or $F(1)$ has the smallest number of vertices (and the smaller to the larger one). This way the colors are divided evenly and when the forests F_1 and F_2 are completely exhausted the difference between the number of vertices in $F(0)$ and the number of vertices in $F(1)$ is at most the number of vertices in T_*. We can then make the number of vertices in $F(0)$ and $F(1)$ differ by at most 1 by partitiong T_*. (The partition is done using either Lemma 1 or 2.)

We note that the two types of colored vertices, which Lemma 4 refers to, are (a) the neighbors of fixed vertices and (b) the neighbors of parent-fixed vertices. Lemma 4 only needs to describe how to partition these two types of vertices and not grandparent-matched vertices, as Lemma 3 is used earlier in the algorithm to partition them. (Furthermore, the nearly matched pair of sets S_0 and S_1, created in Lemma 4 to split T_*, are divided with S_0 going to one side of the partition and S_1 going to the other.) In fact, our construction creates trees which may either have one or two of the indicated colored vertices. However, it is not difficult to show the desired separation result holds also in this case.

The theorem will now follow from simply a calculation of the number of nodes that are mapped to each point in the collision tree. Let f(r) denote the number of vertices in the binary tree that are mapped to a level r node in the collision tree. As indicated in step (a) of the procedure, the vertices mapped to a node at level r are the grandparent-fixed and grandparent-matched vertices. There are clearly $f(r+2)$ grandparent-fixed vertices, as these are the vertices mapped to a $r+2$ level node of the collision tree. If we begin with a binary tree T with at most 2^s vertices then we embed into a height $s+1$ binary collision tree. If we count levels in increasing order from the leaves, then at level r we have at each node a forest of 2^r vertices. By Lemma 1, therefore, we need at most a set of size $\lceil \log_3 (2^r+1) \rceil$ for a partition of some tree. So, there are at most $\lceil \log_3(2^{r+2}+1) \rceil$ grandparent-matched vertices. As nearly matched sets of vertices at the grandparent are of the same size, and one goes to each child and then each are divided in half again to reach the grandchildren, the result is that there are $\frac{1}{2}\lceil \log_3(2^{r+2}+1) \rceil$ grandparent-matched vertices at γ.

Note also that there are at most two remaining neighbors for every vertex when it is fixed, as one neighbor will have already been matched earlier. As we have seen, there are $f(r+2)$ grandparent-fixed vertices. So, there are at most $2 f(r+2)$ neighbors of these grandparent-fixed vertices. These are evenly split between the grandchildren, so each one gets $\frac{1}{2}f(r+2)$ grandparent-fixed vertices. Thus, we get $f(r) \leq \lceil \frac{1}{2}(f(r+2) + \lceil \log_3(2^{r+2}+1) \rceil) \rceil + 1$ and the solution of this satisfies $f(r) \leq \lceil \log_3 2^{r-1} + 1 \rceil + 6$. This follows, because $(\lceil \frac{1}{2}\lceil \log_3 (2^{r+1} - 1) \rceil + \lceil \log_3(2^{r+2} + 1) \rceil) \rceil \leq \lceil \log_3(2^{r-1} + 1) \rceil + 2.)$ Furthermore, $\lceil \log_3(2^{r-1} + 1) \rceil + 6 \leq 2r - 4 = 2(r-2)$ holds, for all $r \geq 7$. So, there are enough points to map the binary tree into the collision tree of the second type of height $s+1$.

Thus, we get a dilation 6 embedding into the hypercube when we don't care to which hypercube locations at a particular node in the collision tree the binary tree vertices are mapped. By caring, as indicated earlier, we get a smaller dilation embedding. However, there are some additional *tricks* that must be done.

For instance, to get dilation 3 we have to define a recurrence relation that in addition to the already mentioned nodes fixed at each level r allows for the inclusion of additional ones. That is, for grandparent-fixed nodes which have a common neighbor we reserve a place for the common neighbor. We call the common neighbor a *center vertex*. The embedding to achieve dilation 3, which we described in section II, is only possible if: (a) the four neighbors of two grandparent-fixed vertices, say assigned to $\gamma 1\alpha 0$ and $\gamma 1\alpha 1$, respectively, are mapped to four different "quadrants", i.e. the two bits after γ in locations assigned these vertices are distinct, which we ensure using "forbidden pairs" in our construction and (b) when two grandparent-fixed vertices have a center vertex in common, then the center vertex is also laid out at the same level of the collision tree. Let $g(r)$ denote the total number of nodes fixed at level r, including the neighbors of grandparent-fixed vertices, the grandparent-matched vertices, and these center vertices. We can show that: $g(r) \leq \lceil \frac{5}{3} \lceil \log_3(2^{r-1}+1) \rceil \rceil + 12$, i.e. $g(r) \leq 2(r-2)$, for all $r \geq 18$. Thus, if all binary trees of size at most 2^{18} can be embedded into their optimal hypercube with dilation 3, then all binary trees can be so embedded. This inequality also shows that, with no further work at all, we can show that all binary trees can be embedded with dilation 3 into the binary collision tree of some extra height (and hence more hypercube locations at the top levels). So, this yields an embedding of arbitrary binary trees with dilation 3 and O(1) expansion into hypercubes.

If, on the other hand, we do not assign center vertex to the same level of the collision tree as we do the two grandparent-fixed vertices they lie between, then to get a dilation 5 embedding we do the following: Map the four neighbors of two grandparent-fixed vertices assigned to $\gamma 1\alpha 0$ and $\gamma 1\alpha 1$, respectively, to four different "quadrants" as described above and a center vertex adjacent to grandparent-fixed vertices assigned to $\gamma 100\alpha b_1$ and $\gamma 100\beta b_2$ is mapped to a location $\gamma b_3 b_4 \alpha b_2$, for some b_3 and b_4 in $\{0,1\}$.

All of these considerations lead to our result:

Theorem. Every binary tree can be embedded with dilation 3 and O(1) expansion into a hypercube. (In fact, every binary tree can be embedded into its optimal hypercube with dilation 3, provided that all binary trees with no more than B vertices, for some fixed number B, can be embedded with dilation 3.) Also every binary tree can be embedded into its optimal hypercube with dilation 5.

VI. Conclusion and Open Questions

We believe that our techniques can in fact show that all binary trees are embeddable into their optimal hypercubes with dilation at most three. What remains is to show that this in fact is true for all binary trees of size at most B, for some fixed constant B (B is about 2^{18}).

We should also note that an earlier version of our work also contained results about embedding complete binary trees with small dilation into small cube connected cycle networks. Later we became aware of a recent paper by Bhatt, Chung, Hong, Leighton, and Rosenberg [BCHLR] which has as one of its main results the theorem that all complete binary trees are embeddable with dilation 4 and expansion 4 into a butterfly network. As the butterfly network B(k) is embeddable with dilation 2 into the cube connected cycle network of dimension k, denoted by CCC(k), it follows from their result that complete binary trees can be embedded into cube connected cycle networks with dilation at most 8 and expansion 4. (That is, the dilation of the composition of two embeddings is the product of their dilation.) A small dilation and simultaneous small expansion embedding of a complete binary tree into the butterfly (or cube connected cycle network) is well worth learning, although all known embeddings involve rather complex constructions. We refer the interested reader to [BCHLR].

References

[AR]   R. Aleliunas and A. L. Rosenberg, "On Embedding Rectangular Grids in Square Grids", *IEEE Trans. on Computers*, C-31,9 (1982), pp. 907-913.

[BMS]   S. Bettayeb, Z. Miller, and I. H. Sudborough, "Embedding Grids into Hypercubes", *these proceedings*.

[BI]   S. N. Bhatt, and I. C. F. Ipsen, "How to Embed Trees in Hypercubes", *Research Report YALEU/DCS/RR-443, Yale Univeristy, Dept. of Computer Science*, 1985.

[BCLR] S. Bhatt, F. Chung, T. Leighton, and A. Rosenberg, "Optimal Simulation of Tree Machines", *Proc. 27th Annual IEEE Symp. Foundations of Computer Sci.*, Oct. 1986, pp. 274-282.

[BCHLR] S. Bhatt, F. Chung, J.-W. Hong, T. Leighton, A. Rosenberg, "Optimal Simulations by Butterfly Networks", *Proc. 20th Annual ACM Theory of Computing Symp.*, 1988, to appear.

[C]   M. Y. Chan, "Dilation 2 Embedding of Grids into Hypercubes", Tech. Report, Computer Science Program, Univ. Texas at Dallas, 1988.

[CC]   M. Y. Chan and F. Y. L. Chin, "On Embedding Rectangular Grids in Hypercubes", to appear in *IEEE Trans. on Computers*.

[E]   J. A. Ellis, "Embedding Rectangular grids into Square Grids", *these proceedings*.

[HL]   I. Havel and P. Liebl, "One Legged Caterpillars Span Hypercubes", *J. Graph Theory*, 10 (1986), pp. 69-76.

[H]   I. Havel, "On Hamiltonian Circuits and Spanning Trees of Hypercubes", *Cas. Pest. Mat.* (in Czech.), 109 (1984), pp. 135-152.

[MSUW] B. Monien, G. Spenner, W. Unger, and G. Wechsung, "On the Edge Length of Embedding Caterpillars into Various Networks", manuscript, Dept. of Math. and Computer Science, Univ. Paderborn, Paderborn, W. Germany, 1988.

[N]   L. Nebesky, "On Cubes and Dichotomic Trees", *Cas. Pest. Mat.* (in Czech.), 99 (1974), pp. 164-167.

EMBEDDING RECTANGULAR GRIDS INTO SQUARE GRIDS

John A. Ellis
University of Victoria
Victoria, British Columbia, Canada

Abstract. *We show that 2-dimensional rectangular grids of large aspect ratio can be embedded into rectangles of smaller aspect ratios with small expansion and dilation. In particular, width can be reduced by a factor of up to 2 with optimal expansion and dilation (2). A factor of 3 can be obtained with dilation 3. In general, any rectangular grid can be embedded into a square grid that is no more than unity larger on the side than the minimum possible, with dilation no more than 3. These results improve on those previously obtained, in which dilation better than 18 could not be guaranteed. They might be applied to more complex grid embedding problems, such as embedding multi-dimensional grids into hyper-cubes.*

1. Introduction

An *embedding* of a guest graph $G = (V_G, E_G)$ into a host graph $H = (V_H, E_H)$ is a one-to-one mapping ϕ from V_G to V_H. We measure the quality of an embedding ϕ with two cost functions: *dilation* and *expansion*. The *dilation* of an edge $\{u,v\} \in E_G$ is the length of a shortest path in H that connects $\phi(u)$ and $\phi(v)$. The dilation of an embedding is the maximum dilation, over all edges in E_G. The *expansion* of an embedding is $|V_H| / |V_G|$. The dilation of an embedding measures the worst case stretching of an edge, the expansion measures the relative size of the guest graph. One can also define a *load factor*, which measures the worst case sharing of an edge in the host by paths from the guest. We do not consider load factors in this paper, except to note that, in the embeddings under consideration, it is small.

Embedding problems can model many questions of interest in Computer Science, such as VLSI circuit layout, simulating one parallel processing architecture by another, assigning processes to processors in a distributed processing system and simulating one data structure by another. Aleliunas and Rosenberg [1] considered the particular problem of embedding 2-dimensional rectangular grids of aspect ratio greater than one into smallest possible grids of square aspect. They point out that the problem has application to the design of large scale integrated circuits since the natural design of a circuit may be rectangular but the circuit must eventually be manufactured on a square chip. The critical factors of area and wire length are represented by expansion and dilation in the model just described.

A solution to this two dimensional problem may be used to solve more complex problems. For example, Bettayeb, Miller and Sudborough [2] discuss the embedding of multi-dimensional grids into hypercubes, which has immediate application to the efficient simulation of grid-like parallel processing architectures by hypercube architectures.

Aleliunas and Rosenberg were unable, in general, to optimise both expansion and dilation simultaneously and conjectured that there may be an inherent expansion-dilation tradeoff. If expansion is minimized, i.e., we embed a rectangular grid into the smallest possible square grid, then the smallest dilation they obtain depends on the aspect ratio of the rectangle. For some aspect ratios, dilation is optimal, i.e., 2, but for others dilation better than 18 could not be guaranteed.

In Section 2 we summarise the results obtained in this paper. In Section 3 we describe an improved compression technique suitable for instances of the problem with smaller aspect ratios. In Section 4 we show that the new compression technique can be combined with an earlier technique called "folding" to obtain a close to optimal result for any aspect ratio.

2. Summary of New Results

Our results show that there is in fact no significant tradeoff between expansion and dilation, for this problem. In many cases optimal expansion and optimal dilation can be achieved simultaneously and in the worst case dilation 3 is obtainable with almost optimal expansion.

Let the rectangular grid be of height h and width w and let $h < w$. Let $h' < h$ and w' be the smallest integer such that $h'w' \geq hw$. We call the $h' \times w'$ rectangle the *ideal rectangle* and the ratio w/w' the *compression ratio*. We will call the square grid of side s where $s = \lceil (hw)^{0.5} \rceil$ the *ideal square grid* and the square of side $s+1$ the *nearly ideal* square grid. The following theorems will be proved.

Theorem 1: If the compression ratio is ≤ 2, then a rectangular grid can be embedded into any of its ideal rectangular grids with dilation 2.

Corollary 1: If the compression ratio is ≤ 2, then a rectangular grid can be embedded into its ideal square grid with dilation 2.

Theorem 2: If the compression ratio is ≤ 3, then a rectangular grid can be embedded into any of its ideal rectangular grids with dilation ≤ 3.

Corollary 2: If the compression ratio is ≤ 3, then a rectangular grid can be embedded into its ideal rectangular grid with dilation ≤ 3.

Theorem 3: Any rectangular grid can be embedded into its nearly ideal square grid with dilation ≤ 3.

3. Small Compression Ratios

In this section we justify Theorems 1 and 2 and hence their corollaries. We will consider compression ratios up to 3 in this section, but we illustrate the general technique in terms of the smallest of these, i.e., in the range 1 through 3/2. Other cases are then covered by generalising at the end of the section.

3.1. Primitive Tiles

We first define a simple compression pattern that embeds any $(n-1) \times n$ grid into a $n \times (n-1)$ grid. This pattern, which for a particular n we call a primitive, light 2-tile, is defined in Figure 3.1.a The tile shown there represents an embedding of a 4×5 grid into a 5×4. The figure shows the position of the edges that were horizontal in the guest grid, after the embedding into the host. A diagonal edge indicates that adjacent nodes have been embedded across the diagonal of a unit square in the host. There are two shortest paths between the images of these nodes, of length 2. The vertical edges are not shown but their position is easily deduced. We note that the dilation of no edge, vertical or horizontal, is greater than two. Note that the pattern can be extended *ad infinitum*.

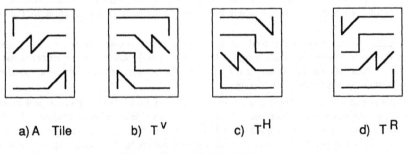

a) A Tile   b) T^V c) T^H d) T^R

Figure 3.1   Light 2-Tiles

3.2. Tile Operations

The general technique constructs *composite* tiles out of smaller tiles. The construction uses three operations on both primitive and composite tiles. These operations are: reflection about a vertical axis, reflection about a horizontal axis, and reversal of orientation. By the *orientation* of a vertical line in the tile diagrams, we refer to the direction, up or down, that is implied by a left to right traversal of the row containing the vertical. We note that in a primitive tile the orientation of the vertical line in each row reverses from row to row. We denote the operations by V, H and R respectively. In Figure 3.1, item b is formed from item a by the V operation, item c is obtained from a by the H operation and item d is obtained from a by the R operation. It is evident that these operations are commutative.

3.3. Iterating Tiles

Suppose that a tile T is placed side by side with T^V. Connecting, horizontal edges suffer no dilation. No vertical edges cross the join. Suppose a tile T is placed above T^{VR}. Connecting, vertical edges suffer dilation no greater than two. No horizontal edges cross the horizontal join. Finally we note that T^{VRH} is identical to T^{HVR}, because the operations are commutative. Hence T^{VRH} can form the bottom right tile of a four tile pattern, since it matches its neighbours both above and to the left.

We can make rows of tiles by repeatedly applying the V operation, and columns by repeatedly applying HR. It is evident that $T^{HRHR} = T = T^{VV}$. Consequently, every other tile along a row or down a column is identical. This process of tile construction we call *iteration*. The ideas are illustrated in Figure 3.2.

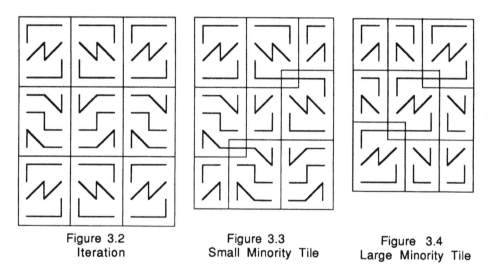

Figure 3.2	Figure 3.3	Figure 3.4
Iteration	Small Minority Tile	Large Minority Tile

3.4. Forming Tiles by Diagonalisation

We use a technique, called *diagonalisation*, to combine tiles of different sizes. To help us describe the process, we make some definitions. A *tile* is either a *primitive tile* or is formed from other tiles by *diagonalisation*, a procedure which is defined below. The *level* of a tile is the number of times the procedure has been used in constructing the tile. An *element* of a primitive tile is as defined for each type of primitive tile. An *element* of a compound tile is a tile of current level minus one. A tile element is a *majority* element if it occurs more than once across a row in the tile structure, and a *minority* element if it occurs exactly once. For example, the elements of a light 2-tile are single nodes (the *majority* elements) and single edges (the *minority* elements). The *dimension* of a tile is the number of majority elements across the height or width of the tile. A *subtile* is a tile contained entirely within another tile.

The diagonalisation process allows us to combine tiles of the same level that differ in dimension by one element into a new tile of the next level. Each row in the new tile contains one element of the *minority* type, and $n \geq 1$ tiles of the *majority* type, where n is the dimension of the new tile. Figures 3.3 and 3.4 illustrate the use of the technique. Note that the majority tile can be larger or smaller than the the minority tile.

We wish to show that all tiles have the following properties:

(1)　The dilation of the embedding defined by the tile is the same as that of the primitive tiles from which it is composed.

(2)　If any subtile is shifted by the extent of two majority elements both horizontally and vertically, it covers an identical sub-tile.

(3)　For all tiles, $T = T^{HRHR}$ and $T = T^{VV}$.

(4)　For all tiles, T placed above T^{HR} yields a dilation for the connecting vertical edges as defined in property 1.

(5)　For all tiles, T placed side by side with T^V yields a dilation for the connecting horizontal edges as defined in property 1.

We now describe the construction by which a tile of dimension n and level k is constructed from elements which are tiles of level $k-1$. Build_large works with minority elements that are larger than the majority elements. Its operation is illustrated in Figure 3.5. Build_small works in the opposite situation.

procedure build_large;
{Builds a tile of dimension n in which the minority tile is the larger. The ith step creates a tile of dimension i from a tile of dimension $i-1$. The 0th tile is a large, minority tile whose diagonal orientation is from top left to bottom right. The procedure builds a new tile whose diagonal orientation is from top right to bottom left. Of course, a mirror image procedure is obtained by inverting left and right.}
for $i = 1$ **to** n **do**

(1)　Add a new bottom row composed of i majority tiles each of which is obtained from its neighbour above by the HR operations. Note that the left end tile is obtained from the majority tile contained within the minority tile.

(2)　Add a new left column composed of i majority tiles, each of which is obtained from its right neighbour by the V operation. Note that the bottom end tile is obtained from the majority tile contained within the minority tile.

(3)　Create a new minority tile for the bottom left corner by:
a) Adding a left neighbour to the left end of the new bottom row which is obtained from the end element by the V operation.
b) Adding a neighbour to the bottom end of the new leftmost column which is obtained from the bottom element by the HR operations.
c) Let the top right element of the minority tile be T^{HR}, where T is the element above it. Let the bottom left element of the minority tile be T^{HR}, where T is the element above it.

endfor endprocedure

We need to show that the procedure is self consistent and that the tiles produced preserve the tile properties. An easy inductive argument can be based on the following observations.

First we note that the tile properties do hold for primitive tiles. Then we note that the procedure may not be self consistent if the overlapping portions of the elements defined in Steps 3.a and 3.b are

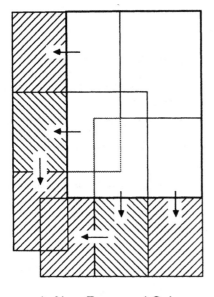

a) New Rows and Columns

Figure 3.6
A Level 2 Tile

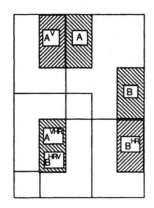

b) Identical Overlap

Figure 3.5 The Diagonalisation Process

not identical. These overlapping portions are A^{VHR} and B^{HVR} by definition of the construction. See Figure 3.5.a. But A is identical to B by property 2 and the operations are commutative. So the overlapping portions are identical and Step 3 is consistent.

The dilation of the composite tile is still 2 because the new elements added match their neighbours at the seams by properties 4 and 5. The second property is preserved because of property 3. Finally we note that alternate pairs of minority elements are identical because their components are likewise 2 operations apart. It is evident that properties 3, 4 and 5 are preserved.

We need a similar procedure for the case in which the minority tile is small relative to the majority tile. In this procedure we refer to the *expansion* and *contraction* of a tile. By the *expansion* of a tile, we mean a tile of dimension one more than the given tile, by *contraction*, one less.

procedure build_small;
{Builds a tile of dimension n in which the minority tile is the smaller. The ith step creates a tile of dimension i from a tile of dimension $i-1$. The 0th tile is a small, minority tile whose diagonal orientation is from top left to bottom right. The procedure builds a new tile whose diagonal orientation is from top right to bottom left. Of course, a mirror image procedure is obtained by inverting left and right.}
for $i = 1$ **to** n **do**

(1)   Add a new bottom row composed of $i-1$ majority tiles each of which is obtained from its neighbour above by the HR operations. Note that this step is null for $i = 1$.

(2)   Add a new left column composed of $i-1$ majority tiles, each of which is obtained from its right neighbour by the V operation. Note that this step is null for $i = 1$.

(3)   Add a neighbour R to the left end of the new bottom row such that R is the expansion of T^V where T is the minority tile in the bottom left corner of the current tile.

(4)   Add a neighbour R to the bottom end of the new leftmost column such that R is the expansion of T^V where T is the minority tile in the bottom left corner of the current tile.

(5)   Create a new minority tile R for the bottom left corner such that R is T^H where T is the lower contraction of the leftmost minority tile in the bottom row.

endfor endprocedure

One can use Figure 3.3 to illustrate the operation of this procedure. Starting from the small tile in the top right corner, a dimension n tile is constructed by adding a left column and bottom row n times. The justification for this procedure is similar to that just given for the build_large procedure.

3.5. Repeated Diagonalisation

It remains to describe the method whereby these techniques can be applied to any instance of the problem of compressing a rectangle of width w down to w', so long as the compression ratio is no more than 3/2. Then we compute the number of tiles required to achieve the required absolute compression $\delta = w-w'$. Each tile achieves an absolute compression of 1, so we require δ tiles.

An example illustrates the fact that the problem is not necessarily solved by iteration of a single tile, because tiles of different sizes are required. Further, one application of the diagonalisation procedure may not be sufficient. Consider for example the problem of compressing a width 23 down to 16, which is the subject of Figure 3.6. We need 7 tiles to achieve the absolute compression, but 7 does not divide 23. In fact we need 5 tiles of width 3 and 2 of width 4, (before compression). The diagonalisation technique only permits us to mix one minority tile with majority tiles. So we make composite tiles. One contains 1 primitive tile of width 4 and 2 of width 3, the other contains 1 of width 4 and 3 of width 3.

The entire process begins from one minority tile, width 4. Using the diagonalisation process we create a dimension 3, level 1 tile from the primitive tile. This level 1 tile is used as the seed for a dimension one, level 2 tile by another application of the diagonalisation process.

The general rule is that if j is the number of majority tiles at level i and n is the number of minority tiles, then the dimension of the minority tile at level $i+1$ is $\lceil j/n \rceil$. We may then define a complete construction process as follows.

procedure construct;
{Compute the number (j and n) of the majority and minority primitive tiles and the size of the minority tile from w and δ}
$j \Leftarrow$ remainder(w div δ); $n \Leftarrow \delta - j$; size $\Leftarrow \lfloor w/\delta \rfloor$; **if** $j < n$ **then** swap(j,n); size $\Leftarrow$ size + 1 **endif**;
Choose a primitive tile of appropriate size; Diagonal orientation $\Leftarrow$ orientation of primitive tile;

loop
    Reverse the diagonal direction;
    Invoke build_large or small according as to whether the minority tile is large or small,
    the dimension is defined by $\lceil j/n \rceil$;
exitif $n = 1$ **or** n divides j
    $t \Leftarrow n$; $n \Leftarrow$ remainder(j div n); $j \Leftarrow t-n$;
    if $j < n$ **then** swap(j,n); Invert the type of the minority tile (small/large) **endif**;
endloop

if j divides n **then** complete the construction by simple iteration of the current tile **endif**
end procedure;

The methods described define, an embedding of a $(w-\delta) \times w$ grid into a $w \times (w-\delta)$ grid. In general we want a solution the problem: Embed a $h \times w$ grid into an $h' \times w'$ grid where h' is the smallest integer such that $h'w' \geq hw$. We do need to be assured that our method embeds the first h rows within the h' rows of the guest grid.

This property is easily established by noting that it holds for the primitive tiles, whether rows are counted from the top or from the bottom of the tile, and that the property is preserved by any application of the building procedures. Consequently Theorems 1 and 2 are justified.

3.6. Larger Compression Ratios

We have considered so far ways of realising a compression ratio in the range 1 through 3/2. This was done by employing the primitive tile which we called the light 2-tile. We also define three other primitive tiles which can be used to obtain compression ratios up to three. They are illustrated in Figure 3.7. The heavy 2-tile yields compressions in the range 3/2 through 2 with dilation 2. The light 3-tile yields compressions in the range 2 through 5/2 with dilation 3, and the heavy 3-tile yields compressions in the range 5/2 through 3, also with dilation 3.

These tiles can be used to construct composite tiles in just the same way as the light 2-tiles. It may be verified, by arguments similar to those just described for the light 2-tiles, that the tile properties hold for all these primitive tiles and that the properties are preserved through the tile building procedures. Figure 3.8 shows a composite tile whose elements are light 3-tiles. Consequently, Theorems 1 and 2, and their corollaries, have been justified.

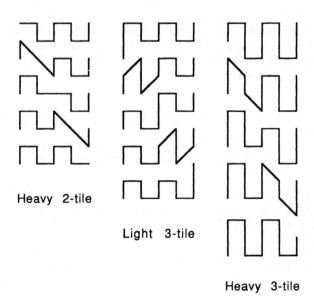

Heavy 2-tile

Light 3-tile

Heavy 3-tile

Figure 3.7   Larger Compression Ratios

Figure 3.8   Composing Light 3-Tiles

4. Embedding Rectangles with Large Aspect Ratios

So far we have considered aspect ratios not exceeding 3. We now want to justify Theorem 3, namely that any rectangle can be embedded into its nearly ideal square with dilation no more than 3. This can be accomplished by combining two separate techniques, folding and compression.

4.1. The Folding Technique

The folding technique was defined by Rosenberg and Aleliunas [1] and is illustrated in Figure 4.1. A long, skinny rectangle can be folded several times so as to reduce its aspect ratio considerably. The significant properties of this method are:

(1)   The dilation is exactly 2.

(2)   If a horizontal edge is dilated it maps onto a path between opposing corners of a unit square in the host grid.

(3)   If a vertical edge is dilated, it maps onto a path comprising two adjacent vertical edges in the host grid.

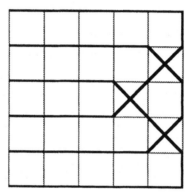

Figure 4.1   Folding

4.2. Dilation Produced by Folding and Compression

Since the folding operation and compression by no more than a factor of 2 each create dilation 2, combining these operations creates dilation no more than 4. We will now show that if a folding operation is followed by a compression based on light 2-tiles, i.e., with a compression ratio no more than 3/2, then the combined dilation is no more than 3. Let the *fold axis* of a fold operation be along the width of the subject grid, e.g., in Figure 4.1, the fold axis is horizontal. We achieve the claimed result by compressing along the axis that is orthogonal to the fold axis, e.g., in Figure 4.1, a horizontal fold would be followed by a vertical compression.

We need to consider three grids: the original grid P which is the guest for a folding embedding into the host Q and the final host R which accepts Q as a guest by way of a compression. Consider two adjacent nodes, a and b, in P. Suppose folding dilates an edge $\{a,b\}$. Let a' and b' be the images of a and b in the host Q. We will show that there always exists a node x in Q and some path (a',x,b') from a' to b' such that compression will not dilate both the edges $\{a',x\}$ and (x,b'), and hence that there is a path between the images of a and b in R of length no more than 3. There are two cases, $\{a,b\}$ is either horizontal or vertical in P.

Suppose $\{a,b\}$ is horizontal in P. From the properties of folding, a' and b' must be situated across the diagonal of a unit square of Q, i.e., their are two paths of length 2 from a' to b' around opposite sides of the square. The compression pattern defined by using light 2-tiles never dilates both of a pair of horizontal edges belonging to a unit square. This observation is justified by inspection for primitive tiles. When tiles are combined, we note that there is no dilation of horizontal edges across vertical seams. For horizontal seams we note that if an edge is diagonal in a tile, then one of its adjacent horizontal edges is vertical in an adjacent tile. The corresponding vertical on the other side of the seam is, by construction, of the opposing orientation. Hence the edge paired with the diagonal one remains horizontal, i.e., is not dilated.

Finally suppose $\{a,b\}$ is a vertical edge in P. From the properties of folding, if this edge is dilated, then there is a node x such that $\{a',x\}$ and $\{x,b'\}$ are both vertical edges in Q. We supposed that the compression occurs along the vertical axis, so these are horizontal edges from the point of view of the compression. Note that the compression never dilates two adjacent horizontal edges. The observation is justified by inspection for the primitive tiles. For composite tiles, note that there is no dilation of horizontal edges across the vertical seams, and that the nodes in the overlap between large primitive tiles are not adjacent of any diagonal edges.

4.3. Maximum Reduction of Large Aspect Ratios

The strategy just discussed suggests that to achieve compression ratios greater than 3 we can first use folding to reduce the problem to one that is within a factor of 3/2 of the desired result. Compression along the axis orthogonal to the folding then completes the embedding. We will investigate the maximum achievable width reduction that can be obtained with dilation 3.

If the folding operation used n folds, we will call $n+1$ the *fold factor*. We defined an *ideal* grid with respect to an $h \times w$ guest grid to be a square grid of side $s = \lceil (hw)^{0.5} \rceil$. If there exists a k such that $kh \le s$ and $\lceil w/k \rceil \le s$ then folding produces an optimal, minimum expansion, dilation 2, embedding. In general such a k may not exist, but we show that there is always a fold factor k such that the height of the rectangle after folding, h', is given by $h' = kh$ and $s < kh \le 3s/2$ and $h'w' < (s+1)^2$. Consequently, any grid can be embedded into its *nearly ideal* square grid with dilation no greater than 3, so justifying Theorem 3.

The fold factor we need is $k = \lceil s/h \rceil$. To show that this value of k implies that $h' < 3s/2$ we note that there exists a d, $0 \le d < 1$, such that $h(k-d) = s$. By assumption, $w > 3s$, and hence $s > 3h$, which implies that $hd < s/3$. Hence $hk-s/3 < s$, which yields the desired result since $hk = h'$. To show that $h'w' < (s+1)^2$, we note that $w' = \lceil w/k \rceil$ and so $h'w' = \lceil w/k \rceil \times wk$. If $k|w$ then $h'w' = hw$. Otherwise, $h'w' = \lfloor w/k \rfloor \times hk \le hw+hk$. But $hw \le s^2$ and $hk < 4s/3$ so $h'w' < (s+1)^2$.

For example, a 7×82 rectangle requires a fold factor of 4, according to the above equations. The rectangle will be folded into a 28×21 rectangle and from thence compressed into 25×25 square. The ideal square is however of side 24.

References

[1] Aleliunas, R. and Rosenberg, A. L., "On Embedding Rectangular Grids in Square Grids", IEEE Trans. on Computers, C-31, 9, pp. 907 - 913, 1982.

[2] Bettayeb, S., Miller, Z. and Sudborough, I. H., "Embedding Grids into Hypercubes", Manuscript, Computer Science Program, University of Texas at Dallas, 1987.

Efficient Reconfiguration of VLSI Arrays [*]
(Extended Abstract)

Bruno Codenotti [1] *and Roberto Tamassia* [2]

[1] Istituto di Elaborazione dell'Informazione
Consiglio Nazionale delle Ricerche
56100 Pisa, Italy

[2] Coordinated Science Laboratory
University of Illinois at Urbana-Champaign
Urbana, Illinois 61801

Abstract   We consider the problem of reconfiguring a 2-dimensional VLSI array with faulty cells. A network flow model of the problem is formulated and an algorithm is presented for interconnecting the functional cells of the array so that they simulate a fault-free array of smaller size. Experimental results on the practical performance of this algorithm and of other techniques previously proposed in the literature are reported.

1. Introduction

The technique of *wafer scale integration* for VLSI circuits has received considerable attention in recent years. The basic idea is to assemble a large system of processors, or *cells*, on a single silicon wafer so that the chip packaging costs are dramatically reduced. Due to the physical and technological limits of the integration process, some cells can be defective, or *dead*. Therefore, the problem arises of reconfiguring the interconnection network using the *live* cells.

The reconfiguration of 1- and 2-dimensional arrays of cells, a system typical of VLSI systolic architectures, has been recently investigated by Leighton and Leiserson [5] and by Greene and El Gamal [2, 3] In both works, a probabilistic model of cell failure is adopted and algorithms for minimizing the maximum wire length in the reconstructed array are given. Channel width and area penalties are also considered. For related issues in fault tolerant VLSI design, see [4].

---

[*] Research partially supported by the Joint Services Electronics Program under Contract N00014-84-C-0149. Portions of this work were done while the first author was visiting the Coordinated Science Laboratory of the University of Illinois.

In this paper we present a new approach to the problem of reconfiguring 2-dimensional VLSI arrays. We model the reconfiguration problem as a *simulation* of a *virtual* fault-free array performed by the live cells of the physical defective array, and provide a network flow characterization of the simulation mapping. This approach yields a practical heuristics for the reconfiguration problem.

We also report on the results of a comparative study on the practical performance of our algorithm and the algorithms of Leighton-Leiserson and Greene-El Gamal. This experimental study has been conducted by implementing the above algorithms and using a variety of random tests. The results have shown that our algorithm consistently produces better results.

It is interesting to notice that a network flow approach is also used by Greene [2]. However, his construction is rather different from ours. In fact, the flow network adopted by Greene is essentially obtained by connecting the live cells lying in the same row and column.

2. Preliminaries

We use the same model for the reconfiguration problem as in [5]. The n cells are positioned in a $\sqrt{n} \times \sqrt{n}$ square array. The strips between two consecutive rows or columns of cells are called *channels*. Each channel contains a fixed number of tracks on which the interconnection wires used for the reconfiguration are routed.

Suppose that some cells of the array are dead. We investigate the problem of interconnecting the remaining m live cells into a smaller square array of size $\sqrt{m} \times \sqrt{m}$. (For simplicity, we assume that the number m of live cells is a perfect square, the extension to the general case is straightforward.) The main quality measures for a solution of this problem are the *maximum wire length*, the *total wire length*, and the *channel width* (number of tracks), all of which should be minimized. The maximum wire length is often considered the most important parameter because of its direct connection with the propagation delay in the reconfigured array. See an example in Fig. 1 (taken from [5]).

The reconfigured array can be viewed as a virtual array free of faults that is simulated by the given defective array. Each cell of the virtual array is simulated by a distinct live cell of the original array.

Consider the distribution of faults shown in Fig. 2.a. It is easy to see that the wiring of Fig. 2.b provides an optimal reconfiguration. More generally, when the set of dead cells consists of an equal number of rows and columns, the optimal simulation of the virtual array is trivial. This observation suggests the following reconfiguration strategy:

(1) Select an array whose dead cell distribution consists of exactly $\sqrt{n} - \sqrt{m}$ rows and $\sqrt{n} - \sqrt{m}$ columns. This array is called *target*.

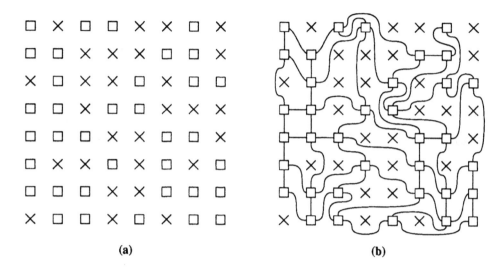

Figure 1 (a) A square array of 64 cells; the live cells are represented by a square and the dead cells by a cross. (b) Example of reconfiguration of the array in part (a) so that the live cells form a square array of 36 cells. The maximum wire length is 4 and the total wire length is 115. For simplicity, a detailed routing is not shown.

(2) Interconnect the live cells of the original array in order to simulate the target.

(3) Wire the target by connecting each of its live cell to the closest live cells on the same row or column.

Notice that the simulation of the virtual array results from the composition of two simulation mappings: the simulation of the target by the original array, and the simulation of the virtual array by the target.

We overview in the next section the main features of our reconfiguration algorithm, based on the above strategy. Experimental results on the performance of our technique and comparisons with the ones of Leighton-Leiserson and Greene-El Gamal are provided in Section 4.

We now recall the basic terminology on network flows, to be used in the next section.

A *flow network N* is a 6-tuple $N = (U,A,b,c,s,t)$ where:

(1) (U,A) is a digraph with vertex set U and arc set A, called the *underlying digraph* of N.

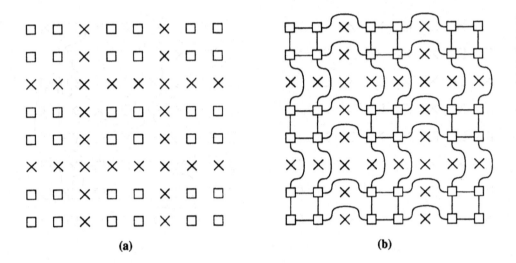

Figure 2 (a) An array of 64 cells whose dead cells are arranged into two rows and two columns. (b) Optimal reconfiguration of the array in part (a).

(2) $b : U \cup A \to N$ associates with all nodes and arcs a nonnegative integer *capacity*;

(3) $c : A \to N$ associates with each arc a nonnegative integer *cost*;

(4) s and t are two designated nodes called the *source* and the *sink* of N, respectively.

A *flow* for N is an integer-valued function $f : A \to Z$ that satisfies the following conditions:

(1) $0 \le f(u,v) \le b(u,v)$ for all $(u,v) \in A$;

(2) $\displaystyle\sum_{u : (u,v) \in A} f(u,v) = \sum_{w : (v,w) \in A} f(v,w) \le b(v)$ for all $v \in U - \{s,t\}$.

The *cost* and the *value* of flow f are defined as follows:

$$\text{COST}(f) = \sum_{(u,v) \in A} c(u,v) \cdot f(u,v); \quad \text{VALUE}(f) = \sum_{v : (s,v) \in A} f(s,v).$$

A *maximum flow* for N is a flow with maximum value. A *minimum cost flow* of value ϕ for N is a flow with value ϕ and minimum cost. Efficient algorithms for computing maximum and minimum cost flows in networks are surveyed in [6].

3. Overview of the Network Flow Technique

We identify the cells of the 2-dimensional array with the integer grid points whose $x-$ and y-coordinates are in the range $[1,\sqrt{n}\,]$ and denote by L and D the subsets of live and dead cells, respectively. Also, we assume that the target is given and we denote by L_T and D_T the subsets of its live and dead cells, respectively. Notice that $|D_T| = |D|$. We denote by τ the number of cells that are dead in the original array but live in the target, i.e. $\tau = |D \cap L_T| (= |L \cap D_T|)$.

The simulation between the cells of the original array and the target is expressed by a function σ mapping vertices into vertices, where $\sigma(u) = v$ iff cell v of the target is simulated by cell u of the original array. The function σ must satisfy the following properties:

(1)   no cell is simulated by more than one cell;

(2)   only the live cells can perform simulations;

(3)   all cells in L_T have to be simulated.

This can be formally expressed by defining a *simulation function* as a bijection $\sigma: L \rightarrow L_T$. We associate with σ a digraph $G_\sigma = (V_\sigma, A_\sigma)$, called *simulation graph*, where $V_\sigma = L \cup L_T$, and $(u,v) \in A_\sigma$ whenever $\sigma(u) = v$ and $u \neq v$. Notice that the isolated vertices of G_σ correspond to the live cells that simulate themselves.

Lemma 1 A function σ is a simulation function if and only if the digraph G_σ consists of disjoint paths that are either

(1)   cycles of one or more live cells, or

(2)   chains starting at a vertex in $L \cap D_T$, ending at a vertex in $D \cap L_T$, and having the intermediate vertices in L.

Furthermore, the number of these chains is exactly τ. $\qquad\qquad\qquad\qquad\square$

Let $d(u,v)$ denote the Manhattan distance between cells u and v , i.e.

$$d(u,v) = |x(v) - x(u)| + |y(v) - y(u)|.$$

The maximum wire length l_σ in the interconnection of the target generated by the simulation function σ is equal to the maximum Manhattan distance $d(u,v)$ between two cells u and v that simulate adjacent cells of the target, i.e.

$$l_\sigma = \max \{d(u,v): d(\sigma(u),\sigma(v)) = 1\}.$$

The total wire length L_σ is given by:

$$L_\sigma = \sum_{(u,v): d(\sigma(u),\sigma(v))=1} d(u,v).$$

We now define two new quantities and show their relation with the maximum and total wire length. The *maximum simulation distance* d_σ is the maximum Manhattan distance between a live cell u and the cell $\sigma(u)$ simulated by u, i.e.

$$d_\sigma = \max \{d(u, \sigma(u)): u \in L\}.$$

The *total simulation distance* D_σ is defined as the sum of the Manhattan distances between a live cell u and $\sigma(u)$, i.e.

$$D_\sigma = \sum_{u \in L} d(u, \sigma(u))$$

If we assign to each arc (u, v) of G_σ a length equal to $d(u, v)$, we have that D_σ is equal to the sum of the lengths of the arcs of G_σ.

Proposition 1 Let σ be a simulation function, then $l_\sigma \leq 2d_\sigma + 1$ and $L_\sigma \leq 2D_\sigma + 2n$.

Sketch of Proof: A consequence of the triangular inequality $d(u, v) \leq d(u, w) + d(w, v)$.□

Given a positive integer k, we can test whether $d_\sigma \leq k$ by constructing an auxiliary digraph $G^{(k)} = (U, A^{(k)})$, defined as follows (see Fig 3):

(1) $U = L \cup D \cup \{s, t\}$, where s and t are two new vertices.

(2) $A^{(k)} = A_s \cup A_m^{(k)} \cup A_t$, where

$$A_s = \{(s, u): u \in L \cap D_T\},$$

$$A_m^{(k)} = \{(u, v): u \in L, v \in L_T, \text{ and } d(u, v) \leq k\}, \text{ and}$$

$$A_t = \{(u, t): u \in D \cap L_T\}.$$

Proposition 2 There exists a simulation function σ with maximum simulation distance $d_\sigma = k$ if and only if the digraph $G^{(k)}$ contains τ vertex-disjoint chains from s to t. Also, the union of such chains minus the vertices s and t is the simulation graph G_σ of σ.

Sketch of Proof: Follows from the definition of $G^{(k)}$ and Lemma 1. □

Proposition 3 The minimum value d^* for the maximum simulation distance of a simulation function σ can be computed in time $O((d^*)^2 n^{1.5} \log n) = O(n^{2.5} \log n)$.

Proof: Let $N^{(k)}$ be the flow network whose underlying graph is $G^{(k)}$ and such that each node and arc has unit capacity. The maximum number of vertex-disjoint chains from s to t in $G^{(k)}$ is equal to the value of a maximum flow in $N^{(k)}$, which can be computed in time

$$O(\sqrt{|U|} |A^{(k)}|) = O(\sqrt{n} \cdot k^2 n) = O(k^2 n^{1.5}) \text{ [6]}.$$

Hence, we can find the minimum value k^* of k for which $G^{(k)}$ has τ vertex-disjoint chains from s to t by performing a binary search in the interval $1 \leq k \leq 2\sqrt{n} - 1$ with total time complexity $O((k^*)^2 n^{1.5} \log n) = O(n^{2.5} \log n)$. By Proposition 2, $k^* = d^*$, the minimum value of the maximum simulation distance. □

Now, assume that $N^{(k)}$ admits a flow of value τ. Assign cost $c(u, v) = d(u, v)$ to each arc $(u, v) \in A_m^{(k)}$, and null cost to the remaining arcs. Given a minimum cost flow f^* for

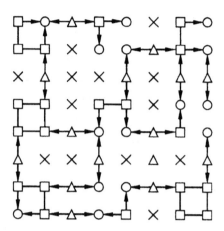

Figure 3 The digraph $G^{(1)}$ corresponding to the array of Fig. 1.a and to the target of Fig. 2.a. The cells in $D \cap L_T$ are represented by a circle. The cells in $L \cap D_T$ are represented by a triangle. Only the arcs in $A_m^{(1)}$ are shown, where an arc without arrow stands for a pair of symmetric arcs.

$N^{(k)}$ of value τ, the set of arcs of $N^{(k)}$ such that $f^*(u,v)=1$ consists of exactly τ vertex-disjoint chains from s to t. Also, the cost of f^* is equal to the the minimum total simulation distance D_σ that can be achieved with $d_\sigma \le k$. Since a minimum cost flow computation in $N^{(k)}$ can be performed in time $O(\tau(|A^{(k)}| + n \log n))$ [6], we conclude:

Proposition 4 A simulation function σ with minimum $d_\sigma = d^*$ and such that D_σ has minimum value (for that value of d_σ) can be computed in time: $O((d^*)^2 n^{1.5} \log n + \tau n((d^*)^2 + \log n)) = O(n^3)$. $\square$

The algorithm for the reconfiguration of the target is based on Propositions 1, 3, and 4. We now show how to tune the cost function in the network $N^{(k)}$ to heuristically improve the practical performance of the reconfiguration algorithm.

We observe that l_σ can reach its maximum value $2d_\sigma + 1$ only if there are two adjacent cells of the target, $\sigma(u)$ and $\sigma(v)$, such that $d(u,\sigma(u))=d(v,\sigma(v))=d_\sigma$. In terms of the graph G_σ, this means that there are two vertices at distance 1 that have an incoming arc of length d_σ. Hence, in order to decrease the probability that l_σ attains its upper bound, we want to have the fewest possible "long" arcs in the chains of G_σ. To this extent, we assign to each arc $(u,v) \in A_m^{(k)}$ a cost $c(u,v)=\chi(d(u,v))$, where the function χ satisfies the following two conditions:

(1) $\chi(x_1)+\chi(x_2)<\chi(x_1+x_2)$;

(2) $x_1+\cdots+x_p<y_1+\cdots+y_q \Rightarrow \chi(x_1)+\cdots+\chi(x_p)<\chi(y_1)+\cdots+\chi(y_q)$,
for $p,q \leq n$, and $x_i,y_j<2\sqrt{n}$.

For example, we can use the function $\chi(x)=x^{1+\varepsilon}$, where ε is a (sufficiently) small positive constant. Condition (1) implies that two "short" arcs of length x_1 and x_2 are preferred to one "long" arc of length x_1+x_2. Condition (2) implies that the new cost function still minimizes the total simulation distance.

To complete the overview of the reconfiguration algorithm, we observe that the following heuristic criteria are used to choose the target:

(1)   minimization of τ;

(2)   uniform distribution of the cells in $L \cap L_T$.

We show in Fig. 4 an example of the application of our reconfiguration algorithm to the array of Fig. 1. It is interesting to compare the reconfigurations of Figs. 1.b and 4.b.

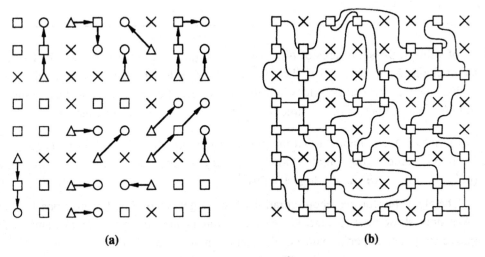

(a)   (b)

Figure 4 (a) Minimum cost flow in the network $N^{(2)}$ corresponding to the array of Fig. 1.a and to the target of Fig. 2.a. Only the arcs in $A_m^{(2)}$ with nonzero (unit) flow are shown. Note that $k=2$ is the minimum value of k for which $N^{(k)}$ admits a flow of value $\tau=15$. (b) Reconfiguration of the array in Fig. 1.a produced by our algorithm. The maximum wire length is 4 and the total wire length is 103. For simplicity, a detailed routing is not shown.

This technique can also be extended in order to take into account the channel width. The basic idea is to modify the flow network so that the node and arc capacities represent an upper bound on the channel width.

4. Experimental Results

We have conducted an experimental study on the performance of our algorithm (CT) and the ones of Leighton-Leiserson (LL) and Greene- El Gamal (GE). The three algorithms have been implemented in Pascal and run on a VAX 11/780 under the Unix operating system [1].

The input data have been randomly generated using a uniform distribution of the dead cells. Namely, we have adopted the probabilistic model used in [3, 5], where each cell has an independent probability p of being dead. The target has been selected using a simple heuristic that takes into account the two criteria mentioned in the previous section.

Several tests have been performed with various values of n and p. Examples of the results obtained are shown in Fig. 5 that gives the average value of the maximum wire length achieved by the three algorithms over about 250 trials. It is important to notice that algorithm GE always reconstructs an array of smaller size than $\sqrt{m} \times \sqrt{m}$. Typically,

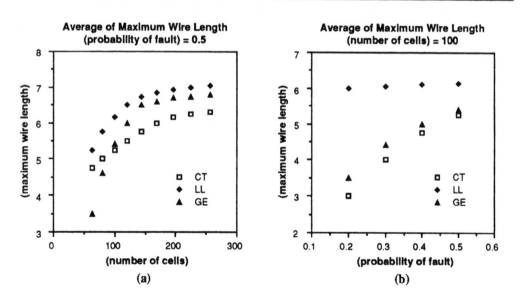

Figure 5 Experimental results on the performance of algorithms CT, GE, and LL.

the array reconstructed has between $0.5m$ and $0.65m$ elements.

In 93% of the tests our algorithm has found a reconfiguration with smaller or equal maximum wire length than the other two algorithms. In fact, the only case when this does not always happen is for $n \leq 100$ and $p \geq 0.5$, for which algorithm GE is the best, followed by CT and LL. However, it should be recalled that algorithm GE reconstructs an array with only a fraction of the live cells available.

We also observe that algorithm LL has better performance for higher values of p than for smaller ones. Namely, for p in the range [0.2 , 0.4], the maximum wire length is almost constant (see Fig. 5.b).

Acknowledgment

We wish to thank Franco Preparata for valuable discussions.

References

[1] D. Baldassarri and V. Carlino, "Implementazione di Algoritmi per la Riconfigurazione di Array VLSI," Tesi di Laurea, Dipartimento di Informatica, Univ. of Pisa, Pisa, Italy, 1985.

[2] J.W. Greene, "Configuration of VLSI Arrays in the Presence of Defects," Ph.D. Thesis, Department of Electrical Engineering, Stanford Univ., California, 1983.

[3] J.W. Greene and A. El Gamal, "Configuration of VLSI Arrays in the Presence of Defects," *J. ACM*, vol. 31, no. 4, pp. 694-717, 1984.

[4] I. Koren and D.J. Pradhan, "Modeling the Effect of Redundancy on Yeld and Performance of VLSI Systems," *IEEE Trans. on Computers*, vol. C-36, pp. 344-355, 1987.

[5] F.T. Leighton and C.E. Leiserson, "Wafer-Scale Integration of Systolic Arrays," *IEEE Trans. on Computers*, vol. C-34, pp. 448-461, 1985.

[6] R.E. Tarjan, "Data Structures and Network Algorithms," *CBMS-NSF Regional Conference Series in Applied Mathematics*, vol. 44, Society for Industrial Applied Mathematics, 1983.

EMBEDDING GRIDS INTO HYPERCUBES

by

Said Bettayeb
Computer Science Program
Univeristy of Texas at Dallas
Richardson, TX 75083-0688

Zevi Miller
Dept. of Math and Statistics
Miami University
Oxford, Ohio 45056

I. Hal Sudborough
Computer Science Program
University of Texas at Dallas
Richardson, TX 75083-0688

Abstract

We consider efficient simulations of mesh connected networks by hypercube machines. In particular, we consider embedding a mesh or grid G into the smallest hypercube that has at least as many points as G, called *the optimal hypercube* for G. In order to minimize simulation time we derive embeddings, i.e. one-to-one mappings of points in G to points in the hypercube, which minimize *dilation*, i.e. the maximum distance in the hypercube between images of adjacent points of G. Our results are:

(1) There is a dilation 2 embedding of the $[m \times k]$ grid into its optimal hypercube, under conditions described in Theorem 2.1.

(2) For any $k < d$, there is a dilation $k+1$ embedding of a $[a_1 \times a_2 \times \cdots \times a_d]$ grid into its optimal hypercube, under conditions described in Theorem 3.1.

(3) A lower bound on dilation in embedding multi-dimensional meshes into their optimal hypercube as described in Theorem 3.2.

I. Introduction and Overview

In this paper we study embeddings of grids of arbitrary dimension into hypercubes. We are motivated by problems arising in VLSI design and the design of optimal parallel architectures. The problem we consider is illustrated in Figure 1. The goal is to find the best dilation over all possible embeddings. The problem of finding good embeddings for various graphs, especially those graphs directly related to using parallel computer architectures and VLSI, has been well studied in the literature. Examples are [AR, BCLR, CMST, E, HLN, Y].

Let a_i, $1 \leq i \leq d$, be a set of d nonnegative integers. Let $[a_1 \times a_2 \times \cdots \times a_d]$ denote the grid with vertex set $\{(x_1, x_2, \ldots, x_d) \mid$ all x_i integers, $0 \leq x_i \leq a_i \}$. Two vertices $(x_1, x_2, \ldots, x_d)$ and $(y_1, y_2, \ldots, y_d)$ are joined by an edge if and only if for some j, $1 \leq j \leq d$, we have $y_i = x_i$ for all i$\neq$j, and $|y_j - x_j| = 1$. A hypercube of dimension n, denoted by Q(n), is an undirected graph of 2^n vertices labelled by the integers between 0 and 2^n-1. There is an edge between two vertices when the binary representation of their labels differ in exactly one bit-position. Now for any graph G the smallest hypercube having at least as many points as G, namely Q($\lceil \log|G|\rceil$) (all logarithms being to base 2), is denoted by *Optimal-hypercube(G)* and is referred to as the optimal cube for G. Our main results are the construction of small dilation embeddings of $[a_1 \times a_2 \times \cdots \times a_d]$ into its optimal hypercube.

Dilation 2 embeddings of some 2-dimensional grids into their optimal hypercube have been described in a paper by Chan and Chin [CC]. We describe a general technique for embedding multi-dimensional grids into their optimal hypercube. Our results improve upon those of [CC].

Remark: For any t, $0 \leq t \leq n$, Q(n) contains a subgraph isomorphic to the grid $[2^t \times 2^{n-t}]$.

Our basic result will be that under certain conditions on m and k, there is a dilation 2 embedding of G=$[m \times k]$ into Optimal-hypercube(G). This is accomplished by constructing a dilation 2 embedding of G into $[\, 2^{\lceil \log m\rceil} \times 2^p \,]$, for some p$\geq$0.

An overview of how G is mapped into Optimal-hypercube(G) is as follows.

Step 1: Construct a dilation 2 embedding of $[m \times 2^{\lceil \log m\rceil}\,]$ into $[2^{\lceil \log m\rceil} \times m]$. We call this embedding a $(m, 2^{\lceil \log m\rceil})$-*tile*.

Step 2: Paste together enough such tiles end to end to obtain an embedding of G into $[2^{\lceil \log m\rceil} \times p]$, for some p, which is a subgraph of Optimal-hypercube(G).

The construction of these tiles and the manner of their pasting will guarantee that the embedding of G into Optimal-hypercube(G) has dilation 2.

Example 1. Consider the $[3 \times 10]$ grid. As this grid has 30 points, its optimal hypercube is Q(5) with 32 points. We use a 3-fold concatenation of the (3,4)-tile using only the first two points in each chain of the last tile. This concatenation is therefore an object with 4 rows and 8 columns, and will therefore represent an embedding of $[3 \times 10]$ into $[4 \times 8]$. The latter grid is a subgrid of Q(5), which is the optimal hypercube for $[3 \times 10]$. Tile T is shown in Figure 3(d). The technique to paste together the tile and its reverse image to achieve a dilation 2 embedding of the entire grid is shown in Figure 2.

Tiles, similar to those we describe, have been previously defined by Ellis [E]. Ellis used tiles for embedding with small dilation rectangular grids into square grids of optimal or near optimal size.

II. Dilation 2 Embedding

In this section we construct for m,k$\geq$1, a dilation 2 embedding of the $[m \times k]$ grid into the hypercube Q($\lceil \log mk\rceil$) under certain conditions. The embedding is carried out in two main steps. The first is to construct, for all $i \geq 1$ and all m, $(2^{i-1} \leq m \leq 2^i)$, dilation 2 embeddings of the $[m \times 2^i]$ grid into the $[2^i \times m]$ grid such that rows of $[m \times 2^i]$ are horizontal chains in $[2^i \times m]$. For example, we do not simply rotate $[m \times 2^i]$ by 90 degrees. This would embed rows of $[m \times 2^i]$ into columns of $[2^i \times m]$. Examples of desired $(m, 2^i)$ tiles are shown in Figure 3.

The basis step of the construction is given in Figure 3, with the following interpretation. Figure 3(d), for example, shows an embedding of a $[3 \times 4]$ grid into a $[4 \times 3]$ grid. There are three chains, each with four points, and these represent the three rows of the $[3 \times 4]$ grid. These chains are placed in a $[4 \times 3]$ array. It is easy to check that the embedding has dilation 2. For example, the second points in the second and third chains are adjacent as points in the $[3 \times 4]$ grid and are at distance 2 in the $[4 \times 3]$ grid. In Figure 3(c), we have embedded the $[2 \times 4]$ grid into the $[4 \times 2]$ grid. The figures representing an embedding of the $[m \times 2^i]$ grid into the $[2^i \times m]$ grid are examples of $(m, 2^i)$-tiles.

The inductive construction of the $(m, 2^i)$ tiles, where $2^{i-1} \leq m \leq 2^i$, from $(\lceil m/2 \rceil, 2^{i-1})$ and $(\lfloor m/2 \rfloor, 2^{i-1})$ tiles is given in Figure 4. The two not necessarily distinct $(m, 2^i)$ tiles are called *standard* and *alternate*.

We call the image of the ith row of $[m \times 2^i]$ under the embedding represented by the standard (alternate) $(m, 2^i)$-tile the ith *chain of the standard* (respectively, *alternate*) $(m, 2^i)$-tile.

Lemma 2.1. For all $i \geq 1$ and all m ($2^{i-1} \leq m \leq 2^i$), both the standard and alternate $(m, 2^i)$-tiles have dilation 2.

Proof: (by induction on i) The basis step is accomplished by inspection. That is, all $(m, 2^i)$-tiles, for $i \leq 2$, have dilation 2.

(Inductive Step) Assume that all standard and alternate $(p, 2^k)$-tiles have dilation 2, for all p $(2^{k-1} \leq p \leq 2^k)$. We show that standard and alternate $(m, 2^{k+1})$-tiles have dilation 2, for all m $(2^k \leq m \leq 2^{k+1})$. The argument is divided into eight cases based on the congruence class of m modulo 8. In all cases we show that the recursive construction results in dilation 2 by showing that the "seams" (where the four smaller component tiles join) are parts of already constructed $(p, 2^k)$ tiles and hence, by the inductive hypothesis, must have dilation 2. Clearly, this is sufficient, as the internal portions of each of the four component tiles used in the recursive construction must have dilation 2 by the inductive hypothesis. A horizontal seam, created by placing one component tile over another, preserves dilation 2, if, for all j, the jth point in the last chain in the upper tile is within distance 2 from the jth point in the first chain of the tile below. Our proof shows this to be the case. Similarly, a vertical seam, created by placing one component tile in front of another, preserves dilation 2, if the last point in the ith chain of the first tile is within distance 2 of the first point of the ith chain in the following tile, for all i. We show that the tiles are indeed constructed so that all seams preserve dilation 2.

case (1) m≡1 (mod 8)

The standard and alternate (m, 2^{k+1})-tiles, say T_1 and T_2, are constructed as shown in Figure 4(b), where the component tiles are: (1) A = standard ($\lceil m/2 \rceil$, 2^k)-tile, (2) B = alternate ($\lceil m/2 \rceil$, 2^k)-tile, and (3) D = alternate ($\lfloor m/2 \rfloor$, 2^k)-tile. As $\lceil m/2 \rceil \equiv 1$ (mod 4) and $\lfloor m/2 \rfloor \equiv 0$ (mod 4), the tiles T_1 and T_2 are in fact constructed as shown in Figure 5.

Observe that the horizontal and vertical seams (as shown in Figure 4(b)) must have dilation 2, as Figure 5 shows that these seams are parts of copies of either A, B, or D. That is, the middle of the tile shown in Figure 5(a) is a copy of A = standard ($\lceil m/2 \rceil$, 2^k)-tile, the middle of the tile shown in Figure 5(b) is a copy of B = alternate ($\lceil m/2 \rceil$, 2^k)-tile, and the remaining parts of the seams are parts of the tile D = alternate ($\lfloor m/2 \rfloor$, 2^k)-tile. More specifically, the middle of the tile consists of subtiles A_1 and A_2 and is identical to the tile A in the northwest quadrant. This means that the portions of the horizontal and vertical seams between A and B in the middle of the tile must have dilation 2, as these seams already appear inside A. Furthermore, the remaining parts of the horizontal and vertical seams consist of adjacent copies of A_2, which appear in the same way in the tile D. Hence these parts of the seams must also have dilation 2.

(The other cases are similarly handled and are not included here due to space limitations.) □

Let T be a (m,2^i)-tile (either standard or alternate). For any j ($1 \leq j < 2^i$) and k ($1 \leq k \leq m$), let $column_T(j,k)$, or simply $column(j,k)$ when T is understood, denote the column in which the jth point in chain k of T appears in. Let $left(T,j)=\min \{column(j,k) \mid 1 \leq k \leq m\}$ and $right(T,j) = \max \{column(j,k) \mid 1 \leq k \leq m\}$. That is, left(T,j) and right(T,j) are the leftmost column and rightmost column that a jth point in any chain of T appear. Let $offset(T,j) = right(T,j) - left(T,j)$ and $offset(T) = \max \{ offset(T,j) \mid 1 \leq j \leq 2^i \}$.

We provide a bound on the size of offset(T,j), for any (m,2^i) tile T. First we explain the purpose of bounding offset(T,j). Recall that to embed a [m×k] grid G into its optimal hypercube we first embed G into a [$2^{\lceil \log m \rceil} \times p$] grid G', for some integer p. To do this we partition G into a number of [m×$2^{\lceil \log m \rceil}$] subgrids and use a (m,$2^{\lceil \log m \rceil}$) tile T or its reversal (either standard or alternate) to embed each of these subgrids into a [$2^{\lceil \log m \rceil} \times m$] subgrid of G'. However, as k is often not an integer multiple of $2^{\lceil \log m \rceil}$, say k = $2^{\lceil \log m \rceil} + j$, the last subgrid of G is embedded into G' by a truncation of T to the first j points in each of its chains. Usually not all of these j^{th} points are located in the same column of T, so extra columns in G' beyond the minimum number of columns needed to receive the points of G may be required. We need to bound the number of columns in which these first j points in each chain of T occur in order to bound the size of the grid G' that contains the image of G. The minimum number of columns needed in a grid with $2^{\lceil \log m \rceil}$ rows just to have enough room to hold the points from the truncation of T to j points in each chain, denoted by $\mu(T,j)$, is $\lceil mj/2^{\lceil \log m \rceil} \rceil$. As the tile T may not achieve the optimum packing of points into the smallest available region, an upper bound for the number of columns actually used by T is the sum of $\mu(T,j)$ and offset(T,j). In general this is an overly generous estimate, but it seems too difficult to exactly compute right(T,j) and thereby the exact number of columns needed. The following helpful lemma can easily be shown.

Lemma 2.2. For any $i \geq 2$ and any $(m,2^i)$ tile T, offset(T) $\leq \lfloor i/2 \rfloor$.

We note that this upper bound on offset(T) is tight; it can be actually achieved.

Using our tiles and the preceeding result on offset we can now give the algorithm 2-EMBED which takes $[m \times k]$ as input and gives as output an embedding of $[m \times k]$ into the grid $L = [2^{\lceil \log m \rceil} \times H(m,k)]$, where $H(m,k) = \lceil mk/2^{\lceil \log m \rceil} \rceil + \lfloor \lceil \log m \rceil / 2 \rfloor$.

The algorithm 2-EMBED is described informally as follows:

1. Let T be the $(m, 2^{\lceil \log m \rceil})$ tile.

2. Let $j = \lfloor k/2^{\lceil \log m \rceil} \rfloor$, and let $k \equiv r \pmod{2^{\lceil \log m \rceil}}$.

   2a) If r=0, let the embedding of G be the j-fold concatenation of T.

   2b) If $r \neq 0$, let the embedding of G be the object obtained from the (j+1)-fold concatenation of T by deleting all but the first r points in the $(j+1)^{st}$ tile, i.e. the rightmost tile. Increase the number of columns to H(m,k).

We can now give our main result for dimension 2.

Theorem 2.1. Algorithm 2-EMBED produces a dilation 2 embedding of $G = [m \times k]$ into its optimal cube provided the following condition (*) holds.

(*)
$$\lceil \log m \rceil + \lceil \log \left(\lceil \frac{mk}{2^{\lceil \log m \rceil}} \rceil + \lfloor \frac{\lceil \log m \rceil}{2} \rfloor \right) \rceil \leq \lceil \log mk \rceil .$$

Some remarks concerning the sufficient condition (*) are in order. First we note that the full offset(T) $= \lfloor \lceil \log m \rceil / 2 \rfloor$ number of columns have been added to insure enough room. Clearly this is in general an overestimate. The number of additional columns (beyond the theoretical minimum $\mu = \lceil mk/2^{\lceil \log m \rceil} \rceil$) need only be the difference right(j) - μ. Since in general $\mu >$ left(j), offset(T) is an overly generous estimate of the required additional number of columns. We note also that the problem of adding additional columns can often be eliminated by an ad hoc and straightforward redesign of the last tile using additional edges of the hypercube.

III. Higher Dimensional Grids.

Under certain conditions [$a_1 \times a_2 \times \cdots \times a_d$] is a subgraph of its optimal hypercube and hence embeds into it with dilation 1. As shown in ([BS], [CC]) the smallest dimension of any hypercube containing $G_d = [a_1 \times a_2 \times \cdots \times a_d]$ as a subgraph is $\sum_{i=1}^{d} \lceil \log a_i \rceil$. It follows that G_d is a subgraph of its optimal hypercube precisely when

$$(**) \qquad \sum_{i=1}^{d} \lceil \log a_i \rceil = \lceil \log \prod_{i=1}^{d} a_i \rceil = \lceil \sum_{i=1}^{d} \log a_i \rceil .$$

We proceed to consider embeddings of G_d when this condition does not hold, i.e., when dilation at least 2 is required.

We begin by isolating certain 2-dimensional subgrids of G_d. For any i,j, where $1 \leq i < j \leq d$, $G_d(i,j)$ is the 2-dimensional subgrid of G_d consisting of the subgraph of G_d induced by the set of points $\{(0,0,\cdots,x,0,0,y,0\cdots 0) \mid 0 \leq x \leq a_i, 0 \leq y \leq a_j\}$, where x occupies position i and y occupies position j. We refer to $G_d(i,j)$ as a *face* of G_d. Now for a fixed set of d-2 integers $z_1, z_2, ..., z_{i-1}, z_{i+1}, z_{i+2}, ... , z_{j-1}, z_{j+1}, z_{j+2}, ..., z_d$ we call the subgraph of G_d induced by $\{(z_1, z_2, ... , z_{i-1}, x, z_{i+1}, ... , z_{j-1}, y, z_{j+1}, ... , z_d) \mid 0 \leq x \leq a_i, 0 \leq y \leq a_j \}$ a *face of G_d parallel to $G_d(i,j)$*. For any point x in a d–dimensional grid, let x_i denote the ith coordinate of x.

Given our grid G_d, we describe a corresponding grid G_d' of the same dimension satisfying condition $(**)$. The algorithm which follows this description will construct a small dilation embedding $f: G_d \rightarrow G_d'$. By condition $(**)$, G_d' is a subgraph of its optimal hypercube. Under certain additional conditions to be described later, this hypercube will also be optimal for G_d. The end result will then be a small dilation embedding of G_d into its optimal hypercube.

Pick an integer k, where $0 \leq k \leq d-1$. We begin by calling on 2-EMBED(a_1, a_d). Recall that it outputs a dilation 2 embedding $g: [a_1 \times a_d] \rightarrow [2^{\lceil \log a_1 \rceil} \times H(a_1, a_d)]$. Let $b^{(1)} = H(a_1, a_d)$, and call on 2-EMBED($a_2, b^{(1)}$). Let $b^{(2)} = H(a_2, b^{(1)})$. In the inductive step, having the number $b^{(i-1)}$, we call on 2-EMBED($a_i, b^{(i-1)}$) and let $b^{(i)} = H(a_i, b^{(i-1)})$. Finally let $b = b^{(k)}$. Then G_d' is the d-dimensional grid $[2^{\lceil \log a_1 \rceil} \times 2^{\lceil \log a_2 \rceil} \times \cdots \times 2^{\lceil \log a_k \rceil} \times a_{k+1} \times a_{k+2} \times \cdots \times a_{d-1} \times 2^{\lceil \log b \rceil}]$. We now present the algorithm described above.

EMBED(k, G_d)

for each $z \in G_d$ **do** $f(z) := z$; $\qquad$ /*f is computed after several iterations. Initially, it is the identity function.* /

$B := a_d$;

for $i = 1$ to k **do**

begin $\quad$ 2–EMBED(a_i, B) $\qquad$ /*Remark: new values of dimension d are given by 2-EMBED and are returned as the variable B.* /

$\qquad$ **for** each point $z \in G_d$ **do**

$\qquad$ **begin** $\quad$ Let (Δ_1, Δ_2) be such that the image of (a,b) under

$\qquad \qquad$ the embedding given by 2-EMBED(a_i, B) is $(a + \Delta_1, b + \Delta_2)$

$\qquad \qquad f(z)_i := f(z)_i + \Delta_1$ $\qquad$ /*This alters the ith and d^{th} coordinates of each point z in the same way*

$\qquad \qquad \qquad \qquad \qquad \qquad$ *as in the 2D embedding defined by 2–EMBED(a_i, B). Thus, it embeds.*

$\qquad \qquad f(z)_d := f(z)_d + \Delta_2$ $\qquad$ *all 2D faces of G_d parallel to $G_d(i,d)$ in the same way* /

$\qquad$ **end**

end

Example 2. Embedding $[3 \times 3 \times 13]$ grid into $Q(7)$. Let g denote this embedding.

First embed $[3 \times 3 \times 13]$ into $[4 \times 3 \times 10]$ using 2–EMBED(3,13) as shown in Figure 6(a); let g^1 denote this embedding. Then embed $[4 \times 3 \times 10]$ into $[4 \times 4 \times 8]$ using 2–EMBED(3,10) as shown in Figure 6(b); let g^2 denote this embedding. In Figure 6(c), we show the resulting embedding which is the composition of the previous steps, i.e. $g = g^2 g^1$.

We note that the composition of k dilation 2 embeddings in general yields dilation 2^k. However, in our case, the dilation after k compositions of 2D embeddings on pairwise orthogonal faces is k+1. Basically this is due to the fact that any single 2D embedding used, maps points that differ only by 1 in the dth dimension to points that differ by at most 1 in the dth dimension (although their images may differ in the other currently considered dimension by 1 as well). As the dth dimension is the only dimension that is considered repeatedly and the other k dimensions are considered once, the result is that each point is mapped to a new point in which at worst k+1 dimensions have changed ... each by at most 1. Consequently, the dilation is at most k+1.

Theorem 3.1. For any k $(1 \leq k \leq d{-}1)$, the algorithm EMBED(k,G_d) constructs a dilation k+1 embedding of $[a_1 \times a_2 \times a_3 \times \cdots \times a_d]$ into its optimal hypercube provided that the following condition C_k holds:

$$(C_k): \quad \sum_{i=1}^{d-1} \lceil \log a_i \rceil + \lceil \log B_k \rceil \leq \lceil \sum_{i=1}^{d} \log a_i \rceil, \quad \text{where} \quad B_k = \frac{a_d \prod_{i=1}^{k} a_i}{\prod_{i=1}^{k} 2^{\lceil \log a_i \rceil}} + \sum_{i=1}^{k} \lfloor \frac{\lceil \log a_i \rceil}{2} \rfloor.$$

The condition C_k is derived from the fact that several 2-dimensional embeddings of Section II are composed in the algorithm EMBED(k,G_d) and each contributes a possible offset. Note also that for any fixed ordering, condition C_k becomes progressively harder to satisfy as k decreases. This makes sense as with decreasing k we get a stronger result, i.e. a smaller dilation embedding of G_d into its optimal hypercube.

Finally, we remark that in the worst case, i.e. k=d, we obtain a dilation d embedding provided (C_d) is satisfied. The next theorem shows that this is off by a little more than a factor of log d from a lower bound for embeddings of grids with equal sides. As notation, let $[q]^d$ denote the d - dimensional grid $[q \times q \times \cdots \times q]$ (where there are d factors q). Of course the optimal cube for $[q]^d$ is $Q(\lceil d \log q \rceil)$. For a function h(n) we let $\Omega(h(n))$ denote any function which is bounded below by a constant times h(n) for sufficiently large n and as usual let $O(h(n))$ (resp. $o(h(n))$) denote a function f(n) such that $\lim_{n \to \infty} |f(n) / h(n)| \leq c$ (resp. $\lim_{n \to \infty} |f(n) / h(n)| = 0$,) for some constant c. Finally, we let $\Theta(h(n))$ denote a function f(n) such that $f(n) = O(h(n))$ and $f(n) = \Omega(h(n))$.

Theorem 3.2: Let $f = f(k) \geq k$ be a function of k and let $g : [f(k)]^k \to Q([k \log f])$ be an embedding. Then dilation$(g) = \Omega(\, k \log (k + \log f) \, / \, \log f \log (k \log f) \,)$, as $k \to \infty$.

Proof: Let dilation$(g) = D$.

Consider the sphere S of radius r in $[f(k)]^k$ centered at $(0,0,\ldots,0)$, that is, the set of points in $[f(k)]^k$ that are at graph distance at most r from the origin. Let $s(r,k) = |S|$. Clearly, g must map S into a sphere of radius at most rD in $Q = Q([k \log f])$. It follows that the latter sphere has at least as many points as S. For a properly chosen r we will see that the resulting inequality implies the theorem.

We first calculate the number $n(r,k)$ of points in $[f(k)]^k$ at distance exactly r from the origin. Clearly $n(r,k)$ is related to partitions in number theory. Thus consider the number $N(d,k)$ of ordered k-tuples $(x_1, x_2, \ldots, x_k)$, each coordinate a nonnegative integer, such that $\sum_{i=1}^{k} x_i = d$. By letting each such k-tuple correspond in the natural way to a way of distributing d indistinguishable objects into k distinguished cells, we see $N(d,k) = \binom{d+k-1}{k-1}$. Now for $r \leq f(k)$, we have $n(r,k) = N(r,k)$. We later choose r to be $O(\log f) << f$, so that we may in fact assume $n(r,k) = \binom{r+k-1}{k-1}$. It now follows that $s(r,k) = \sum_{j=0}^{r} n(j,k) = \sum_{j=0}^{r} \binom{j+k-1}{k-1} = \binom{r+k}{k}$.

Now the number of points in the sphere of radius rD in Q is $\sum_{i=0}^{rD} \binom{q}{i}$, where $q = [k \log f]$. We now bound this from above as follows. Observe that we may suppose $rD \leq O(k)$, since otherwise with $r = \log f$ we obtain $D = \Omega(k \, / \, \log f)$ and there is nothing left to prove. We can then use standard estimates on the tail of the binomial distribution [Fe] which imply that for $s = o(n)$ we have $\sum_{i=0}^{s} \binom{n}{i} = O\left(\binom{n}{s}\right)$. Since $rD = O(k) = o(q)$, it follows in our case that $\sum_{i=0}^{rD} \binom{q}{i} = O\left(\binom{q}{rD}\right)$. We are therefore led to $\binom{q}{rD} \geq \binom{r+k}{k}$, with constant factors implicit but not indicated. Now using $\binom{n}{i} = \Theta(n^i / i!)$ and Stirling's formula $n! = \Theta(\sqrt{n}(\frac{n}{e})^n)$, we obtain $q^{rD} = \Omega\left(\sqrt{rD} \; e^{k-rD} \, (rD)^{rD} \, \dfrac{(r+k)^k}{\sqrt{k} \, k^k}\right)$. On taking logarithms we get

$$D = \Omega\left(\frac{1}{r \log q}\left[k - rD + \frac{\log rD}{2} - \frac{\log k}{2} + rD \log (rD) + k \log (r+k) - k \log k \right]\right).$$

Clearly the first three terms in the brackets are dominated by the others. Since $rD = O(k)$, we see that $rD \log (rD)$ and $k \log k$ are dominated by $k \log (r+k)$ for large k. Hence with $r = \log f$ we get the stated result.

Corollary: If $f(k)$ is any polynomial bounded function of k such that $f(k) > k$ then any embedding $g : [f(k)]^k \to Q([k \log f])$ satisfies dilation$(g) = \Omega(k \, / \, \log (k \log k))$.

IV Conclusion

We note that Chan [Ch] has recently described a technique for embedding all 2D meshes into their optimal hypercubes with dilation at most 2. Thus the 2D embedding problem is solved. The principal open problem left is to describe embeddings of grids of arbitrary dimensions that are provably optimal. In particular, are there grids G whose embeddings into Optimal-hypercube(G) have dilation which is within a constant factor of the lower bound given in Theorem 3.2 ? The interested reader should also consult other recent work on embeddding meshes into hypercubes ([G], [HJ]).

References

[AR]   R. Aleliunas and A. L. Rosenberg, "On Embedding Rectangular Grids in Square Grids", *IEEE Trans. on Computers*, C-31,9 (1982), pp. 907-913.

[BCLR]   S. Bhatt, F. Chung, T. Leighton, and A. Rosenberg, "Optimal Simulation of Tree Machines", *Proc. 27th Annual IEEE Symp. Foundations of Computer Sci.* , Oct. 1986, pp. 274-282.

[BS]   J. E. Brandenburg and D. S. Scott, "Embeddings of Communication Trees and Grids into Hypercubes", *Intel Scientific Computers Report* #280182-001, 1985.

[CC]   M. Y. Chan and F. Y. L. Chin, "On Embedding Rectangular Grids in Hypercubes", to appear in *IEEE Trans. on Computers*. (Tech. Report TR-B2-87, February, 1987, Centre of Computer Studies and Applications, University of Hong Kong, Pokfulam Road, Hong Kong)

[Ch]   M. Y. Chan, " Dilation 2 Embedding of Grids into Hypercubes", Technical Report, Univerity of Texas at Dallas, Computer Science Program, 1988.

[CMST]   M.-J. Chung, F. Makedon, I. H. Sudborough and J. Turner, "Polynomial Algorithms for the Min-Cut Linear Arrangement Problem on Degree Restricted Trees", *SIAM J. Computing* 14,1 (1985), pp. 158-177.

[E]   J. A. Ellis, "Embedding Rectangular grids into Square Grids", *Proc. of Aegean Workshop On Computing*, Corfu, Greece 1988.

[Fe]   W. Feller, "An Introduction to Probability Theory and Its Applications",

[G]   D. S. Greenberg, "Optimum Expansion Embeddings of Meshes in Hypercube", Technical Report YALEU/CSD/RR-535, Yale University, Dept. of Computer Science.

[HLN]   J. Hastad, T. Leighton, and M. Newman, "Reconfiguring a Hypercube in the Presence of Faults", *Proc. 19th Annual ACM Symp. Theory of Computing*, May 25-27, 1987.

[HJ]   Ching-Tien Ho and S. Lennart Johnson, "On the Embedding of Arbitrary Meshes in Boolean Cubes with Expansion Two Dilation Two", *Proc. 1987 International Conference on Parallel Processing*, pp. 188-191.

[Y]   M. Yannakakis, "A Polynomial Algorithm for the Min Cut Linear Arrangement of Trees", *J. ACM*, 32,4 (1985), pp. 950-959.

f(1,1) = 0000   f(2,3) = 1011
f(1,2) = 0001   f(2,4) = 1111
f(1,3) = 0111   f(2,5) = 0000
f(1,4) = 0101   f(3,1) = 1000
f(1,5) = 0100   f(3,2) = 1010
f(2,1) = 0010   f(3,3) = 1001
f(2,2) = 0011   f(3,4) = 1101
        f(3,5) = 1100

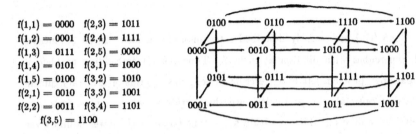

Figure 1. A dilation 2 embedding f of the [3×5] grid into the hypercube with 16 points. The adjacent points (1,3) and (2,3) in the grid, for example, are mapped to 0111 and 1011, respectively, which are at distance 2 in the hypercube.

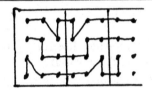

Figure 2. A partition of a [3×10] grid into two [3×4] subgrids and one additional [3×2] subgrid and the embedding of each of these [3×4] subgrids into a [4×3] subgrid of the [4×8] grid via the tile T or its reverse image T^R

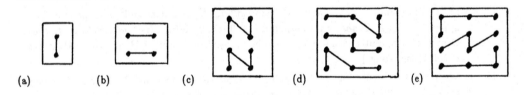

(a)   (b)   (c)   (d)   (e)

Figure 3. (a) the standard and alternate (1,2)-tile, (b) the standard and alternate (2,2)-tile, (c) the standard and alternate (2,4)-tile, (d) the standard (3,4)-tile, (e) the alternate (3,4)-tile

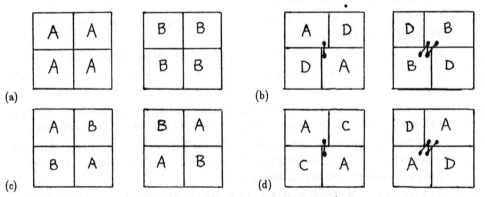

(a)   (b)   (c)   (d)

Figure 4. Constructing the standard and alternate $(m,2^i)$-tile when (a) $m \equiv 0 \pmod 4$, (b) $m \equiv 1 \pmod 4$, (c) $m \equiv 2 \pmod 4$, and $m \equiv 3 \pmod 4$, where A is the standard ($\lceil m/2 \rceil$, 2^{i-1})-tile, B is the alternate ($\lceil m/2 \rceil$, 2^{i-1})-tile, C is the standard ($\lfloor m/2 \rfloor$, 2^{i-1})-tile, and D is the alternate ($\lceil m/2 \rceil$, 2^{i-1})-tile.

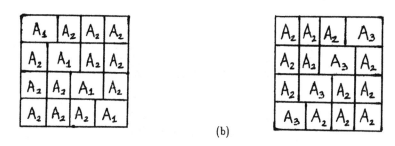

Figure 5. (a) Construction of a standard $(m, 2^{k+1})$-tile, when $m \equiv 1 \pmod 8$, and (b) construction of an alternate $(m, 2^{k+1})$-tile, when $m \equiv 1 \pmod 8$, where each A_1 is a standard $(\lceil m/4 \rceil, 2^{k-1})$-tile, A_2 is an alternate $(\lfloor m/4 \rfloor, 2^{k-1})$-tile, and A_3 is an alternate $(\lceil m/4 \rceil, 2^{k-1})$-tile.

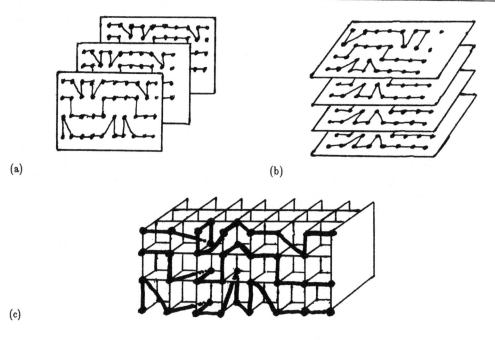

Figure 6. (a) Embedding of parallel faces of the $[3 \times 3 \times 13]$ grid into $[4 \times 3 \times 10]$ grid; (b) embedding of parallel faces of the $[4 \times 3 \times 10]$ grid into $[4 \times 4 \times 8]$ and (c) example of chains produced by composition to obtain the embedding of $[3 \times 3 \times 13]$ grid into $[4 \times 4 \times 8]$ grid.

Compaction on the Torus[*]
(Extended Abstract)

K. Mehlhorn, W. Rülling
Fachbereich 10, Informatik
Universität des Saarlandes
D-6600 Saarbrücken

Abstract: In this paper we introduce a general framework for compaction on a torus. This problem comes up whenever an array or row of identical cells has to be compacted. We instantiate our framework with several specific compaction algorithms: one-dimensional compaction without and with automatic jog insertion and two-dimensional compaction.

I. Introduction

A *compactor* takes as input a VLSI-Layout and produces as output an equivalent layout of smaller area. An effective compaction system frees the designer from the details of the design rules and hence increases his productivity and on the other hand produces high quality layouts. For these reasons, compaction algorithms have gained widespread attention in the VLSI-Literature [Hsueh, Kedem/Watanabe, Leiserson/Pinter, Lengauer, Maley, Mehlhorn/Näher], and are the basis for several computer-aided circuit design systems [Dunlop, Hsueh, Lengauer/Mehlhorn, Rülling, Williams].

Regular layouts composed of rows or arrays of identical cells arise frequently in practice, e.g., bit slice architecture or systolic arrays. Let S be the cell to be replicated. We adress the following problem:

> Compact S into a cell S' such that cell S' still can be used to tile the plane (or an infinite strip)

This problem is called the compaction problem on the torus, because a layout S can be used to tile an array iff its left and right, and top and bottom boundaries are compatible, i.e. if the cell S can be drawn on a torus, cf. figure 1. The compaction problem on a torus is interesting for three reasons.

(1) Row-like and array-like arrangements of a single cell arise frequently in practice. In such an arangement it is desirable to compact all instances of the cell identically to

---

[*] This research was supported by the DFG, grant SFB 124, TP B2

(2) guarantee identical electrical behavior of all instances and to

(3) allow further hierarchical processing.

Our own interest in compaction on a torus was stimulated by a Kulisch-arithmetic-chip designed by P. Lichter [Lichter]. A central component of this design is an accumulator consisting of 1152 identical cells which are arranged into a 36 by 32 array. The fully instantiated layout overstrained the compactor of the HILL-system (although it can handle 100 000 rectangles) and so compaction on a torus was called for. Several simple-minded approaches failed. Compacting a single cell does not guarantee tileability, compacting the layout using the algorithm for hierarchical compaction by Lengauer does not guarantee that the instances stay identical, and compacting a single cell and insisting that the boundary stays rectangular wastes area, although it guarantees tileability. Finally, the approach of Eichenberger/Horowitz works only for constraint based compaction without jog insertion.

In this paper we describe a framework for compaction on a torus. It can be combined with several known compaction algorithms, e.g. one-dimensional compaction [Hsueh,Lengauer], one-dimensional compaction with automatic jog insertion [Maley, Mehlhorn/Näher], and two-dimensional compaction [Kedem/Watanabe], to yield specific compaction algorithms.

Our approach is very simple. Let the cell S have length L and height H. We draw S on a cylinder of circumference L and height H. If S is supposed to tile the plane (instead of a strip) then we also identify the upper and lower rim of the cylinder and obtain a torus. We now let the circumference shrink. In this way the features of the cell will move closer together until a tight cut, i.e., two features reaching their minimum separation, arises. These two features are kept at their mimimum separation from now on. We continue in this fashion until a cycle of saturated cuts around the cylinder (or torus) arises. At this point we have minimized the x-width of the cell but still guarantee that it can tile a strip (or the plane).

In section II we describe our approach in more detail and fill in some algorithmic details. We stay however on the generic level. In section III we instantiate our framework in three specific cases: one-dimension compaction without jog insertion (section III.1), two-dimensional compaction (section III.2), one-dimensional compaction with jog insertion (section III.3).

II. Definitions and Results

We give a precise definition of the cylindrical compaction problem. A *cylindrical sketch* is a quadruple (F, W, P, L) consisting of a cylinder Z of circumference L, a finite set F of *features*, which are points (= point feature) and open straight line segments (= line feature) on the surface of Z, a finite set W of *wires*, which are simple paths on the surface of Z, and a partition P of the features F. Each block of the partition is called a *module*. figure 1 shows an example of a sketch. When the partition P and the period L are understood we will refer to a pair (F, W) as a sketch. The features and wires of a sketch must satisfy the following conditions:

(1) Distinct features do not intersect and the endpoints of each line feature are point features.
(2) No wire may cross itself.
(3) Each wire touches exactly two features, which are point features lying at the endpoints of the wire. They are called the *terminals* of the wire.

A *point in a sketch* is a point lying on a feature. Modules form the rigid part of a layout and wires represent the flexible interconnections.

Sketches comprise the information of placement and global routing. A (detailed) *routing* of a sketch (F, W, P, L), $W = \{p_1, \ldots, p_m\}$, is a sketch $(F, W', P, L), W' = \{q_1, \ldots, q_m\}$, such that q_i is *homotopic* to p_i, i.e., p_i and q_i have the same endpoints and p_i can be transformed continuously into q_i without moving its endpoints and without allowing its interior to touch a feature in F, and such that the q_i's satisfy the constraints of the particular wiring model used. We consider only the grid model in this extended abstract; our results extend however to any polygonal wiring norm. In the *grid model* wires are rectilinear paths with a minimum vertical and horizontal separation of 1.

A *cut* C is any open line segment connecting two points of the sketch, say p and q, and not intersecting any feature. The *density* of cut C is the number of crossings of C by wires which are enforced by the topology of the sketch, cf. figure 2. Crossings of C which can be removed by deforming the wires do not contribute to the density. The *capacity* of a cut in the grid model is given by $\max\{x\text{-length}(C), y\text{-length}(C)\} - 1$, where $x\text{-length}(C)$ ($y\text{-length}(C)$) is the length of C in horizontal (vertical) direction, cf. figure 3. A cut is called *safe* if its density does not exceed its capacity and it is called *tight* or *saturated* if its density is equal to its capacity. The following theorem was proved by Cole/Siegel and Leiserson/Maley for the grid model.

Theorem 1. *A sketch has a routing iff all cuts of the sketch are safe.*

Actually, the results are slightly stronger. Let us call a cut $\overline{pq}$ *critical*, if either p and q are point features or at most one of them lies on a line feature and the line segment $\overline{pq}$ is perpendicular to that line feature. Then a sketch is routable iff all critical cuts are safe.

With every cylindrical sketch S we can associate an infinite planar sketch S^∞ as follows. Let R be any vertical line on the cylinder. Then S^∞ is obtained by unrolling the cylinder and then tiling a strip with the unrolled cylinder, cf. figure 3.

We are now ready to define the one-dimensional cylindrical compaction problem. The goal of compaction is to displace the modules in x-direction such that the resulting sketch is routable and has minimal period. Let $S = (F, W, P, L)$ be a *routable cylindrical sketch* and let S^∞ be the associated planar sketch. We denote the different instances of a feature f by $f_i, i \in \mathbf{Z}$. A displacement (or configuration) of S^∞ is given by a vector $d \in \mathbb{R}^{F \times \mathbf{Z}}$; $d(f_i)$ is the displacement of feature f_i. Of course, not all displacements make sense. Firstly, features in the same module must be displaced by the same amount and therefore we must have $d(f_i) = d(g_j)$ for any two feature instances in the same module. Secondly, features should not cross over during compaction and we therefore must have $x_p + d(f_i) < x_q + d(g_j)$ for any two points $p = (x_p, y_p)$ and $q = (x_q, y_q)$ where $x_p < x_q$ and $y_p = y_q$ and p lies on a feature f_i and q lies on a feature g_j. Let d be a configuration satisfying the two constraints above. We can now define the sketch $S^\infty(d)$ in a natural way. A point p on feature instance f_i with coordinates (x_p, y_p) in the sketch S^∞ has coordinates $(x_p + d(f_i), y_p)$ in $S^\infty(d)$ and the wires in $S^\infty(d)$ have the "same" homotopies as in S^∞; cf. [Maley 85] for a more precise definition. The *configuration space* $C(S) \subseteq \mathbb{R}^F \times \mathbb{R}$ of a cylindrical sketch S consists of all pairs $(d, \delta), d \in \mathbb{R}^F, \delta \in \mathbb{R}$, such that the configuration $\tilde{d} \in \mathbb{R}^{F \times \mathbf{Z}}$ with $\tilde{d}(f_i) = d(f) + i\delta, f \in F$ and $i \in \mathbf{Z}$, satisfies the two constraints above and $S^\infty(\tilde{d})$ is routable. Note that the pair $(0, L)$, where 0 is the zero-vector, belongs to $C(S)$, since the sketch S is assumed to be routable. Also note that the sketch $S^\infty(\tilde{d})$ can be wrapped around a cylinder of circumference δ and hence gives rise to a cylindrical sketch which we denote $S(d, \delta)$. The *essential configuration* space $C_0(S)$ of a sketch S consists of that connected component of $C(S)$ which contains the pair $(0, L)$, i.e., a configuration (d, δ) belongs to $C_0(S)$ if the cylindrical sketch $S(d, \delta)$ can be obtained from $S = S(0, L)$ by continuously shrinking the cylinder and deforming the layout drawn on its surface whilst maintaining the routability of the sketch.

Definition: One-dimensional cylindrical compaction problem

    Input: A routable cylindrical sketch $S = (F, W, P, L)$

    Output: A configuration $(d, \delta) \in C_0(S)$ such that δ is minimal.

Theorem 2. *Let* $S = (F, W, P, L)$ *be a routable cylindrical sketch. Then the essential configuration space* $C_0(S)$ *of the sketch* S *is a convex polyhedron.*

Proof: [Maley 85] proved the analogous result for planar sketches. Because of the correspondence between cylindrical sketches and periodic planar sketches described above the result carries over.

Before we can state our results we need one additional notation. Let us assume that the wires in a sketch are given by polygonal paths. For a wire $w \in W$, let b_w be the number of line segments in the polygonal path for w, let $b = \sum_{w \in W} b_w$, and let $m = |F|$ be the number of features.

Theorem 3 (Cylindrical Compaction with Automatic Jog Insertion) .
In the grid model the cylindrical compaction problem can be solved in time
$O(m^3 W_{max}^2 \log m + K \log m) = O(m^4 W_{max}^2 \log m)$ *where* $W_{max} = 1 + \lfloor H/\Delta_{min} \rfloor$,
H *is the height of the sketch* S, Δ_{min} *is the period of the compacted layout and* K
is the number of times a feature moves across a critical cut during compaction. ∎

The quantity K is a measure of how much the sketch changes during compaction.
We believe that the bound $K \leq m^4 W_{max}^2$ which we derive in section IV is overly
pessimistic.

Compaction without jog insertion is a special case of theorem 3. Let us assume
that wires are specified as rectilinear polygonal paths; view vertical wire pieces as
modules and horizontal wire pieces as wires in the sense of the definition of a sketch,
cf. figure 4. Then the compaction of such a sketch is tantamount to compaction
without jog insertion.

Theorem 4. *Cylindrical compaction without jog insertion can be solved in time*
$O(m^2 \log m)$.

Cylindrical compaction without jog insertion was previously considered by Eichen-
berger/Horowitz. They did not analyse their algorithm.

Our approach can also handle maximum and minimum distance constraints which
are specified by the user as long as the constraints are satisfied by the initial layout S. Since *toroidal compaction* amounts to cylindrical compaction in the presence of equality constraints between the upper and the lower cell boundary our
algorithms carry over to toroidal compaction with unchanged running time. Finally, we want to mention that the algorithm underlying theorem 4 can be used for
Kedem/Watanabe-like two-dimensional compaction.

III. Compaction on a Torus: the Framework

In this section we describe the general framework for compaction on a torus. For
simplicity, we deal only with the cylinder in this extended abstract. Let $S = (F, W, P, L)$ be the cylindrical sketch to be compacted; let $\Delta = L$.

The central concept of our approach is shrinking which we define next. Let $S(\Delta)$ be
the sketch obtained for the circumference Δ of the cylinder and let R be a vertical
line on the cylinder which we call the reference line. For a feature f let $p(f, \Delta)$
be the distance from f to the reference line when going to the left starting in f.
The local meaning of "left" and "right" is defined by viewing the cylinder from the
outside. We refer to $p(f, \Delta)$ as the position of f in the sketch $S(\Delta)$. For a cut C let
the wrapping number $w(C, \Delta)$ be the number of intersections between C and the
reference line R. We extend the concept of wrapping number to features as follows.

Consider an auxiliary digraph G_A with vertex set $F \cup \{R\}$. For every feature f there is an edge (R, f) of cost 0 and for every saturated cut C with endpoints f and g, where the left-to-right orientation is from f to g, there is an edge (f, g) of cost $w(C, \Delta)$. We denote such a cut C by $\overline{fg}$; note that this notation is ambiguous since only the endpoints together with the wrapping number identify a cut. Let us assume for the moment that the auxiliary graph G_A is acyclic; the other case is treated in the proof of lemma 3 below. Let $T(\Delta)$ be a longest path tree with root R in the auxiliary graph G_A, let $w(f, \Delta)$ be the length of a longest path from R to f in G_A and let $d(f, \Delta) = \Delta \cdot w(f, \Delta) + p(f, \Delta)$, cf. figure 5 for an illustration. We refer to $w(f, \Delta)$ as the wrapping number of f and to $d(f, \Delta)$ as the distance from R to f in $S(\Delta)$. With these concepts it is now easy to define the sketch $S(\Delta - \epsilon)$. The position $p(f, \Delta - \epsilon)$ of feature f in $S(\Delta - \epsilon)$ is given by $d(f, \Delta) \bmod (\Delta - \epsilon)$. The wire homotopy in $S(\Delta - \epsilon)$ is defined in the natural way by considering the continuous transformation (ϵ grows starting at 0) from the positions $p(f, \Delta)$ to the positions $p(f, \Delta - \epsilon), f \in F$.

Lemma 1. a) Let $C = \overline{fg}$ be a cut and let $x(C, \Delta) := p(g, \Delta) - p(f, \Delta) + w(C, \Delta) \cdot \Delta$ be the x-length of C in $S(\Delta)$. Then the x-length of C in $S(\Delta - \epsilon)$ is given by $x(C, \Delta - \epsilon) = x(C, \Delta) + \epsilon(w(g, \Delta) - w(f, \Delta) - w(C, \Delta))$.
b) If $T(\Delta)$ exists then $S(\Delta - \epsilon)$ is legal for $\epsilon > 0$ sufficiently small.

Proof: a) Let $\Delta' = \Delta - \epsilon$. Then

$$x(C, \Delta') = p(g, \Delta') - p(f, \Delta') + w(C, \Delta') \cdot \Delta'$$
$$= (p(g, \Delta) + w(g, \Delta) \cdot \Delta) \bmod \Delta' - (p(f, \Delta) + w(f, \Delta) \cdot \Delta) \bmod \Delta'$$
$$+ w(C, \Delta') \cdot \Delta'$$
$$= (p(g, \Delta) + w(f, \Delta) \cdot \epsilon) \bmod \Delta' - (p(f, \Delta) + w(f, \Delta) \cdot \epsilon) \bmod \Delta'$$
$$+ w(C, \Delta') \cdot \Delta - w(C, \Delta') \cdot \epsilon$$

Let us assume for simplicity that $p(g, \Delta) + w(g, \Delta) \cdot \epsilon < \Delta'$ and $p(f, \Delta) + w(f, \Delta) \cdot \epsilon < \Delta'$; the other case is similar and left to the reader. Then $w(C, \Delta') = w(C, \Delta)$ and hence $x(C, \Delta') = x(C, \Delta) + \epsilon(w(g, \Delta) - w(f, \Delta) - w(C, \Delta))$.

b) Let $C = \overline{fg}$ be any cut. Clearly, if C is not tight in $S(\Delta)$ then C is not tight in $S(\Delta - \epsilon)$ for ϵ sufficiently small. If C is tight in $S(\Delta)$ then there is an edge (f, g) of cost $w(C, \Delta)$ in the auxiliary graph and hence $w(g, \Delta) \geq w(f, \Delta) + w(C, \Delta)$ by the definition of wrapping number. Thus $x(C, \Delta - \epsilon) \geq x(C, \Delta)$. Since the y-coordinates of the features do not change during shrinking the capacity of C does not go down when passing from $S(\Delta)$ to $S(\Delta - \epsilon)$. Also, the density of C does not change for ϵ sufficiently small. Thus $S(\Delta - \epsilon)$ is legal for ϵ sufficiently small. ∎

Lemma 1 is the basis for our compaction algorithm. If $T(\Delta)$ exists, then the shrinking process yields a legal sketch of smaller period. This leads to the following algorithm.

$\Delta \leftarrow w$; $S(\Delta) \leftarrow S$ (* the initial sketch S has period w *)

while $T(\Delta)$ exists

do let $\epsilon > 0$ be maximal such that $S(\Delta - \epsilon)$ is legal;

    compute $S(\Delta - \epsilon)$ and $T(\Delta - \epsilon)$;

    $\Delta \leftarrow \Delta - \epsilon$;

od;

It remains to prove termination (lemma 2) and correctness (lemma 3). Let $W_{max}(\Delta) = \max\{w(C,\Delta); C = \overline{fg}$ is a tight cut and there is no sequence $f_0, \ldots, f_k$ such that $f = f_0, g = f_k, C_i = \overline{f_i f_{i+1}}$ is tight for $0 \leq i < k$, and $w(C,\Delta) = \sum_i w(C_i,\Delta)\}$, i.e., $W_{max}(\Delta)$ is the maximal wrapping number of any cut which must be used in $T(\Delta)$. We prove upper bounds for W_{max} in various compaction models in section IV. In particular, $W_{max}(\Delta) = 1$ for compaction without jog insertion and $W_{max}(\Delta) \leq 1 + \lfloor k/\Delta \rfloor$ for compaction with jog insertion in the grid model.

Lemma 2. a) $0 \leq w(f,\Delta) \leq m \cdot W_{max}(\Delta)$ for all f and Δ and $w(f,\Delta)$ is non-decreasing for every f.

    b) The number of iterations is bounded by $m^2 W_{max}(\Delta_{min})$ where Δ_{min} is the period of the final sketch.

Proof: a) The bounds $0 \leq w(f,\Delta)$ and $w(f,\Delta) \leq m \cdot W_{max}(\Delta)$ follow immediately from the definition of wrapping numbers. We show next that $w(f,\Delta)$ is non-decreasing for every f. Consider any $S(\Delta)$ and let ϵ be maximal such that $S(\Delta - \epsilon)$ is legal. Then there must be a cut $C = \overline{fg}$ which is tight in $S(\Delta - \epsilon)$ and oversaturated in $S(\Delta - \epsilon - \delta)$ for $\delta > 0$. Thus $w(g,\Delta) - w(f,\Delta) - w(C,\Delta) < 0$ by lemma 1a.

Consider the layouts $S(\Delta - \delta)$ where $0 \leq \delta < \epsilon$. In these layouts exactly the cuts $D = \overline{hk}$ with $w(k,\Delta) = w(h,\Delta) + w(D,\Delta)$ are tight. This follows from lemma 1a and the definition of ϵ. In particular, all cuts in $T(\Delta)$ stay tight. Moreover, the tree $T(\Delta - \delta), 0 \leq \delta < \epsilon$, is independent of δ. This can be seen as follows. As we increase δ from 0 to ϵ, the wrapping number of a feature h increases by one whenever h moves across the reference line R on the cylinder. Note that in this case the wrapping number of all cuts incident to h and leaving h to the left goes up by one and the wrapping number of the cuts leaving h to the right goes down by one. Thus the longest path tree does not change. The argument also shows that the quantity $w(g, \Delta - \delta) - w(f, \Delta - \delta) - w(C, \Delta - \delta)$ is a constant independent of δ.

In $S(\Delta - \epsilon)$ the cut C becomes tight and is added to the auxiliary graph. Since $w(g, \Delta - \delta) - w(f, \Delta - \delta) - w(C, \Delta - \delta) = w(g,\Delta) - w(f,\Delta) - w(C,\Delta) < 0$ and hence $w(f, \Delta - \delta) + w(C, \Delta - \delta) > w(g, \Delta - \delta)$ for all $\delta, 0 \leq \delta < \epsilon$, there is now a longer path to g in the auxiliary graph. Thus we obtain $T(\Delta - \epsilon)$ by replacing in

$T(\Delta)$ the edge (= cut) currently ending in g by an edge corresponding to C. Also,
$w(g, \Delta - \epsilon) = w(f, \Delta - \epsilon) + w(C, \Delta - \epsilon)$.

b) We have shown in part a) that the values $w(f, \Delta)$ are non-decreasing, that
at least one such value is increased in each iteration and that $0 \leq w(f, \Delta) \leq m \cdot W_{max}(\Delta_{min})$. Thus the number of iterations is bounded by $m^2 W_{max}(\Delta_{min})$. ∎

Lemma 3. *The algorithm constructs a sketch of minimum period.*

Proof: Let S_{fin} be the final sketch. It is clearly reachable (the algorithm shows
how) from the initial sketch S by a continuous transformation which only passes
through legal configurations. Thus $S_{fin} = S(d_0, \Delta_0)$ for some $(d_0, \Delta_0) \in C_0(S)$.
In S_{fin} there must be a sequence $f_0, \ldots, f_k$ of features such that $f_0 = f_k$, and
the cuts $C_i = \overline{f_i f_{i+1}}, 0 \leq i < k$, are tight. Let x_i be the x-length of C_i and
let w_i be its wrapping number. Then $x_0 + \ldots + x_{k-1} = \Delta_0(w_0 + \ldots + w_{k-1})$.
Let $S' = S(d, \delta)$ with $(d, \delta) \in C_o(S)$ be arbitrary. For $0 \leq \lambda \leq 1$ consider the
configuration $(d(\lambda), \delta(\lambda))$ with $d(\lambda) = (1 - \lambda)d_0 + \lambda d$ and $\delta(\lambda) = (1 - \lambda)\Delta_0 + \lambda\delta$.
Then $(d(\lambda), \delta(\lambda)) \in C_0(S)$ since $C_0(S)$ is convex by theorem 1. Next observe that
the cuts C_i exist in $S(d(\lambda), \delta(\lambda))$ for λ sufficiently small and that their density is
the same as in S_{fin}. Thus their capacity must be no smaller than in S_{fin} and
hence their x-length must be no smaller than in S_{fin}. Their total x-length is
$\delta(\lambda)(w_0 + \ldots + w_{k-1})$ and hence $\delta(\lambda) \geq \Delta_0$ for λ small. Thus $\delta \geq \Delta_0$ and S_{fin} has
minimal period. ∎

At this point we have proved termination and correctness of our generic compaction
algorithm. We fill in some more algorithmic detail next. The data structures are
the longest path tree $T = T(\Delta)$ and the set $A = A(\Delta)$ of cuts. With every cut
$C \in A$ we associate the minimal value $\epsilon(C)$ of ϵ such that C becomes tight in
$S(\Delta - \epsilon)$. The value $\epsilon(C)$ is easily computed from the density and the capacity
of C in $S(\Delta)$ using lemma 1a. Let $\epsilon_1 = min\{\epsilon(C); C \in A\}$. For every feature f
let $\epsilon(f)$ be the minimal value of ϵ such that $p(f, \Delta - \epsilon) \bmod (\Delta - \epsilon) = 0$, i.e. f
moves across the reference line in $S(\Delta - \epsilon)$. Let $\epsilon_2 = min\{\epsilon(f); f \in F\}$. Finally,
let $\epsilon_3 = min\{\epsilon; A(\Delta - \epsilon) \neq A(\Delta)\}$ and let $\epsilon = min(\epsilon_1, \epsilon_2, \epsilon_3)$. We distinguish
three cases according to whether $\epsilon = \epsilon_1, \epsilon = \epsilon_2$ or $\epsilon = \epsilon_3$. The three cases are not
mutually exclusive.

Case 1, $\epsilon = \epsilon_1$: Let $\epsilon = \epsilon(C)$ and $C = \overline{fg}$. Let T_g be the subtree of T rooted at
g. We perform the following actions:

1) If $f \in T_g$ then **STOP**, $T(\Delta - \epsilon)$ does not exist.

2) Increase $w(h)$ by $w(f) + w(C) - w(g)$ for every feature $h \in T_g$; here $w(h)$
   denotes the current wrapping number of feature $h \in T_g$.

3) Delete the current edge ending in g from T and add the edge (f, g).

4) Recompute $\epsilon(D)$ for every cut D incident to a feature $h \in T_g$, recompute $\epsilon(h)$
   for every $h \in T_g$ and recompute ϵ_1, ϵ_2 and ϵ_3.

Case 2, $\epsilon = \epsilon_2$: Let $\epsilon = \epsilon(f), f \in F$.

5) Increase $w(f)$ and $w(C)$ by one for all cuts C leaving f to the left and decrease $w(C)$ by one for all cut leaving f to the right. Update $\epsilon(f)$ and ϵ_2.

Case 3, $\epsilon = \epsilon_3$: A either grows or shrinks at $\Delta - \epsilon$.

Case 3.1. A shrinks, say cut C disappears, cf. figure 6.

6) Delete C from A. Update ϵ_1 and ϵ_2.

Case 3.2. A grows, say cut C appears, cf. figure 6.

7) Add C to A, compute $\epsilon(C)$ and update ϵ_1 and ϵ_3.

Remark: In the high level description of the algorithm cases 2 and 3 did not appear because case 2 does not change the longest path tree and the values $\epsilon(C)$. Case 3.1 either removes an unsaturated cut or occurs together with case 1. Case 3.2 creates only unsaturated cuts, cf. figure 6.

IV. Specific Compaction Algorithms

In this section we derive specific compation algorithms from the general framework of section III.

IV.1. One-dimensional Compaction without Jog Insertion

We assume that wires are specified as rectilinear polygonal paths in the input sketch S. We treat vertical wire segments as modules and horizontal wire segments as wires in the sense of the definition of sketch. The only cuts which have to be considered are horizontal cuts connecting pairs of features which are visible from each other. Thus there are only $O(m)$ critical cuts, the set A of cuts does not change and $W_{max} = 1$ since no cut can wrap around twice.

For a feature h let $deg(h)$ be the number of cuts incident to h. Then $\sum_{h \in F} deg(h) = O(m)$ since the set of cuts defines a planar graph on the set of features. Also, actions 4 and 5 are executed at most m times for each h by lemma 2a and hence the total cost is $\sum_{h \in F} m \, deg(h) \cdot \log m = O(m^2 \log m)$; the $\log m$ factor results from the fact that a change of $\epsilon(D)$ requires a heap operation. This proves theorem 4.

IV.2. Two-dimensional Compaction without Jog Insertion

Kedem/Watanabe describe a branch-and-bound approach to the two-dimensional compaction which uses one-dimensional compaction to compute the lower bounds in the bound step. This readily extends to toroidal compaction by virtue of theorem 4. The details will be given in the full paper.

IV.3. One-dimensional Compaction with Jog Insertion

In order to apply our generic algorithm we need a bound on W_{max} and we need a way to manage the set A of cuts.

Lemma 4. *Let S be a cylindrical sketch of height H and period Δ. Let W_{max} be the maximal wrapping number of any tight cut which cannot be replaced by a sequence of shorter tight cuts. Then $W_{max} \leq 1 + \lfloor H/\Delta \rfloor$ in the grid model.*

Proof: We use the concept of shadowing, cf. [Cole/Siegel, Maley 87]. Let $C = \overline{fg}$ be a saturated cut with $w(C, \Delta) > 1 + H/\Delta$, cf. figure 7. Consider the straight line segments $C' = \overline{fg'}$ and $C'' = \overline{g'g}$ with $w(C', \Delta) = w(C, \Delta) - 1$ and $w(C'', \Delta) = 1$. The triangle with sides C, C' and C'' must contain a feature h ($h = g'$ is possible) such that the line segments $D' = \overline{fh}$ and $D'' = \overline{hg}$ with $w(D', \Delta) + w(D'', \Delta) = w(C, \Delta)$ are both cuts and the triangle with sides C, D' and D'' contains no features. Then the capacity of C is equal to the sum of the capacities of D' and D'' plus 1, since D' and D'' have both slope at most one, and the density of C is at most the sum of the densities of D' and D'' plus 1. Thus D' and D'' are tight and can replace C. We conclude $W_{max} \leq 1 + h/\Delta$.   ∎

We store the set of cuts as in [Mehlhorn/Näher], i.e., for each point feature f we store the set of critical cuts incident to f in clockwise order in a balanced tree. The results of [Mehlhorn/Näher] imply that actions 1 to 5 take time $O(deg(h) \log m)$ for each feature h whose wrapping number is increased where $deg(h)$ is the number of cuts incident to h and that actions 6 and 7 take time $O(\log m)$ each. Let K be the number of times actions 6 and 7 are performed. Then the running time is $O(K \log m + m \cdot (mW_{max}) \cdot (mW_{max} \log m))$ since the maximal wrapping number of any feature is $m \cdot W_{max}$ by lemma 2 and since the maximal number of critical cuts incident to any feature is mW_{max}; note that for each feature g there can be up to W_{max} different cuts with endpoints f and g. Finally, note that $K \leq 3m^4 W^2_{max}$. This can be seen as follows. Consider a pair (f, g) of features and one of the W_{max} possible cuts C with endpoints f and g. Any feature h can cross C at most $3mW_{max}$ times since between consecutive crossings the wrapping number of at least one of the three features must have been increased. Thus the running time is $O(K \log m + m^3 W^2_{max} \log m) = O(m^4 W^2_{max} \log m)$ and theorem 3 is proved. We believe that our bound on K is overly pessimistic.

Acknowledgment: We thank M. Jerrum, St. Näher and Th. Schilz for many helpful discussions. M. Jerrum suggested that toroidal compaction is the adequate approach to compacting regular structures.

V. References

[1] Cole, R., and Siegel, A., "River routing every which way, but loose", Proceedings of the 25th Annual Symposium on Foundations of Computer Science, October 1984, pp. 65–73

[2] Dunlop, A.E., "SLIP: symbolic layout of integrated circuits with compaction", Computer Aided Design, Vol. 10, No. 6, 1978, pp. 387–391

[3] Eichenberger, P., and Horowitz, M., "Toroidal Compaction of Symbolic Layouts for Regular Structures", 1987 IEEE, ICCAD, pp. 142–145

[4] Hsueh, M.Y., "Symbolic Layout and Compaction of Integrated Circuits", Ph.D. thesis, EECS Division, University of California, Berkeley, 1979

[5] Gao, S., Jerrum, M., Kaufmann, M., Mehlhorn, K., Rülling, W., Storb, C., "On homotopic river routing", BFC 87, Bonn, 1987

[6] Kaufmann, M., and Mehlhorn, K., "Local Routing of Two-Terminal Nets", 4th STACS 87, LNCS 247, pp. 40–52

[7] Kedem, G., and Watanabe, H., "Optimization techniques for IC layout and compaction", Technical Report 117, Computer Science Department, University of Rochester, 1982

[8] Leiserson, C.E., and Maley, F.M., "Algorithms for routing and testing routability of planar VLSI layouts", Proceedings of the 17th Annual ACM Symposium on Theory of Computing, 1985, pp. 69–78

[9] Leiserson, C.E., and Pinter, R.Y., "Optimal placement for river routing", SIAM Journal on Computing, Vol. 12, No. 3, 1983, pp. 447–462

[10] Lengauer, T., "Efficient algorithms for the constraint generation for integrated circuit layout compaction", Proceedings of the 9th Workshop on Graphtheoretic Concepts in Computer Science, 1983

[11] Lengauer, T., and Mehlhorn, K., "The HILL system: a design environment for the hierarchical specification, compaction, and simulation for integrated circuit layouts", Proceedings, Conference on Advanced Research in VLSI, 1984

[12] Lengauer, T., "On the solution of inequality systems relevant to IC layout", Journal of Algorithms, Vol. 5, No. 3, 1984, pp. 408–421

[13] Lengauer, T., "The Complexity of Compacting Hierarchically Specified Layouts of Integrated Circuits", 1982 FOCS, pp. 358–368

[14] Lichter, P., "Ein Schaltkreis für die Kulischarithmetik", Diplomarbeit, Universität des Saarlandes, 1988, in preparation

[15] Maley, F.M., "Compaction with Automatic Jog Insertion", 1985 Chapel Hill Conference on VLSI

[16] Maley, F.M., "Single-Layer Wire Routing", Ph.D. thesis, MIT, 1987

[17] Mehlhorn, K., and Näher, St., "A Faster Compaction Algorithm with Automatic Jog Insertion", Proceedings of the 1988 MIT VLSI Conference

[18] Pinter, R.Y., "The Impact of Layer Assignment Methods on Layout Algorithms for Integrated Circuits", Ph.D. thesis, MIT Department of Electrical Engineering and Computer Science, 1982

[19] Rülling, W., "Einführung in die Chip-Entwurfssprache HILL", Techn. Bericht 04/1987, SFB124, Univ. des Saarlandes, 1987

[20] Williams, J.D., "STICKS — a graphical compiler for high level LSI design", National Computer Conference, 1978, pp. 289–295

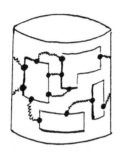 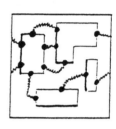

Figure 1 :
A typical cylindrical sketch (F, W, P, L) and its representation in the plane. Dark points and line segments are features, light lines are conceptual module boundaries. Formally, the connected components formed by dark and light lines are the blocks of the partition P. Wires are shown as wiggled lines.

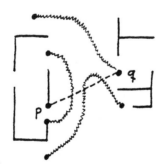

Figure 2 :
A portion of a sketch with cut $\overline{pq}$. The flow across $\overline{pq}$ is 1.

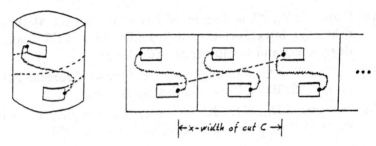

Figure 3 :
The unrolled version S^∞ of a cylindrical sketch S. The shown cut C between f and g has density 3.

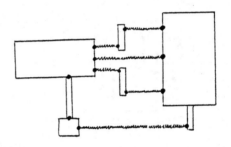

Figure 4 :
A sketch for compaction without jog insertion. Vertical wire segments are treated as modules.

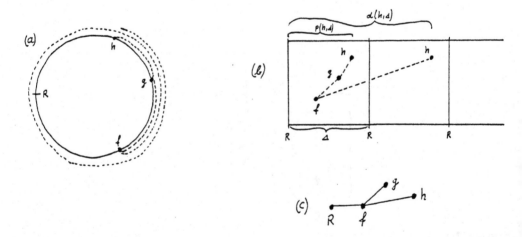

Figure 5 :
Part a) shows a top view of the cylinder. There are saturated cuts $\overline{fg}, \overline{gh}$ and $\overline{fh}$. The latter cut has wrapping number 1 and the former two have wrapping number 0. Part b) shows the unrolled picture. Part c) shows the tree $T(\Delta)$.

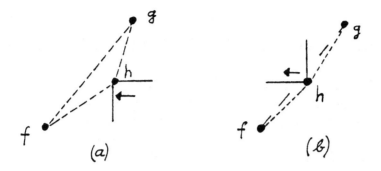

(a)   (b)

Figure 6 :
In (a) h moves to the left relative to the cut $\overline{fg}$ and hits this cut in $S(\Delta - \epsilon)$. The cut $\overline{fg}$ will then disappear. If the cut $\overline{fg}$ is tight then both cuts $\overline{fh}$ and $\overline{fg}$ will be tight in $S(\Delta - \epsilon)$ and hence cases 1 and 3.1. arise together. In b) h also moves to the left with respect to the line segment $\overline{fg}$. Thus the cut $\overline{fg}$ will arise, say in $S(\Delta - \epsilon)$. The cut $\overline{fg}$ cannot be saturated because o.w. the cuts $\overline{fh}$ and $\overline{hg}$ would be saturated in $S(\Delta - \epsilon)$ and hence $\overline{fh}$ would be oversaturated in $S(\Delta - \epsilon - \delta)$ for $\delta > 0$ small. Thus $(f, h) \in T(\Delta - \epsilon)$ and g would never become visible from f.

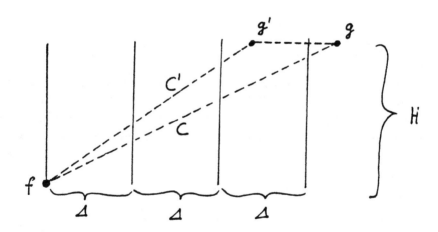

Figure 7 :
The cut $C = \overline{fg}$ has wrapping number $\geq 1 + H/\Delta$. g' is equal to g and the line segment C' has wrapping number one less than C.

CHANNEL ROUTING WITH SHORT WIRES

Michael Kaufmann

FB 10, Informatik

Universität des Saarlandes

6600 Saarbrücken

West Germany

Ioannis G. Tollis

Department of Computer Science

P.O. Box 830688, FN 3.3

The University of Texas at Dallas

Richardson, TX 75083 - 0688

Abstract.

Channel routing is a fundamental problem in VLSI layout and has received much attention recently ([F],[MPS],[PL],[BBL],[BBBL],[GK]..). Unfortunately, most algorithms produce good layouts only with respect to the channel width, but not with respect to the length of the wires, which is a second important measure for the performance of the layout. Here, we consider the problem of routing a given channel with short wires. We show that our algorithms produce routings with minimum total wire length.

1. Introduction.

Channel routing is a fundamental problem in VLSI layout and has received much attention recently ([F],[MPS],[PL],[BBL],[BBBL],[GK]..). Unfortunately, most algorithms produce good layouts only with respect to the channel width, but not with respect to the lenght of the wires, which is a second important measure for the performance of the layout. Here, we consider the problem of routing a given channel with short wires. We show that the routings produced by our algorithms achieve the minimum total wire length.

The channel routing problem is defined as follows:

A *channel* is a rectangular grid G of *tracks* (numbered from 0 to $w + 1$) and *columns* (numbered from 1 to m), where w is the *width* and m the *length* of the channel. A *net* $N = (\{t_1, \ldots, t_u\}, \{b_1, \ldots, b_l\})$ is a collection of *terminals*, where *lower terminal* b_j is located at grid point $(b_j, 0)$ on track 0 (the *lower boundary*) and *upper terminal* t_j at $(t_j, w+1)$ on track $w+1$ (the *upper boundary*). N is called a *2-terminal net* if $u = l = 1$. The other nets are called *multiterminal nets*. A *channel routing problem* (*CRP*) is a set of pairwise disjoint nets $N = \{N_1, \ldots, N_n\}$. Its solution (also called *layout*) is a set $W = \{w_1, \ldots, w_n\}$ of subgraphs of G

(called *wires*) such that w_i connects the terminals of N_i, for $1 \leq i \leq n$, under the condition of the corresponding wiring model.

The wiring model can be formulated as restrictions on how wires share grid points and edges of the grid G. In this paper we consider the *knock-knee* model where any two wires should be edge-disjoint. Wires of a layout are physically located on one or more layers so that no two wires run through the same grid point on the same layer (such a situation is called a short circuit). Layouts in the knock-knee model may require four layers ([BB], [T]).

The goal of the classical channel routing is to find a layout with minimal channel width w for a given *CRP*. A basic parameter of the problem is its *density d*, defined as the maximum, over all columns c, of the density $d(c)$ at column c:

$$d(c) := | \{ N \mid N \text{ is a net with } min\{t_1, b_1\} \leq c \text{ and } max\{t_u, b_l\} \geq c + 1 \} |.$$

The channel routing problem has been intensively studied. In particular, many provably good algorithms have been developed for problems which only consist of 2-terminal nets ([BBBL], [BBL], [RBM], [PL]). But for the general multiterminal channel routing problem (*MCRP*) there are only some simple generalizations of the algorithms for 2-terminal problems ([BBBL],[MPS],[SP]). Recently, Gao and Kaufmann have shown that the channel widths for multiterminal problems are a factor of approximately 3/2 worse than those for 2-terminal problems.

The density of a problem is a trivial lower bound for the channel width. For 2-terminal problems in knock-knee model one can find optimal solutions with respect to the channel width. But a good layout should not only be area-efficient, an other important measure for the performance is the length of the wires.

Since it is NP-hard to find a solution for a multiterminal channel routing problem with the minimum number of tracks ([S]), it is clearly NP-hard to route a multiterminal channel routing problem with minimum total wire length. Hence, we consider only two-terminal channel routing problems. Furthermore, we restrict our attention to problems where all the grid points at the upper and lower boundaries are terminals of some (two-terminal) net, and nets do not have both terminals on the same boundary. We call a problem trivial, if each net starts and ends in the same column. It is obvious that any trivial problem can be solved optimally with respect to the area and the total wire length. If the problem is not trivial, then we need an additional column to find any solution for the problem. We know from previous papers that one additional column is always sufficient to solve the problem. Let us assume that we add this column to the right of the channel. Hence we ensure that our layouts will be area-optimal in a certain sense (minimal length and width).

We present algorithms that route a given two-terminal channel routing problem using the minimum total wire length.

First important notions will be the *conflict graph* and the *cycles* of the problem. Let $P = N_1, N_2, ...$ be a (two-terminal) channel routing problem. From P we derive a directed graph $G = (V, E)$, called *conflict graph*, as follows: The vertices of G are the nets of P, $N_1, N_2, ...$. There is an arc $e = (N_i, N_j)$ from vertex N_i to vertex N_j if and only if the upper terminal of net N_i lies at the same column as the lower terminal of net N_j, for all i, j. We say that nets $N_1, N_2, ..., N_l$ form a cycle of length l if there exist arcs (N_i, N_{i+1}) for $1 \leq i < l$ and the arc (N_l, N_1). We consider those cycles since each of them will produce a detour of wires.

2. A Special Case.

We first describe our algorithm and prove its optimality on a special case of a channel routing problem. In the next section we will show how to extend the ideas presented here for a general case of a channel routing problem.

Consider the following problem:
A channel consists of d nets which form $d/2$ nested cycles, each of length two, such that there is a column of density d. An example of such a channel is shown in Figure 1(a) for $d = 6$, and it is called a *block*.

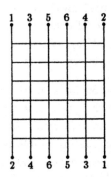

Fig. 1(a): A block of 3 nested cycles

Consider now a channel routing problem that consists of n such blocks $B_1, ..., B_n$ and one additional free column to the right of the rightmost block, B_n, see Figure 1(b).

Note that each block is divided into two parts by its column of density d, a left part and a right part.

Our algorithm consists of two different layout steps:

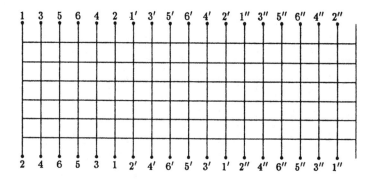

Fig. 1(b): An example for the special case (3 blocks)

(a) In the first step we show how to layout everything that lies to the left of the rightmost column of density d. Each cycle C will produce a detour that lies to the right side of the rightmost terminals of the two nets of C. We layout the nets of each cycle in two adjacent tracks, such that the nets of i-th depth of the nesting always use tracks $2i - 1$ and $2i$.

(b) The part to the right of the rightmost column of density d is laid out as follows:

Assume that we have already constructed the layout for the left part. Let N_s be the net with the upper terminal in the previous column. We route net N_s as follows: the lower terminal of N_s is routed on the track 1 to the rightmost column of the channel, then, using the rightmost column, we route N_s until we reach track d, and finally, to the left along track d until its upper terminal is reached. The remaining nets are routed in the obvious way without any detours from the left to their terminals.

Figure 2 shows a layout constructed by our algorithm.

The total wire length produced by our algorithm is equal to the total number of edges in the channel minus the edges not used by the layout. Clearly, there are $n \cdot d \cdot (d + 1) + (d + 1)$ vertical and $n \cdot d \cdot d$ horizontal edges in the channel. Hence, there are $n \cdot d \cdot (2d + 1) + d + 1$ edges in total.

From Figure 2 one can see that there are $1 + 2 \cdot \sum_{i=1}^{d/2}(2i - 1)$ unoccupied edges in the layout, and exactly $d + 1$ of them are vertical.

Now we show that the layout constructed by our algorithm is optimal. We will think of the channel routing problem as a switchbox routing problem. We first recall some definitions and results from the switchbox routing theory [F], [MP], [KM]. A switchbox routing problem P is called *even* if for every vertex v the

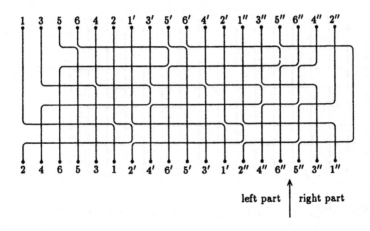

Fig. 2: An optimal layout for the example

degree of v, $deg(v)$, plus the number of nets having v as their terminal, $ter(v)$, is even. Otherwise, P is called *odd*. We call $deg(v) + ter(v)$ the *extended degree of* v.

Theorem 1. *An even problem is solvable if and only if no straight cut is oversaturated.*

Lemma 1. *If an odd problem is solvable then there exists an extension of the problem to an even solvable problem, by inserting some fictitious nets.*

If we consider our channel routing problem as a switchbox routing problem then the only vertices with odd extended degree lie on the first and last column, see Figure 3. We extend the problem by pairing up neighboring vertices of odd extended degree by fictitious nets, F_i, $i = 1, ..., d + 1$. Note that this extension is the only possible one, since otherwise we would produce some oversaturated cuts.

We know that each of these fictitious nets have to occupy at least one vertical edge in the layout of the extended switchbox routing problem. Furthermore, it is clear that each of these nets may use only one vertical edge, since there are $n \cdot d$ real nets which must use exactly $d + 1$ vertical edges each. Hence the remaining $d + 1$ vertical edges can be used by the $d + 1$ fictitious nets.

Notice that $d/2$ of the fictitious nets lie on the leftmost column of the leftmost block and $d/2 + 1$ of them lie on the rightmost (additional) column. Now we have to estimate the number of horizontal edges that will be used by the fictitious nets. Because of parity arguments [KM], we know that all unused edges in the layout of the original (channel routing) problem must be used by the fictitious nets.

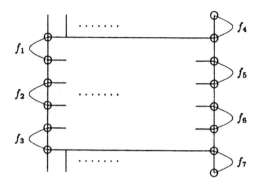

Fig. 3: The extension to an even problem

Lemma 2. *The number of unused horizontal edges is* $\geq 2 \cdot \sum_{i=0}^{d/2-1} i + 2 \cdot \sum_{i=1}^{d/2-1} i.$

Proof Sketch:

At least one fictitious net must be routed on the leftmost column, since nets N_1 and N_2 use one horizontal edge each to the right of this column. The same happens for columns $2, 3, ..., d/2$. Hence, the number of unused horizontal edges on the left side of the channel is $\geq 2 \cdot \sum_{i=0}^{d/2-1} i$. If we apply a similar argument to the right side of the channel we will find that the number of unused horizontal edges on the right side of the channel is $\geq 2 \cdot \sum_{i=1}^{d/2-1} i$. ∎

Since there must be $d + 1$ unused vertical edges, we have the following result:

Lemma 3. *The number of unused edges in any solution of the special case of a channel routing problem is*

$$\geq 2 \cdot \sum_{i=0}^{d/2-1} i + 2 \cdot \sum_{i=1}^{d/2-1} i + d + 1.$$

But notice that this is exactly the number of unused edges in the analysis of our algorithm. Therefore, we have the main theorem of this section:

Theorem 2. *The solution obtained by our algorithm is optimal for the special case of a channel routing problem.*

3. The General Case.

Let us assume now that the cycles within a block could have length greater than two, and that they might or might not be nested. Furthermore, we assume that the cycles could have a free shape. We distinguish three configurations of one or two cycles and show how to solve them optimally. We can simulate any shape of a cycle by a combination of these three configurations.

1. Let C be a cycle of length $l > 2$ such that the number of nets of C crossing any column is at most two, see Figure 4(a).

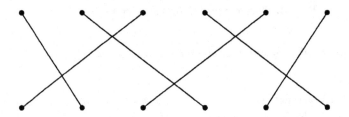

Fig. 4(a): A cycle of length 6 as described above

We modify the problem by replacing cycle C by a cycle of length two and $l - 2$ trivial nets between the endpoints of the new cycle as shown in Figure 4(b). The layout of the modified problem is easy to do: Simply layout the cycle as discussed before. The trivial nets are laid out in the obvious way, along their column.

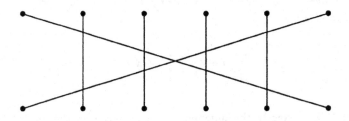

Fig. 4(b): The layout of this cycle

This layout can be converted to a layout for the original problem by replacing some crossings by knock-knees. Deciding which crossings must be replaced by knock-knees is rather obvious: In the original problem each rising net (left terminal on the lower channel boundary) is crossed by exactly one falling net (left terminal at the upper boundary) at every column. In the modified problem however, there are two crossings per column. It is easy to determine which crossing must be replaced by a knock-knee.

The layout for the modified problem is optimal with respect to the total wire length, since the trivial nets are laid out in an optimal way. Clearly, the layout generated by our technique is optimal for the original problem, since the lower bound is the same as for the modified problem.

2. Assume that the cycles have length two. Let C_1 and C_2 be two cycles such that the left end of C_1 lies to the left of the left end of C_2 and the right end of C_1 lies between the ends of C_2, see Figure 5(a). Such cycles are called *intersecting*.

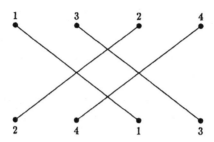

Fig. 5(a): Two intersecting cycles

We modify the problem by replacing C_1 and C_2 by two nested cycles C_1' and C_2', as shown in Figure 5(b). We now layout the cycles as discussed in the previous section.

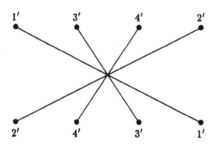

Fig. 5(b): The layout of these two cycles

Again, this layout can be converted to a layout for the original problem by replacing some crossings by knock-knees. More specifically, for the example of Figure 5(b), the crossing between nets 1' and 3' is replaced by a knock-knee. By replacing the crossing between nets 2' and 4' by a knock-knee we obtain a layout for the original problem. Again, the lower bound for the original problem is the same as for the modified problem. Hence, the obtained solution is optimal.

3. Assume that there is a cycle C such that there exists a column that is crossed by more than two nets of C, see Figure 6(a).

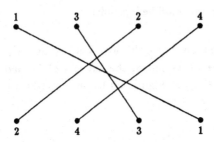

Fig. 6(a): An example for a cycle described above

We modify the problem by replacing cycle C by nested cycles of length two, as shown in Figure 6(b). We now layout the cycles as discussed in the previous section.

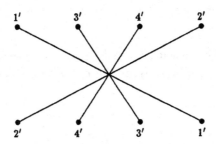

Fig. 6(b): The layout for the example in figure 6(a)

Again, this layout can be converted to a layout for the original problem by replacing some crossings by knock-knees. More specifically, for the example of Figure 6(b), the crossing between nets $3'$ and $4'$ is replaced by a knock-knee. By similar arguments we can show that the obtained solution is optimal.

Hence, we have an algorithm which constructs optimal layouts for channel routing problems consisting of nested cycles of length two, and some operations that convert general problems to those of nested cycles of length two.

4. Remarks.

If the densities of the blocks are different we will always use the lowest available tracks. As discussed above, there is only one vertical edge for each fictitious net. Furthermore, all fictitious nets lie to the left of the leftmost column of maximal density or to the right of the rightmost column of maximal density.

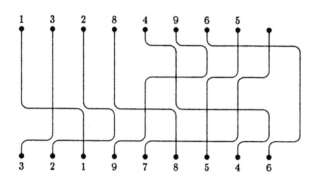

Fig. 7: A layout for a more complicated problem

The following problems still remain open:

1. How can we layout a problem optimally if nets are allowed to have terminals on the same side, or, if some boundary points are free?

2. How can this technique be extended for switchbox routing?

References.

[BBBL] B. Berger, M. Brady, D. Brown, and F. T. Leighton: "Nearly Optimal Algorithms and Bounds for Multilayer Channel Routing". Manuscript.

[BBL] B. S. Baker, S. N. Bhatt, and F. T. Leighton: "An Approximation Algorithm for Manhattan Routing". Proc. of the 15th Ann. ACM Symposium on Theory of Computing, pp. 477-486 (1983).

[BB] M. Brady and D. Brown: "VLSI Routing: Four Layers Suffice". MIT VLSI Conference 1984.

[BP] D. Brown and F. P. Preparata: "Three-Layer Channel Routing of Multi-terminal Nets". Technical Report, Coordinated Science Lab., University of Illinois at Urbana-Champaign (Oct. 1982).

[F]   A. Frank "Disjoint Paths in Rectilinear Grids" Combinatorica, Vol. 2. No. 4, 1982, pp. 361-371

[GK]   S. Gao and M. Kaufmann: "Channel Routing of Multiterminal Nets". Proc. of 28th Ann. Symposium on Foundations of Computer Science, pp. 316-325 (1987).

[KM]   M. Kaufmann and K. Mehlhorn: "Routing through a Generalized Switchbox". Journal of Algorithms 7, pp. 510-531 (1986).

[MPS] K. Mehlhorn, F. P. Preparata, and M. Sarrafzadeh: "Channel Routing in Knock-Knee Mode: Simplified Algorithms and Proofs". Algorithmica 1(2), pp. 213-221, Springer-Verlag (1986).

[PL]   F. P. Preparata and W. Lipski: "Optimal Three-Layer Channel Routing". IEEE Trans. on Computers C-33, pp. 427-437 (1984).

[RBM] R. L. Rivest, A. Baratz, and G. Miller: "Provably Good Channel Routing Algorithms". Proc. CMU Conf. on VLSI, pp. 153-159 (Oct. 1981).

[S]   M. Sarrafzadeh: "On the Complexity of the General Channel Routing Problem in the Knock-Knee Mode". to appear in IEEE Transactions on Computer-Aided Design

[SP]   M. Sarrafzadeh and F. P. Preparata: "Compact Channel Routing of Multiterminal Nets". Annals of Discrete Math., North Holland (Apr. 1985)

[T]   I. G. Tollis: "A New Algorithm for Wiring Layouts". Proceedings of AWOC, (1988)

Simple Three-Layer Channel Routing Algorithms ‡

Teofilo Gonzalez and Si-Qing Zheng‡‡
University of California, Santa Barbara

Abstract. In this paper we present a simple three-layer assignment algorithm for planar layouts generated by a class of layout algorithms. This class of algorithms includes simple variations of the currently best algorithms for the three layer channel routing problem (CRP). More specifically, this class includes algorithms "equivalent" to the following algorithms (i-iii) developed by Mehlhorn, Preparata and Sarrafzadeh [7].

(i) The algorithm that generates planar layouts for the two-terminal net CRP with d_{max} tracks.

(ii) The algorithm that generates planar layouts for the two- and three-terminal net CRP with at most $\lfloor 3d_{max}/2 \rfloor$ tracks.

(iii) The algorithm that generates planar layouts for the multi-terminal net CRP with at most $2d_{max} - 1$ tracks.

The planar layouts generated by these algorithms and by their "equivalent" algorithms are three-layer wirable by the layer assignment algorithm given in [8]. Our approach is different. We make simple modifications to these layout algorithms and incorporate a simple wire assignment strategy to generate three-layer wirings under the knock-knee model. Consequently, we obtain simpler and faster algorithms that generate three-layer wirings with layouts similar to the ones generated by algorithms (i) - (iii). Our algorithms are faster and conceptually simpler because there is no need to construct diagonal diagrams and legal partitions. The channel width of the wiring generated by our algorithm is identical to that of the corresponding planar layout generated by algorithms (i) - (iii).

1. Introduction

The channel routing problem (CRP) has been recognized as one of the most important problems in VLSI design automation. The CRP problem is defined over the rectangular grid formed by lines $\{x = i \mid i \in Z\}$ and $\{y = j \mid 0 \le j \le h+1\}$. The horizontal grid lines $y = 0$ and $y = h+1$ are called the *boundaries* of the channel, and the horizontal grid lines $y = j$, $1 \le j \le h$, are called *tracks* of the channel. All the vertical grid lines are called *columns* of the channel. A channel routing problem consists of a collection of pairwise disjoint sets of grid points, $N = \{N_1, N_2, ..., N_m\}$, located on the channel boundaries. Each N_i in N is called a *net* and each point in N_i is called a *terminal* of net N_i. Terminals in each net need to be connected by wires running along the grid lines. It should be noted that the terms "layout" and "wiring" are frequently interchangeable. In this paper we follow Preparata and Lipski's [8] convention and give these two terms a different connotation. A *planar layout* (or simply a layout) of a channel routing problem is a collection of edge disjoint connected subgraphs $W = \{W_1, W_2, ..., W_m\}$ of the channel grid, such that each subgraph W_i connects all the terminals in net N_i. This definition implies that at each

‡ This research was supported in part by the National Science Foundation under Grant DCR - 8503163.

‡‡ Dr. Si-Qing Zheng new address is: Department of Computer Science, Louisiana State University, Baton Rouge, Louisiana, 70803-4020.

grid point in a planar layout there can be at most two wires, and for i ≠ j, W_i and W_j do not share any grid line segment. There are several different wiring models for the channel routing problem. In this paper, we consider the CRP under the knock-knee model. In the knock-knee model, when two different wires share a grid point they either *cross* or form a *knock-knee*. The two types of knock-knees are given in figure 1.1. The horizontal (vertical) portions of a knock-knee are called the *horizontal (vertical) arms of the knock-knee*.

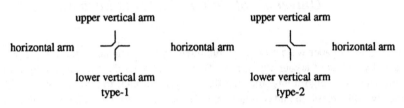

Figure 1.1: Knock-knees.

There are k ≥ 2 conducting layers $L_1, L_2, ..., L_k$ available and L_{i+1} is stacked on top of L_i for 1 ≤ i < k. A *wiring* of a given layout W = $\{W_1, W_2, ..., W_m\}$ is a mapping that associates each edge of W_i, 1 ≤ i ≤ n, to a layer such that for every i ≠ j if edges (p_1, p_2) and (p_2, p_3) in W_i are assigned to L_s and L_t, respectively, and edge (p_2, p_4) in W_j is assigned to L_u then either u > max{s, t} or u < min{s, t}. Contact cuts (called *vias*) can be established only at grid points. Vias allow a wire to change from one layer to the another. The objective of the channel routing problem consists of finding an *optimal wiring*, i.e., a wiring on a grid with least number of horizontal grid lines. We also refer to this criteria as *minimum channel width* or minimum number of tracks.

We call the open interval (c,c+1) a *vertical cut*, where c and c+1 are two adjacent columns of the channel. We define the *channel density* d_{max} for a CRP as $d_{max} = max\{d(c)\}$, where d(c), the *local density* for the vertical cut, (c,c+1), is the number of nets in N whose leftmost terminal is located to the left of vertical cut (c,c+1) and the rightmost terminal is located to the right of vertical cut (c,c+1). Clearly, the channel density d_{max} is a lower bound for the channel width in an optimal wiring for the CRP problem.

Preparata and Lipski [8] developed an efficient algorithm to generate a three-layer optimal wiring for the two-terminal net CRP. Their algorithm consists of two phases. In the first phase a minimum-track planar layout that satisfies certain properties is constructed. In the second phase the planar layout is transformed into a three-layer wiring by a powerful transformation, legal partition of the diagonal diagram induced by the layout. Recently, Mehlhorn, Preparata and Sarrafzadeh [7] developed another algorithm to construct a planar layout for this problem. The algorithm is conceptually simpler and the planar layout can also be three-layer wired by the algorithm in [8]. Another algorithm that finds a planar layout with d_{max} tracks for this problem appears in [10]; however, it is not known whether or not the layouts generated by this algorithm are three-layer wirable.

For the case when each net consists of at most three terminals, the algorithms in [7] and [9] generate a planar layout with no more than $\lfloor 3d_{max}/2 \rfloor$ tracks. For multi-terminal net channel routing problems, the algorithms in [4], [7] and [11] generate planar layouts with at most $2d_{max}-1$ tracks. The layouts generated by these algorithms are three-layer wirable by the algorithm given in [8]. The bounds $\lfloor 3d_{max}/2 \rfloor$ and $2d_{max}-1$ are believed to be best possible when one restricts to three layer wirable planar layouts. Recently, Gao and Kaufmann [2] showed that every multiterminal net CRP has a planar layout with $3d_{max}/2 + O(\sqrt{d_{max}\log d_{max}})$ tracks. However, it is not known whether these layouts are three layer wirable or not.

Lipski [5] showed that there are planar layouts that are not three-layer wirable. Gonzalez and Zheng [3] showed that there even exist six-row planar layouts which are not three-layer wirable. In general, the problem of determining whether a given planar layout is three-layer wirable is NP-complete([5]). Brady and Brown ([1]) showed that every planar layout is four-layer wirable by finding a legal partition (that satisfy some additional properties) of the diagonal diagram induced by the layout.

In this paper we present a simple three-layer assignment algorithm for planar layouts generated by a class of layout algorithms. This class of algorithms includes simple variations of the currently best algorithms for the three layer channel routing problem. More specifically, this class includes algorithms "equivalent" to the following algorithms developed by Mehlhorn, Preparata and Sarrafzadeh [7].

(i)   The algorithm that generates planar layouts for the two-terminal net CRP with d_{max} tracks.

(ii) The algorithm that generates planar layouts for the two- and three-terminal net CRP with at most $\lfloor 3d_{max}/2 \rfloor$ tracks.

(iii) The algorithm that generates planar layouts for the multi-terminal net CRP with at most $2d_{max} - 1$ tracks.

The planar layouts generated by these algorithms and by their "equivalent" algorithms are three-layer wirable by the layer assignment algorithm given in [8]. Our approach is different. We make simple modifications to these layout algorithms and incorporate a simple wire assignment strategy to generate three-layer wirings. Consequently, we obtain simpler and faster algorithms that generate three-layer wirings with layouts similar to the ones generated by algorithms (i) - (iii). Our algorithms are faster and conceptually simpler because there is no need to construct diagonal diagrams and legal partitions. The channel width of the wiring generated by our algorithm is identical to that of the corresponding planar layout generated by algorithms (i) - (iii). As mentioned before, algorithms (i) - (iii) are currently the best algorithms for the three layer CRP problem.

2. Three-Layer Wiring Algorithms

Before we define the class of layouts algorithms we are interested in and our layer assignment strategy, we need to introduce additional notation. Let A be a layout algorithm that generates a layout by a single left-to-right column-by-column sweeping (scanning) of the terminals. For column c, we define the *strip area S(c) around column c* as the area delimited by the two vertical lines c - 1/2 and c + 1/2, and the top and bottom boundaries. When algorithm A process column c it generates the layout, W(c), for the strip area S(c). The horizontal wires leaving W(c) from the right of S(c) are called *the output wires of W(c)*. There are no input wires for W(1) and for c > 1 *the input wires of W(c)* are the output wires for W(c-1). A horizontal wire that is both an input wire and an output wire on the same track in W(c) is called a *continuing wire*. A horizontal wire that is only an output wire on some track k in W(c) is called a *beginning wire*. We say that a wire W_i is a *k-stranded input (output) wire in W(c)* if the vertical line x = c-1/2 (x = c+1/2) intersects k times wire W_i in layout W(c). We say that vector $V = (v_1, v_2, ..., v_m)$ is an *input (output) strand vector for W(c)* if wire W_i is a v_i-stranded input (output) wire in W(c). We use osv(W(c)) (isv(W(c))) to denote the output (input) strand vector for W(c).

The planar layout algorithms which can be modified by our strategy to generate three-layer wirings are called *conservative planar layout algorithms*. An algorithm A is said to be a conservative layout algorithm if and only if it satisfies the following three properties.

(1) Algorithm A generates a layout by a single left-to-right column-by-column sweep (scan). When column c is being considered the algorithm generates its final layout $W(c)$, i.e., once $W(c)$ is generated, the layouts $W(1), W(2), ..., W(c)$ will not be modified.

(2) For column c > 1 every layout $W'(c-1)$ such that $osv(W'(c-1)) = osv(W(c-1))$ (remember that $W(c)$ is the layout generated by algorithm A for column c), algorithm A generates a layout $W'(c)$ with $osv(W'(c)) = osv(W(c))$.

(3) For any layout $W'(c-1)$ with $osv(W'(c-1)) = osv(W(c-1))$ algorithm A generates a layout $W'(c)$ with no more than two knock-knees. If there are two knock-knees, the knock-knees are of different types, and the knock-knee of type-1 is below the knock-knee of type-2. A type-1 (type-2) knock-knee has its lower (upper) vertical arm intersect the bottom (top) boundary.

Without loss of generality, assume that every conservative algorithm is initially assigned h empty tracks and throughout the execution of the algorithm the value h is never increased nor decreased. An algorithm A that does not satisfy this restriction can be easily simulated by executing A once to obtain the value of h. Once this value is computed, algorithm A can be easily modified to satisfy this additional property. One can easily develop algorithms "equivalent" to algorithms (i) - (iii) that satisfy properties (1) - (3). By "equivalent" we mean that the new algorithm has the same (asymptotic) worst case time complexity, and it never generates a layout with a number of tracks that exceeds the bounds stated in (i) - (iii). The planar layout algorithms given in [8] and [10] do not satisfy some of these properties and it seems that these algorithms do not have "equivalent" algorithms that satisfy properties (1) - (3).

Let A be any algorithm that satisfies properties (1) - (3). In what follows we define our algorithm, A^*, to construct a layout (similar to the one constructed by A) and find its layer assignment simultaneously. Algorithm A^* constructs the layout as algorithm A in a single left-to-right scan of the columns. When column c is being considered, we first take $W^*(c-1)$, the layout generated by algorithm A^* at the (c-1)th iteration (if c = 1, $W^*(0) = \varnothing$) and mimic algorithm A on this input. Let $W'(c)$ be the layout obtained by this process. Note that the input wires in $W'(c)$ are identical to the output wires in $W^*(c-1)$. Depending on the knock-knees in $W'(c)$ and the layer assignment for $W^*(c-1)$, $W^*(c)$ is defined as either $W'(c)$ or a slightly modified version of $W'(c)$. In either case the output strand vector for $W^*(c)$ and $W'(c)$ are identical. Let $W(1)$, $W(2), ... (W^*(1), W^*(2), ...)$ be the layout constructed by algorithm A (A^*) for some CRP problem instance N. From this brief description and the assumption that algorithm A satisfies property (2) one can easily prove that at each step, the output strand vector for $W(c)$ is identical to the output strand vector for $W^*(c)$. When algorithm A^* is processing column c, the layer assignment for each wire segment in $W^*(c)$ is determined. The layer assignment rules are quite simple: horizontal wires are always assigned to either the top layer or the bottom layer, whereas the vertical wires are assigned to the middle layer in "normal" regions and to either the top or bottom layer in other regions. Vias are introduced whenever necessary. For column c, we use $[k_1, k_2]$, where $k_1 \le k_2$, to represent the vertical grid segments from track k_1 to track k_2. Note that it is a closed interval. For open intervals we use parentheses instead of square brackets. Remember that the bottom (top) boundary is track 0 (h+1). We define the *strip area S(c,I) (S(c) restricted to I)* as the set of all points in S(c) with y-coordinate value $y \in I$, where $I = [k_1, k_2]$ for some $k_1 <$

k_2. Similarly, the layout $W^*(c,I)$ $(W'(c,I))$ is defined as $W^*(c)$ $(W'(c))$ restricted to the strip area $S(c,I)$. Depending on the knock-knees in $W'(c)$, each of the vertical grid segments in column c is labeled, R_1, R_N, or R_2. A region R_N is a normal region, and the other two regions, R_i, contain exactly one type-i knock-knee. The labeling procedure is given below.

procedure LABELING
 case
  :there is no knock-knee in $W'(c)$ /* fig. 2.1(a) */:
   the interval [0,h+1] is labeled R_N;
  :there is exactly one knock-knee in $W'(c)$ and its type is type-1 /* fig. 2.1(b) */:
   let k be the track where the knock-knee is located;
   the interval [0,h+1] is partitioned and labeled as follows: [0,k-1) is labeled R_N, [k-1,k] is
   labeled R_1 and (k,h+1] is labeled R_N.
  :there is exactly one knock-knee in $W'(c)$ and its type is type-2 /* fig. 2.1(c) */:
   let k be the track where the knock-knee is located;
   the interval [0,h+1] is partitioned and labeled as follows: [0,k) is labeled R_N, [k,k+1] is labeled R_2, and (k+1,h+1] is labeled R_N.
  :there are two knock-knees in $W'(c)$ /* fig. 2.1(d) */:
   /* From property (3) we know that the knock-knees are of different types and that the
   knock-knee of type-1 is below the one of type-2 */
   let k_1 (k_2) be the track where the knock-knee type-1 (type-2) is located;
   /* From property (3) we know that $k_1 < k_2$ */
   the interval [0,h+1] is partitioned and labeled as follows: [0,k_1-1) is labeled R_N, [k_1-1,k_1]
   is labeled R_1, (k_1,k_2) is labeled R_N, [k_2,k_2+1] is labeled R_2, and (k_2+1,h+1] is labeled
   R_N.
 endcase
end of procedure LABELING

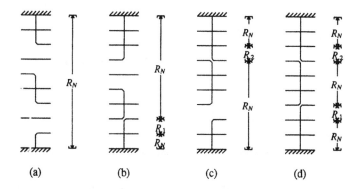

Figure 2.1: Labeling examples.

 In the following figures the region inside the wiggled lines is a three-layer wiring and the region not enclosed by wiggled lines is a planar layout. In a three-layer wiring we use solid lines to represent wires assigned to the top layer, dashed lines for wires assigned to the bottom layer and dotted lines for the wires assigned to the middle layer. Let us now explain how the layout $W^*(c)$ and its layer assignment are generated by algorithm A^* from $W'(c)$. Algorithm A^*

generates $W^*(c)$ by constructing sublayouts $W^*(c,I)$ for each interval I in $W'(c)$ uniformly labeled (i.e., with the same label). Let us now consider any uniformly labeled interval I. Initially we set the input wires in $W^*(c,I)$ to be the same as the output wires in $W^*(c-1,I)$ which are identical to the set of input wires in $W'(c,I)$. The layer assignment for the input wires in $W^*(c,I)$ is identical to the layer assignment for the output wires in $W^*(c-1,I)$. Depending on the label assigned to an interval, I, the remaining part of the layout $W^*(c,I)$ and the layer assignment for it is constructed as follows.

PROCEDURE LAYOUT AND LAYER ASSIGNMENT

case 1: Interval I is labeled R_N.
Algorithm A^* generates the layout $W^*(c,I) = W'(c,I)$. The layer assignment in this case is defined as follows. (Remember that the input wires in $W^*(c,I)$ are assigned to the same layers as the output wires in $W^*(c-1,I)$). All vertical wires are assigned to the middle layer, all the continuing wires remain in the layer assigned to their input portion, and the beginning wires are assigned to the top layer (note that they could have also be assigned to the bottom layer).

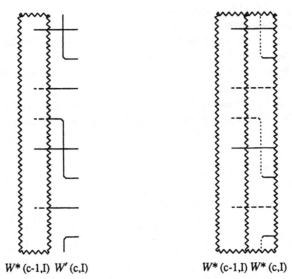

$$W^*(c-1,I) \quad W'(c,I) \qquad\qquad W^*(c-1,I) \quad W^*(c,I)$$

Figure 2.2: Example for case 1.

case 2: Interval I is labeled R_1.
Clearly, the interval is of the form $I = [k-1,k]$ for some track k, there is only one knock-knee in $W'(c,I)$, the type of knock-knee is type-1, and the knock-knee is located at grid point (c,k). There are two cases.

subcase 2.1: There is no input wire in $W'(c,I)$ assigned to track k-1 or the input wire in $W'(c,I)$ assigned to track k-1 is electrically common with the lower vertical arm of the knock-knee.

Algorithm A^* generates the layout $W^*(c,I) = W'(c,I)$. The layer assignment in this case is defined as follows. If the input wire in track k of $W^*(c,I)$ is in the top (bottom) layer, then the only vertical wire and all the output wires in $W^*(c,I)$ are assigned to the bottom (top) layer.

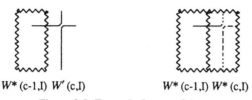

$W^*(c\text{-}1,I)\ \ W'(c,I)$ $W^*(c\text{-}1,I)\ W^*(c,I)$

Figure 2.3: Example for case 2.1.

subcase 2.2: The input wire in $W'(c,I)$ assigned to track k-1 is not electrically common with the lower arm of the knock-knee.
There are two cases depending on whether the output wires in $W^*(c\text{-}1,I)$ assigned to tracks k-1 and k are in the same layer or not.

subcase 2.2.1: The output wires in $W^*(c\text{-}1,I)$ assigned to tracks k-1 and k are in the same layer.
Algorithm A^* generates the layout $W^*(c,I) = W'(c,I)$. The layer assignment in this case is defined as follows. If the output wires in tracks k-1 and k of $W^*(c\text{-}1,I)$ are in the top (bottom) layer, then the only vertical wire and the beginning horizontal wires in $W^*(c,I)$ are assigned to the bottom (top) layer, and the continuing wire in $W^*(c,I)$ remains in the same layer as its input portion.

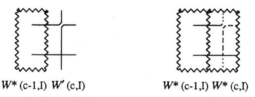

$W^*(c\text{-}1,I)\ \ W'(c,I)$ $W^*(c\text{-}1,I)\ W^*(c,I)$

Figure 2.4: Example for case 2.2.1.

subcase 2.2.2: The output wires assigned to tracks k-1 and k in $W^*(c\text{-}1,I)$ are in different layers.
In this case the layout $W^*(c,I) \neq W'(c,I)$. $W^*(c,I)$ is $W'(c,I)$ after performing the transformation shown in figure 2.5.

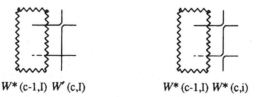

$W^*(c\text{-}1,I)\ \ W'(c,I)$ $W^*(c\text{-}1,I)\ W^*(c,i)$

Figure 2.5: Layout transformation for case 2.2.2.

The layer assignment in this case is defined as follows. If the output wire in track k of W^*(c-1,I) is in the top (bottom) layer, then the only vertical wire and the output wire in track k of W^*(c,I) are assigned to the bottom (top) layer. The output wire in track k-1 of W^*(c,I) is assigned to the top (bottom) layer.

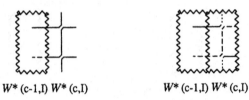

W^* (c-1,I) W^* (c,I) $\qquad$ W^* (c-1,I) W^* (c,I)

Figure 2.6: Example for case 2.2.2.

case 3: Interval I is labeled R_2.
Since case 3 is similar to case 2, it will be omitted.

end of procedure LAYOUT AND LAYER ASSIGNMENT

In figure 2.11 we give a planar layout constructed by algorithm (i) in [7]. The corresponding layout and the wiring constructed by our procedure is given in figure 2.12.

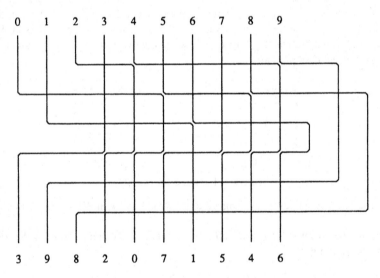

Figure 2.11: Layout constructed by algorithm (i) in [7].

Theorem 2.1: Let N, A, and A^* be as defined above. Algorithm A^* constructs a planar layout and its three layer wiring for N. Furthermore, the number of tracks in the three-layer wiring constructed by A^* and the number of tracks in the planar layout constructed by A are identical.
Proof: Since algorithm A satisfies property (1) we know that it constructs the planar layout in a single left-to-right scan of the columns. This implies that once the layout for W(c) is constructed none of the layouts for W(1), W(2), ..., W(c) is modified. Algorithm A^* mimics this process. If the input strand vectors of W^*(c) and W(c) are identical, we know by property (2) and our

construction rules that the output strand vectors for $W^*(c)$ and $W'(c)$ are also identical. Therefore, it follows inductively (after a trivial proof for the base) that $W^*(c)$ and $W(c)$ have identical output strand vectors for every c. One can easily verify that W^* is a planar layout for N and that the three-layer wiring is valid. Therefore, Algorithm A^* constructs a planar layout and its three layer wiring for N; furthermore, the number of tracks in the three-layer wiring constructed by A^* and the number of tracks in the planar layout constructed by A are identical. This completes the proof of the theorem. □

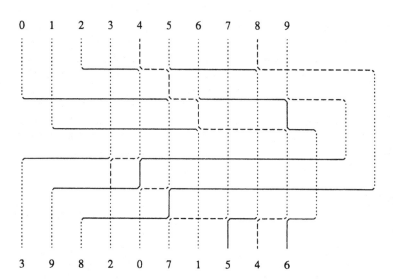

Figure 2.12: Layout and wiring constructed by our procedure.

3. Discussion

There are two major approaches to solve three-layer routing problems: the two-phase approach and the single-phase approach. In the two-phase approach, a planar layout is constructed in the first phase. In the second phase a three-layer wiring for the planar layout obtained in the first phase is constructed through a transformation, e.g., legal partition of the diagonal diagram induced by the layout. In the single-phase approach, layout construction and the three-layer assignment of the layout are performed simultaneously.

In this paper we presented a simple three-layer assignment algorithm for planar layouts generated by conservative layout algorithms. This class of algorithms includes simple variations of well known algorithms for the channel routing problem. Our approach consists of making simple modifications to the layout algorithm and incorporating a simple wire assignment strategy to generate three-layer wirings. Consequently, we obtain simpler and faster algorithms that generate three-layer wirings for layouts similar to the ones generated by algorithms (i) - (iii). Our algorithms are faster and conceptually simpler because there is no need to construct diagonal diagrams and legal partitions. The channel width of the wirings generated is identical to that of the planar layouts (i) - (iii). Algorithms (i) - (iii) are currently the best algorithms for the three layer channel routing problem.

We believe that if the structure of the planar layouts generated by a layout algorithm are simple, a three-wiring for the layout may be found by using diagonal diagrams. On the other hand, if a layout algorithm generates planar layouts with simple structures, it is not unlikely that this layout algorithm can be transformed into a single-phase routing algorithm.

There is a broader class of algorithms for which transformations similar to ours can generate three-layer wirings in a single phase. Property (2) defined in this paper is too restrictive. We defined conservative algorithms this way in order to have a simple equivalence proof. One may relax property (2) and only require that the number of extended nets, paired nets, etc., have identical counts at the end of each step. Equivalence proofs can also be obtained for these cases. For brevity we did not include the broader class of layout algorithms.

In general, wiring generated through legal partitions of the diagonal diagrams tend to have a large number of vias. For the layouts whose diagonal diagrams satisfy certain properties, some techniques can be used to reduce the number of vias. For example, the layouts generated by the three algorithms given in [7] can be wired in three layers by using the layer assignment algorithm given in [8]. This layer assignment algorithm finds a legal partition of the diagonal diagram corresponding to the layout. Special techniques are used to minimize the number of vias in the three layer wiring. It is easy to show that our layer assignment algorithm has similar performance with respect to the number of vias. We should point out that the time complexity for our algorithm is identical to that of the procedures given in [7]. Note that one does not need to output at each step unit wire segments in each of the tracks.

Bibliography

[1] Brady, M. L. and D. J. Brown, "VLSI Routing: Four Layers Suffice", Advances in Computing Research, vol. 2, 1984.

[2] Gao, S. and M. Kaufmann, "Channel Routing of Multiterminal Nets", Proceedings of the 28th Symposium on Foundations of Computer Science, pp 316-325, 1987.

[3] Gonzalez, T. and Zheng, S.-Q., "Wirability of Planar Layouts", Technical Report, # 87-11, CS Dept., UC Santa Barbara, Aug. 1987.

[4] Gonzalez, T. and Zheng, S.-Q., "Three-Layer Channel Routing of Multi-terminal Nets", Technical Report, # 87-13, CS Dept., UC Santa Barbara, Aug. 1987.

[5] Lipski, W. Jr, "An NP-complete Problem Related to Three-layer Channel Routing", Advances in Computing Research, vol. 2, 1984.

[6] Mehlhorn, K. and F. P. Preparata, "Routing Through a Rectangle", J. ACM, vol. 33, no. 1, 1986.

[7] Mehlhorn, K., F. P. Preparata and M. Sarrafzadeh, "Channel Routing in Knock-Knee Mode: Simplified Algorithms and Proofs", Algorithmica, no. 1, 1986.

[8] Preparata, F. P and W. Lipski, Jr, "Optimal Three-layer Channel Routing", IEEE Transaction on Computer., vol .33, no. 5, 1984.

[9] Preparata, F. P. and M. Sarrafzadeh, "Channel Routing of Nets Bounded Degree", VLSI: Algorithms and Architectures, North-Holland, 1984.

[10] Rivest, R. L., A. Baratz and G. Miller, "Provably Good Channel Routing Algorithms," in Proc. 1981 Carnegie-Mellon Conf on VLSI, Oct. 1981, pp. 153-159.

[11] Sarrafzadeh, M. and F. P. Preparata, F. P., "Compact Channel Routing of Multiterminal Nets", Annals of Discrete Mathematics, no. 25, April 1985, pp. 255-279.

Applying the Classification Theorem for Finite Simple Groups
to Minimize Pin Count in Uniform Permutation Architectures

Extended Abstract

Larry Finkelstein[1]
College of Comp. Science
Northeastern University
Boston, MA 02115

Daniel Kleitman[2]
Mathematics Department
M.I.T.
Cambridge, MA 02139

Tom Leighton[2]
Mathematics Dept. and
Lab. for Computer Science
M.I.T.
Cambridge, MA 02139

Abstract

In this paper, we show how to efficiently realize permutations of data stored in VLSI chips using bus interconnections with a small number of pins per chip. In particular, we show that permutations among chips from any group G can be uniformly realized in one step with a bus architecture that requires only $\Theta(\sqrt{|G|})$ pins per chip. The bound is within a small constant factor of optimal and solves the central question left open by the recent work of Kilian, Kipnis and Leiserson on uniform permutation architectures [8]. The proof makes use of the Classification Theorem for Finite Simple Groups to show that every finite group G of nonprime order contains a nontrivial subgroup of size at least $\sqrt{|G|}$. The latter result is also optimal and improves the old pre-Classification Theorem lower bound of $|G|^{1/3}$ proved by Brauer and Fowler [1] and Feit [4].

Key Words: architecture, busses, Classification Theorem, difference cover, group theory, pins, simple groups, VLSI.

[1]Research supported in part by NSF grant DCR-8603293.

[2]Research supported in part by Air Force Contract OSR-86-0076, DARPA Contract N00014-80-C-0622, Army Contract DAAL-03-86-K-0171 and an NSF Presidential Young Investigator Award with matching funds from AT&T and IBM.

1. Introduction

In this paper, we consider the problem of efficiently realizing permutations of data stored in VLSI chips. Given a collection of N chips and a permutation π from a fixed group G, our task is to set up dedicated directed links from chip i to chip $\pi(i)$ for all i. Then the data in the chips can be permuted according to π in a single step.

Communication between chips will be handled by busses. A single bus can connect to an arbitrary number of chips, but can only serve as a dedicated link between a single pair of chips. To *realize* a permutation π, we must have a collection of N distinct busses $b_1, \ldots, b_N$ so that bus b_i is linked to chip i and chip $\pi(i)$. Then chip i can send data to chip $\pi(i)$ by simply writing on bus i and having chip $\pi(i)$ read on bus i.

Given a group of permutations G, the problem is to construct a set of busses with the minimum amount of hardware so that any permutation in the group can be realized. The most important measure of hardware is the total number of pins on the chips. *Pins* are the connection points between the chips and the busses and are the commodity which is scarcest in practical settings.

A set of busses is said to *uniformly realize* a permutation π if the permutation can be realized by every chip writing to the same pin (say the third pin) and every chip reading from the same pin (say the tenth). Uniform permutation architectures are much easier to implement in practice because the control structure is so simple (e.g., it is SIMD). To perform a permutation π, each chip is simply told to write to pin W_π and to read from pin R_π. Without uniformity, the local decisions about where to write and read might be a lot costlier in terms of time and/or space.

In this paper, we consider only schemes that realize permutations in a uniform fashion. We show that for any permutation group G, there is a collection of busses with $O(\sqrt{|G|})$ pins per chip that can uniformly realize any permutation in the group. The architecture is easily constructable in polynomial time, and is guaranteed to be within a small constant factor of optimal for any group. The result improves the previous bound of $O(\sqrt{|G|\log|G|})$ pins per chip by Kilian, Kipnis and Leiserson [8].

The key to our construction lies in our ability to find a large subgroup of any group of nonprime order. Applying the Classification Theorem for Finite Simple Groups, we prove that every finite group G of nonprime order contains a nontrivial subgroup of size at least $\sqrt{|G|}$. The result is optimal and substantially improves the pre-Classification Theorem lower bound of $|G|^{1/3}$ due to Brauer and Fowler [1] and Feit [4]. The proof is not difficult, and essentially consists of checking that the known finite simple groups all have large subgroups. That such a result might be true has been noticed by others, including Suzuki [11, p. 125].

The paper is divided into six sections. We start with the construction of a uniform permutation architecture in Section 2. The necessary group theory is presented in Section 3. We conclude with some remarks, acknowledgements and references in Sections 4–6.

This paper was motivated by and naturally follows the foundational work of Kilian, Kipnis and Leiserson [8]. For more information on the VLSI aspects of our work, we refer the reader to that paper. For more information on the group theoretic aspects of our work, we refer the reader to the book by Gorenstein [6] on the classification of finite simple groups.

2. Constructions

In this section, we show how to configure a collection of busses so that every permutation from a group G can be uniformly realized, and so that each chip contains at most $3\sqrt{|G|}$ pins. The construction makes use of the fact that every group has a small difference cover. A *difference cover* for a permutation set Π is a set $\Phi = \{\phi_0, \phi_1, \ldots, \phi_{K-1}\}$ of permutations such that for each $\pi \in \Pi$, there exist $\phi_i, \phi_j \in \Phi$ such that $\pi = \phi_j^{-1}\phi_i$.

In [8], Kilian, Kipnis and Leiserson prove that there is a uniform architecture of busses for a permutation set Π with K pins per chip if and only if Π has a K-element difference cover. The configuration of the busses given the difference set is quite straightforward. Each chip u is connected via its ith pin to bus $\phi_i(u)$. To see that this configuration uniformly realizes any permutation π in G, let $\phi_i, \phi_j \in \Phi$ be such that $\pi = \phi_j^{-1}\phi_i$ and define the write-read operation for π to be "all chips write on pin i and read on pin j." Then chip u will write on bus $\phi_i(u)$ and will read on bus $\phi_j(u)$. Since ϕ_i is a permutation, no two chips will write on the same bus. Moreover, the data written by chip u on bus $\phi_i(u)$ will be read by chip v where $\phi_i(u) = \phi_j(v)$. Hence, $v = \phi_j^{-1}\phi_i(u) = \pi(u)$, and the data written by chip u is read by chip $\pi(u)$ as desired.

We now need only prove that every permutation group has a small difference cover. This is accomplished in the following theorem. The result is a simple consequence of the fact that every finite group G of nonprime order has a nontrivial subgroup of size at least $\sqrt{|G|}$. This fact is proved as Theorem A in Section 3. A similar result on minimal bases of groups was independently proved by Pyber [10].

Theorem B. *Every group G has a difference cover of size at most $3\sqrt{|G|}$.*

Proof. We first prove by induction on $|G|$ that for any x, $1 \leq x \leq |G|$, every group G can be factored into subsets L and R of sizes at most $\frac{c|G|}{x}$ and x such that $G \subseteq L \times R$. In other words, L and R are such that every element of G can be written as the product of an element of L times an element of R. The optimal value of c depends on G, but is always at most 2. The proof is concluded by selecting $x = \sqrt{G}$ and observing that $L^{-1} \cup R$ is a difference cover for G.

There are two cases for the induction. If $|G|$ is prime (case 1), then G is cyclic with some generator g. The factorization is then given by

$$L = \{g^i \mid 0 \leq i < |G|, \ i \equiv 0 \bmod \lfloor x \rfloor\}$$
$$R = \{g^i \mid 0 \leq i < \lfloor x \rfloor\}.$$

Note that $|L| = \lceil \frac{|G|}{x} \rceil$ and $|R| = \lfloor x \rfloor$. Hence the claim is verified in this case, provided that $c \geq 2$.

If $|G|$ is nonprime (case 2), then G contains a subgroup H of size $|H| \geq \sqrt{|G|}$ by Theorem A. If $x \leq \sqrt{|G|}$, then use induction to factor H into subsets L' and R' of sizes at most $\frac{c|H|}{x}$ and x, respectively. Let L'' be any set of left-coset representatives for G/H. Then $G = L'' \times H$ and $|L''| = \frac{|G|}{|H|}$. Next set $L = L'' \times L'$ and $R = R'$. Then $G = L \times R$, $|L| = |L''| \cdot |L'| \leq \frac{c|G|}{x}$ and $|R| = |R'| \leq x$.

If $x > \sqrt{|G|}$, then the argument is similar except that we factor H into subsets of size $\frac{c|H|}{x'}$ and x' where $x' = \frac{x|H|}{|G|}$, and we let $R'\prime$ be a set of right-coset representatives for G/H. Then $L = L'$, $R = R' \times R'\prime$ and $G = L \times R$ where $|L| \leq \frac{c|G|}{x}$ and $|R| \leq x$, as desired. $\square$

The constant 3 in Theorem B is simply the value of $c+1$ where c is as defined in the proof. The value of c must be close to 2 for some groups but can be smaller for others. In fact, c is nearly one for large groups for which $|G|$ does not have small prime factors. The extra additive 1 can also be reduced in cases when $L^{-1} \cup R$ contains pairs of elements that are inverses of each other. In no event can the overall constant be reduced below 1, however, since a difference cover must have at least $\sqrt{|G|}$ elements in order to cover all $|G|$ permutations in G.

3. Group Theory

The main result of this section is a proof of the following theorem.

Theorem A. Let G be a finite group which is not cyclic of prime order. Then G has a proper subgroup H such that $|H| \geq \sqrt{|G|}$.

The proof of Theorem A reduces in a fairly straightforward manner to a verification that the conclusion holds in all cases where G is a non-abelian simple group. This will be done in a uniform fashion using the identification of the infinite families of simple groups with Chevalley groups, excluding of course the alternating series. The sporadic groups will be treated individually.

We start by reducing the general result to the case of simple groups.

Lemma A. Suppose the conclusion of Theorem A holds when G is a non-abelian simple group. Then Theorem A is true in general.

Proof. The proof is by induction on the order of G. The base case occurs when G is isomorphic to $Z_2 \times Z_2$ and the result is clearly true in this case. Now let G be an arbitrary group which satisfies the hypotheses of Theorem A. If G is simple, then the conclusion of Theorem A holds by assumption. Otherwise, let K be a maximal normal subgroup of G so that G/K is simple. If G/K is non-abelian as well, then by assumption, there exists a proper subgroup H of G containing K, so that $|H/K|^2 \geq |G/K|$. But then $|H|^2 \geq |G|$ and so H satisfies the conclusion of Theorem A.

If G/K is cyclic of order p, p a prime, then K satisfies the conclusion of Theorem A unless $|K| < p$. But then if σ is an element of order p of G, then $\langle \sigma \rangle$ satisfies the conclusion of Theorem A. $\square$

We will assume for the remainder of this discussion that G is a finite non-abelian simple group.

According to the classification of finite simple groups, ([5], Table 2.4) G is isomorphic to a Chevalley group, an alternating group A_n, ($n \geq 5$), or one of the 26 sporadic simple groups.

If G is isomorphic to A_n, then the conclusion of Theorem A is true since G has order $n!/2$ and a subgroup of index n. Therefore, we may assume that G is isomorphic to a Chevalley group or a sporadic simple group. We shall organize the proof along these lines.

3.1. G Isomorphic to a Chevalley Group

The two primary references for this section are the book by Carter (*Simple Groups of Lie Type* [2]) and the paper by Curtis, Kantor and Seitz [3]. Each Chevalley group G has an underlying field of definition, $GF(q)$ where q is a power of a prime p and is associated with an indecomposable root system Δ of Euclidean space E_n. G is generated by certain p-subgroups U_α, $\alpha \in \Delta$, called *root* subgroups.

The Dynkin diagrams for the indecomposable root systems are given in ([3], p.4). Let Π be a set of fundamental roots $\alpha_1, \ldots, \alpha_n$ for Δ. We will denote the root subgroup $U_{\pm \alpha_i}$ by $U_{\pm i}$ respectively.

Many important structural properties of G can be deduced through the existence of a (B, N) pair ([2], Sections 8.2 and 13.4). The subgroup B of G, called the Borel subgroup, has the form $B = UH$ where U, called the unipotent subgroup, is a Sylow p-subgroup of G, and H is the Cartan (or diagonal) subgroup. N contains H as a normal subgroup and $W = N/H$ may be identified with a group generated by the reflections $s_1, \ldots, s_n$ corresponding to the set Π of fundamental roots in Δ. We call n the *rank* of G.

For each subset $I \subseteq \{1, \ldots, n\}$, define

$$G_I = \langle B, U_{-j} \mid j \notin I \rangle.$$

Then G_I is a parabolic subgroup, that is a subgroup of G containing B and has the form $G_I = Q_I L_I$ where

$$Q_I = \langle U_\alpha \mid \alpha = \sum m_j \alpha_j, m_j \neq 0, \text{ for some } j \in I \rangle, \text{ and}$$
$$L_I = \langle U_j, U_{-j} \mid j \notin I \rangle.$$

L_I is called a Levi Complement and is usually a covering group of a Chevalley Group or the direct product of covering groups of Chevalley groups. The structure of L_I can be determined from the Dynkin Diagram for G. When $I = \{i\}$, we will denote G_I by G_i. In this case, it can be shown that G_i is a maximal parabolic subgroup.

We will now show, case by case, that every Chevalley group has a subgroup of the form G_i which satisfies the conclusion of Theorem A.

3.1.1. G Isomorphic to a Linear Group

The Chevalley Groups which are isomorphic to linear groups are given in Table 1. In each case, we may choose G_1 to be the appropriate subgroup which satisfies the conclusion of Theorem A. Table 1 gives the Chevalley groups which are isomorphic to Linear Groups.

Table 1   Linear Groups

G	Linear Group Notation	$|G|$	d
$A_n(q)$	$PSL(n+1, q)$	$q^{n(n+1)/2} \prod_{i=1}^n (q^{i+1} - 1)$	$(n+1, q-1)$
$B_n(q), n > 1$	$PSO(2n+1, q)$	$q^{n^2} \prod_{i=1}^n (q^{2i} - 1)$	$(2, q-1)$
$C_n(q), n > 2$	$PSp(2n, q)$	$q^{n^2} \prod_{i=1}^n (q^{2i} - 1)$	$(2, q-1)$
$D_n(q), n > 3$	$PSO^+(2n, q)$	$q^{n(n-1)}(q^n - 1) \prod_{i=1}^{n-1} (q^{2i} - 1)$	$(4, q^n - 1)$
$^2A_n(q), n > 1$	$PSU(n+1, q)$	$q^{n(n+1)/2} \prod_{i=1}^n (q^{i+1} - (-1)^{i+1})$	$(n+1, q+1)$
$^2D_n(q), n > 3$	$PSO^-(2n, q)$	$q^{n(n-1)}(q^n + 1) \prod_{i=1}^{n-1} (q^{2i} - 1)$	$(4, q^n + 1)$

Theorem 3.1.1. *If G is isomorphic to a linear group, then the conclusion of Theorem A holds.*

Proof. We will show that the maximal parabolic subgroup G_1 has the appropriate order. This in turn will follow from the equation $[G : G_1]^2 < |G|$. We refer the reader to ([3], Table 4) where the indices of $[G : G_1]$ are tabulated.

(i) $G \cong A_n(q)$; $[G : G_1] = \frac{q^{n+1} - 1}{q - 1}$.

Suppose first that $n > 1$. Then $[G : G_1]^2 \leq q^{n+1}(q^{n+1} - 1)$ and $q^{n(n+1)/2}(q^{n+1} - 1)$ is a factor of $|G|$ from Table 1. If $n = 1$, then $[G : G_1] = q + 1$ and $|G| = q(q+1)(q-1)/d$ where $d = (q-1, 2)$. Since $q > 3$ in order for G to be simple, it follows that $[G : G_1]^2 \leq |G|$ in this case as well.

(ii) $G \cong B_n(q), C_n(q)$; $[G : G_1] = \frac{q^{2n} - 1}{q - 1}$.

Since $[G : G_1]^2 \leq q^{2n}(q^{2n} - 1)$, it suffices to show that $q^{2n}(q^{2n} - 1)$ is a factor of $|G|$. But this follows from the fact that $n > 1$.

(iii) $G \cong D_n(q), {}^2D_n(q)$; $[G : G_1] = \frac{(q^n - 1)(q^{n-1} + 1)}{q - 1}$, $\frac{(q^n + 1)(q^{n-1} - 1)}{q - 1}$, respectively.

First note that

$$\prod_{i=1}^{n-1}(q^{2i}-1) \geq \prod_{i=1}^{n-1}(q^i+1) \geq q^{n(n-1)/2}.$$

Thus $|G| \geq q^{3n(n-1)/2}$. Furthermore, in either case, $[G:G_1] \leq q^{2n}$. Thus it suffices to show that $q^{4n} \leq q^{3n(n-1)/2}$. But this follows from the fact that $n \geq 4$.

(iv) $G \cong {}^2A_n(q)$; $[G:G_1] = \frac{(q^{n+1}-(-1)^{n+1})(q^n-(-1)^n)}{q^2-1}$.

It is easy to prove directly that $[G:G_1]^2 \leq |G|$ when $n = 2, 3$, keeping in mind that $q > 2$ when $n = 2$. Otherwise, assume $n \geq 4$. Then $|G|/[G:G_1]^2 \geq q^{n(n+1)/2}/[G:G_1]$. But $[G:G_1] \geq q^{2n+2}$, and so $|G|/[G:G_1]^2 \geq q^{n(n+1)/2}/q^{2n+2} \geq 1$. The last inequality uses the fact that $n \geq 4$. □

3.1.2. G Isomorphic to an Exceptional Chevalley Group of Rank > 2.

The following table gives the orders of the Chevalley groups of exceptional type of rank > 2.

Table 2 Exceptional Groups of rank > 2

G	$|G|$	d
$F_4(q)$	$q^{24}(q^{12}-1)(q^8-1)(q^6-1)(q^2-1)$	1
$E_6(q)$	$q^{36}(q^{12}-1)(q^9-1)(q^8-1)(q^6-1)(q^5-1)(q^2-1)$	$(3, q-1)$
$E_7(q)$	$q^{63}(q^{18}-1)(q^{14}-1)(q^{12}-1)(q^{10}-1)(q^8-1)(q^6-1)(q^2-1)$	$(2, q-1)$
$E_8(q)$	$q^{120}(q^{30}-1)(q^{24}-1)(q^{20}-1)(q^{18}-1)(q^{14}-1)(q^{12}-1)(q^8-1)(q^2-1)$	1
${}^2E_6(q)$	$q^{36}(q^{12}-1)(q^9+1)(q^8-1)(q^6-1)(q^5+1)(q^2-1)$	$(3, q+1)$

Theorem 3.1.2. *If G is isomorphic to an exceptional Chevalley group of rank > 2, then the conclusion of Theorem A holds.*

Proof. As in the proof of Theorem 3.1.1, we will show that there is a maximal parabolic subgroup G_i, such that $[G:G_i]^2 \leq |G|$. We will refer to [3]. Table 4, to find a suitable G_i. It suffices then to find an upper bound u for $[G:G_i]$ and a lower bound ℓ for $|G|$ such that $u^2 \leq \ell$. The following lower bounds on the orders of the groups in this class are obtained by using the inequalities $q^n - 1 \geq q^{n-1}$ and $q^{n+1} \geq q^n + 1$.

$$|F_4(q)| \geq q^{48} \quad |E_6(q)| \geq q^{72} \quad |E_7(q)| \geq q^{126} \quad |E_8(q)| \geq q^{240} \quad |{}^2E_6(q)| \geq q^{73}$$

The upper bound u on $[G:G_i]$ such that $u^2 \leq \ell \leq |G|$ may be computed directly from $[G:G_i]$ and is given as follows.

(i) $G = F_4(q)$; $[G:G_1] = \frac{(q^4+1)(q^{12}-1)}{q-1}$:

$u = q^{17}$.

(ii) $G = E_6(q)$; $[G:G_2] = \frac{(q^4+1)(q^9-1)(q^{12}-1)}{(q-1)(q^3-1)}$:

$u = q^{27}$.

(iii) $G = E_7(q)$; $[G:G_1] = \frac{(q^{14}-1)(q^{12}-1)(q^{18}-1)}{(q^6-1)(q^4-1)(q-1)}$:

$u = q^{44}$.

(iv) $G = E_8(q)$; $[G:G_8] = \frac{(q^{10}+1)(q^{24}-1)(q^{30}-1)}{(q^6-1)(q-1)}$:

$u = q^{65}$.

(v) $G = {}^2E_6(q)$; $[G:G_1] = \frac{(q^4+1)(q^9+1)(q^{12}-1)}{(q^3+1)(q-1)}$:

$u = q^{27}$. $\square$

3.1.3. G Isomorphic to a Chevalley Group of Rank ≤ 2

The remaining Chevalley groups have rank ≤ 2. An exceptional case occurs for ${}^2F_4(q)$ when $q = 2$. In this case, the commutator subgroup has index 2 and is simple. In the case of the rank-1 groups, a Borel subgroup is maximal and the action of G on the cosets of B is 2-transitive. For the rank 2-groups (excluding ${}^2F_4(2)'$), the set Π of fundamental roots consists of a short root and long root. Let P be the maximal parabolic subgroup corresponding to a short root (a long root would do as well). Then the Levi Factor L of P is isomorphic to $SL(2,q)$ except in the case of ${}^3D_4(q)$ in which case L is isomorphic to $SL(3,q^3)$ ([6], Section 3.5.). We may then use the formulas $|SL(2,q)| = q(q^2-1)$ and $|SL(3,q^3)| = q^3(q^3-1)(q^2-1)$ to compute the order of P and hence $[G:P]$. We present in Table 3, the orders of the groups in this class together with the index of a maximal parabolic subgroup P. The structure of ${}^2F_4(2)'$ is discussed in [12].

Table 3 Exceptional Groups of rank ≤ 2

G	$\|G\|$	$[G:P]$	d
$G_2(q)$, $q > 2$	$q^6(q^6-1)(q^2-1)$	q^6-1	1
${}^2F_4(q)$, $q = 2^{2m+1}$	$q^{12}(q^6+1)(q^4-1)(q^3+1)(q-1)$	$(q^6+1)(q^3+1)(q^2+1)(q-1)$	1
${}^2F_4(2)'$	$2^{11} \cdot 3^3 \cdot 5^2 \cdot 13$	$3^3 \cdot 5 \cdot 13$	1
${}^3D_4(q)$	$q^{12}(q^8+q^4+1)(q^6-1)(q^2-1)$	$(q^8+q^4+1)(q^2-1)$	1
${}^2G_2(q)$, $q = 3^{2m+1}$, $q > 3$	$q^3(q^3+1)(q-1)$	q^3+1	1
${}^2B_2(q)$, $q = 2^{2m+1}$, $q > 2$	$q^2(q^2+1)(q-1)$	q^2+1	1

Theorem 3.1.3. If G is isomorphic to an exceptional Chevalley group of rank ≤ 2, then the conclusion of Theorem A holds.

Proof. It is easy to see from Table 3 that $[G:P]^2 \leq |G|$ for all groups in this class. For example if $G = {}^2F_4(q)$, then after some elementary calculations, we have

$$\begin{aligned}
|G|/[G:P]^2 &= \frac{q^{12}(q^4+1)}{(q^6+1)(q^3+1)(q^2+1)^2(q-1)} \\
&\geq \frac{(q^{12}-1)(q^4-1)}{(q^6+1)(q^3+1)(q^2+1)^2(q-1)} \\
&= \frac{(q^6+1)(q^3+1)(q^3-1)(q^2+1)(q^2-1)}{(q^6+1)(q^3+1)(q^2+1)^2(q-1)} \\
&= \frac{(q^3-1)(q+1)}{q^2+1} \\
&\geq q+1.
\end{aligned}$$

The remaining cases follow in a similar way. $\square$

3.2. G a Sporadic Simple Group

There are 26 sporadic simple groups. In the following table, we describe the structure of a maximal subgroup M for each sporadic group and give its structure. We use the notation $H \cdot Z_2$ to denote an extension of a simple group H

by an automorphism of order 2 and $\hat{H}$ to denote a central extension of H by Z_2. In addition, we will use the notation $E_{2^n} \cdot H$ to denote an extension of an elementary abelian group of order 2^n by H Most of the information in Table 4 may be obtained from ([6], Chapter 2). An additional reference is [7] for the existence of a $PSL(2,11)$ subgroup of J_1. Missing entries in the table are easily derived from other entries.

Table 4 The Sporadic Simple Groups

G	$|G|$	M	$[G:M]$
M_{11}	$2^4 \cdot 3^2 \cdot 5 \cdot 11$	M_{10}	11
M_{12}	$2^6 \cdot 3^3 \cdot 5 \cdot 11$	M_{11}	11
M_{22}	$2^7 \cdot 3^2 \cdot 5 \cdot 7 \cdot 11$	$PSL(3,4)$	22
M_{23}	$2^7 \cdot 3^2 \cdot 5 \cdot 7 \cdot 11 \cdot 23$	M_{22}	23
M_{24}	$2^{10} \cdot 3^3 \cdot 5 \cdot 7 \cdot 11 \cdot 23$	M_{23}	24
J_1	$2^3 \cdot 3 \cdot 5 \cdot 7 \cdot 11 \cdot 19$	$PSL(2,11)$	$2 \cdot 17 \cdot 19$
J_2	$2^7 \cdot 3^3 \cdot 5^2 \cdot 7$	$PSU(3,3)$	$2^2 \cdot 5^2$
J_3	$2^7 \cdot 3^5 \cdot 5 \cdot 17 \cdot 19$	$PSL(2,16) \cdot Z_2$	$2^2 \cdot 3^4 \cdot 19$
J_4	$2^{21} \cdot 3^3 \cdot 5 \cdot 7 \cdot 11^3 \cdot 23 \cdot 29 \cdot 31 \cdot 37 \cdot 43$	$E_{2^{11}} \cdot M_{24}$	$11^2 \cdot 29 \cdot 31 \cdot 37 \cdot 43$
HS	$2^9 \cdot 3^2 \cdot 5^3 \cdot 7 \cdot 11 \cdot 23$	M_{22}	$2^2 \cdot 5^2$
Mc	$2^7 \cdot 3^6 \cdot 5^3 \cdot 7 \cdot 11 \cdot 23$	$PSU(3,4)$	$5^2 \cdot 11$
Suz	$2^{13} \cdot 3^7 \cdot 5^2 \cdot 7 \cdot 11 \cdot 13$	$G_2(4)$	$2 \cdot 3^4 \cdot 11$
Ru	$2^{14} \cdot 3^3 \cdot 5^3 \cdot 7 \cdot 13 \cdot 29$	$^2F_4(2)$	$2^2 \cdot 5 \cdot 7 \cdot 29$
He	$2^{10} \cdot 3^3 \cdot 5^2 \cdot 7^3 \cdot 17$	$SP(4,4) \cdot Z_2$	$2 \cdot 3 \cdot 7^3$
Ly	$2^8 \cdot 3^7 \cdot 5^6 \cdot 7 \cdot 11 \cdot 31 \cdot 37 \cdot 67$	$G_2(5)$	$2^2 \cdot 3^4 \cdot 11 \cdot 37 \cdot 67$
On	$2^9 \cdot 3^4 \cdot 5 \cdot 7^3 \cdot 11 \cdot 19 \cdot 31$	$PSL(3,7) \cdot Z_2$	$2^3 \cdot 3^2 \cdot 5 \cdot 11 \cdot 31$
$\cdot 1$	$2^{21} \cdot 3^9 \cdot 5^4 \cdot 7^2 \cdot 11 \cdot 13 \cdot 23$	$\cdot 2$	$2^3 \cdot 3^3 \cdot 5 \cdot 7 \cdot 13$
$\cdot 2$	$2^{18} \cdot 3^6 \cdot 5^3 \cdot 7 \cdot 11 \cdot 23$	$PSU(6,2) \cdot Z_2$	$2^2 \cdot 5^2 \cdot 23$
$\cdot 3$	$2^{10} \cdot 3^7 \cdot 5^3 \cdot 7 \cdot 11 \cdot 23$	$Aut(Mc)$	$2^2 \cdot 3 \cdot 23$
$M(22)$	$2^{17} \cdot 3^9 \cdot 5^2 \cdot 7 \cdot 11 \cdot 13$	$U_6^{\hat{}}(2)$	$2 \cdot 3^3 \cdot 5 \cdot 13$
$M(23)$	$2^{18} \cdot 3^{13} \cdot 5^2 \cdot 7 \cdot 11 \cdot 13 \cdot 17 \cdot 23$	$M(\hat{2}2)$	$3^4 \cdot 17 \cdot 23$
$M(24)'$	$2^{21} \cdot 3^{16} \cdot 5^2 \cdot 7^3 \cdot 11 \cdot 13 \cdot 17 \cdot 23 \cdot 29$	$M(23)$	$2^3 \cdot 3^3 \cdot 7^2 \cdot 29$
F_5	$2^{15} \cdot 3^{10} \cdot 5^3 \cdot 7^2 \cdot 13 \cdot 19 \cdot 31$	$E_{2^5} \cdot PSL(5,2)$	$3^8 \cdot 5^2 \cdot 7 \cdot 13 \cdot 19$
F_3	$2^{14} \cdot 3^6 \cdot 5^6 \cdot 7 \cdot 11 \cdot 19$	A_{12}	$2^5 \cdot 3 \cdot 5^4 \cdot 19$
F_2	$2^{41} \cdot 3^{13} \cdot 5^6 \cdot 7^2 \cdot 11 \cdot 13 \cdot 17 \cdot 19 \cdot 23 \cdot 31 \cdot 47$	$E_6^{\hat{}}(2) \cdot Z_2$	$2^3 \cdot 3^4 \cdot 5^4 \cdot 23 \cdot 31 \cdot 47$
F_1	$2^{46} \cdot 3^{20} \cdot 5^9 \cdot 7^6 \cdot 11^2 \cdot 13^3 \cdot 17 \cdot 19 \cdot 23 \cdot 29 \cdot 31 \cdot 41 \cdot 47 \cdot 59 \cdot 71$	$\hat{F}_2$	

Theorem 3.1.4. *If G is isomorphic to a sporadic simple group, then the conclusion of Theorem A holds.*

Proof. It is straightforward to check from Table 4 that $[G:M]^2 \leq |G|$ for each sporadic group G. $\square$

4. Remarks

The permutation architecture described in this paper can be quickly and easily constructed once the structure of the underlying group is understood and/or a tower of large subgroups is found. Fast polynomial time algorithms for finding group towers are well-known. For example, see [9].

We are currently trying to generalize our results to include t-step architectures, e.g., configurations of busses that can uniformly realize permutations in t steps where $t > 1$. We hope to be able to show that any permutation from a fixed group can be so realized using $O(t|G|^{1/2t})$ pins per chip, which is within a constant factor of optimal for constant t.

5. Acknowledgements

We are indebted to Walter Feit, Daniel Gorenstein, Laszlo Pyher, Bill Reynolds, and others for their helpful comments concerning the Classification Theorem for Finite Simple Groups.

6. References

1 R. Brauer and K. Fowler, "On groups of even order", *Ann. Math.*, **62** (1955), 565-583.

2 R.W. Carter, "Simple Groups of Lie Type", John Wiley & Sons, New York, 1972.

3 C. Curtis, W. Kantor and G. Seitz, "The 2-transitive permutation representations of the finite Chevalley groups", *Trans. Amer. Math. Soc.*, **218**, (1976), 1-59.

4 W. Feit, *Proc. National Academy of Sciences*, Vol. 48, pp. 968–970, 1962.

5 D. Gorenstein, "Finite Simple Groups", Plenum Press, New York, 1982.

6 D. Gorenstein, "The Classification of Finite Simple Groups, Volume 1: Groups of Noncharacteristic 2 Type", Plenum Press, New York, 1983.

7 Z. Janko, "A new finite simple group with abelian 2-Sylow subgroups and its characterization," *J. Algebra*, **3** (1966), 147-186.

8 J. Kilian, S. Kipnis and C. Leiserson, "The Organization of Permutation Architectures with Bussed Interconnections," *Proc. 1987 IEEE Conf. on Foundations of Computer Science*, pp. 305–315, 1987.

9 E. Luks, "Computing the Composition Factor of a Permutation Group in Polynomial Time," *Combinatorica*, Vol. 7, 1987, pp. 87–100.

10 L. Pyber, personal communication, 1988.

11 M. Suzuki, *Group Theory, Volume II*, Springer-Verlag, 1986.

12 J. Tits, "Algebraic and abstract simple groups," *Ann. of Math.*, **80** (1964), 313-329.

A NEW ALGORITHM FOR WIRING LAYOUTS [*]

Ioannis G. Tollis

Department of Computer Science
The University of Texas at Dallas
P.O. Box 830688, FN 3.3
Richardson, TX 75083-0688

ABSTRACT

In this paper we present a new algorithm for wiring layouts in the square grid. This algorithm provides a rather intuitive proof that the number of layers required to wire a given layout W is at most four, and it constructs a two-layer wiring if and only if W requires two layers. The algorithm runs in $O(N)$ time, where N is the number of grid points occupied by the layout. The technique used here can be extended in order to construct wirings of layouts in the tri-hexagonal grid that use at most five layers.

1. Introduction

Wiring is the problem of converting a wire layout in the plane into an actual three-dimensional configuration of wires, and is a fundamental and classical problem in VLSI layout [1-9]. Wiring is formulated as follows: Given a layout W in the plane, assign each edge of W to a unique layer such that the connectivity between the terminals is preserved, and edges of distinct wires do not share a point on the same layer.

Typically, the process of creating a VLSI layout consists of two phases: *placement* and *routing*. Routing consists of two steps: *wire layout* and *wiring*. In the first step, a planar (wire) *layout* is obtained; all wires connecting the corresponding terminals are projected to the plane. In the wiring step, the layout in the plane is converted into an actual three-dimensional configuration of wires. There is a constant number (usually small) of conducting layers, and each edge of every wire is assigned to a unique layer, so that all edges of a wire are electrically connected and no two edges belonging to distinct wires are connected. Adjacent edges of the same wire lying on different layers are connected with vertical connections, called *vias*.

All of the past research on wiring is based on the concept of *two-colorable maps*, introduced by Preparata and Lipski [7]. They showed that three layers are sufficient to solve a two-

---

[*] This research was performed while the author was with the Coordinated Science Lab., University of Illinois, Urbana, IL 61801, and was supported in part by the Semiconductor Research Corporation under contract RSCH 84-06-049-6.

terminal channel routing problem with the minimum number of tracks. Subsequently, Lipski [4] showed that it is NP-complete to decide whether an arbitrary planar wire layout in the square grid can be wired using three layers. Brady and Brown showed that every layout in the square grid can be wired using four layers [1]. A systematic approach to layout wirability for layouts in the square grid appears in [5]. Finally, we recently developed a systematic approach to layout wirability for layouts in any *uniform grid* [8, 9].

The Brady-Brown algorithm [1] constructs a wiring for a given layout W in the square grid as follows: (i) First, it converts W into an "equivalent full layout" W^* by inserting new (fictitious) wires, so that every edge of the grid is used by some wire of W^*. (ii) It then produces a two-colorable map that provably corresponds to a four-layer wiring for W^*. (iii) Finally, it constructs a four-layer wiring for W from the wiring of W^*. Hence, its disadvantages are the following: (a) It only works for full layouts. (b) It always constructs a four-layer wiring of W^*, even if layout W can be wired using fewer layers. (c) The method used to construct a wiring of W from the wiring of W^* uses the same number of layers.

In this paper, we present a new algorithm for constructing wirings of layouts in the square grid also using at most four layers. However, a given layout W is wired using two layers if W is wirable with two layers. This is the first wiring technique that is not based on two-colorable maps. Our algorithm takes advantage of non-full layouts and therefore in most instances it constructs wirings using at most three layers. Also, it runs in time linear with respect to the area occupied by the layout. The technique presented here extends to the design of a linear-time algorithm for wiring layouts in the tri-hexagonal grid (see Figure 1) using at most five layers. It is an interesting open problem to extend this wiring technique to other uniform grids.

2. Preliminaries

The most widely used grid is the square grid. The tri-hexagonal grid is interesting for its attractive symmetries, although, to our knowledge, it has not yet been applied in practice. The lines of a grid are called *grid lines*. The intersection point of two grid lines is called a *grid point*, and an edge of the grid is called a *grid edge*. The length of a grid edge and the distance between a grid point and any not incident grid edge are at least one unit. The square grid has been the subject of many investigations [1, 5, 7]. In this paper we will develop a new technique for wiring layouts in the square grid.

A *wire w* is a connected subgraph of the grid R such that every edge of w is a grid edge of R, and is called *(wire) edge*. Grid points of R with exactly one wire edge of w incident with them are called *terminals*. A *(planar) wire layout*, or simply *layout*, is a finite collection

$W = \{w_1, w_2, ..., w_n\}$ of wires. If the wires of W are edge disjoint then the layout is a *knock-knee* layout. Let w_1 and w_2 be two wires that share a grid point p. If we scan the wire edges incident with p, say, counterclockwise around p, then we distinguish two cases:

(i) the wire edges of w_1 and w_2 alternate. This is called a *crossing*, see Figure 2(a).

(ii) the wire edges of w_1 are adjacent and so are the ones of w_2. This configuration is called a *knock-knee*, see Figure 2(b).

If the wire edges incident with some grid point p belong to at most one wire, then p is called *trivial* (with respect to W). Similarly, any grid point p whose incident wire edges belong to two wires is called *nontrivial*.

A *conducting layer*, or simply *layer*, is a graph isomorphic to the grid R. The conducting layers are stacked on top of each other and are denoted by $L_1, L_2, ..., L_\nu$. L_1 is the bottommost layer, L_2 is above L_1, ..., and L_ν is the topmost layer. A contact between two layers, called a *via*, can be placed only at a grid point of R.

A (ν-*layer*) *wiring* of a given layout $W = \{w_1, ..., w_n\}$ is an assignment of each edge of every wire in W to a layer such that: if adjacent edges of a wire are assigned to different layers at a grid point p, then no other wire can use that grid point on any of the intermediate layers. More formally, if in a wiring of W $\phi(e, w_k) \in \{L_1, ... L_\nu\}$ is the layer to which edge e of wire w_k has been assigned, $k = 1, ..., n$, then the following condition must be satisfied: If edges e_1, e_2 of w_k incident with grid point p are assigned to layers L_i and L_j, respectively, then $\phi(e, w_l) \neq L_m$, $i \leq m \leq j$, for every edge e of w_l incident with p, $l \neq k$.

The *layer number* $\nu(W)$ of a layout W is the minimum number of layers needed to wire W, i.e., the minimum number k such that a k-layer wiring of W exists.

3. Relative Order and the Layer-Graph

In this section we will see that any wiring of a given layout corresponds to some vertical ordering of the wires on all the grid points. Our goal is to characterize the vertical orderings that induce legal wirings, and to develop the necessary tools that will allow us to find wirings that require a small number of layers. Although our discussion is based on layouts in the square grid, it is clear that the following results hold also for other grids, e.g., the tri-hexagonal grid and the uniform grids.

Let W be a layout in the square grid, and w_1 and w_2 be two wires in W that pass through grid point p_1. Assume that e_1 is an edge of wire w_1 connecting the grid points $p_1 = (x, y)$ and

$p_2 = (x, y + 1)$, and e_2 is an edge of wire w_2 connecting $p_1 = (x, y)$ and $p_3 = (x + 1, y)$, see Figure 3. Any wiring ϕ for W must assign distinct layers to edges e_1 and e_2, i.e., $\phi(e_1, w_1) \neq \phi(e_2, w_2)$. There are two cases:

(1) If $\phi(e_1, w_1) < \phi(e_2, w_2)$ then wire w_1 is *below* wire w_2 at grid point p_1. Furthermore, we say that edge e_1 is *below* (and we label e_1 with the symbol **b**) at grid point p_1, and edge e_2 is *above* (labeled with the symbol **a**) at grid point p_1.

(2) If $\phi(e_1, w_1) > \phi(e_2, w_2)$ then e_1 is above (labeled **a**), and e_2 is below (labeled **b**) at p_1.

Hence, any wiring ϕ for W induces a vertical *relative order* between the wires at every grid point p, and this implies a relative order r between the edges used by the wires at p. We associate a (relative order) label $r(e, p)$ with every edge e used by some wire at grid point p. The labels of all the edges incident with p that are used by some wire induce a relative order on the wires that pass through p. Namely, $r(e, p)$ is a label associated with edge e at point p, and it is equal to **a** if e is above at grid point p; it is equal to **b** if edge e is below at grid point p. If p is trivial then all the edges of w incident with p could be either above or below at p. We denote this situation by $r(e, p) = \mathbf{x}$.

Assume that $e = (p_1, p_2)$ is an edge of some wire w. If $r(e, p_1) = \mathbf{a}$ and $r(e, p_2) = \mathbf{b}$ then e is oriented from p_1 to p_2 and it is called a *layer-increasing arc*, see Figure 4. It means that at grid point p_2 we are forced to use one more layer than the number of layers necessary at grid point p_1. More formally, if e is assigned to layer L_i at grid point p_1, and since a wire can change layers only at grid points, e is also on layer L_i at grid point p_2. But since $r(e, p_2) = \mathbf{b}$, the other wire passing through p_2 must use some layer L_j, where $j \geq i + 1$.

For any given relative order r of the edges of a layout W we define the *layer-graph G^r* induced by r as follows:

(1) G^r is a directed graph whose vertices are the grid points of W.

(2) The arcs of G^r are the layer-increasing arcs (induced by r).

Hence, it is clear that any given relative order r induces a layer-graph G^r.

Conversely, any directed graph G induces a relative order r_G on the edges of layout W as follows: For every arc $e = (p_1, p_2)$ in G, directed from p_1 to p_2, set $r(e, p_1) = \mathbf{a}$ and $r(e, p_2) = \mathbf{b}$. For every edge $e = (p_1, p_2)$ of W not in G do the following:

(1) If either p_1 (or p_2) is a trivial point then $r(e, p_1) = \mathbf{x}$ (or $r(e, p_2) = \mathbf{x}$).

(2) Find an arc $e' = (p_1', p_2')$ of G that belongs to the same wire w as e, and such that e' is the edge closest to e. If p_1' is closer (on w) to e than p_2' then

$r(e,p_1) = r(e,p_2) = r(e',p_1')$, else $r(e,p_1) = r(e,p_2) = r(e',p_2')$, see Figure 5.

(3)   If no edge of G belongs to wire w, then all edges of wire w have the same label, i.e., for every edge $e = (p_1,p_2)$ of w, $r(e,p_1) = r(e,p_2) = \mathbf{a}$ or $\mathbf{b}$. Whether it is $\mathbf{a}$ or $\mathbf{b}$ is determined very easily by checking the label of an edge of any wire that passes through the same grid point as w, at that point.

We have seen that any relative order induces a unique layer-graph and the converse. As a matter of fact, in the following a relative order and its corresponding layer-graph will be treated as one. However, a more important question to be asked is whether any given relative order induces a wiring. The following theorem gives a negative answer.

Theorem 1 Let r be a relative order of the edges of a layout W. There is a wiring of W that induces r if and only if the layer-graph G^r does not contain a (directed) cycle. $\square$

An example of a relative order whose layer-graph has a cycle is shown in Figure 6. It is easy to see that there is no way of assigning layers to the wires which is consistent with the relative order shown. The number of layers needed to wire a layout with respect to some relative order r can also be determined from the layer-graph.

Theorem 2 A relative order r for a layout W is induced by a k-layer wiring for W if and only if the layer-graph G^r is acyclic and the length of a longest path in G^r is $\leq k - 2$.

Proof sketch: (sufficiency): Let r be a relative order for W, and let $k - 2$ be the length of a longest path in G^r. Suppose the layers are labeled as B below L_1 below L_2 below $\ldots L_{k-2}$ below T. The following algorithm constructs a k-layer wiring for W that induces r (i.e., is consistent with the layer-graph G^r).

Algorithm WIRE

Input: A layout W, and an acyclic layer-graph G^r.

Output: A wiring of W.

for every edge e in G^r **do**

   $l(e) = $ length of the longest path in G^r ending at e

   $\phi(e) = L_{l(e)}$

end-for

for every edge e not in G^r **do**

   if the label of e is $\mathbf{a}$ **then** $\phi(e) = T$

   else $\phi(e) = B$

   end-if

end-for $\square$

Algorithm *WIRE* spends a constant time for every edge of the layout. Hence, its time complexity is linear with respect to the number of grid points occupied by the layout. Therefore, we have the following:

Theorem 3 Algorithm *WIRE* constructs a wiring for W in time $O(|N|)$, where N is the number of grid points occupied by W. $\square$

Notice that Theorems 2 and 3 hold also for layouts in other grids, e.g., the tri-hexagonal grid and the uniform grids.

4. Constructing Layer-Graphs in the Square Grid

We will present an algorithm that, for any given layout in the square grid, constructs a layer-graph for which the length of the longest path is at most two. Hence, by Theorem 2, this allows us to wire any layout with at most four layers.

Informally, the algorithm visits the grid points of a row from left to right, starting from the bottommost row that contains some edge of the layout. When the grid points of a row have all been visited, then it starts visiting the grid points of the row immediately above the current one in the same fashion. At each grid point p the algorithm simply assigns a relative order r to the edges incident upon p in such a way that the length of a longest path of the layer-graph induced by r is bounded. The labels of the edges incident to the leftmost grid point of the bottommost row are assigned arbitrarily. Before we give a formal description of the algorithm, we need to introduce some more terminology and discuss related results.

Let W be a given layout. Assume that grid point $p = (x,y)$ is incident to edges $e_1 = (p,p_1)$, $e_2 = (p,p_2)$, $e_3 = (p,p_3)$, $e_4 = (p,p_4)$ of W, where $p_1 = (x-1,y)$, $p_2 = (x,y-1)$, $p_3 = (x+1,y)$ and $p_4 = (x,y+1)$, as shown in Figure 7. Further assume that the algorithm has visited all grid points (x',y') with $y' < y$ or, if $y' = y$, with $x' < x$. Hence, the edges e_1 and e_2 are already assigned labels at the grid points p_1 and p_2, $r(e_1,p_1), r(e_2,p_2)$, respectively. We say that p has a *conflict* if p is *not* trivial, and (a) e_1 and e_2 belong to the same wire w and $r(e_1,p_1) \neq r(e_2,p_2)$, or, (b) if e_1 and e_2 belong to different wires and $r(e_1,p_1) = r(e_2,p_2)$. In this case p is called a *conflict point*. Otherwise, p does not have a conflict and is called a *conflict-free point*. Notice that if p is a trivial grid point then it is conflict-free and the labels of the edges incident with p could be either **a** or **b** at p.

If p is a nontrivial conflict-free point, then the action to be taken is straightforward. The labels of e_1 and e_2 at p must be such that a layer-increasing arc is never introduced. In other words $r(e_1,p) = r(e_1,p_1)$ and $r(e_2,p) = r(e_2,p_2)$. Furthermore, the labels of e_3 and e_4 at p

are assigned in such a way that edges of the same wire have the same label at p. For example, if e_1 and e_4 belong to the same wire, then $r(e_4,p)=r(e_1,p)$ and $r(e_3,p)=r(e_2,p)$. This action is called *propagate*.

If p is trivial then the edges incident to p are assigned a relative order at p such that no layer-increasing arc is introduced. This might cause the labels of the edges (of the same wire) incident to p to be different, but since p is trivial a via can be introduced at p in order to connect all the edges.

If p is a conflict point, then we need to introduce layer-increasing edges in order to resolve the conflict. However, as is shown in the following, the introduction of only one layer-increasing arc suffices.

Lemma 1 Let p be a conflict point with incident edges $e_1=(p,p_1)$, $e_2=(p,p_2)$, $e_3=(p,p_3)$, $e_4=(p,p_4)$. If we have assigned labels to edges e_1 and e_2 at p_1 and p_2, respectively, then the introduction of one layer-increasing edge is enough to resolve the conflict. Furthermore, we can choose where to introduce it, at e_1 or at e_2. □

More specifically, if we need to introduce a horizontal layer-increasing arc, i.e., between grid points $p=(x,y)$ and $p_1=(x-1,y)$, we perform the following:

Action-conflict-h

let $e_1=((x,y),(x-1,y))$

let $e_2=((x,y),(x,y-1))$

$r(e_2,(x,y))=r(e_2,(x,y-1))$

if $r(e_1,(x-1,y))=$ **a then**

   $r(e_1,(x,y))=$ **b**

else { $r(e_1,(x-1,y))=$ **b** }

   $r(e_1,(x,y))=$ **a**

end-if □

The introduction of a vertical layer-increasing arc, i.e., between grid points $p=(x,y)$ and $p_2=(x,y-1)$, is done in a similar fashion:

Action-conflict-v

let $e_1=((x,y),(x-1,y))$

let $e_2=((x,y),(x,y-1))$

$r(e_1,(x,y))=r(e_1,(x-1,y))$

if $r(e_2,(x,y-1))=$ **a then**

$$r(e_2,(x,y)) = \mathbf{b}$$

else { $r(e_2,(x,y-1)) = \mathbf{b}$ }

$$r(e_2,(x,y)) = \mathbf{a}$$

end-if □

The labels of the edges e_3 and e_4 at p, $r(e_3,p)$ and $r(e_4,p)$, are assigned in such a way that edges of the same wire have the same label at p. For example, if e_1 and e_3 belong to the same wire, then $r(e_3,p) = r(e_1,p)$ and $r(e_4,p) = r(e_2,p)$.

Let G^r be the layer-graph that is being constructed by our algorithm. With each grid point $p = (x,y)$, we associate a number $l(x,y)$, which denotes the length of the longest directed path of the layer-graph G^r that starts, or ends at grid point (x,y). However, because of the nature of the algorithm, only the l function of the grid points in the current and previous row are ever used. Therefore, the algorithm will maintain up to date only the l function of the aforementioned grid points. We claim that for any layout W, our algorithm constructs a layer-graph such that $l(x,y) \leq 2$, for every grid point (x,y).

Algorithm LAYER-GRAPH

Input: A layout W with n rows and m columns.

Output: A layer-graph G^r for W such that length of any (directed) path in G^r is at most two.

for $y = 1,n$ **do**

for $x = 1,m$ **do**

let $p = (x,y)$

if p is a conflict point **then**

  if $l(x,y-1) = 2$ **then**

    introduce a layer-increasing arc at the edge $e = ((x,y),(x-1,y))$

  else { $l(x,y-1) \leq 1$ }

    let $p^* = (x^*,y^*)$ be the lexicographically closest grid point with $y^* \geq y-1$ and $l(x^*,y^*) = 2$

    if p^* exists and $|x-x^*| + |y-y^*|$ is odd **then**

      introduce a layer-increasing arc at the edge $e = ((x,y),(x-1,y))$

    else { p^* does not exist, or $|x-x^*| + |y-y^*|$ is even }

      introduce a layer-increasing arc such that $(x+1,y)$ becomes a conflict-free point

        { see below }

    end-if

  end-if

update the l function of the at most two grid points affected that lie in the current or previous row

else { p is conflict-free }

    propagate

end-if

end-for

end-for. ☐

    The introduction of a layer-increasing arc such that $(x+1,y)$ becomes a conflict-free point is done as follows: Since $(x+1,y)$ must be a conflict-free point, the label of edge $e = ((x+1,y),(x+1,y-1))$ at grid point $(x+1,y-1)$, $r(e,(x+1,y-1))$, clearly determines the label of edge $e_3 = ((x+1,y),(x,y))$ at point $(x+1,y)$, $r(e_3,(x+1,y))$ in a unique way. Furthermore, $r(e_3,(x,y)) = r(e_3,(x+1,y))$. Hence, the label $r(e_3,(x,y))$ determines which layer-increasing arc needs to be introduced.

Theorem 4 Let W be a layout in the square grid. Algorithm *LAYER-GRAPH* produces a layer-graph G^r for W such that the length of any directed path in G^r is at most two.

Proof sketch: Let $p = (x,y)$ be the current grid point. The proof of the theorem is based on showing that the following two invariants are maintained [8]:

I1   Let $p' = (x',y')$ be any grid point which is visited before p, with $y' \geq y-1$. If $|x-x'| + |y-y'|$ is odd and $l(x,y) = 2$ then $l(x',y') < 2$.

I2   If $p = (x+1,y)$ has a conflict then $l(x,y) + l(x+1,y-1) \leq 2$. ☐

    Figure 8 shows an example of a layout and the layer-graph produced by Algorithm *LAYER-GRAPH*. Notice that the length of a longest path in the produced layer-graph is one. Algorithm *LAYER-GRAPH* spends a constant amount of time for resolving conflicts or propagating at each grid point. Furthermore, by Theorem 4, the operation of updating the l function takes also constant time.

Theorem 5 Let W be a layout in the square grid which occupies N grid points. Algorithm *LAYER-GRAPH* computes a layer-graph G^r for W in $O(N)$ time. ☐

    Combining Theorems 2, 3, 4 and 5 we obtain the following:

Corollary 1 Let W be a layout which occupies N grid points. A four-layer wiring of W can be constructed in $O(N)$ time. ☐

5. Extensions

    For a given layout W let G^r be the layer-graph constructed by our algorithm. If there is no layer-increasing arc in G^r, then W can be wired using exactly two layers. The following theorem shows that the converse is also true [8].

Theorem 6 A layout W is two-layer wirable if and only if the layer-graph G^r constructed by Algorithm *LAYER-GRAPH* contains no layer-increasing arcs. □

For layouts in the tri-hexagonal grid we have the following [8]:

Theorem 7 Let W be a layout in the tri-hexagonal grid which occupies N grid points. There exists an $O(N)$-time algorithm that produces a layer-graph G^r for W such that the length of any path in G^r is at most three. □

Combining Theorems 2, 3, and 7 we obtain the following:

Corollary 2 Let W be a layout in the tri-hexagonal grid which occupies N grid points. A five-layer wiring of W can be constructed in $O(N)$ time. □

ACKNOWLEDGEMENT

I would like to thank Franco Preparata for his guidance and support.

REFERENCES

[1] M.L. Brady and D.J. Brown, "VLSI ROUTING: Four Layers Suffice," pp. 245-257 in *Advances in Computing Research, vol. 2*, ed. F.P. Preparata, JAI Press Inc., 1984.

[2] M.L. Brady and M. Sarrafzadeh, "Layout Streching to Ensure Wirability," *Proceedings of the 21st Annual Conference on Information Sciences and Systems*, Johns Hopkins University, Baltimore MD, 1987.

[3] T. Gonzalez and S.Q. Zheng, "Streching and Three-Layer Wiring Planar Layouts," Technical Report TRCS87-10, Univeristy of California at Santa Barbara, 1987.

[4] W. Lipski, Jr, "On the Structure of Three-Layer Wirable Layouts," pp. 231-243 in *Advances in Computing Research, vol. 2*, ed. F.P. Preparata, JAI Press Inc., 1984.

[5] W. Lipski, Jr and F.P. Preparata, "A Unified Approach to Layout Wirability," *Mathematical Systems Theory*, vol. 19, pp. 189-203, 1987.

[6] K. Mehlhorn and F.P. Preparata, "Routing Through a Rectangle," *J. ACM*, vol. 33, no. 1, pp. 60-85, 1986.

[7] F.P. Preparata and W. Lipski, Jr, "Optimal Three-Layer Channel Routing," *IEEE Trans. on Computers*, vol. 33, no. 5, pp. 427-437, 1984.

[8] I.G. Tollis, "Algorithms for VLSI Layout," Ph.D. Thesis, Department of Computer Science, Technical Report ACT-85, Coordinated Science Laboratory, University of Illinois at Urbana-Champaign, 1987.

[9] I.G. Tollis, "Wiring Layouts in the Octo-Square Grid and Two-Colorable Maps," Manuscript, Urbana, 1987.

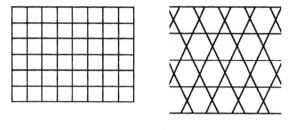

(a)   (b)

Figure 1. (a) The square grid; (b) the tri-hexagonal grid.

Figure 7. A grid point with its incident edges.

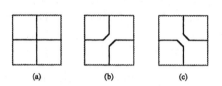

(a)   (b)   (c)

Figure 2. (a) Crossing, (b), (c), knock-knees in the square grid.

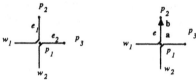

Figure 3. Two wires form a knock-knee.   Figure 4. A layer-increasing edge.

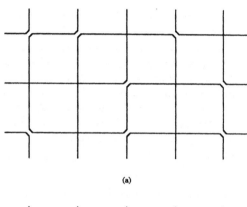

(a)

Figure 5. A layer-increasing edge of wire w

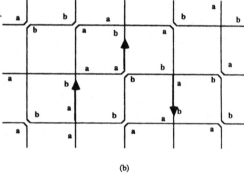

(b)

Figure 8. Part (b) shows the layer-graph obtained for the layout of part (a).

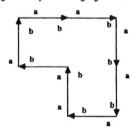

Figure 6. A directed cycle of layer-increasing edges.

Input Sensitive VLSI Layouts for Graphs of Arbitrary Degree

Deepak D. Sherlekar
University of Michigan

Joseph JáJá
University of Maryland

Abstract: A general method to find area-efficient VLSI layouts of graphs of arbitrary degree is presented. For graphs of maximum degree Δ, the layouts obtained are smaller by a factor of Δ^2 than those obtained using existing methods.

Optimal planar layouts, and near-optimal nonplanar layouts are also derived for planar graphs of arbitrary degree and gauge. The results span the spectrum between outerplanar graphs (*gauge 1*), and arbitrary planar graphs (*gauge $O(n)$*). Optimality is established by developing families of planar graphs of varying gauge and degree, and proving lower bounds on their layout area. These techniques can be combined to exhibit a trade-off between area and the number of contact cuts. The resulting scheme is sensitive to all three parameters that affect the area: the *maximum degree*, the *gauge*, and the number of *contact cuts*.

1 Introduction

An examination of previous results on VLSI graph layouts [10,3,15,18] indicates that the partition tree of a graph and its maximum vertex degree are the dominant factors affecting its layout area. In case of planar graphs, the important parameters are the degree of 'outerplanarity' of the graph [5], the degree of planarity of the layout [1], and the maximum vertex degree. The first two parameters have been examined separately in the past [5,1]. The objective of this paper is to present a scheme which is *simultaneously* sensitive to all these parameters.

The degree of resemblance of a planar graph to an outerplanar graph is expressed by its *gauge* [2,1,5]. A planar graph has gauge F if there exists a drawing of the graph in the plane for which no vertex is at a distance greater than $F - 1$ from the outer face. Thus outerplanar graphs have gauge 1. There exist planar graphs [19,16], which have gauge $O(n)$. The number of *contact cuts* in a layout is a measure of its degree of planarity. Although degree-sensitive layouts have not been studied before, layouts schemes for planar graphs of maximum degree 4 have addressed parameters such as gauge and number of contact cuts. [10,19,5,1].

Under the classical model for VLSI layouts [17], vertices occupy unit area, edges have unit thickness, and no vertex has degree greater than 4. Thus existing schemes consider circuits which satisfy these assumptions. However, these assumptions may be violated in practice. If vertices have varying area, it would be reasonable to represent larger vertices by regions of appropriate area. The constraint on vertex degree can relaxed by representing a vertex of degree d by a square region of perimeter d. This representation was originally suggested in [17], and is consistent with the motivation behind this paper. Avoiding vertices of arbitrary degree by transforming into an equivalent trivalent graph [7] prevents more than 3 signals to reach a vertex simultaneously, and ignores the fact that vertices of high degree may have high area as well. Allowing vertices of arbitrary degree also opens up the possibility of formulating a hierarchical scheme to design layouts. The graph at the bottom level of the hierarchy is the circuit graph. At other levels, vertices correspond to modules, instead of elementary circuit elements.

A solution to the problem of vertices of high degree would also solve the problem of vertices of varying area. Thus the discussion below is confined to addressing the former problem alone. The problems addressed in this paper are described in Section 2. Section 3 presents a scheme to find layouts of nonplanar graphs. Section 4 deals with *gauge* and *degree* sensitive nonplanar layouts of planar graphs. Similar results are presented in Section 5 for planar layouts. These results are combined in Section 6 to exhibit a trade-off between area and number of contact cuts. This results in a scheme which is sensitive to all three parameters mentioned above. Concluding remarks appear in Section 7.

2 Problem Definitions and Results

Dissection Mechanism	Maximum Degree (Δ)	Gauge F	Layout Area	Reference
$(F, \sqrt{2})$ bifurcator	4	—	$O\left(F^2 \log^2 \frac{n}{F}\right)$	[3]
$(F, \sqrt{2})$ weighted bifurcator	PARAMETER	—	$O\left(F^2 \log^2 \frac{n}{F}\right)$	*Theorem 3.2*
Various Planar Separators	4	$O(n)$	$O(n \log^2 n)$	[10,19]
	4	1	$O(n)$	[10,19]
	4	PARAMETER	$O(n \log^2 F)$	*Theorem 4.1*
	PARAMETER	$O(n)$	$O\left(\Delta n \log^2 \frac{n}{\Delta}\right)$	*Theorem 4.3*
	PARAMETER	PARAMETER	$O\left(\Delta n \log^2 \frac{F}{\Delta}\right)$	*Theorem 4.4*

Table 1: Nonplanar Layouts of Planar and Nonplanar Graphs

The method presented here to derive nonplanar layouts was introduced in [15]. The layouts in [15] were obtained via an intermediate embedding of the graph in a binary tree. This technique is reformulated here as one that uses the tree of meshes as the host graph for the intermediate embedding, as was done in [3]. However unlike [3], the dissection mechanism here satisfies a weaker balancing property, called the *external edge balancing condition (EEBC)*, which leads to smaller layout areas. These results are described in Table 1 along with relevant results from the literature.

For planar layouts, the idea behind the layouts in [19] and [5] is extended to

Maximum Degree (Δ)	Gauge F	Layout Area	Reference
$O(n)$	$O(n)$	$O(n^2)$	[19,1]
4	1 (Tree)	$O(n)$	[19]
4	PARAMETER	$O(Fn)$	[5]
PARAMETER	PARAMETER	$O(\Delta Fn)$	Theorem 5.1

Table 2: Planar Layouts of Planar Graphs

find layouts of area $O(\Delta Fn)$ for graphs of gauge F and maximum degree Δ such that $\Delta F \leq n$. This result and other known results appear in Table 2. The results for planar and nonplanar layouts are combined to exhibit a trade-off between the layout area and the number of contact cuts in Section 6.

The gap of $O(\log n)$ between the lower and upper bounds for nonplanar layouts of planar graphs of maximum degree 4 [9] is shown to exist as a gap of $O(\log F)$ for graphs of any specified gauge F. The planar layouts of Table 2 are also shown to be optimal. This extends the lower bound of $\Omega(Fn)$ in [5] to graphs of arbitrary degree.

3   A Scheme for Nonplanar Layouts

The problem with extending the approach in [19,10] to vertices of arbitrary degree is the marriage step [15]. When reinserting an edge to connect vertices in two layouts, all vertices blocking the path of the edge must be moved. For a vertex of degree d, this increases the width and height of the layouts by a factor of $O(d)$. The resulting layout are larger in area by a factor of $O(\Delta^2)$ for graphs of maximum degree Δ [15]. The displacement of vertices to route wires can be avoided by embedding the vertices of the graph in the leaf meshes of an appropriate truncated *tree of meshes* [3], and using the internal meshes for routing the edges. However, the dissection mechanism used to obtain the embedding has to satisfy certain balancing properties in order to ensure that the internal meshes that perform the routing are not too large. For graphs having an F bifurcator and maximum degree Δ, the size of the *fully balanced* bifurcators used in [3] are of size $O(\Delta F)$. Thus the layout area increases quadratically with Δ.

A weaker balancing property can correct the problem of large bifurcator size [15]. Consider an internal mesh in the tree of meshes. The number of edges routed through this mesh is the sum of the size of the bifurcator at that level, and the number of external edges routed through it. These external edges correspond to the edges connecting the subgraph embedded in the tree rooted at the mesh, and the rest of the graph. Following the argument in [15], merely ensuring equal sized partitions, such that each partition has at most a fixed fraction of the external edges incident with it is sufficient. This condition is called the *external edge balancing condition (EEBC)*. A bifurcator satisfying the *EEBC* can be derived efficiently using a *weighted* bifurcator. If a graph has nonnegative weights on its vertices summing to at most 1, then a weighted F-bifurcator splits it into two parts, each of weight at most $1/2$, by removing at most F edges, such that both parts recursively satisfy weighted $F/\sqrt{2}$-bifurcators.

A *weighted* bifurcator can be used to derive a two-weight bifurcator on the same lines as two-weight separators [6,15]. Consider a graph with two independent sets of nonnegative vertex weights, each summing to at most 1. Then a two weight bifurcator splits the graph into two parts, each of weight $1/2$ of the first kind and $(1 + \epsilon)/2$ of the second kind, for some fixed $\epsilon < 1$. Assigning the first weight to correspond to number of vertices, and the second to the number of external edges incident with the vertex leads to the following theorem.

Theorem 3.1 *Any graph having a weighted F-bifurcator also has a weighted F-bifurcator satisfying the* external edge balancing condition. $\qquad\square$

The above theorem can now be used to find an embedding in a truncated tree of meshes whose layout corresponds to a layout of the graph.

Theorem 3.2 *Any graph having a $(F, \sqrt{2})$ weighted bifurcator has a layout of area $O(F^2 \log^2(n/F))$ regardless of its maximum vertex degree.* $\qquad\square$

Use the F bifurcator satisfying the *EEBC* to embed the graph in a truncated tree of meshes, such that the completely disconnected partitions at the bottom of the recursive decomposition are mapped to the leaf meshes. Let ϵ be the constant of the two-weight bifurcator used to derive the bifurcator satisfying the *external edge balancing condition*. Then it can be shown that the edges handled by the mesh at level i is

$$O\left(F/2^{i/2}\sum_{j=0}^{i}\left((1+\epsilon)/\sqrt{2}\right)^j\right) = O\left(F/2^{i/2}\right) \quad if \;\; \epsilon < \sqrt{2}-1$$

Thus the graph can be embedded in a truncated tree of meshes having root mesh of size $O(F \times F)$ and depth $2\log(n/F) + \Theta(1)$. The layout of this tree of meshes found using the H-tree scheme is $O(F^2 \log^2(n/F))$. $\qquad\square$

4 Non-Planar Layouts of Planar Graphs

4.1 Bounded Degree Vertices

Two separator theorems are of relevance. The first is the $\sqrt{n}$-separator for planar graphs [12,4]. The other is the F-separator for planar graphs of gauge F [5]. The counterparts of these separators which satisfy the *EEBC* can be derived on the same lines as Theorem 3.1. Thus planar graphs of maximum degree 4 and gauge F satisfy an $F \log n$-separator theorem and a $\sqrt{n}$-separator theorem, both of which satisfy the *EEBC*. Hence the partition tree of the graph is best predicted as a combination of these two separators. This leads to the following theorem.

Theorem 4.1 *Let G be a planar graph having vertices of degree at most 4 and gauge F. Then G has a nonplanar layout of area $O(n \log^2 F)$.*

Proof: Apply the $F \log n$-separator recursively up to a level L such that the size of the separator at level L is no smaller than that found by the $\sqrt{n}$-separator theorem. Thus $L = \log(n/(F^2 \log^2 F)) + \Theta(1)$. Then switch to the $\sqrt{n}$-separator. Recall the parameter ϵ which decides the accuracy to which the external edges are split over the partitions. Select a value of ϵ in the range $(0,1)$ for the $F \log n$-separator, and in the range $(0, \sqrt{2} - 1)$ for the $\sqrt{n}$-separator.

The meshes at any level $i \leq L$ of the *tree of meshes* for G have sides of length $O(F(\log n - i))$. The subtrees rooted at level L are essentially trees of meshes corresponding to $\sqrt{n}$-separable graphs of bounded degree having $n/2^L$ vertices. This yields the following recurrence for the height H_{np4} of the layout.

$$H_{np4}(L - i) = 2H_{np4}(L - (i + 2)) + O(F \log(n/2^i)), \quad 0 \leq i < L$$
$$H_{np4}(0) = O\left(\sqrt{n/2^L} \log(n/2^L)\right)$$

which solves to $H_{np4}(L) = O(\sqrt{n} \log F)$. Similarly, The width of the layout may be shown to be $O(\sqrt{n} \log F)$. Hence the area of the layout is $O(n \log^2 F)$. $\square$

The resolution of the $O(\log n)$ gap between the upper and lower bounds on the layout area of planar graphs is an important open problem [9]. The above result may be placed in context of this problem by the following theorem.

Theorem 4.2 *There exist planar graphs of gauge F which require $\Omega(n \log F)$ area.*

Proof: The *tree of meshes* T_p introduced in [9] has $n = 2p^2 \log p + p^2$ vertices and gauge $p = \Theta(\sqrt{n/\log n})$. T_p requires $\Omega(p^2 \log^2 p) = \Omega(n \log n)$ area [9]. For any $F > n/\log n$, construct the following graph G of gauge F. G consists of $\Theta(n/F^2 \log F)$ disjoint graphs, each of which is a *tree of meshes* T_F. Then from [9], it follows that the area of G is $\Omega\left((n/F^2 \log F)(F^2 \log^2 F)\right) = \Omega(n \log F)$. $\square$

4.2 Arbitrary Degree Vertices

For planar graphs of arbitrary degree, it is possible to infer edge separators smaller than $O(\Delta \sqrt{n})$. This requires the use of a cycle separator which partitions a graph having weights assigned to its vertices as well as faces. An edge separator can be inferred from such a cycle separator in the dual graph using the technique of [8]. The separator in [13] is a cycle separator of size $O(\sqrt{dn})$, where d is the maximum face size. Given a planar graph G of maximum degree Δ, consider its dual graph G'. The size of the largest face in G' is Δ. Thus G has a $\sqrt{\Delta n}$ edge separator. The derivation of a $\sqrt{\Delta n}$ edge separator satisfying the $EEBC$ follows from the proof of Theorem 3.1. Applying the procedure in Theorem 3.2 yields the following result.

Theorem 4.3 *Let G be a planar graph having vertices of degree at most Δ. Then G possesses a VLSI layout of area $O(\Delta n \log^2(n/\Delta))$.* $\square$

Finally, the layout algorithm can be made sensitive to the gauge of the graph using the same strategy as Theorem 4.1. Consider the dual graph G' of G with assignments of weights to its faces as in Theorem 4.3. The diameter of G' is F (the gauge of G), and the maximum face size of G' is Δ (the maximum vertex degree of G). Find a spanning tree T' of G' having diameter F. Then by Theorem 6 of [13], G' has a cycle separator of size at most $F + \Delta$. This yields an edge separator for G of size at most $O(F + \Delta)$, which yields a $(\Delta + F) \log n$ edge separator satisfying the *EEBC*.

Theorem 4.4 *Let G be a planar graph of gauge F and maximum degree Δ. Then G has a layout of area $O(\Delta n \log^2(F/\Delta))$.*

Proof: Adopt the strategy of Theorem 4.1, applying the $(\Delta + F) \log n$ edge separator as long as it is smaller than the $\sqrt{\Delta n}$ edge separator. The level of recursion L at which the separators are switched may be shown to be $\log \left(n / \left(\frac{F+\Delta}{\sqrt{\Delta}} \right)^2 \log^2 \left(\frac{F+\Delta}{\sqrt{\Delta}} \right)^2 \right)$

The tree of meshes corresponding to the graph is made up of the tree of meshes corresponding to the $(\Delta + F) \log n$ edge separator up to level L, each of whose leaves is the layout of an $n/2^L$ vertex graph having a $\sqrt{\Delta n}$ edge separator. The recurrence for the height of the layout can be formulated and solved on the same lines as Theorem 4.1, yielding an area of $O(\Delta n \log^2(F/\Delta))$. $\square$

5 Gauge and Degree-Sensitive Planar Layouts

5.1 Upper Bounds

The solution is obtained by modifying the approach in [19] and [5] for graphs of maximum degree 4 to handle vertices of high degree. The scheme starts with a drawing of the graph in the plane which minimises the maximum distance of any vertex to the outer face. The layout produced is topologically equivalent to this drawing of the graph. However, the separator should provide an incision which allows the layouts of the partitions to be placed side by side during the marriage step [5]. The type of incision required for the purpose, and an algorithm to find it are described in [5]. Any planar graph of gauge F can be split into two parts, each having at most 2/3 of the vertices of G, by a simple, non-circular separating path of size $O(F)$ [5]. The corresponding edge separator is inferred by deleting all edges incident with the vertices in the separating path.

The 'marriage' phase reinserts the deleted edges in such a way that the topology of the drawing G_ϕ is preserved in the layout. The above separator ensures that this can be done. The reinserted wires snake around the boundary of the layouts to avoid crossing over other vertices or edges. It can be shown that the number of *kinks* [19] or bends that any wire needs to make to emerge out of a layout increases by $O(1)$ at each level as the recursion unwinds [19,5]. For the base case, the convex boundary of the vertices ensures that the number of *kinks* required to connect one vertex to another is $O(1)$. The recursion bottoms out upon encountering graphs of size $O(\Delta F)$ which can be laid out in area $O(\Delta^2 F^2)$ [19,1]. The number of *kinks* in a wire inserted at the top level is $O(\log(n/\Delta F))$. Hence the recurrence giving the area A of the graph is

$$A(n) = \max_{1/3 \le x \le 2/3} \left(\sqrt{A(xn) + A((1-x)n)} + c\Delta F \log(n/\Delta F) \right)^2 \quad n > \Delta F$$

$$A(n) = O(n^2) \quad n \le \Delta F$$

This recurrence can be solved along the same lines as [10], by defining an auxiliary linear recurrence B such that $A(n) = O(nB^2(n))$. Proceeding on these lines leads to the following theorem.

Theorem 5.1 *Let G be any planar graph of gauge F and maximum degree Δ. Then G has a planar layout of area $O(\min(\Delta Fn, n^2))$.* □

5.2 Lower Bounds

The planar layouts obtained in Theorem 5.1 are optimal in area for any Δ and F such that $\Delta F = O(n)$. This is proved in the following theorem.

Theorem 5.2 *For any Δ and F such that $\Delta F = O(n)$, there exist planar graphs of maximum degree Δ and gauge F whose layouts require $\Omega(\Delta Fn)$ area.*

Proof: Define a family of graphs, called the Δ-*starred diamond graphs*, over the range $0 < \Delta < n$. The graph is illustrated in Figure 1. It is a derivative of the *diamond graph* [19], also shown in Figure 1. If the horizontal and vertical edges of the *diamond graph* are grouped into disjoint sets of $\Delta/2$ consecutive edges and 'pinched' in the middle to be joined at a vertex, then the resulting graph is a Δ-*starred diamond graph*. For an n_d vertex *diamond*, the Δ-*starred diamond* has $n = n_d + \lceil n_d/4\Delta \rceil = \Theta(n_d)$ vertices. The distance of the vertex at the center of the Δ-*starred diamond* to the outer face is $\Theta(n_d/\Delta) = \Theta(n/\Delta)$. Thus the gauge of the Δ-*starred diamond* is $\Theta(n/\Delta)$.

The layout area of a Δ-*starred diamond* is $\Omega(n^2)$. This may be shown by reducing the problem of finding a planar layout of the corresponding *diamond graph* to that of the Δ-*starred diamond*. Consider any layout of the Δ-*starred diamond* having area A. Then a planar layout of the corresponding *diamond graph* can be derived from this layout by replacing every vertex in the layout which has degree Δ by the gadget illustrated in Figure 1. The gadget links up every pair of edges which correspond to a single edge in the corresponding *diamond graph*. The area of the gadget is at most four times that of the original vertex. Thus the area of the transformed layout is $O(A)$. The area of the *diamond graph* is known to be $\Omega(n^2)$. Hence the area of the Δ-*starred diamond graph* is also $\Omega(n^2)$.

For $\Delta F = \Omega(n)$, define a graph, called the (Δ, F)-*starred diamond graph*. This graph consists of $\Theta(n/\Delta F)$ disjoint graphs, each of which is a Δ-*starred diamond* of $\Theta(\Delta F)$ vertices. Then from the above argument, the area of the (Δ, F)-*starred diamond graph* is $\Omega\left(\frac{n}{\Delta F}(\Delta F)^2\right) = \Omega(\Delta Fn)$. □

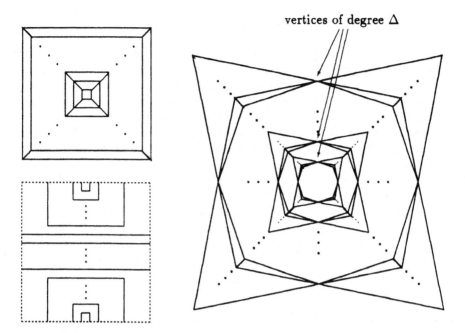

Figure 1: Clockwise, from top left: the Diamond Graph, the Δ Starred Diamond Graph, and an illustration showing replacement of a degree Δ vertex by $\Delta/2$ wires

6   Area-Contact Cut Trade-Offs

A trade-off between the layout area $A(n)$ and the number of permissible contact cuts $C(n)$ for planar graphs of bounded degree was shown in [1] to be $A(n)C(n) = \Theta(n^2)$, $1 \leq C(n) = O(n/\log^2 n)$. In view of the results in the previous sections, these trade-offs can be extended as shown below.

Theorem 6.1 *Let G be any planar graph of gauge F and maximum degree 4. Then for any number C, $n/F \leq C = O(n/\log^2 F)$, G has a layout with C contact cuts and area $O(n^2/C)$.* □

In view of Theorem 4.4 and Theorem 5.1, the area-cut trade-off of Theorem 6.1 can be extended to graphs having vertices of arbitrary degree as well. From the discussion in [1], it is clear that one can remove $O(\Delta n/g)$ from a planar graph of maximum degree Δ so that the resulting components have gauge at most g. The Δ-starred diamond graph can be used to prove that there exist planar graphs which require removal of $\Omega(\Delta n/g)$ edges to reduce the gauge to g. The trade-off between area and number of contact cuts for planar graphs of gauge F and maximum degree Δ is given by the following theorem.

Theorem 6.2 *Let G be any planar graph of maximum degree Δ and gauge F. Then for any number C, $(\Delta n/F) \leq C = O(\Delta n/\log^2((F + \Delta)))$, G has a layout with C contact cuts and area $O(\Delta^2 n^2/C)$.* □

7 Concluding Remarks

A general, area-efficient layout strategy was developed for planar and other separable graphs of arbitrary degree. The bounds on the layout area obtained are better by a factor of $O(\Delta^2)$ than corresponding bounds obtained by extending existing techniques to handle vertices of high degree. The techniques described in [3] can be adapted to minimise the maximum wire length of these layouts as well.

Considering that graphs with bisection width $o(n)$ have only $O(n)$ edges, it is unlikely that when Δ is large, there are a large number of vertices of degree Δ. The analysis of the preceding sections assumed that all the edges in the graph were 'covered' by vertices of degree Δ. It is possible to show better results when the graph has only a few vertices of degree $\Theta(\Delta)$. Such distributions of edges over vertices have been considered, and shall be discussed in a later paper.

The techniques developed here also have other applications. For example [14], dissection mechanisms satisfying the *external edge balancing condition* are also useful for solving the problem of task allocation in tree-connected multiprocessors or multiprocessors based on Fat Trees [11].

Acknowledgements: The research reported here was conducted while the first author was pursuing his doctoral studies at the Department of Computer Science, University of Maryland. The first author was supported by the National Science Foundation grants NFSD-CR-83-05992 and CDR-8500108, and the Air Force Office of Scientific Research grant AFOSR-82-0303. The second author was supported by the National Science Foundation grant CDR-8500108 and NASA grant NAG5-776.

References

[1] A. Aggarwal, M. M. Klawe, D. Lichtenstein, N. Linial, and A. Wigderson. Multi-layer grid embeddings. *Proc. 26th Annual IEEE Symposium on Foundations of Computer Science*, 186–196, 1985.

[2] B. S. Baker. Approximation algorithms for np-complete problems on planar graphs. *Proc. 24th Annual IEEE Symposium on Foundations of Computer Science*, 265–273, 1983.

[3] S. N. Bhatt and F. T. Leighton. A framework for solving vlsi graph layout problems. *Journal of Computer and System Sciences*, 28:300–343, 1984.

[4] H. N. Djidjev. On the problem of partitioning planar graphs. *SIAM Journal on Algebraic and Discrete Methods*, 3(2):229–240, June 1982.

[5] D. Dolev, F. T. Leighton, and H. W. Trickey. Planar embedding of planar graphs. In *Advances in Computing Research, vol. 2*, pages 147–161, JAI Press, 1984.

[6] J. R. Gilbert, D. J. Rose, and A. Edenbrandt. A separator theorem for chordal graphs. *SIAM Journal on Algebraic and Discrete Methods*, 5(3):306–313, 1984.

[7] F. Harary. *Graph Theory*. Addison-Wesley, Reading, MA, 1969.

[8] D. B. Johnson and S. M. Venkatesan. Parallel algorithms for minimum cuts and maximum flows in planar networks. *Proc. 23rd Annual IEEE Symposium on Foundations of Computer Science*, 244–254, 1982.

[9] F. T. Leighton. New lower bound techniques for vlsi. *Proc. 22nd Annual Symposium on Foundations of Computer Science*, 1–12, 1981.

[10] C. E. Leiserson. *Area Efficient VLSI Computation*. PhD thesis, Carnegie Mellon University, 1981.

[11] C. E. Leiserson. Fat-trees: universal networks for hardware-efficient supercomputing. *IEEE Transactions on Computers*, C-34(10):892–901, Oct 1985.

[12] R. J. Lipton and R. E. Tarjan. A separator theorem for planar graphs. *SIAM Journal on Applied Mathematics*, 36(2):177–189, April 1979.

[13] Gary Miller. Finding small simple cycle separators for 2-connected graphs. *Proc. 16th Annual ACM Symposium on Theory of Computing*, 376–382, 1984.

[14] D. D. Sherlekar. *Graph Dissection Techniques for VLSI and Algorithms*. PhD thesis, University of Maryland, College Park, Maryland, 1987.

[15] D. D. Sherlekar and J. Ja'Ja' Layouts of graphs of arbitrary degree. *Proceedings of the 25th Annual Allerton Conference*, September 1987.

[16] Y. Shiloach. *Linear and Planar Arrangement of Graphs*. PhD thesis, Weizmann Institute, Rehovot, Israel, 1976.

[17] C. D. Thompson. *A Complexity Theory for VLSI*. PhD thesis, Carnegie Mellon University, 1980.

[18] J. D. Ullman. *Computational Aspects of VLSI*. Computer Science Press, Rockville, MD, 1984.

[19] L. G. Valiant. Universality considerations in vlsi circuits. *IEEE Transactions on Computers*, C-30(2):135–140, Feb 1981.

Fast Self-Reduction Algorithms for
Combinatorial Problems of VLSI Design

Michael R. Fellows[1]
Michael A. Langston[2]

Abstract.

In a recent series of papers [FL1, FL2, FL3], we have proven the existence of decision algorithms with low-degree polynomial running times for a number of well-studied VLSI layout, placement and routing problems. These results make use of the powerful Robertson-Seymour theorems on the well-partial-ordering of graphs under both the minor and immersion orders. In the present paper, we study the complexity of *construction* versions of these problems, focusing on efficient self-reduction strategies. We introduce a notion of *fast* self-reduction in this setting and develop a general technique, which we term *scaffolding*, that is useful in the design of fast self-reduction algorithms.

1. Introduction

In the design and manufacturing of VLSI systems, practical problems are often characterized by fixed-parameter instances. The parameter may represent the number of tracks on a chip, the number of processing elements, the number of channels required to connect circuit elements or the load on communications links.

We have recently employed Robertson-Seymour posets to prove small-degree polynomial-time decision complexity for a variety of fixed-parameter combinatorial problems of VLSI design (see, for example, [FL1, FL2, FL3]). Some were previously known to be in P only by way of dynamic programming formulations with unboundedly high-degree polynomial running times (see, for example, [EST, GS, MaS]).

In this paper, we explore the complexity of associated construction problems. Our primary focus is on self-reduction, the process whereby a decision algorithm is used as a subprogram by a construction algorithm. Although we have previously identified straightforward self-reduction algorithms for a wide assortment of problems [BFL, FL3], we herein develop more efficient self-reduction strategies aimed at layout permutation problems. To accomplish this, we introduce a notion of *fast* self-reduction and a general technique that we call *scaffolding*. We show how scaffolding can be the basis for fast self-reduction algorithms for a number of fixed-parameter problems, including gate matrix layout, cutwidth, vertex separation number, and several others.

In the next section, we briefly outline the advances from graph theory and graph algorithms that motivate and make possible these results. Fast self-reduction is defined in Section 3. In Sections 4 and 5, respectively, we present and apply the scaffolding technique. A property shared by several types of self-reductions is defined in Section 6, where we illustrate with an example that even more efficient self-reductions may be possible. The final section contains a discussion of several practical issues relevant to this general topic.

---

[1]Department of Computer Science, University of Idaho, Moscow, ID 83843. This author's research is supported in part by the Sandia University Research Program and by the National Science Foundation under grant MIP–8603879.
[2]Department of Computer Science, Washington State University, Pullman, WA 99164–1210. This author's research is supported in part by the Washington State Technology Center and by the National Science Foundation under grants ECS–8403859 and MIP–8603879.

2. Background

All graphs we consider are finite and undirected, but may have loops and multiple edges. A graph H is less than or equal to a graph G in the *minor* order, written $H \leq_m G$, if and only if a graph isomorphic to H can be obtained from G by a series of these two operations: taking a subgraph and "contracting" edges. For example, the construction that follows shows that $W_4 \leq_m Q_3$ (although $W_4 \not\leq_t Q_3$ where $\leq_t$ denotes less than or equal to in the topological order).

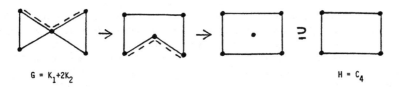

$G = Q_3$ $H = W_4$

— — — contract

Note that the relation $\leq_m$ defines a partial ordering on graphs. A family F of graphs is said to be *closed* under the minor ordering if the facts that G is in F and that $H \leq_m G$ together imply that H must be in F. The *obstruction set* for a family F of graphs is defined to be the set of graphs in the complement of F that are minimal in the minor ordering. Therefore, if F is closed under the minor ordering, it has the following characterization: G is in F if and only if there exists no H in the obstruction set for F such that $H \leq_m G$.

Theorem 1. [RS6] (formerly known as Wagner's Conjecture [Wa]) Any set of finite graphs contains only a finite number of minor-minimal elements.

Theorem 2. [RS5] For every fixed graph H, the problem that takes as input a graph G and determines whether $H \leq_m G$ is solvable in polynomial time.

Remarkably, Theorems 1 and 2, when applicable, guarantee only the *existence* of a polynomial-time decision algorithm. We are assured of the existence of a finite obstruction set for F without being given (by the arguments that establish the theorem) a means for identifying the elements of the set, the cardinality of the set, or even the order of the largest graph in the set (see [FRS]) .

Another interesting feature of Theorems 1 and 2 is the low degree of the polynomials bounding the decision algorithms' running times. Letting n denote the number of vertices in G, the general bound is $O(n^3)$. If F excludes a planar graph, then the bound is $O(n^2)$. These polynomials possess enormous constants of proportionality, rendering them impractical [Jo]. Therefore, Theorems 1 and 2 can be viewed largely as tools for determining problem complexity. Whether the algorithms promised by these theorems can be made effective in practice is an important open question.

A graph H is less than or equal to a graph G in the *immersion* order, written $H \leq_i G$, if and only if a graph isomorphic to H can be obtained from G by a series of these two operations: taking a subgraph and "lifting" pairs of adjacent edges [Ma]. For example, the construction that follows shows that $C_4 \leq_i K_1 + 2K_2$ (although $C_4 \not\leq_m K_1 + 2K_2$ and $C_4 \not\leq_t K_1 + 2K_2$).

$G = K_1 + 2K_2$ $H = C_4$

The relation $\leq_i$, like $\leq_m$, defines a partial ordering on graphs with the associated notions of closure and obstruction sets.

Theorem 3. [RS3] (formerly known as Nash-Williams' Conjecture [Na]) Any set of finite graphs contains only a finite number of immersion-minimal elements.

Theorem 4. [FL3] For every fixed graph H, the problem that takes as input a graph G and determines whether $H \leq_i G$ is solvable in polynomial time.

Theorems 3 and 4, like Theorems 1 and 2, guarantee only the existence of a polynomial-time decision algorithm for any immersion-closed family F of graphs. The method we use in proving Theorem 4 yields a general time bound of $O(n^{h+6})$, where h denotes the order of the largest graph in F's obstruction set. For the results in this paper, however, all bounds will be $O(n^2)$, since the graph families of interest exclude a planar graph of maximum degree three as a minor, and arguments special to this situation allow us to prove bounds similar to those known for minor-closed families.

For an application of Theorems 1 and 2, consider gate matrix layout (GML for short), a combinatorial problem arising in several VLSI layout styles. Formally, we are given an $n \times m$ Boolean matrix M and an integer k, and are asked whether we can permute the columns of M so that, if in each row we change to $*$ every 0 lying between the row's leftmost and rightmost 1, then no column contains more than k 1s and $*$s.

Although the general problem is NP-complete [KF], we have shown that for any fixed value of k, an arbitrary instance of GML can be mapped to an equivalent instance with only two 1s per column, then modeled as a graph whose family of "yes" instances is closed under the minor order and excludes a planar graph.

Theorem 5. [FL1] For any fixed k, GML can be decided in $O(n^2)$ time.

For an application of Theorems 3 and 4, consider two-dimensional grid load factor (2DGLF for short), a two-dimensional analog of the min cut linear arrangement problem [GJ]. An *embedding* of a graph G into a graph H is a one-to-one map $f: V(G) \to V(H)$ together with an assignment, to each edge uv of G, of a path from $f(u)$ to $f(v)$ in H [Ro]. The *minimum load factor* of G relative to H is the minimum, over all embeddings of G in H, of the maximum over all edges e of H of the number of paths of the embedding that contain e. In the 2DGLF problem, we are given a graph G and integers k and w, and are asked whether the minimum load factor relative to an infinite-length, width-w grid is less than or equal to k. Although the general problem is NP-complete even when w is fixed, when both k and w are fixed the family of "yes" instances is closed under the immersion order.

Theorem 6. [FL3] For any fixed k and w, 2DGLF can be decided in polynomial time.

In Section 5, we shall show how the phrase "polynomial time" can be replaced with "$O(n^2)$ time."

3. Fast Self-Reduction

Observe that complexity results based on the Robertson-Seymour theorems are nonconstructive at another level as well. Even if known, such algorithms do not decide based on finding, or failing to find, natural *evidence* (such as a layout). There are now many problems for which polynomial-time decision algorithms can be proven to exist, but for which there are no known algorithms to construct evidence in polynomial time [Fe]. These developments call into question the previous intuitive wisdom that only the complexity of decision problems need be addressed (see, for example, [KUW]).

A natural computational setting in which to consider this issue is that of oracle computations for solving problems of construction. We assume that an *oracle algorithm*, A, to solve a construction problem has access to some decision problem oracle, O. The *overhead* of A is the time required by A to produce a solution, if any exist, where each invocation of O is charged with only a unit-time cost. Thus the overhead of A is the time required by A outside of the running time of the oracle. For example, we have previously established the following.

Theorem 7. [BFL] For any fixed k, a satisfactory solution to GML can be constructed, if any exist, by an oracle algorithm with overhead $O(n^2)$ that makes $O(n^2)$ calls to a decision oracle for GML.

If, as in Theorem 7, the oracle consulted by A answers the decision version of the same problem A solves, then A is said to be a *self-reduction*. A novelty of our approach is that we obtain efficient oracle algorithms using oracles for closely-related, but different decision problems, with no increase in the asymptotic time complexity of the oracle.

How good is Theorem 7? More generally, what computational behavior constitutes a best oracle reduction? Clearly, there may be tradeoffs between the power (and computational cost) of the oracle, the amount of overhead time required, and the number of oracle calls.

Definition. An oracle algorithm A for solving a construction problem is *fastest*, relative to a function $f(n)$, if A has overhead $O(f(n))$ and if the number of oracle calls is within a constant factor of the minimum number required by any oracle algorithm (one free to use any oracle) with overhead $O(f(n))$ solving the same problem.

Because the decision problem oracles available for the layout problems we consider all have time complexity $O(n^2)$, it is natural to focus in this paper on oracle algorithms that are fastest relative to $f(n) = n^2$.

Theorem 8. The oracle algorithm of Theorem 7 is not fastest relative to n^2.

Proof. We argue that there is an oracle algorithm with overhead $O(n^2)$ that asks only $O(n \log n)$ questions (of an unrestricted, "omniscient" oracle) in order to discover a satisfactory solution to GML, when any exist.

The first question to ask is the decision version of GML. If the answer is "yes," then, bit by bit, a sequence of $\log_2 n! = O(n \log n)$ questions can reveal the "index" to a least-numbered satisfactory permutation of the columns of the matrix. The overhead required for this scheme is easily bounded by $O(n^2)$. $\square$

In [FL3], we proved the existence of polynomial-time self-reductions for several other fixed-parameter layout problems by a strategy that proceeds in two stages. In the first stage, using the decision problem oracle once on each pair of nonadjacent vertices, it is determined whether an edge can be added to the graph without turning a "yes" instance into a "no" instance. Thus, in the first stage, the graph is "saturated" with edges. In the second stage, a satisfactory solution is easily determined from the structure of the saturated graph. Since each of these problems asks for a permutation of vertices or edges, the type of oracle algorithm used in the proof of Theorem 8 can be applied.

Corollary 9. None of the self-reductions described above is fastest relative to n^2. $\square$

Note that we are up against an information-theoretic upper bound on the number of oracle calls that a fastest oracle-reduction can make, namely, the number of bits required to write down a solution, or, in other words, the logarithm of the size of the search space. Since all of the problems we consider in this paper are layout permutation problems on graphs with a linear number of edges, this bound is in every case $O(n \log n)$.

Definition. An oracle algorithm A solving a construction problem is *fast*, relative to $f(n)$, if A has overhead $O(f(n))$ and makes $O(s)$ oracle calls, where s is the number of bits required to write down a solution.

This notion of fast is more impressive if $f(n)$ is small. The oracle algorithm of Theorem 7 is not even fast relative to n^2. We shall show how a technique that we term *scaffolding* can be used along with Robertson-Seymour oracles to obtain fast oracle-reductions relative to n^2. Whether these algorithms are truly fastest, relative to n^2, is unknown.

4. Scaffolding

In this section we describe a general method for the design of self-reduction algorithms that are fast in the sense defined in Section 3. This method is particularly applicable to layout permutation problems concerning width metrics on graphs and hypergraphs, such as cutwidth [MiS], modified cutwidth [EST], pathwidth [RS3], gate matrix layout [DKL], vertex separation number [Le], search number [MHGJP], node search number [KP], two dimensional grid load factor [FL3] and others.

A *width* metric on graphs or hypergraphs is a measure $w(G)$ defined as the minimum, over all permutations of the vertex or edge set, of an objective function defined on permutations. Thus the size of the solution space is at least $n!$

The approach that we describe makes sense for problems that satisfy the following *uniformity condition* concerning the complexity of the associated fixed-parameter decision problem for the metric w: there is a constant c such that for *all* k, it can be decided in time $O(n^c)$ for an arbitrary graph G whether $w(G)$ is at most k.

We exploit this uniformity condition to solve width k layout problems by devising an oracle algorithm that efficiently employs an oracle for the width k' decision problem, for an appropriately chosen k' greater than k. This can therefore be viewed as a generalized form of self-reduction that incurs a penalty of only a multiplicative constant, while allowing us to quickly find a width k permutation, when any exist, by making at most $O(n \log n)$ oracle calls.

A key ingredient in our method is to imitate the binary insertion sort algorithm [Kn], by attaching to G a *scaffolding component* that encodes the data structure required by the sorting algorithm. Assuming a vertex-permutation problem, the vertices of G are attached one by one to the *template level* of the scaffold as a permissible sorted order of the attached vertices is progressively extended. (Such an order on a subset of $V(G)$ is one that can be extended to a permutation of $V(G)$ with width k or less.) The determination of where, among the already attached vertices, a newly-chosen unattached vertex may be inserted in a permissible order is made by means of the *probe level* of the scaffold. Attachments at this level permit a call to the decision algorithm (for width k') to determine if v can be inserted in any permissible order between (not necessarily consecutive) vertices x and y that are already attached to the template. Thus, with at most $O(\log n)$ oracle calls, v can be inserted with a binary search. When all vertices of G have been attached to the template in this fashion, the order of attachment describes a permutation of $V(G)$ of width less than or equal to k. The figure below illustrates the general construction. A specific example will be presented in the next section.

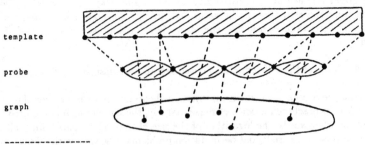

5. Sample Applications

Our scaffolding technique is applicable to problems on width metrics that satisfy the afore-mentioned uniformity condition. We have recently shown that this condition holds for virtually all of the width metrics that have been studied in the context of VLSI layout. In most cases, these results are a dramatic improvement on the best previously-known time-complexity bounds.

The next theorem shows that 2DGLF satisfies the uniformity condition and illustrates a special argument that can sometimes be applied to width problems closed under the immersion order.

Theorem 10. For any fixed k and w, 2DGLF can be decided in $O(n^2)$ time.

Proof Sketch. It can be shown that, for any k and w, there is a (large) binary tree T having minimum load factor greater than k relative to the infinite-length, width-w grid. An easy lemma shows that any graph that is greater than or equal to T in the minor order is greater than or equal to T in the immersion order, and therefore must be a "no" instance (because the "yes" instances are closed in the immersion order). Thus, by the Robertson-Seymour theorems, the graphs of minimum load factor at most k have bounded branch width, and a branch decomposition can be found in $O(n^2)$ time. Testing for the obstructions in the immersion order can be done in linear time for graphs of bounded branch width, given the branch decomposition. □

Note that, as a special case of the above theorem with $w = 1$, min cut linear arrangement (MCLA for short) can be decided in $O(n^2)$ time for all k, improving on the $O(n^4)$ bound of [FL3]. (The cutwidth of a graph is the metric of MCLA, that is, the minimum load factor of G relative to the infinite path.)

Theorem 11. For any fixed k, a satisfactory solution to MCLA can be constructed, if any exist, by an oracle algorithm that is fast relative to n^2.

Proof Sketch. We set $k' = 3k$. A scaffolding of a particular graph G is illustrated below as it might appear at some stage of the algorithm, with $k = 2$.

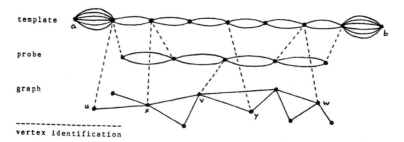

A call to the decision problem oracle (for cutwidth $k' = 6$) will determine if inserting v between x and w is a permissible extension of the currently established permissible order on the vertices u, x, y, w that are already attached to the template.

The main arguments that establish the proof are outlined as follows:

1) If the scaffolded graph G' of G is connected, then vertices a and b must be first and last (or vice versa) in any permutation of $V(G')$ with cutwidth at most k'.

2) In any permutation p of $V(G')$ of cutwidth at most k', the template component and the probe component cannot be "kinked."

3) If the sequence of attachments of the vertices of G to the template is permissible, then there is a permutation p of $V(G')$ of cutwidth at most k'.

4) In any permutation p of $V(G')$ of cutwidth at most k', p restricted to $V(G)$ must have cutwidth at most k. □

The choice of k' and the details of the scaffold components are specific to the particular layout problem considered. Attachments can take the form of edge additions, vertex identifications, or more complex graph gadgets. By arguments and constructions similar to those above we have obtained the following result on the complexity of other well-known layout permutation problems.

Theorem 12. For any fixed k, a satisfactory solution to modified cutwidth, pathwidth, gate matrix layout, vertex separation number, search number, node search number and two-dimensional grid load factor can be constructed, if any exist, by an oracle algorithm that is fast relative to n^2.

6. Blind Oracle Reductions

There is a certain attribute of the fast self-reductions developed in Sections 4 and 5 that may point to even more efficient oracle-reductions that merit exploration.

Definition. An oracle algorithm A that solves a construction problem is *blind* if it knows only the size and format of the input, and otherwise has access to the input only through queries to the oracle.

Thus, in the case of graph problems, we may assume that the algorithm knows only that the vertex set is $\{1, \ldots, n\}$. (Knowing the format, in this case, means having names for the vertices, so that a solution can be expressed.) This arrangement readily generalizes to other kinds of problems.

Surprisingly, even straightforward self-reductions are often blind without any modifications. Indeed, many construction problems (including problems such as Hamiltonian circuit and k-colorability with natural evidence) have blind self-reductions that are asymptotically as fast as any other blind oracle algorithm relative to the number of oracle calls, for fixed overhead $f(n)$.

Theorem 13. The fast oracle algorithms based on scaffolding as described in Section 5 are blind.

Proof. At no point do the algorithms need to know more than the size of the graph. The algorithms need only set up the scaffold and determine how the vertices of the graph are to be attached to it, asking the same questions that would be asked if the graph were visible. □

This prompts the following definition of a complexity metric that is well-defined on all construction problems, and may provide some insight into the nature of fast oracle reductions. The notion we offer seems to generalize, in some ways, the concept of *evasiveness* of graph properties [Bol].

It is also related to *oracle complexity* in constructivist foundations of computational complexity [AFLM].

Definition. The *blind oracle complexity* of a construction problem Π is the minimum number of calls to any omniscient, trustworthy oracle that is required for any blind oracle algorithm to solve Π (producing the required evidence, when it exists).

Note that, unlike fast oracle algorithms, we do not define blind oracle complexity relative to overhead. This is because blind oracle algorithms cannot even access the input without consulting the oracle. Using the upper bounds of Theorems 11 and 12 and standard adversary arguments for lower bounds, we achieve the following result.

Theorem 14. The blind oracle complexity of the layout permutation problems considered in this paper is $\Theta(n \log n)$.

Thus the scaffolding technique yields oracle-reductions that are optimal (to within a constant factor) for the class of blind oracle algorithms.

These results also point out that, since the scaffolding technique does not analyze the particular structure of an input graph, more efficient self-reductions may be possible for algorithms that are able to take advantage of the input. In fact, such improvements have already been made for other problems amenable to a Robertson-Seymour approach to decision problem complexity. Consider, for example, k-feedback vertex set [GJ] for fixed k. Although the logarithm of the size of the search space is $O(\log n)$, we have devised an algorithm to analyze the input graph's structure (which therefore is not blind) that needs only $O(1)$ oracle calls.

Theorem 15. For any fixed k, there is a fastest oracle algorithm relative to n^2 that finds a k-feedback vertex set, when any exist, and makes at most $(2k)^k$ oracle calls.

7. Toward Practical Algorithms

The powerful new tools offered by the recent theorems of Robertson and Seymour on well-partially-ordered sets enable straightforward proofs of the existence of decision algorithms with small-degree polynomial-time bounds for a wide range of layout problems of VLSI design. Herein we have explored the bounds that can be obtained by fast self-reduction strategies for the problem of constructing a permutation of width at most k when any exist.

Because of the nonconstructive nature of the theorems and the astronomical constants of proportionality involved, our results must, at this time, be viewed mainly as establishing complexity bounds for these problems—bounds that are theoretically achievable, rather than approachable in practice. An important, open question is whether these bounds can be attained with practical algorithms.

There are, however, grounds for optimism. Among the approaches that may ultimately yield reasonable algorithms with performance bounds as good as we have shown to be theoretically achievable are the following.

1) Be inspired and find a direct algorithm. This is the approach advocated in [Jo] where vertex cover is used as an example. Since there are several obvious planar obstructions to having a k-vertex cover, the Robertson-Seymour theorems give a bound of $O(n^2)$ for all k. Unfortunately, this involves enormous constants and one must know the (also possibly enormous) obstruction sets in order to know the algorithms. A direct approach with time complexity $O(n^2)$ based on matching was found [BR] as was an improvement to $O(n)$ time based on fast self-reduction [BF].

2) Find a direct algorithm based on bounded branch-width. As suggested in [Jo], the fact that there are planar obstructions for vertex cover implies that all "yes" instances have bounded branch-width (by the Robertson-Seymour theorems). Moreover, in $O(n^2)$ time a branch decomposition can be found, in case the graph might be a "yes" instance. In this setting, the vertex cover problem can be solved directly by a linear-time dynamic programming algorithm. In fact, in this setting, many problems (including many that are NP-complete in general) can be solved in linear time [AP, BLW, Se, SS, WHL].

It should be noted that this is essentially the method of the Robertson-Seymour complexity bounds for minor-closed families of graphs having a planar obstruction. The difference is that in the bounded branch-width setting one is now solving the problem directly, rather than looking for obstructions. At least one need only find a single planar obstruction and not the entire set. This approach is still impractical unless one can prove a decent bound on branch-width [Bod], since the bound supplied by the Robertson-Seymour theorems is the principal astronomical constant precluding the practicality of a direct application of these immensely powerful and general results. As discussed in [Wi], however, this approach may not work for layout permutation problems.

3) Employ approximate obstruction sets. It seems likely that, in many cases, a small subset of the obstruction set may contain most of the needed information [BFKL]. In conjunction with fast minor tests, hand-crafted to the particular graphs in question, this may provide a viable approach.

4) Utilize path-shape. Note that for all of the width metrics we have considered, the graphs of bounded width are "path-shaped." Technically, for any two of these width metrics w and w', it should be possible to obtain reasonable bounds of the form $w(G) = \Theta(w'(G))$. By a trivial modification of the Robertson-Seymour quadratic-time algorithm for finding an approximate branch decomposition of a graph, when any exist, a path-shaped branch-decomposition can be found in $O(n^2)$ time, or the graph determined to be a "no" instance. Next, one wants an efficient algorithm to convert this path-shape into the path-shape corresponding to the width metric under consideration.

References

[AFLM] K. Abrahamson, M. R. Fellows, M. A. Langston and B. Moret, "Constructive Complexity," to appear.

[AP] S. Arnborg and A. Proskurowski, "Linear Time Algorithms for NP-hard Problems on Graphs Embedded in k-trees," TRITA–NA–8404, The Royal Institute of Technology (1984).

[BF] S. R. Buss and M. R. Fellows, "Achieving the Robertson-Seymour Bounds: k-Feedback and Related Vertex Sets," to appear.

[BFKL] R. L. Bryant, M. R. Fellows, N. G. Kinnersley and M. A. Langston, "On Finding Obstruction Sets and Polynomial-Time Algorithms for Gate Matrix Layout," *Proc. 25th Allerton Conf. on Communication, Control, and Computing* (1987), 397–398.

[BFL] D. J. Brown, M. R. Fellows and M. A. Langston, "Polynomial-Time Self-Reducibility: Theoretical Motivations and Practical Results," Computer Science Technical Report CS-87-171, Washington State University, 1987.

[BLW] M. W. Bern, E. L. Lawler and A. L. Wong, "Why Certain Subgraph Computations Require Only Linear Time," *Proc. 26th IEEE Symposium the Foundations of Computer Science* (1985), 117–125.

[Bod] H. L. Bodlaender, "Classes of Graphs with Bounded Tree-Width," Technical Report RUU–CS–86–22, Department of Computer Science, University of Utrecht, 1986.

[Bol] B. Bollabás, Extremal Graph Theory, Academic Press, New York, 1978.

[BR] M. Blum and S. Rudich, private communication.

[DKL] N. Deo, M. S. Krishnamoorthy and M. A. Langston, "Exact and Approximate Solutions for the Gate Matrix Layout Problem," *IEEE Trans. on Computer-Aided Design* 6 (1987), 79–84.

[EST] J. Ellis, I. H. Sudborough and J. Turner, "Graph Separation and Search Number," to appear.

[Fe] M. R. Fellows, "Applications of the Robertson-Seymour Theorems: A Survey," to appear.

[FL1] M. R. Fellows and M. A. Langston, "Nonconstructive Advances in Polynomial-Time Complexity," *Info. Proc. Letters* 26 (1987), 157–162.

[FL2] ——, "Nonconstructive Tools for Proving Polynomial-Time Decidability," *J. of the ACM*, to appear.

[FL3] ——, "Layout Permutation Problems and Well-Partially-Ordered Sets," *Proc. 5th MIT Conf. on Advanced Research in VLSI* (1988), to appear.

[FRS] H. Friedman, N. Robertson and P. D. Seymour, "The Metamathematics of the Graph Minor Theorem," in Applications of Logic to Combinatorics, American Math. Soc., Providence. RI. to appear.

[GJ] M. R. Garey and D. S. Johnson, Computers and Intractability: A Guide to the Theory of NP-Completeness, Freeman, San Francisco, CA, 1979.

[GS] E. M. Gurari and I. H. Sudborough, "Improved Dynamic Programming Algorithms for Bandwidth Minimization and the Min Cut Linear Arrangement Problem," *J. of Algorithms* 5 (1984), 531–546.

[Jo] D. S. Johnson, "The Many Faces of Polynomial Time," in The NP-Completeness Column: An Ongoing Guide, *J. Algorithms* 8 (1987), 285–303.

[KF] T. Kashiwabara and T. Fujisawa, "NP-completeness of the Problem of Finding a Minimum-Clique-Number Interval Graph Containing a Given Graph as a Subgraph," *Proc. IEEE Symp. on Circuits and Systems* (1979), 657–660.

[Kn] D. E. Knuth, The Art of Computer Programming, Vol 3: Sorting and Searching, Addison-Wesley, Reading, MA, 1973.

[KP] M. Kirousis and C. H. Papadimitriou, "Searching and Pebbling," Technical Report, National Technical University, Athens, Greece, 1983.

[KUW] R. M. Karp, E. Upfal and A. Wigderson, "Are Search and Decision Problems Computationally Equivalent," *Proc. 17th ACM Symp. on Theory of Computing* (1985), 464–475.

[Le] T. Lengauer, "Black-White Pebbles and Graph Separation," *Acta Informatica* 16 (1981), 465–475.

[Ma] W. Mader, "A Reduction Method for Edge-Connectivity in Graphs," *Annals of Disc. Math* 3 (1978), 145–164.

[MaS] F. S. Makedon and I. H. Sudborough, "On Minimizing Width in Linear Layouts," to appear.

[MHGJP] N. Megiddo, S. L. Hakimi, M. R. Garey, D. S. Johnson and C. H. Papadimitriou, "On the Complexity of Searching a Graph," IBM Research Report RJ 4987, 1986.

[MiS] Z. Miller and I. H. Sudborough, "Polynomial Algorithms for Recognizing Small Cutwidth in Hypergraphs," to appear.

[Na] C. Nash-Williams, "On Well-Quasi-Ordering Infinite Trees," *Proc. Cambridge Phil. Soc.* 61 (1965), 697–720.

[Ro] A. Rosenberg, "Issues in the Study of Graph Embeddings," *Lecture Notes in Computer Science* 100 (1981), 150–176.

[RS1] N. Robertson and P. D. Seymour, "Disjoint Paths-a Survey," *SIAM J. Alg. Disc. Meth.* 6 (1985), 300–305.

[RS2] ———, "Graph Minors—a Survey," in Surveys in Combinatorics (I. Anderson, ed.), Cambridge Univ. Press, 1985.

[RS3] ———, "Graph Minors I. Excluding a Forest," *J. Comb. Th. Ser. B* 35 (1983), 39–61.

[RS4] ———, "Graph Minors IV. Tree-Width and Well-Quasi-Ordering," to appear.

[RS5] ———, "Graph Minors XIII. The Disjoint Paths Problem," to appear.

[RS6] ———, "Graph Minors XVI. Wagner's Conjecture," to appear.

[Se] D. Seese, "Tree-Partite Graphs and the Complexity of Algorithms," Preprint P–Math–08/86, Karl-Weierstrass-Institut für Mathematik, Akademie der Wissenschaften der DDR, 1986.

[SS] P. Scheffler and D. Seese, "Graphs of Bounded Tree-Width and Linear-Time Algorithms for NP-Complete Problems," manuscript, 1986.

[Wa] K. Wagner, "Uber Einer Eigenshaft der Ebener Complexe," *Math. Ann.* 14 (1937), 570–590.

[WHL] T. V. Wimer, S. T. Hedetniemi and R. Laskar, "A Methodology for Constructing Linear Graph Algorithms," *Congressus Numerantium* 50 (1985), 43–60.

[Wi] T. V. Wimer, "Linear Algorithms on *k*-Terminal Graphs," Ph.D. Dissertation, Clemson University, 1987.

Regular Structures and Testing: RCC-Adders

(Extended Abstract)

Bernd Becker, Uwe Sparmann

Fachbereich Informatik, Universität des Saarlandes

D-6600 Saarbrücken, West Germany

Abstract. *We confine ourselves to one of the basic problems of testing, the test pattern genera-tion problem for combinational circuits, and study the relation between structural properties and test complexity.*

Here, this relation is looked at in detail for Recursive Carry Computation adders. The class of RCC-adders has been introduced in [BeKo] and contains a wide range of different adder realizations (e.g., optimal time adders such as the carry look-ahead adder of [BrKu] and the conditional carry adder of [BeKo]). We show that symbolic computation can be used to define this class and at the same time offers a uniform test approach which can be applied at an early stage of the design process. The class of RCC-adders itself splits into several subclasses which are specified by structural properties of the overall computation scheme and functional properties of the basic cells. Optimal complete test sets with respect to two commonly used fault models, the single stuck-at fault model and the single cellular fault model, are developed for these RCC-subclasses. The cardinality of the test sets depends on the choice of the fault model and on structural properties of the RCC-subclass.

To summarize our results, we finally obtain two tables with upper and lower bounds characterizing the test complexity of classes of RCC-adders. The upper bounds are obtained by the effective construction of complete test sets. The cardinality of these sets varies between a logarithmic or linear number of patterns for an n-bit RCC-adder.

1. Introduction

As a result of technological improvements VLSI electronic circuitry may nowadays contain hundreds of thousands of transistors on a single silicon chip. Even if the chips are correctly designed, a certain fraction of them will have physical defects caused by imperfections occurring during the manufacturing process (e.g., open connections induced by dust particles). Therefore there has to be a test phase in which the "good" chips are sorted from the "bad" ones. No precise statistical data about the extent of the test problems exist, but it is assumed at the moment that they contribute to at least 25% of the total product costs ([Mi]). With further increasing chip complexity these costs may even increase. Thus, the necessity of new methods and efficient algorithms for the test phase is evident. Specialists in the field of testing agree that testability questions have to be considered from the very beginning of the design process in order to develop a test concept which guarantees the testability of the circuit at the end of the manufacturing process.

In this paper we confine ourselves to one of the basic problems of testing, the test pattern generation problem for combinational circuits. (With the introduction of scan design and built-in test methods ([EiWi], [KMZ]) the test of a complete VLSI chip can be reduced to testing its combinational sub-circuits.) We show that, under certain circumstances, the construction of test sets for combinational circuits can be successfully performed on a relatively high design level.

Universal test pattern generation methods based on the D-algorithm or 'related' algorithms are be-coming very costly or even computationally infeasible in general. (The computation time may be

This work was supported by DFG, SFB 124, TP B1, VLSI Entwurfsmethoden und Parallelität, and by BMFT Grant 413-5839-ITS8501A7, Kooperationsprojekt TESUS der Firmen Nixdorf und Siemens exponential in the size of the circuit!) Another method is random pattern generation. In this case, the generation costs are normally very low, but the quality of the tests is unknown. Thus, fault simulation is necessary to verify the quality of the test set. Depending on the size of the circuit, a very large number of random patterns has to be simulated to obtain a satisfactory 'fault coverage'. This again leads to extensive computation times. Universal methods use no structural information about the circuits under test. However, there are in practice no large scale circuits with random logic. There will always be some sort of 'structure'. And these structural properties turn out to be very useful for test pattern generation. This is shown by several authors for important classes of circuits with regular structures, such as PLA's, memories or arithmetical units. (See e.g. [Sm], [DaMu], [AbRe], [Ka], [Fr], [FeSh]). Whereas these authors mainly use methods which are based on linear models like "iterative logic arrays", the basic structure of 'fast' circuits for addition and multiplication usually is a tree-like structure. Linear models can no longer be used. Nevertheless, complete tests based on structural properties can be constructed. Work in this direction has been done in [Be1], [BeSo] on an optimal time multiplier and in [AbGa], [MoAb] and [Be2] on two types of optimal time adders, the carry look-ahead adder and the conditional sum adder.

In this paper we study the test generation problem for **Recursive Carry Computation adders, RCC-adders**, for short. This is a class of (optimal time) adders, which has been introduced in [BeKo] and contains a wide range of different adder realizations, e.g., the carry look-ahead adder as presented in [BrKu] and the conditional carry adder of [BeKo]. We show in section 2 that symbolic computation can be used to define the class of RCC-adders. (In [BeKo] this is done with methods from Boolean algebra.) Single elements of the RCC-class are then specified by a choice of an overall computation structure and a transformation of symbolic basic cells into digital basic cells. A choice of an encoding and of some redundant function values is necessary for this transformation. The results of the paper concerning testability of RCC-adders are presented in section 3 and can be summarized as follows: Two different fault models, both of which are well-known, are considered, the cellular fault model (= cf-model) and the stuck-at fault model (= sa-model). The cf-model works on a functional level, whereas the sa-model checks values on digital wires. Thus it is not at all clear that the test patterns are constructible on the symbolic level in the latter case. In a first lemma we therefore "translate" faults on digital wires to faults on the symbolic level. Now all constructions can be performed on the symbolic level. This makes them independent of details of a specific digital realization, such as the encoding of the symbolic values.

We derive uniform complete test sets for classes of RCC-adders, i.e., complete test sets which work for each element of a given class. The cardinality of the tests depends on the fault model and the computation structure of the considered subclass. In a first step the cardinality is bounded from below: Theorem 1 presents lower bounds for the class RCC and several subclasses. For the proof of this theorem the "depth" of the circuit is related to the cardinality of a complete test in lemma 2.

Then we give upper bounds for the size of uniform complete test sets in theorem 2. These bounds are obtained by effective construction of tests and vary between $4log(n) + 22$ and $12n$ for an n-bit adder. Even if no detailed properties of the structure are known, a complete test (with a linear number of patterns) can be specified. If the overall computation scheme of the device under test is known, smaller test sets can be derived. This is shown for two well-known and often used structures, which result in (asymptotically) optimal time adders: the complete binary tree (= CBT) structure, which mainly consists of n overlapping binary trees, and the BK-structure as introduced in [BrKu]. In the stuck-at fault model the test sets are independent of the redundancies, in the cellular fault model some properties of the redundancies are required. (Carry look-ahead adders and conditional carry adders fulfill these properties.) Because of lack of space all constructions necessary for a complete proof of theorem 2 can not be given in this paper. To demonstrate our methodology we more or less restrict ourselves to the detailed construction of one uniform complete test set in lemma 3. Comprehensive proofs for all results can be found in the full paper [BeSp].

The quality of the constructions can be judged by a comparison of the lower and upper bounds for the cardinality of complete test sets. It turns out that all tests except one are (asymptotically) optimal with very small constants.

2. RCC-Adders – A Class of (Optimal Time) Adders

The class of RCC-adders is defined in [BeKo] by using methods of boolean algebra as introduced in [Ho]. In the following we present a different method to derive this class, which is based on symbolic computation and turns out to be well-suited for the development of a uniform test approach for the whole class. We start with a description of the basic addition method.

Let $a := a_{n-1} \ldots a_0$, $b := b_{n-1} \ldots b_0$ and $c := c_{-1}$ be binary encodings of non-negative numbers. (For simplicity of presentation assume $n = 2^m$.) We want to compute the binary encoding $s := s_n \ldots s_0$, of the sum $\sum_{j=0}^{n-1} a_j 2^j + \sum_{j=0}^{n-1} b_j 2^j + c_{-1}$. The usual way of achieving this is to compute the carry bits c_i for $i = 0, \ldots, n-1$. Then $s_i := a_i \oplus b_i \oplus c_{i-1}$ for $i = 0, \ldots, n-1$ and $s_n := c_{n-1}$ give the sum bits.

Thus, a fast carry computation is the main problem which has to be solved with the design of an (optimal time) adder. In the following we present an algorithm which computes the carries by computation of 'types' of subnumbers of a and b. For this we partition the set $\{0,1\}^{2k}$ ($k \in \{1, \ldots, n\}$) into three subsets:

$$A_k := \{(\alpha_{k-1} \ldots \alpha_0, \beta_{k-1} \ldots \beta_0) | \alpha_l, \beta_l \in \{0,1\} \text{ and } \sum_{l=0}^{k-1} \alpha_l 2^l + \sum_{l=0}^{k-1} \beta_l 2^l < 2^k - 1\}$$

$$P_k := \{(\alpha_{k-1} \ldots \alpha_0, \beta_{k-1} \ldots \beta_0) | \alpha_l, \beta_l \in \{0,1\} \text{ and } \sum_{l=0}^{k-1} \alpha_l 2^l + \sum_{l=0}^{k-1} \beta_l 2^l = 2^k - 1\}$$

$$G_k := \{(\alpha_{k-1} \ldots \alpha_0, \beta_{k-1} \ldots \beta_0) | \alpha_l, \beta_l \in \{0,1\} \text{ and } \sum_{l=0}^{k-1} \alpha_l 2^l + \sum_{l=0}^{k-1} \beta_l 2^l > 2^k - 1\}$$

Figure 1 illustrates these sets for the case $k = 3$. The elements of A_k (G_k) correspond to the points in the upper (lower) triangle and denote pairs $(\alpha_{k-1} \ldots \alpha_0, \beta_{k-1} \ldots \beta_0)$, whose sum **absorbs** an incoming carry (**generates** a carry). The elements of P_k correspond to the points of the diagonal and denote pairs, whose sum **propagates** an incoming carry.

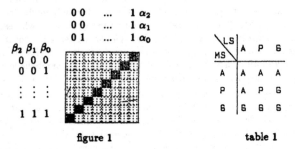

figure 1   table 1

In the following the subnumber $\gamma_{[i,j]}$ of a binary encoding $\gamma := \gamma_{n-1} \ldots \gamma_0 \in \{0,1\}^n$ is given by $\gamma_{[i,j]} := \gamma_i \ldots \gamma_j$ ($i, j \in \{0, \ldots, n-1\}, i \geq j$). For a pair $(a_{[i,j]}, b_{[i,j]})$ of subnumbers of a and b we then define: $(a_{[i,j]}, b_{[i,j]})$ has **type** T or $type(a_{[i,j]}, b_{[i,j]}) = T$ ($T \in \{A, P, G\}$), iff $(a_{[i,j]}, b_{[i,j]}) \in T_k$ for $k := i - j + 1$. Since A_k, P_k, G_k forms a partition of $\{0,1\}^{2k}$, the type of any pair $(a_{[i,j]}, b_{[i,j]})$ in $\{0,1\}^{2k}$ is well-defined. Note that the computation of the carry bits is reducible to the computation of $type(a_{[i,0]}, b_{[i,0]})$ for $i = 0, \ldots, n-1$ because of the following equivalence:

$$c_i = 1 \iff type(a_{[i,0]}, b_{[i,0]}) = \begin{cases} G & \text{if } c_{-1} = 0 \\ G \text{ or } P & \text{if } c_{-1} = 1 \end{cases} \qquad (*)$$

A divide and conquer approach is appropriate for the computation of the types: the type $T_{[i,j]}$ of a pair $(a_{[i,j]}, b_{[i,j]})$ ($i, j \in \{0, \ldots, n-1\}$, $i > j$) is determined by the types $T_{[i,h+1]}$ and $T_{[h,j]}$ of two "adjacent" pairs $(a_{[i,h+1]}, b_{[i,h+1]})$ and $(a_{[h,j]}, b_{[h,j]})$ (for any h with $h \in \{i - 1, \ldots, j\}$). For example $T_{[i,j]} = A$, iff $T_{[i,h+1]} = A$ or $[T_{[i,h+1]} = P$ and $T_{[h,j]} = A]$. We summarize these rules in table 1 by defining an operation '$\triangleright$' on the set of types. Column "MS" (=most significant) gives the type of the pair $(a_{[i,h+1]}, b_{[i,h+1]})$, row "LS" (=least significant) gives the type of the pair $(a_{[h,j]}, b_{[h,j]})$ and the entry specifies the type of the "composed" pair $(a_{[i,j]}, b_{[i,j]})$. '$\triangleright$' is non-commutative (that is the reason why we do not use a symmetric operator symbol), but has the following nice properties, which will be of great importance for the construction of the test patterns:

- It follows directly from the definition of type that $'\triangleright'$ is an associative operation. (Thus, we can omit brackets in expressions of the form $T_1 \triangleright T_2 \triangleright \ldots \triangleright T_l$ and the result of this expression is independent of the order in which the expression is evaluated.)
- P is a unit element with respect to $'\triangleright'$.
- A and G are **left-stable** or **left-0-elements** of $'\triangleright'$, since $A \triangleright T = A$, $G \triangleright T = G$ for any $T \in \{A, P, G\}$. Thus, P is the only element which is **left-injective**, i.e., $P \triangleright T_1 = P \triangleright T_2$ implies $T_1 = T_2$.

We are now ready to give a preliminary scheme of an adder with Recursive Carry Computation (= RCC-adder) on an "algorithmic" level. For illustration see figure 2. $T_{[i,i]}$ is abbreviated as $T_{[i]}$.

- Compute the type of (a_i, b_i) for $i = 0, \ldots, n-1$ ("**Preparation**"). Then use a divide and conquer approach and the operation $'\triangleright'$ to compute the type of $(a_{[i,0]}, b_{[i,0]})$ for $i = 0, \ldots, n-1$ ("**Recursive Carry Computation**"). Finally compute s_i for $i = 0, \ldots, n$ ("**Final Addition**").

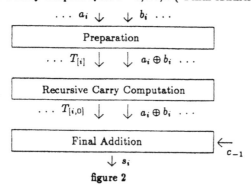

figure 2

In the following this preliminary scheme is realized with basic cells on a symbolic level. For the Preparation Part and the Final Addition Part this is easily done in a bitslice mode by use of basic cells α, γ and $EXOR$. (See figure 3 below and note that the γ-cells use the equivalence (*) to compute the carries.)

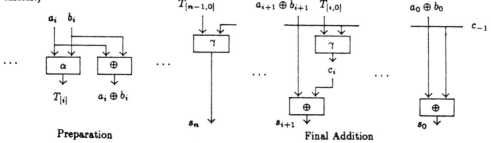

Preparation   Final Addition

figure 3

We omit any further details of these components of the adder and now concentrate on the crucial part of the realization and the test as well, the Recursive Carry Computation Part (= RCC-part). The types $T_{[i]}$ ($i \in \{n-1, \ldots, 0\}$) are inputs to this part. (The outputs $a_i \oplus b_i$ of the Preparation Part are passed through the RCC-part without modification.) We use binary trees tr_i ($i \in \{n-1, \ldots, 0\}$) to compute $T_{[i,0]}$ according to the divide and conquer method proposed above. Thereby the nodes of the trees tr_i correspond to a further basic cell type, the β-cell, which is used to compute the operation $'\triangleright'$. A β-cell has a left input "MS", a right input "LS" and computes the type MS$\triangleright$LS according to table 1. A node in the tree tr_i is denoted by $\beta_{[l,k]}$, iff the type $T_{[l,k]}$ of the pair $(a_{[l,k]}, b_{[l,k]})$ is computed at this node ($n-1 \geq l > k \geq 0$). The inputs to $\beta_{[l,k]}$ are outputs of β-cells $\beta_{[l,h+1]}$ and $\beta_{[h,k]}$ (h fixed with

$l > h \geq k$); i.e., the input edges of $\beta_{[l,k]}$ correspond to the types $T_{[l,h+1]}$ and $T_{[h,k]}$ of the two adjacent pairs $(a_{[l,h+1]}, b_{[l,h+1]})$ and $(a_{[h,k]}, b_{[h,k]})$. Thus, the computation $T_{[l,k]} = T_{[l,h+1]} \rhd T_{[h,k]}$ is performed at cell $\beta_{[l,k]}$. The trees tr_i are interleaved in such a way that no type $T_{[l,k]}$ is computed more than once. (Cells computing the same type are identified, and their output is diagonalized to the cells which use this type as an input.) The structure of the binary trees tr_i depends on the selection of the parameter h for the inputs of any cell $\beta_{[l,k]}$. Thus the general method described in 2.1 provides a large number of possible carry computation structures (see e.g. [LaFi]). The following two structures are often used and their impact on testability will be discussed in section 3.

The **CBT-Structure** (= Complete Binary Tree Structure) consists of complete binary trees tr_i, if $i = 2^x - 1$ for some $x \in \mathbb{N}$. If $i \neq 2^x - 1$, then tr_i is an incomplete tree "cut off on its right hand side". Figure 4 shows the trees for the case $n = 4$, and how they are overlapped to yield the CBT-Structure.

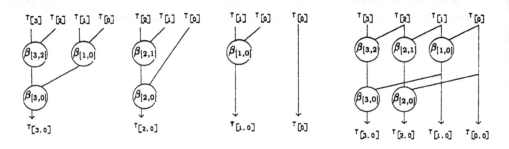

figure 4

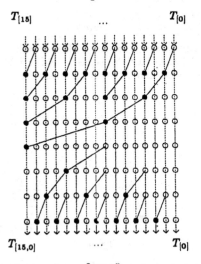

figure 5

The **BK-Structure** (= Computing Scheme of [BrKu]) consists of two parts. First of all, the type $T_{[n-1,0]}$ is computed in a complete binary tree of depth $log(n) = m$. (Note that this includes the computation of the types $T_{[l,k]}$ at the inner nodes of the tree.) The remaining types are computed in the second part, which again uses the complete binary tree structure, this time in the reverse order (and for subtrees of size $2^{m-1}, 2^{m-2}, \ldots, 2^1$). See figure 5 for an illustration of this computation scheme in the case $m = 4$. (The black nodes correspond to β-cells, the white nodes are introduced to emphasize the regularity of the structure and simply transmit or diagonalize data.)

Until now we have only given a description of the class of RCC-adders on a symbolic computation level without regard to a physical realization. To obtain a physical realization of an RCC-adder we have to transform the symbolic representation and computation of types into a representation and computation on a digital level. It turns out that various transformations are feasible. First of all, we have to choose a binary **encoding** φ for the symbolic types A, P and G. We restrict our attention to encodings using 2 bits. This is the minimum number of bits possible. There are 24 possibilities to encode A, P, G. The redundant code word is named R. R is also called redundant type in the following. Secondly, we have to specify the Boolean functions which are to be computed by digital realizations of the basic cells α, β and γ. Until now the 'symbolic' functions corresponding to the basic cells have been merely defined for the types A, P, G. We extend these functions to input values containing the redundant value R. Denote the extended 'symbolic' functions by f_α, f_β, f_γ. Together with a fixed encoding, f_α, f_β, f_γ define the Boolean functions of α, β, γ completely. The function f_α is uniquely determined, whereas there are 4^7 (4) possibilities for selecting the outputs of the β-cells (γ-cells) on inputs containing R.

Let RCC denote the set of RCC-adders. An element in RCC is now completely specified by the choice of a computation structure for the RCC-part, the choice of an encoding and the choice of the redundancies for the functions f_β, f_γ. In the following $RCC(\text{struc}, \text{enc}, \text{red})$ denotes a subclass of RCC-adders. 'struc' provides information about the structure of the RCC-part, 'enc' gives an encoding or a set of encodings, 'red' specifies the redundancies, e.g., by defining f_β, f_γ. Thus, the set of all RCC-adders, the β-cells (γ-cells) of which realize the function f_β (f_γ) for a given function $f_\beta(f_\gamma)$ is denoted by $RCC(-, -, (f_\beta, f_\gamma))$. Further examples are presented in the following:

- $-RCC(O(log(n)), -, -)$ denotes the class of all RCC-adders with time complexity $O(log(n))$. From the theoretical point of view this is precisely the subclass of RCC, which contains the optimal time adders.
- $-RCC(BK, -, -)$ or BK denotes the set of RCC-adders with BK-structure in the RCC-part.
- $-RCC(CBT, -, -)$ or CBT denotes the set of RCC-adders with complete binary tree structure.

Since both computation structures (the BK- and the CBT-structure) have depth $O(log(n))$, BK and CBT are subsets of $RCC(O(log(n)), -, -)$. BK and CBT denote classes with fixed computation structure, but with a free choice of the encoding and redundancies. We now describe two subclasses of RCC-adders with a fixed encoding φ and a fixed choice of the redundancies given by functions f_β, f_γ: the class CLA (= Carry Look Ahead Class) containing the well-known carry look-ahead adder as presented in [BrKu] and the class CC (= Conditional Carry Class) containing the conditional carry adder as introduced in [BeKo]. Choose the encoding $\varphi_{CLA}(A) = (0,0)$, $\varphi_{CLA}(P) = (0,1)$, $\varphi_{CLA}(G) = (1,0)$, $\varphi_{CLA}(R) = (1,1)$ and $\varphi_{CC}(A) = (0,0)$, $\varphi_{CC}(P) = (0,1)$, $\varphi_{CC}(G) = (1,1)$, $\varphi_{CC}(R) = (1,0)$, respectively. (Usually the first (second) bit of the encoding given by φ_{CLA} is called generate (propagate) bit. In [BeKo] the first (second) bit of the encoding given by φ_{CC} is called the c^0 (c^1) bit, c^0 (c^1) corresponds to the carry, if the incoming carry is 0 (1).) Figure 6.a (6.b) shows the standard function table of the β-cell denoted by f_β^{CLA} (f_β^{CC}) and a possible digital realization of the β-cell. (The function f_β^{CC} can be realized with two parallel multiplexers. Depending on the technology, this may result in an overall computation time which is shorter than for other known (optimal time) adders, e.g., the CLA-adder. For more details see [BeKo].) Similar definitions can be given for the γ-cells resulting in functions f_γ^{CLA}, f^{CC}. Using the notation from above we then have $CLA = RCC(-, \varphi_{CLA}, (f_\beta^{CLA}, f_\gamma^{CLA}))$ and $CC = RCC(-, \varphi_{CC}, (f_\beta^{CC}, f_\gamma^{CC}))$.

MS\\LS	A	P	G	R
A	A	A	A	A
P	A	P	G	R
G	G	G	G	G
R	G	R	G	R

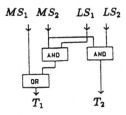

LS / HS	A	P	G	R
A	A	A	A	A
P	A	P	G	R
G	G	G	G	G
R	G	R	A	P

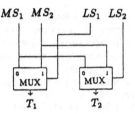

figure 6.b: f_β^{CC} and a possible realization

3. Complete Test Sets for RCC-Adders

As mentioned above we consider two different fault models, the single stuck-at fault model (= sa-model) and the single cellular fault model (= cf-model). First of all, we give an informal description of both models, for more details see e.g. [Br] and [FeSh]. Our aim is to construct test patterns for RCC-adders at an early stage of the design process (i.e., on the symbolic computation level). On the other hand, we have to consider the digital level to give the definition of the **stuck-at fault model** On this level, RCC-adders are circuits which consist of copies of (digital) basic cells $\alpha, \beta, \gamma, MUX, EXOR$, input and output ports, diagonalizations (= fanout points) and signal lines. In the sa-model faults on lines, input or output ports are considered: Assume that exactly one line, input or output port x is fixed to logic 0 (= x stuck-at 0). An input t to the RCC-adder is called a **test** for x **stuck-at 0** (or equivalently a test on x for 1/0) iff the application of t to the correct circuit and to the circuit with the single fault x stuck-at 0 results in different output combinations, i.e., t generates the value 1 on x for the correct circuit, while x has value 0 in the faulty circuit and the difference on x, denoted by 1/0, is propagated to the primary outputs of the circuit. Analogously, a test for x stuck-at 1 can be defined. In the **cellular fault model**, it is assumed that a fault modifies the behaviour of exactly one basic cell ρ and that the modified cell remains combinational. Since this fault can be detected by observing the incorrect output of ρ for one suitable input, it suffices to test for faults of the following kind. A **cellular fault** is a tuple $(\rho, I, X/Y)$, where ρ is the faulty cell, I is the input for which ρ does not behave correctly, and $X (Y)$ is the output of the correct (faulty) cell on input I. A **test** for the cellular fault $(\rho, I, X/Y)$ must apply the input I to ρ and propagate the difference X/Y to a primary output. A fault (in one of both models) is **detectable**, iff there is a test for the fault. A set of input combinations is a **complete test set** for an RCC-adder with respect to a given fault model iff it contains a test for any detectable fault in this model. Let $Cl \subset RCC$. A set of input combinations is a **uniform complete test set** for Cl with respect to a given fault model, iff it is a complete test set for any RCC-adder in Cl. In the following we develop upper and lower bounds for (uniform) complete test sets for classes $CL \subset RCC$.

It follows directly from the definition that tests in the cf-model do not depend on the encoding of the types. In addition, it can be shown that the correct behaviour of an RCC-adder in the cf-model implies the correct behaviour in the sa-model, but not vice versa. Thus, the cf-model seems to be a convenient fault model for the test pattern construction for RCC-adders on a higher design level. On the other hand it turns out that some technical details have to be considered: The set of detectable faults depends on the choice of the redundancies; since (detectable) differences X/R have to be propagated to the outputs, we have to make some assumptions about the redundant entries in the function tables of the β- and γ-cells. In the sa-model we have to deal with differences $1/0, 0/1$ on signal lines bearing digital values. Thus, a test in the sa-model normally has to be constructed on the digital level. In the case of RCC-adders, digital differences on digital lines can be translated to symbolic differences on symbolic lines in a way, which makes the construction of tests totally independent of the encoding and the choice of the redundancies. It can be further shown that all stuck-at faults are detectable (i.e., the set of detectable faults is clearly independent of the encoding and the redundancies). In this sense a complete test in the sa-model requires less assumptions than in the cf-model.

We give results for the test complexity with respect to both fault models. The translation of the differences from the digital level to the symbolic level, which is the starting point for our test pattern construction in the sa-model, is mainly performed in the following lemma. The proof of this lemma is omitted. (It requires a close study of all possible encodings and their properties.)

Lemma 1. Let *Add* be an RCC-adder. i) A test of all (symbolic) lines in the RCC-part of *Add* for the differences P/A, G/A, G/P, A/P, P/G, A/G is sufficient for a complete test of the RCC-part in the sa-model.

ii) A complete test in the sa-model must check at least one difference $X/Y \in \{G/A, A/G, G/P, A/P\}$.

For the rest of the paper we argue on the symbolic computation level. First of all we derive lower bounds. For that define the **depth** of an RCC-adder to be the maximum number of β-cells on a path from one of its inputs to one of its outputs. The **left depth** is the maximum number of left input edges to β-cells on a path from one of the inputs to one of the outputs. Then the following lemma holds

Lemma 2. Let *Add* be an RCC-adder with left depth d_l and depth d. Then d_l and $2d$ are lower bounds on the test size of *Add* with respect to the sa-model and the cf-model respectively.

Proof. We give the proof in the case of the cf-model, i.e., we show that $2d$ is a lower bound. (For the sa-model similar considerations using lemma 1.ii are necessary.) Let p be a path in *Add* with depth d. We attach 2 input combinations to each of the β-cells on p by the following rule: *The path input of a cell is marked by P, and the input which does not belong to the path is marked by G and A.* Denote an input combination to a β-cell by a word of length 2 over the alphabet $\{A, P, G, R\}$. Then 2 input combinations from the set $\{GP, AP, PG, PA\}$ are attached to each β-cell of the path by the above rule. A complete test set must apply these input combinations to the corresponding cells. Since R cannot be generated in the correct adder, the type P can only be computed as $P \triangleright P$. Thus a test pattern which applies P to the output of a β-cell must 'colour' the whole subtree corresponding to this cell with P. We conclude that no two of these input combinations can be applied by a single test. ∎

It is mainly the application of this lemma which leads to the lower bound theorem.

Theorem 1. The following lower bounds hold for classes of RCC-adders

| | **lower bounds** | |
class of RCC-adders	sa-model	cf-model
RCC	$n - 1$ *	$2(n-1)$ *
	2	$2log(n)$
$RCC(O(log(n)), -, -)$		n *
	$\Omega(log(n))$	$2log(n)$
BK	$log(n)$	n
CBT	$log(n)$	$2log(n)$

The following comment is necessary to understand of the above table: A lower bound which is supplied with a "*" stands for a bound on the size of a uniform complete test set for this class. The remaining entries give lower bounds on the size of a (not necessarily uniform) complete test set for any element in a given class. (They are naturally less or equal than the size of a uniform complete test set.)

Proof of theorem 1. The lower bounds for RCC, the $2log(n)$-lower bound for $RCC(O(log(n)), -, -)$, the $log(n)$-lower bound for BK and the lower bounds for CBT follow easily from lemma 2. The n *-lower bound for $RCC(O(log(n)), -, -)$ is directly implied by the n-lower bound for BK. For the proof of the n-lower bound for BK explicit use is made of the BK-structure and of the fact that no two different nodes $\beta_{[i,0]}$ and $\beta_{[j,0]}$ ($i, j \neq 0$ and even) can be tested for the input combination AP and GP at the same time. The $\Omega(log(n))$-lower bound requires the (lengthy) calculation of the left depth of trees with depth $O(log(n))$ and then follows by with lemma 2. The details of both proofs are omitted in this paper. ∎

We give a short interpretation of theorem 1 and point out some questions. The first row of the above table shows that any uniform complete test set for the class RCC must contain at least $n - 1$ tests. On the other hand, the size of a complete test set for an element in RCC can (only) be bounded from below by 2 (in the sa-model) and $2log(n)$ (in the cf-model). Thus, depending on the fault model, RCC-adder with optimal test complexity may have tests with a constant or logarithmic number of patterns. Th leads to the question, as to whether there are any interesting subclasses of RCC, e.g., classes of optima time adders, which achieve these lower bounds with uniform test sets or vice versa, if the lower bounds rise for specific subclasses. (Note that a precise knowledge of the functions f_β, f_γ, i.e., the restriction to a subclass $RCC(-, -, (f_\beta, f_\gamma))$, does not change the lower bounds proven for RCC. We therefore concentrate on subclasses which are specified by structural properties.)

The lower bound of 2 in the sa-model can not be achieved by optimal time adders. This follows directly rom the $\Omega(log(n))$-lower bound. (Optimal time adders do not have minimum left depth in RCC.) The est we can hope to obtain for optimal time adders, e.g., the two well-known classes BK and CBT, are uniform complete test sets with a logarithmic number of patterns. In the cf-model the situation is different. Optimal time adders are candidates for the smallest test sets since they have (asymptotically) minimum depth in RCC. One class of optimal time adders, the class BK, is eliminated by the above table. The size of any complete test set for an element in BK is at least n. We have a lower bound of $2log(n)$ for another class of optimal time adders, the class CBT. Thus, there is some hope that CBT is a subclass which achieves the lower bound $2log(n)$ for RCC with a uniform complete test set.

An answer to all these questions is provided by the effective construction of uniform complete test sets. As mentioned in the introduction we do not give the complete construction for all subclasses in this paper. But we want to give an idea of the most important methods used: Firstly, the problem of testing an RCC-adder is partitioned into subproblems. It can be shown that the Preparation Part and the Final Addition Part are easily testable in both fault models because of their linear structure. (E.g. a set of eight test patterns is sufficient for both parts in the sa-model.) Thus, the test problem for the whole RCC-adder is reduced to the construction of a test set for the RCC-part and the 'embedding' of this test in the whole adder. In the cf-model this is rather complicate and requires the consideration of some technical details. In this paper, we restrict to the sa-model and give the main ideas for the construction of embeddable tests in this case. An input combination to the RCC-part is specified by a word $(T_{n-1} \ldots T_0)$ with $T_i \in \{A, P, G, R\}$. We construct test sets $test_{sa}$ for the RCC-part with the following properties: The test patterns are words over $\{A, P, G\}$, and they check the differences P/A, G/A, G/P, A/P, P/G, A/G on each edge in the RCC-part. It follows from lemma 1 that a set $test_{sa}$ with these properties is a complete test set for the RCC-part. Also the embedding of $test_{sa}$ can now be easily accomplished: Firstly, generate the patterns in $test_{sa}$ by primary input combinations of the adder. (Note that any combination of the types A, P, G can be generated as output of the Preparation Part.) Then combine the resulting patterns with $c_{-1} = 0$ and $c_{-1} = 1$. This guarantees the propagation of the difference to the primary outputs of the adder if there is a fault in the RCC-part. Altogether we obtain a complete test set of size $2|test_{sa}| + 8$ for the whole RCC-adder, if the set $test_{sa}$ for the RCC-part is given.

The construction of complete (uniform) test sets for RCC-parts is heavily based on structural properties of the computation scheme. The details change with different computation schemes and fault models We restrict ourselves to one example which provides some of the main ideas and, at the same time, i not affected by too many technical details.

Lemma 3. For RCC-parts with BK-structure, there exists an embeddable, complete test set (in th sa-model) with $2log(n) + 7$ test patterns.

Proof. Tests for the differences given in lemma 1 have to be constructed. The tests split into 3 groups The first group consists of the pattern (P^n). Since P is a unit element with respect to '$\triangleright$', it follow immediately that the differences P/A and P/G are checked on all edges in the RCC-part.

The second group is responsible for the tests of G/P, A/P, G/A, A/G on left input edges of β-cell and on the output edges. We construct 4 patterns which apply AG and GA to each β-cell in the BK-scheme: Use the equations $A \triangleright G = A$, $G \triangleright A = G$ to generate AG and GA as inputs to $\beta_{[n-1,0}$

in_{AG}.) in_{AG} and in_{GA} do not work in the lower half. Look at figure 7.b to realize that the lower part of the BK-scheme corresponds to an incomplete binary tree of size 2^m, which is reversed and translated one bit position to the right, compared to the complete binary tree in the upper part. From this we may conclude that a reversion of the patterns in_{AG}, in_{GA} and a translation by one bit position to the right applies AG and GA to the β-cells in the lower half. (Note the following: if we denote the reverse of in_{AG} and in_{GA} by in_{AG}^{rev} and in_{GA}^{rev}, respectively, we obtain with induction $\{in_{AG}^{rev}, in_{GA}^{rev}\} = \{in_{AG}, in_{GA}\}$. Thus, two patterns performing the reversion and translation are, e.g., $[A\,in_{AG}]_n$ and $[G\,in_{GA}]_n$. The notation $[\ldots]_n$ is used to define replications of the same input pattern, which are cut off at the right hand side: Let $T := T_1 \ldots T_k$ ($k \in \mathbb{N}$) be a word over $\{A, P, G, R\}$. Then $[T]_n := T^{n\,\mathrm{div}\,k} T_1 \ldots T_{n\,\mathrm{mod}\,k}$.) The application of AG to a β-cell $\beta_{[l,k]}$ yields a test for A/G and A/P on the left input edge in account of the following: A/G remains unchanged by $\beta_{[l,k]}$, A/P is changed to A/G. The difference A/G is visible at output $T_{[l,0]}$ since A and G are left-stable. Analogous arguments prove that an application of GA to $\beta_{[l,k]}$ results in a test for G/A and G/P.

The third group has to test for differences G/P, A/P, G/A, A/G on right input edges of β-cells. With analogous arguments as in the lower bound proof of lemma 2 we know that at least $log(n)$ test patterns are necessary in this case. We show that $2log(n) + 2$ are sufficient. First of all, test patterns for the complete tree in the upper half of the BK-scheme are constructed. Consider level j, $1 \leq j \leq m$, of the complete binary tree and the two patterns $[P^{2^{j-1}} P^{2^{j-1}-1} A P^{2^{j-1}} P^{2^{j-1}-1} G]_n$, $[P^{2^{j-1}} P^{2^{j-1}-1} G P^{2^{j-1}} P^{2^{j-1}-1} A]_n$. These patterns apply PA and PG to the β-cells of level j in alternating order starting with PA or PG at the leftmost cell $\beta_{[n-1,n-1-(2^j-1)]}$. (In figure 7.c the pattern $[P^{2^{j-1}} P^{2^{j-1}-1} A P^{2^{j-1}} P^{2^{j-1}-1} G]_n$ for $j = 2, n = 16$ is represented.) We show that with these patterns the differences A/G, G/A, A/P and G/P are checked on the right input edges of level j.

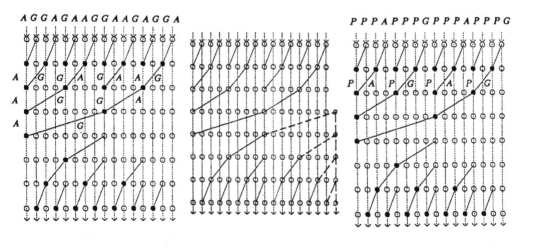

figure 7.a   figure 7.b   figure 7.c

Consider the pattern which applies PA (PG) to cell $\beta_{[l,l-(2^j-1)]}$ of level j. We prove that this pattern is a test for the difference A/G (G/A) at the right input edge of $\beta_{[l,l-(2^j-1)]}$. Since P is the value on the left input edge, the difference A/G (G/A) is passed through to the output of $\beta_{[l,l-(2^j-1)]}$. A and G are left-stable, so A/G (G/A) is propagated identically to the output $T_{[l,0]}$.

Propagation of the difference A/P (G/P) is possible because of the alternating sequence of inputs on level j. A difference A/P (G/P) at the right input of cell $\beta_{[l,l-(2^j-1)]}$ of level j leads to the difference A/P (G/P) on the output of $\beta_{[l,l-(2^j-1)]}$. Two cases have to be considered. Either the difference is directly propagated to the output $T_{[l,0]}$ (without passing through any other β-cell) or several β-cells have to be passed on the path to the output $T_{[l,0]}$. In this case consider the next β-cell $\beta_{[l,k]}$ on this path. We show that the difference A/P (G/P) is changed to A/G (G/A). The input edges of $\beta_{[l,k]}$ are $T_{[l,l-(2^j-1)]}$ and $T_{[l-2^j,k]}$. On the left input edge $T_{[l,l-(2^j-1)]}$ we have the difference A/P (G/P). Since the RCC-part has BK-structure, we know that $(l-2^j) - k \geq 2^j - 1$ (i.e., if two blocks are combined, then the block on the right is at least as long as the block on the left). Thus for the right input we have the equation $T_{[l-2^j,k]} = T_{[l-2^j,l-2^j-(2^j-1)]} \triangleright T_{[l-2^{j+1},k]}$. According to our construction the value of $T_{[l-2^j,l-2^j-(2^j-1)]}$ is G (A). Therefore the value of the right input edge $T_{[l-2^j,k]}$ of $\beta_{[l,k]}$ is also G (A), which implies the difference A/G (G/A) at the output of $\beta_{[l,k]}$. As above we then conclude that A/G (G/A) is propagated to $T_{[l,0]}$.

Altogether we get $2log(n)$ test patterns for the upper tree. The test of the lower tree for the differences $A/G, G/A, A/P$ and G/P requires the two additional test patterns $(P^{n-1}A)$ and $(P^{n-1}G)$. The β-cells of the lower tree only compute types $T_{[i,0]}$. Thus, it is clear that the two patterns from above apply the inputs PA and PG to them. Since the output value of any β-cell in the lower tree is directly propagated to an output of the RCC-part, these two patterns are sufficient to check the differences $A/G, G/A, A/P$ and G/P on right input edges of β-cells in the lower tree of the RCC-part. Thus, $test_{sa}$ consists of $2log(n) + 7$ patterns. Since $test_{sa}$ is embeddable, we obtain a uniform complete test set for BK of size $4log(n) + 22$. ∎

Lemma 3 proves one entry of the table given in the following theorem. This theorem provides the corresponding upper bounds to the lower bounds of theorem 1. In the full paper all entries are verified by constructions similar to that of lemma 3. Here we merely present the list of upper bounds.

Theorem 2. There are uniform complete tests for classes of RCC-adders which imply the following upper bounds

upper bounds

class of RCC-adders	sa-model	cf-model
RCC	$4n + 8$	$12n$ [R]
$RCC(O(log(n)), -, -)$	$4n + 8$	$12n$ [R]
BK	$4log(n) + 22$	$12n$ [R]
CBT	$8log(n) + 8$	$16log(n) + 16$ [R]

The index R on the upper bounds in the cf-model indicates that these entries are only valid if some technical properties about the choice of the redundancies are fulfilled. As mentioned above, these properties are necessary for the propagation of differences X/R and thus allow a generalization of the methods used in the sa-model to the case of the cf-model. The redundancy properties are introduced and discussed in detail in the full paper. In this paper we only state that a large variety of RCC-adders, e.g., all adders in CC and CLA fulfill them.

We finish with a short discussion of the results. From theorem 1 we conclude that all bounds exhibited in theorem 2 (with the exception of one) are asymptotically optimal with very small constants. Even if no information about structural properties of the RCC-part is given, we are able to construct a uniform test of linear size in both fault models. If some structural information is given, we obtain better results. A restriction from RCC to $RCC(O(log(n)), -, -)$ only changes the lower bound in the sa-model. This proves that RCC-adders with optimal time complexity do not have the minimum test sets in the sa-model. (A linear time adder with left depth 1 is testable with a constant number of patterns!) The problem of achieving the $\Omega(log(n))$ bound with a uniform test set for this class is unsolved. In the cellular fault model the linear test (constructed for RCC) still forms an asymptotically optimal uniform test. We give uniform test sets of size $4log(n) + 22$ (sa-model) and $12n$ (cf-model) for the class BK. These tests achieve the lower bounds. Thus, BK is a class in which optimal tests in different fault models have different complexities, i.e., a test in the cf-model is of higher complexity than a test in the sa-model. A second important class of optimal time adders, the class CBT, has different properties. We again construct optimal uniform tests, which achieve the lower bounds for any test of an element in CBT. In this case both tests have cardinality $O(log(n))$, i.e., in an asymptotic sense a test in the sa-model is as hard as a test in the cf-model. If we restrict our attention to the cf-model CBT specifies a subclass in RCC, the elements of which have optimal time complexity and optimal test complexity $O(log(n))$ at the same time. As an example, consider the set of carry look-ahead adders and the set of conditional carry adders with CBT-structure, denoted by $RCC(CBT, -, (f_\beta^{CLA}, f_\gamma^{CLA}))$ and $RCC(CBT, -, (f_\beta^{CC}, f_\gamma^{CC}))$.

Finally, we should at least mention the following interesting points. To keep the paper short, a discussion of these points is omitted:

-If one allows the generation of the redundant type R at the inputs of the RCC-part, it can be shown that there exist constant tests for the classes BK and CBT with respect to the stuck-at fault model. This (again) implies optimal time complexity and optimal test complexity at the same time.

-Our approach can be extended to multiple fault detection in RCC-adders.

-On-chip generation and evaluation of the test patterns is possible in a comfortable way.

References

[AbRe] M.S.Abadir, H.K.Reghbati: 'Functional Testing of Semiconductor Random Access Memories', Computing Surveys 15, 1983, pp. 175-198

[AbGa] J.A.Abraham, D.D.Gajski: 'Design of Testable Structures Defined by Simple Loops', IEEE Tr. C-30, 1981, pp. 875-883

[Be1] B.Becker: 'An Easily Testable Optimal-Time VLSI Multiplier', ACTA INF. 24, 1987, pp. 363-380

[Be2] B.Becker: 'Efficient Testing of Optimal Time Adders', Proc. MFCS 86, Bratislava, LNCS 233, pp. 218-229

[BeKo] B.Becker, R.Kolla: 'On the Construction of Optimal Time Adders', Proc. 5th STACS 88, Bordeaux, LNCS 294, pp. 18-28

[BeSo] B.Becker, H.Soukup: 'CMOS Stuck-Open Self-Test for an Optimal-Time VLSI-Multiplier', The Euromicro J., Microproc. and Microprog. 20, 1987 pp. 153-157,

[BeSp] B.Becker, U.Sparmann: 'A Uniform Test Approach for RCC-Adders', T.R., 8/1987, SFB 124, Saarbrücken 1987

[BrKu] R.P.Brent, H.T.Kung: 'A Regular Layout for Parallel Adders', IEEE Tr. C-31, 1982, pp. 260-264

[Br] M.A.Breuer, Ed.: 'Diagnosis and Reliable Design of Digital Systems', Woodland Hills, CA: Computer Science Press, 1976

[DaMu] W.Daehn, J.Mucha: 'A Hardware Approach to Self-Testing of large PLA's', IEEE Tr. CAS-28, 1981, p. 1033

[EiWi] E.B.Eichelberger, T.W.Williams: 'A Logic Design Structure for LSI Testability', J. Design Autom. Fault-Tolerant Comput. 2, 1978, pp. 165-178

[FeSh] J.Ferguson, J.P.Shen: 'The Design of Two Easily-Testable VLSI Array Multipliers', Proc. 6th Symp. on Comp. Arith., 1983, Aarhus, pp. 2-9

[Fr] A.D.Friedman: 'Easily Testable Iterative Systems', IEEE Tr C-22, 1973, pp. 1061-1064

[Ho] G.Hotz: 'Zur Reduktionstheorie der booleschen Algebra', Colloquium über Schaltkreis- und Schaltwerk-Theorie, (1960) Herausgeber: Unger H., Peschel E., Birkhäuser Verlag 1961

[Ka]   W.H.Kautz: 'Testing for Faults in Cellular Logic Arrays', Proc. of 8th Symp. on Switch. Autom. Th., 1967, pp. 161-174

[KMZ] B.Koenemann, J.Mucha, G.Zwiehoff: 'Built-In Logic Block Observation Technique', Proc. IEEE Test Conf., 1979, pp. 37-41

[LaFi] R.E.Ladner, M.J.Fischer: 'Parallel Prefix Computation', Proc. of Intern. Conf. on Par. Proc., 1977, pp. 213-223

[Mi]   B.Milne: 'Testability, 1985 Technology Forecast', Electronic Design, January 10, 1985, pp. 143-166

[MoAb] R.K.Montoye, J.A.Abraham: 'Built-in Tests for Abitrarily Structured VLSI Carry Look-Ahead Adders', IFIP 1983, pp. 361-371

[Sk]   J.Sklansky: 'Conditional-sum Addition Logic', IRE-EC 9, 1960, pp. 226-231

[Sm]   J.E.Smith: 'Detection of Faults in PLA's', IEEE Tr. C-28, 1979, p. 845

Parallel Simulation and Test of VLSI Array Logic

Pradip Bose

IBM T. J. Watson Research Center, H2-B48
P. O. Box 704, Yorktown Heights, NY 10598

Abstract

We consider the task of accelerating existing array logic CAD software by running them under parallel processing environments. We restrict our discussion to VLSI programmable logic arrays (PLAs); we present results based on fault simulation/test generation software implemented for PLAs. Due to the regular layout of such array logic, considerable parallelism is expected in any data processing problem involving these modules. However, depending on the underlying machine model, and the adopted programming model, varying levels of speed-up are exhibited. We first consider a typical pipelined, vector mainframe like the IBM 3090 VF. Through a combination of novel problem reformulation strategies, and educated coding guidelines for vector machines, we show how an original scalar program can be efficiently speeded up on such machines. The other machine model considered is that of a shared memory parallel processing system. Our results here are based on experiments run under an experimental multi-processor VM environment (VM/EPEX) which has been set up to predict the possible speed-up on a parallel processor such as RP3. The overall processing environment is assumed to be constrained to a scientific/engineering one, with Fortran as the primary coding medium and the hardware biased toward numerically-intensive applications.

I. Introduction

With the growing availability of parallel processing systems, the need and motivation for adapting existing application programs and algorithms to run efficiently on the newer machines, is quite obvious. The fundamental promise of achieving high speed-up of computationally intensive applications through pipelined (vector) processing and/or multi-processing, would be fruitless without educated programming methodologies which adequately exploit the underlying hardware. A poorly coded "parallel" or "parallelizable" program is not unlikely to achieve *less* performance than the sequential version. In the specific domain of applications encompassing CAD and fault simulation/ test generation of VLSI, special-purpose hardware accelerators[1], will make specialized application 'programs run faster. Using parallel computation schemes, hardware approaches have demonstrated the possibility of performance enhancement by a factor of 1000 over conventional software solutions. However, what remains to be fully exploited is the power of *existing* scientific supercomputers via software means. Recently reported work[15] indicates that there is a potential for very high ratios of speed-up, by adapting current CAD solutions to run on vector processors. Most of the modern supercomputers, such as the Cray machines[23], the IBM 3090-VF[24] or the (proposed) RP3 computer[25] are general-purpose *scientific/engineering* machines, but their core hardware is geared toward *numerically* intensive scientific/engineering codes. Thus, when it comes to CAD software, a straightforward mapping from sequential to (possibly) vector or parallel code will not always be able to fully exploit the hardware concurrency available.

In this paper, we restrict our attention to the problem of fault simulation and test generation of programmable logic arrays (PLAs). PLAs are representative of a broader class of array logic modules[2]. Their role as versatile, easy-to-use, programmable modules in large scale logic design is well established. There is also a great body of existing knowledge in the area of test generation and design for testability of PLAs (e.g.[3-13]). Moreover, due to the inherent structure and regularity of PLAs, considerable parallelism is expected in any data processing problem involving PLAs. Thus, it would be interesting to investigate the techniques for accelerating known fault simulation and test generation methodologies by adapting the programming model to the underlying concurrent machine model. As will be brought out by the results in this paper, intelligent exploitation of the hardware and software mechanisms available leads one to the goal of desired speed-up, without having to resort to more difficult tasks like word-oriented, bit-wise parallel simulation methodologies, which usually require lower-level (viz., assembly language), programming aids and models.

We present the results of our investigation in three steps. In section II, we review the basic concepts of PLA fault modeling, fault simulation and test generation. In section III, we consider pipelined, vector processing as the first move up from conventional scalar processing. We present a novel technique of recasting the problem formulation into a computation structure (matrix multiplication) which is extremely well supported in such machines. The speed-up obtained is many times larger than would be obtained otherwise. In section IV, we consider a typical shared memory parallel processing machine model, such as the RP3. Our results in that section are based on parallelized codes run under an experimental multi-processor VM environment (VM/EPEX) to predict the possible speed-up on a parallel processor like RP3. Some performance results depicting the efficacy of techniques described in sections III and IV, are presented in section V. Due to lack of space, the analysis and results are presented in brief; a more detailed treatment is available in a research report[26].

II. PLA Fault Simulation and Test Generation

In this section we briefly review the basic nature of the problems of fault simulation and test generation of PLAs. Based on the insight gleaned from such analysis, we shall present, in sections III and IV, pragmatic yet novel techniques for *acceleration* of these problem solution methodologies on vector and parallel machines.

The Crosspoint Fault Model

In this paper, we restrict ourselves to the crosspoint fault model traditionally used for PLAs, in which a single fault causes a programmed 0(1) in the input or output array to change to a 1(0). This physical fault model is equivalent to a logical fault model in which the product terms are subject to growth (G), shrinkage (S), disappearance (D) and appearance (A) faults[5,6]. As is well-known in the literature, test generation under this fault model is comprehensive enough to cover almost all of the other types of faults of interest in PLAs. Moreover, the general techniques for speed-up under concurrent processing environments (as presented in sections III and IV), are general enough to cover fault models which are broader than the basic crosspoint fault model.

Approaches to Fault Simulation and Test Generation

Conceptually, and logically, a PLA may be viewed as a collection of 2-level AND-OR networks. However, as has been pointed out by several authors, the 2-level AND-OR network model of a PLA does not represent the complete PLA structure when it comes to fault simulation and test generation. In particular, the class of crosspoint faults, which is unique to PLAs is not covered by a minimal stuck-at fault test set derived on the basis of a 2-level AND-OR structure. One possibility is to remodel the logic of a PLA in terms of a combinational network such that each single crosspoint fault in the actual PLA is functionally equivalent to some single stuck-at fault in the modeled network. Agarwal[10] has shown that such a logic remodeling for the purposes of PLA fault analysis and test generation is indeed feasible; the *stuck-at-equivalent* (SAE) network, as referred to by Agarwal, is a *4-level* AND-OR network. The main intent of such logic remodeling in Agarwal's work[10], was to analyze the multiple fault coverage of single fault test sets. However, one approach to fault simulation and test generation for PLAs is to apply traditional[14] techniques to such a functionally equivalent random logic network. Early approaches (e.g.[3]) were precisely along this path. The enormous computational burden presented by such an approach is discussed by Ostapko and Hong[4]. Although some speed-up is obtainable by attempting to vectorize or parallelize such algorithms, the basic nature of this approach is not well-matched to concurrent computational structures of interest to us in this paper (see sections III and IV); in other words, such attempts quickly become computationally intractable with increasing PLA size, even under concurrent processing modes. Approaches such as the one adopted by Ostapko and Hong[4], on the other hand, vectorize and/or parallelize much better, provided suitable bit-wise vector/parallel logic operation primitives are available in hardware, with corresponding software (compiler) support. Typical *supercomputers*[23-25] do not provide such architectural support. While Ostapko and Hong's approach[4] calls for fault simulation at the level of 1's and 0's of the PLA personality, later approaches[5,6] have shown that modeling of the faults at the logical level in terms of G, S, D and A faults, enables simplified fault analysis and test generation procedures. As shown later on in this paper, such an approach lends itself well to concurrent methodologies available under scientific supercomputers.

Brief Review of Our Approach

In the following, we present a brief review of relevant aspects of the theory behind fault simulation and test generation for PLAs, based on our earlier work[6-8]

Notations and Definitions

A PLA with n variables, m product terms and p output functions is referred to here as a (n,m,p)-PLA. This general PLA organization has $m(2n+p)$ crosspoints or contacts. We use an $(n+p)$-tuple notation for the rows· (cubes) of a (n,m,p)-PLA. This notation is of the form $(y_1y_2...y_n, z_1z_2...z_p)$, with $y_i \in [0,1,x]$ and $z_j \in [0,1]$. The n-tuple $(y_1y_2...y_n)$ is the product term (or, "implicant," or "input part," or simply, "term"), and the p-tuple $(z_1z_2...z_p)$ is the corresponding output connection vector (OCV), or "output part." Figure 1 shows the symbolic description of a small (4,4,3)-PLA implementing three switching functions: $f_1 = x_2\bar{x}_3x_4 + \bar{x}_1x_2x_4$, $f_2 = x_1\bar{x}_3x_4 + x_1\bar{x}_2 + \bar{x}_1\bar{x}_2x_4$ and $f_3 = x_2\bar{x}_3x_4 + x_1\bar{x}_2$. We shall use this PLA as a running example for illustrating specific techniques in test generation and logic/fault simulation.

$$
\begin{array}{ll}
\text{1x01} & \text{010} \\
\text{x101} & \text{101} \\
\text{10xx} & \text{011} \\
\text{00x1} & \text{110}
\end{array}
$$

Figure 1. Example PLA Specification

A growth (G) term is an n-tuple $G^{(n)} = (g_1g_2...g_n), g_i \in [0,1,x]$, denoting the set of extra minterms contributed by a G-fault. An appearance (A) term is defined similarly. A shrinkage (S) term is an n-tuple $S^{(n)}$ denoting the set of minterms which disappear on a S-fault. A disappearance (D) term is the n-tuple $D^{(n)}$ denoting the product term which disappears from the Karnaugh map on a D-fault. A growth (G), or shrinkage (S) cube is, respectively, a G or S term augmented (or concatenated) with the OCV of the corresponding source cube of the PLA. For example, if 1x01 011 is a cube (i.e., row) of a PLA (n = 4, p = 3), then, 0x01 011 is a G-cube, where 0x01 is the G-term; 1001 011 is an S-cube, where 1001 is the S-term. A disappearance (D), or an appearance (A) cube is a D or A term augmented with an OCV denoting a propagated single fault. Thus, for the example input cube 1x01 011 the two possible D-cubes are: 1x01 001 and 1x01 010, whereas, the only possible A-cube is 1x01 100. For a given output function, a minterm of a product term is said to be free if it is not covered by any other product term of the function under consideration. Two terms $A^{(n)}$ and $B^{(n)}$ are said to be compatible if there is no i $(1 \leq i \leq n)$ for which $a_i = 0$ and $b_i = 1$ or vice versa. The xcount of a product term is the number of x's in its symbolic specification. Let C denote the set of m cubes, and P the set of m product terms of a (n,m,p)-PLA. We use the symbols G, S, D and A to denote the sets of growth, shrinkage, disappearance and appearance faults, respectively, of a PLA. Also, let G_c, S_c, D_c, and A_c denote, respectively, the set of G, S, D and A cubes of the PLA.

Generation of G, S, D and A Tests

Clearly, any free minterm covered by a G-term qualifies as a test vector for the corresponding G-fault. For a given set of G-terms, a set of free minterms, chosen to cover all the G-terms is a (single) G-fault test set. Similarly, one can define S, D and A test sets. The following two lemmas are representative of the results which follow from an understanding of the basic methodologies required for generating G, S, D and A tests from individual product terms. These results also help in establishing overall bounds on test set size.

Lemma 1: The number of G-tests, n^g, needed for a product term t is bounded as follows:
$$0 \leq n^g \leq k,$$
where, k = number of literals (non-x elements) of t. ∎

An efficient algorithm to generate G-tests from G_c, called the **tabular covering method** has been used in our implemented programs; a brief illustration of this procedure was presented earlier[6], and is described in detail in our report[8].

Lemma 2: The number of S-tests, n^t, needed for a general product term t is bounded as follows:
$$n^t \le n - k + 1,$$
where, n = number of input variables, k = number of literals in t. ∎

The heuristic methods to generate a minimal S-test set for a (n,m,p)-PLA, using the "opposite-adjacent" method has been described earlier[6]. It can be shown that in the vast majority of cases, the upper bound on the number of S-tests for a product term of *arbitrary* size is only 2.

Minimal Test Set

The fundamental steps to generate a minimal T_t constitute a minimal *covering* problem[8]; we refer to this as **Procedure 1.** This problem is known to be NP-complete. Faster, heuristic procedures which generate close to minimal test set sizes are often more desirable than optimal algorithms. Several such methods have been studied and some of them implemented as part of our PLA design and test software[8]. We present below *one* such procedure which has proven to be very efficient.

Procedure 2: Faster, near-minimal T_t generation.

<u>Step A:</u> Derive a minimal (S + D) test set T^{S+D}, using the "opposite-adjacent" method.

<u>Step B:</u> Derive a minimal (G + A) test set T^{G+A}, using the "tabular covering" method referred to earlier.

Then, $T_t = T^{S+D} \cup T^{G+A}$.

Bound on size of a T_t

Theorem 1: For a general (n,m,p)-PLA, the size of a T_t covering all single G, S, D and A faults is bounded as follows:
$$|T_t|^{G+S+D+A} \le m(n + p) .$$

Thus, for a PLA with 16 input variables, 48 product terms and 8 output functions, $|T_t| \le 1152$.

Fault Simulation

As discussed by Ostapko and Hong[4], test generation methodologies for VLSI designs often require the inclusion of some fault simulation capabilities. This is especially true for those testing methodologies which support self-test or pseudo-random test generation algorithms. Also, given a set of test patterns generated by a deterministic methodology, simulation capability helps determine *fault coverage*. In the case of PLA testing, a combination of deterministic test generation and fault simulation is usually used in implemented software (e.g.[4,9]). Even under strictly deterministic techniques[5-8] implied *logic* simulation is repeatedly used to solve covering problems. The basic mechanism behind all fault simulation is the underlying logic simulation of correct and faulty copies of the circuit; hence, we first consider the *logic simulation problem* for PLAs:

In the highest level of description, the PLA logic simulation problem is no different than that for an arbitrary combinational logic circuit: namely, given the Boolean equations describing the output function(s), and a set of input test vectors, we want to determine the set of correct (fault-free, design-intended), output vectors. A simple one-for-one mapping from such a functional description to the corresponding 2-level AND-OR network or the corresponding PLA personality layout, does not change the problem in any way. The only reason one would want to do a gate-level logic simulation (for the AND-OR network) or a bit-level logic simulation (for the PLA) is for verification of the correctness of *actual implementation*. Now, in the case of a PLA, it should be clear from our earlier discussion in section II, that given the nature and extent of the crosspoint fault model, it suffices to consider a symbolic form representation of the PLA (see Figure 1c) for all considerations of logic and fault simulation. In other words, there is no extra advantage or information to be gleaned by resorting to a detailed bit-level personality layout model or an "equivalent" logic network model of the PLA.

Referring to our example PLA (Figure 1), the problem at hand then is: given a (n,m,p)-PLA described symbolically and given t binary n-tuples, representing a set of t input test vectors, we want to determine the corresponding set of t binary p-tuples, representing the fault-free output vectors. Conceptually, the problem can be broken down into the following iterative sequence of operations:

1. Take the next input test vector (n-tuple) and compare it, position-wise, with each of the m input parts (also n-tuples) in sequence.

2. Mark those input parts which are compatible with the test vector under question, and select the corresponding OCVs.

3. Compute the bit-wise OR of the OCVs selected in step 2. This gives the desired (correct) output vector. Add it to the list of output vectors.

4. If there are more input test vectors to be considered, go to step 1; otherwise, we have our set of m output vectors (p-tuples) and the procedure ends.

Clearly, if the underlying model of execution model is strictly sequential, with the lowest operation primitive being a (single, 2-ary) position-wise comparison (for input parts) or a bit-wise logical OR (for OCVs), then the worst case computational complexity involved, in terms of the number of primitive operations is: $\sim O(n.m.t + m.p.t)$. This indicates that the complexity increases linearly with each of the parameters of concern: n, m, p and t; i.e., if one parameter is increased, keeping the others fixed, then the number of operations increases in a linear fashion. Typically, of course, we are interested in certain *kinds* of test vectors, say those derived by a deterministic test generation algorithm. In this case, t depends on n, m and p. For example, if we consider our own test generation methods, highlighted in section II, a realistic upper bound on t is $m(n + p)$ [see Theorem 1 in section II]. In that case, the above complexity becomes: $\sim O(m^2(n + p)^2)$. We now see a polynomial effect in n and m. Obviously, however, there is a lot of potential parallelism in the above steps, which may be exploited. For example, the position-wise comparison operation (step 1) may be done as a pipelined, vector operation or it may be performed in parallel. Similarly, at the next level, each of the n-way comparisons of step 1 may be pipelined or parallelized to achieve speed-up. Acceleration of all the steps via vector/pipelined or parallel computation modes are clearly possible. It is not our intention here to resort to a theoretical analysis of the possible improvement in time complexity under various concurrent processing modes, although such an analysis is clearly possible using known techniques[19]. Instead, in this paper we are interested in dealing with the real constraints imposed by actual scientific machines like the IBM 3090VF or RP3 and devising high-level programming techniques which exploit their underlying hardware concurrency in the best possible manner. The next two sections deal with this objective of achieving best speed-up for the PLA fault simulation/test generation problem under actual vector and parallel machines.

Before proceeding, however, we would like to point out the most usual constraint we have to deal with when it comes to solve CAD problems of our nature on typical scientific supercomputers: namely, the lack of hardware and/or software support for doing logical (Boolean or comparison) operations in vector mode. Such support is usually for numerical (arithmetic) operations only. Thus, our *primary* problem is to map the conceptual tasks and procedures outlined in this section, into well-structured *arithmetic* operations and functions: those which are well-supported by the supercomputer of our choice.

III. Exploiting Concurrency in Scientific Vector Machines

We shall base our discussion in this section on the assumption that the target computer is a typical scientific mainframe like the IBM 3090-VF[24], although the general techniques employed apply to all other machines of this class.

PLA Logic and Fault Simulation

In order to recast the problem definition to suit our machine constraints, our attempt must be to formulate the operations and tasks in terms of vector MULTIPLYs and ADDs: primitives which are well-supported. At the algorithm or high-level coding level, this means we should be thinking in terms of operations like vector dot product, scalar-vector products, matrix-vector or matrix-matrix operations, etc.

Taking this view, the basic (*logic* simulation) problem statement can be reduced to:

1. Given the (m x n) *PLA input matrix* I, and a (n x t) *test vector matrix*, V, derive an (m x t) *activation matrix*, A; this matrix should indicate, in some manner, the product terms activated by each test vector.

2. Given the (t x m) matrix A^T, and the (m x p) *PLA output matrix* O, derive the (t x p) *result matrix*, R.

The result matrix R should give the set of t binary p-tuples representing the set of correct output vectors obtained by applying the t input test vectors to the PLA. In order to use integer *matrix multiplication* to achieve this, we adopt the following technique:

The entries of the PLA input matrix, **I** are mapped onto the set of integers $[0, +1, -1]$ by using the mapping rules:
'0' → -1
'1' → +1
'x' → 0

Similarly, the entries of the test vector matrix **V** are mapped onto the set of integers $[-1, +1]$ using the mapping rules:
'0' → -1
'1' → +1

The (desired) activation matrix **A**, the (desired) result matrix **R**, and the PLA output matrix **O** are all 'bit' matrices, in that their elements carry binary information (0 or 1); hence, we would prefer to process them in that mannner, whether we use matrix multiplication schemes or 'logical' (Boolean) or 'compare' vector operations. Thus, we treat them as integer matrices, using the simple mapping rules:
'0' → 0
'1' → 1

The question facing us now is how to obtain the correct forms of **A** and **R** by using matrix multiplication methods. Clearly, multiplying the mapped (integer) matrices **I** and **V**, will not directly give us the desired (integer) matrix **A**. Instead, we will get a matrix **A'**, whose elements have integral values as given by the following lemma:

Lemma 3: The elements of the matrix **A'** obtained by performing the integer matrix multiplication $A' = I \cdot V$, are integers limited to the set $[-n, -(n-1), ..., 0, +1, ..., +n]$.

<u>Proof:</u> An element of **A'** results as the dot product between a row vector of **I** and a column vector of **V**. The minimum of the range of possible integer values occurs when (a) none of the elements involved in the arithmetic is 0; in other words, each element is either 1 or -1; and (b) if an element in one vector is 1 then the corresponding element of the other vector must be -1: this will ensure that each multiplication operation results in -1. For the maximum value to occur, condition (b) above has to be modified to read: the row and column vectors involved must be pairwise identical. Condition (a) remains unchanged. Since the length of the vectors involved in the dot product is n (the number of input variables of the PLA), the minimum and maximum values possible are clearly -n and +n respectively. That all intermediate integer values, including 0, are also possible, is obvious. ■

In order to get the desired **A** matrix from **A'**, we make use of the following lemma:

Lemma 4: The desired **A** matrix is that obtained by modifying $A'^{(i)}$, the ith row vector of **A'** ($1 \leq i \leq m$) as follows:
$A^{(i)} = (A'^{(i)} + \text{xcount(i)}) \text{ div n}$,
where xcount(i) is a vector of length t (second dimension of **A** or **A'**), each of whose elements is the xcount of the ith row of **I** (see section II for definition of xcount); the **div** operation implies integer divide and truncate (vector); **n** is the integer vector of length t, each of whose elements is the integer n (the second dimnesion of matrix **I**). ■

The formal proof of Lemma 4 is omitted here for brevity.

In order to arrive at the result matrix **R**, we use the next lemma:

Lemma 5: The result matrix is obtained by performing the matrix multiplication $R' = A^T \cdot O$, and then replacing each non-zero (positive) element of **R'** by 1 to get the final **R**.

The proof in this case is quite easy. Essentially, in this step, for each test vector, the OCVs of the corresponding activated product term must be OR-ed, bitwise. This is being simulated through integer matrix multiplication.

Summarizing, therefore, in terms of matrix manipulations, the steps involved to simulate the PLA logic simulation problem are, simply:

1. Compute $A' = I \cdot V$ (* matrix multiplication *).
2. Compute **A** from **A'** using Lemma 4; i.e., by modifying $A'^{(i)}$, the ith row vector of **A'** ($1 \leq i \leq m$) as follows:
   $A^{(i)} = (A'^{(i)} + \text{xcount(i)}) \text{ div n}$.
3. Compute $R' = A^T \cdot O$ (* matrix multiplication *).
4. Obtain **R** by replacing each non-zero (positive) element of **R'** by 1.

Example

With reference to our earlier (4,4,3)-PLA, let us simulate two test vectors (t = 2): 0101 and 1001. The result of applying steps 1-4 above to the suitably mapped integer matrices is shown below.

<u>Step 1.</u>

$$
\begin{bmatrix} 1 & 0 & -1 & 1 \\ 0 & 1 & -1 & 1 \\ 1 & -1 & 0 & 0 \\ -1 & -1 & 0 & 1 \end{bmatrix} \cdot \begin{bmatrix} -1 & 1 \\ 1 & -1 \\ -1 & -1 \\ 1 & 1 \end{bmatrix} \text{gives} \begin{bmatrix} 1 & 3 \\ 3 & 1 \\ -2 & 2 \\ 1 & 1 \end{bmatrix}
$$

$$\uparrow \qquad\qquad \uparrow \qquad\qquad \uparrow$$

$$\mathbf{I} \qquad\qquad \mathbf{V} \qquad\qquad \mathbf{A'}$$

<u>Step 2.</u>

The values xcount(i), for i = 1, 2, 3, 4 for our PLA are 1, 1, 2 and 1 respectively. Thus, simply adding xcount(i) to each of the elements of the ith row (i = 1, 2, 3, 4) of $\mathbf{A'}$, we get the following matrix:

$$
\begin{bmatrix} 2 & 4 \\ 4 & 2 \\ 0 & 4 \\ 2 & 2 \end{bmatrix}
$$

Now, to convert to the desired $\mathbf{A}$ matrix, we replace all elements which are less than n (= 4, in this case) by 0; elements which are equal to n are replaced by 1. Thus, the $\mathbf{A}$ matrix derived is:

$$
\begin{bmatrix} 0 & 1 \\ 1 & 0 \\ 0 & 1 \\ 0 & 0 \end{bmatrix}
$$

To see why this works, note that an element in the $\mathbf{A'}$ matrix acquires the maximum value of $+n$ (see Lemma 3 and its proof), if and only if (a) the xcount of the involved product term (row) (of $\mathbf{I}$ and $\mathbf{V}$ respectively) is 0; and (b) the product term and the test vector are bitwise identical. This is the case in which the product term involved has no 'x' element and the test vector being simulated matches the product term in every position. Hence, this is a case in which the activation matrix should indicate: 1. If the xcount of the product term is non-zero, but the test vector matches the product term in every non-'x' position, then the value of the corresponding converted integer dot product will be n - xcount(product term). Thus, on adding xcount to the dot product, if we get back the maximum value of $+n$, we can be sure that the product term involved is compatible with the test vector and hence the corresponding element in the activation matrix $\mathbf{A}$ must be set to 1.

<u>Steps 3 and 4</u>

The $\mathbf{R'}$ and $\mathbf{R}$ matrices obtained are, respectively:

$$
\begin{bmatrix} 1 & 0 & 1 \\ 0 & 2 & 1 \end{bmatrix} , \begin{bmatrix} 1 & 0 & 1 \\ 0 & 1 & 1 \end{bmatrix}
$$

Thus, the desired output vectors corresponding to the input vectors 0101 and 1001 are 101 and 011 respectively.

For implementation efficiency, it may be shown that the intermediate step 2 above may actually be eliminated and its function merged with the final $\mathbf{R'}$ to $\mathbf{R}$ transform (steps 3 and 4). Similarly, steps 3 and 4 may actually be performed as a single *modified* matrix multiplication routine. So, basically, the procedure involves a cascade of two matrix multiplication/manipulation routines; one computing $\mathbf{A'}$ and the next deriving $\mathbf{R}$. This is actually what has been done in our actual Fortran code, although the steps have been shown as above for conceptual clarity.

To go from logic simulation to *fault simulation*, is but an easy extension of the logic simulation problem. It involves repeated simulation of fault-injected PLAs under the set of vectors considered. In terms of our methodology, this involves, essentially repeated matrix operations (mainly matrix multiplications). Note that in this model, it is no more complex to simulate a *multiple* fault than it is to simulate a single fault. Formulating the basic logic and fault simulation problems in terms of simple matrix and vector operations, automatically points towards its exploitation under efficient vector and parallel architecures typical of scientific supercomputers[23-25] and other VLSI array structures[22].

Test Generation

For brevity, we do not present the full, formal treatment of the (vectorized) PLA test generation procedure. However, with reference to the basic methodology described in section II, we point to the key targets of algorithm reformulation under a scientific vector machine environment. Through such reformulation, the combined fault simulation/test generation procedure exhibits significant speed-up over existing methods (see section VI).

Firstly, the basic opposite-adjacent method for deriving S-tests (and D-tests), is easily implemented in terms of vector/matrix operations and manipulations; the formal treatment is omitted here. Secondly, the basic *covering* test used in the tabular covering method, is a procedure which is easily implemented in terms of matrices as before. Since the above two aspects form the kernel of our overall test generation procedure, it is clear that a neat reformulation (based mainly on vector/matrix operation primitives), of the integrated fault simulation/test generation algorithm is possible.

Fortran Coding Guidelines for High Performance

As is typical of many commercially available scientific vector mainframes and supercomputers, Fortran is the only language supported by a *vectorizing* compiler under the IBM 3090 VF system. Hence our basic challenge was to see how ordinary Fortran could be used to solve our fault simulation and test generation problem by making effective use of the parallelism offered by the underlying machine. The VS Fortran (version 2) compiler is a highly sophisticated state-of-the-art vectorizing compiler. However, in order to take full advantage of the software and the underlying vector hardware, educated programming styles, guided by the knowledge of the compiler and the machine are often necessary. Our research group has been involved in inventing simple, yet effective solutions in tuning computationally intensive Fortran application programs in general. Without dwelling on this aspect of performance enhancement in this paper, we provide two references[20,21] of such work done in our group. We have used such techniques in fine tuning the PLA test and fault simulation programs run on the IBM 3090 VF (see results in section V)

IV. Speed-Up in a Parallel Processing Environment

Our primary motivation for considering parallel implementations of fault simulation and test generation software is the Research Parallel Prototype (RP3) being built at the IBM T. J. Watson Research Center[25]. RP3 is a highly parallel MIMD precessing system intended for research into hardware and software aspects of parallel processing. The machine is capable of being configured as a shared memory or message-passing machine or as a hybrid. In parallelizing this particular application, the shared memory configuration is assumed. An actual, full configuration will provide 512 state-of-the-art microprocessors, with a peak projected performance of 1.3 GIPS. The basic hardware organization should be very effective for a variety of CAD applications in addition to the usual number-crunching parallel applications. However, once again, suitable problem reformulation at the highest level is essential for ensuring optimum usage of the available parallelism.

The EPEX System

Since the RP3 hardware is is not yet available, the application was written and debugged using the experimental mult-processor environment VM/EPEX[18], designed to simulate key aspects of RP3-style processing. EPEX operates on IBM System/ 370 machines under the VM/SP operating system and uses standard facilities of VM/SP to provide either real or virtual parallel processing. Its operation mimics RP3, particularly the memory organization aspects. The system allows a user to spawn a multiple number of virtual machines (processes) to cooperate in executing an application program. The programming model assumed is primarily single program multiple data (SPMD). The multiple virtual machines execute the same program and simulate the shared memory via the use of the shared segment facility of VM/EPEX.

Essentially, the only parallel construct we have made use of in our experiments so far is the *parallel DÔ* loop, in which independent iterations of the loop may be assigned to distinct processors. This is supported under EPEX as an user-specified @DO statement. Process synchronization is achieved via system-supported fetch-and-add, with standard *barrier* type synchronization[18] specified in source code. (See also, Ostapko et al[16]).

V. Some Performance Results

We present a couple of charts to illustrate performance enhancement on the IBM 3090-VF, as a result of applying the techniques presented in section III. The results are based on running randomly generated PLAs through logic/fault simulation programs implemented in VS Fortran (version 2) on the IBM 3090-VF. Figure 2 illustrates the speed-up in virtual CPU time for the basic logic simulation problem. Curve A shows the increase of execution time with increasing values of m (the number of product terms or rows), for a classical implementation (see section II), while curve B shows the results for an implementation based on our technique (section III). Note that in both cases, the programs were compiled using the *same* high-technology VS Fortran vectorizing compiler; so, the original program (curve A) was also vectorized as much as possible by the compiler prior to execution. The number of test vectors (t) simulated for each point was kept equal to m(n + p); i.e., the number of test vectors was increased linearly with increasing m. As is apparent from Figure 2, suitable problem reformulation using our methodology, results in much better performance; the execution time increases *almost* linearly with m, as compared with the distinctly polynomial rate of classical implementation.

Figure 3 shows the results for a particular test generation/fault simulation procedure, run for the same PLA configurations as in Figure 2. This procedure is the same as Procedure 2 of section II, except that an intermediate logic/fault simulation step is introduced between steps A and B, in order to eliminate those G and A faults which are already covered by T^{S+D}. This step decreases the complexity of Step B as well as the overall size of the test set; but as a price, the extra simulation step is required. (The crosspoint single fault coverage in both cases is 100 percent). Figure 3 illustrates the drastic reduction in run-time complexity of a typical fault simulation/test generation program achieved by adopting our methods; as before, curve A depicts performance of the original package, while curve B represents the recoded version.

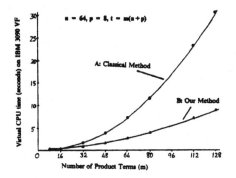

Figure 2. Performance Curves (IBM 3090): PLA Logic Simulation

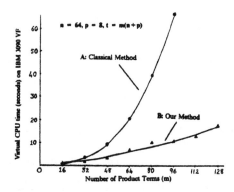

Figure 3. Performance Curves (IBM 3090): PLA Fault Simulation/Test Generation

For programs parallelized under EPEX Fortran, we present a representative speed-up chart (Table 1), for a randomly generated PLA with parameters $n = 64$, $m = 128$ and $p = 8$; the same test generation/fault simulation programs as used for Figure 3 were used for parallelization. Upto 32 virtual processors running under VM/EPEX on 2-way physical machine (IBM 3090-VF) were used. Once again, method A refers to a classical implementation, while method B is our improved version. Table 1 shows that the our method results in a much better speed-up efficiency as the number of processors is increased.

Table 1.

No. of processors →		2	4	8	16	32
Speedup	Method A	1.6	3.0	5.7	7.2	13.7
	Method B	1.8	3.5	7.2	14.5	23.7

VI. Conclusions and Discussion

The primary motivation behind this work was to explore ways to exploit the large degree of concurrency available in modern scientific supercomputers, with respect to design-and-test (CAD) applications. In particular, applications related to well-structured hardware, such as array logic and its variants, are expected to be good candidates for such an exercise. Yet, as our results for PLA fault simulation/test generation indicate, efficient exploitation of available parallelism entails a combination of *problem reformulation* (at the algorithm level) and *coding style*. Of course, problem reformulation is the key to ensure that the architectural primitives and features of the machine are utilized in the best possible manner. The secondary issue of coding style comes in due to the fact that the problem solver must go through the medium of a translator (compiler) to reach the raw hardware. In this paper, our focus has been mainly on the problem reformulation aspects. We have shown, that for array logic applications, a simple, yet novel rethinking in terms of vectors, matrices and their associated *arithmetic* operations, enables us to map the essentially non-numeric domain of fault simulation/ test generation into well-understood numeric problems, solutions for which are very well supported (in hardware and software) on powerful scientific supercomputers.

The main advantages of the basic methodology may thus be summarized as follows:

1. The whole test generation/fault simulation procedure can be programmed very conveniently and compactly using a high-level scientific language like Fortran or APL. In particular, assembly-level coding or bit-parallel manipulations are not needed.

2. Since arithmetic vector and matrix operations are very well supported in scientific supercomputers, this method offers the convenience of standard high-level scientific programming combined with the power of pipelined, vector/parallel architectures provided by these high-end machines. Experiments run under VS Fortran, version 2 (IBM 3090 VF), and under EPEX Fortran (RP3), for medium to large sized PLAs, have demonstrated the significant speed-up advantage afforded by our method.

3. If desired, further improvements in speed are possible, by using or calling hand-optimized assembly language routines for vector/matrix operations; for the IBM 3090 VF, for example, standard Engineering and Scientific Subroutine Library routines could be used easily in the source program. Also, word-oriented, bit-wise parallel simulation[22] could be super-imposed for even more speed-up.

4. The method generalizes easily to a wide class of array logic macros. The general technique, with variations, is likely to be applicable to a wide class of structured logic design/verification problems.

References

1. T. Blank, "A survey of hardware accelerators used in computer-aided design," *IEEE Design & Test of Computers*, Vol. 1, No. 3, pp. 21-39, August 1984.

2. H. Fleisher and L. I. Maissel, "An introduction to array logic," *IBM Journal of Research and Development*, Vol. 19, pp. 98-109, March 1975.

3. E. I. Muehldorf and T. W. Williams, "Optimized stuck fault test pattern generation for PLA macros," *Dig. Semiconductor Test Symp.*, Cherry Hill, NJ, October 25-27, 1977, pp. 88-101.

4. D. L. Ostapko and S. J. Hong, "Fault analysis and test generation for programmable logic arrays," *IEEE Trans. Comput.*, vol. C-28, Sept. 1979, pp. 617-626.

5. J. Smith, "Detection of faults in programmable logic arrays," *IEEE Trans. Computers*, vol. C-28, Nov. 1979, pp. 845-853.

6. P. Bose and J. A. Abraham, "Test generation for programmable logic arrays," *Proc. 19th Design Automation Conf.*, Las Vegas, June 1982, pp. 574-580.

7. P. Bose, "Logical fault analysis and design for testability of programmable logic arrays," *Proc. 23rd Annual Allerton Conf.*, Monticello, October 1985, pp. 158-167.

8. P. Bose, "Functional testing of programmable logic arrays," IBM Research Report RC 10681, Yorktown Heights, NY October 1984.

9. R-S. Wei and A. Sangiovanni-Vincentelli, "PLATYPUS: A PLA test pattern generation tool," *IEEE Trans. on Computer-Aided Design*, Vol. CAD-5, No. 4, pp. 633-643, October 1986.

10. V. K. Agarwal, "Multiple fault detection in programmable logic arrays," *IEEE Trans. on Computers*, vol. C-29, pp. 518-522, June 1980.

11. K. S. Ramanatha and N. N. Biswas, "A design for testability of undetectable crosspoint faults in programmable logic arrays," *IEEE Trans. Comput.*, vol. C-32, June 1983, pp. 551-557.

12. S. M. Reddy and D. S. Ha, "A new approach to the design of testable PLAs," *IEEE Trans. on Computers*, Vol. C-36, No. 2, pp. 201-211, February 1987.

13. H. Fujiwara and K. Kinoshita, "A design of programmable logic arrays with universal test sets," *IEEE Trans. Comput.*, vol. C-30, pp. 823-828, November 1981.

14. J. P. Roth, W. G. Bouricius and P. R. Schneider, "Programmed algorithms to compute tests to detect and distinguish between failures in logic circuits," *IEEE Trans. Electron. Comput.*, vol. EC-16, pp. 567-579, October 1967.

15. N. Ishiura, H. Yasuura and S. Yajima, "High-speed logic simulation on vector processors," *IEEE Trans. on Computer-Aided Design*, Vol. CAD-6, No. 3, May 1987.

16. D. L. Ostapko, Z. Barzilai and G. M. Silberman, "Fast fault simulation in a parallel processing environment," *Proc. Int'l. Test Conf.*, Washington, D.C., September 1987.

17. F. Darema and G. F. Pfister, "Multipurpose parallelism for VLSI CAD on the RP3," *IEEE Design & Test of Computers*, Vol. 4, No. 5, pp. 19-27, October 1987.

18. F. Darema et al., "A Single-Program-Multiple-Data computational model for EPEX Fortran," IBM Research Report RC 11552, Yorktown Heights, NY, October 1986.

19. D. J. Kuck, *The Structure of Computers and Computations*, Vol. I, John Wiley & Sons, New York, 1978.

20. B. Liu and N. Strother, "Peak vector performance from VS Fortran," IBM Research Report RC 12849, June 1987.

21. P. Bose, "A brief status report on EAVE: an Expert Advisor for Vectorization," IBM Research Report (to appear, December 1987).

22. K. Hwang, "Partitioned matrix algorithms for VLSI arithmetic systems," *IEEE Trans. on Computers*, Vol. C-31, No. 12, pp. 1215-1224, December 1982.

23. R. M. Russell, "The CRAY-1 computer system," *Comm. ACM*, vol. 21, no. 1, pp. 63-72, January 1978.

24. Several papers on the IBM 3090 system, architecture and performance, *IBM Systems Journal*, Vol. 25, No. 1, pp. 4-82, 1986.

25. G. F. Pfister et al., "The RP3 Research Parallel Processor Prototype (RP3): introduction and architecture," *Proc. Int'l. Conf. on Parallel Processing*, August 1985, pp. 764-771.

26. P. Bose, "Fast Fault Simulation and Test Generation for PLAs in a Parallel Processing Environment," IBM Research Report RC 13343, Yorktown Heights, NY, December 1987.

Universal Hashing in VLSI

Martin Fürer
Computer Science Department
113 Whitmore Lab.
The Pennsylvania State University
University Park, PA 16802

Abstract. *A problem is presented with deterministic VLSI complexity $AT_{det}^2 = \Omega(N^2)$, but Las Vegas complexity only $AT_{Las\ Vegas}^2 = O(N\ poly(\log N))$. (The Las Vegas algorithm always decides correctly, but T is only the expected running time; A is the area of the chip.) Previously $AT_{Las\ Vegas}^2 = O(N^{3/2}\ poly(\log N))$ has been shown for a similar problem with a more complicated algorithm. Here, we use a simple universal hashing technique based on random linear functions. We hope this will give rise to other applications of universal hashing in VLSI.*

Our algorithm is very practical, because the random bits can even be wired into the chip. For every sequence of inputs during a chip's lifetime, the chances are high that the same short random bit string will always produce the result quickly.

1. Introduction

We investigate the power of randomization in VLSI computations. More precisely, we compare the AT^2-complexity of standard deterministic VLSI chips with the AT^2-complexity of Las Vegas chips solving the same problem.

Las Vegas chips are either assumed to contain switching elements independently producing truly random bits using reasonable amounts of area and time, or more realistically, they receive an additional random input. The algorithms are not allowed to produce wrong results, but the time can be variable. T is the expected running time for worst case inputs.

Improving an earlier result of [Mehlhorn & Schmidt 82], we present a decision problem R with

$$AT_{det}^2(R) = \Omega(N^2),$$

but

$$AT_{Las\ Vegas}^2(R) = O(N\ poly(\log N))$$

where N is the number of input bits.

The method for this efficient Las Vegas algorithm is based on the universal hashing technique of [Carter and Wegman 79]. The connection between communication complexity and random hashing has been discussed in [Fürer 87].

Traditionally, the deterministic AT^2 measure has been considered to be the main complexity measure in VLSI. [Mehlhorn and Schmidt 82], interpret their result as a method being able to provide deterministic lower AT^2 bounds in cases where other methods have not been successful, because the lower bounds are higher than the corresponding non-deterministic upper bounds. We do not touch the deterministic lower bounds here, but improve the Las Vegas upper bounds significantly with a very simple and practical algorithm.

We want to focus attention to the Las Vegas AT^2 measure not just as a natural theoretical tool, but also as a practical complexity measure. Thus, the significance of our algorithm is twofold. First it questions the practical importance of deterministic VLSI lower bounds, because Las Vegas chips might be much more efficient and just as practical, as we shall see in section 5. Second, it exhibits a powerful algorithmic technique.

2. Universal hashing

[Carter and Wegman 79] call a class $H = \{h_j: j \in J, h_j: A \to B\}$ of functions *universal$_2$*, if for all $x,y \in A$ with $x \neq y$

$$prob(h_j(x) = h_j(j)) \leq 1/|B|$$

when $j \in J$ is chosen with uniform distribution.

Thus, e.g. the class of all functions from A to B is *universal$_2$*. This class certainly contains bad hash functions. It is a desirable property of hash functions to take every value with (about) equal frequency.

Definition
A class of functions $H = \{h_j: j \in J, h_j: A \to B\}$ is *uniformly k-universal*, if for every $X \subseteq A$ with $2 \leq |X| \leq k$

$$prob(|h_j(X)|=1) \leq 1/|B|^{|X|-1},$$

when $j \in J$ is chosen with uniform distribution, and for all $j \in J$ and $y \in B$

$$prob(h_j(x)=y) = 1/|B|,$$

when x is chosen from A with uniform distribution, i.e., every h_j in H takes each value equally often.

The following class of functions turns out to be very useful.

Definition
H_0 is the class of all linear functions from the vector space F^n onto F^m where $F = \mathbb{Z}_2$ is the field with two elements.

This class without the restriction to functions onto F^m has been proposed by [Carter and Wegman 79]. They have shown it to be *universal*$_2$. In fact its restriction to linear functions onto F^m is uniformly 3-universal.

Lemma
H_0 is uniformly 3-universal.

Proof
For distinct x, y, z and random j the probability that $h_j(x) = h_j(y) = h_j(z)$ is equal to the probability that $h_j(x-z) = h_j(y-z) = 0$. Note that $x-z$ and $y-z$ are linearly independent.

If $F = \mathbb{Z}_2$ is replaced by another finite field, then the class corresponding to H_0 remains uniformly 2-universal.

The main advantage of H_0 over other classes of uniformly 3-universal hash functions is the extreme simplicity to evaluate $h_j(x)$. In fact it is so simple and useful that it would be a good idea to build it into the instruction set of general purpose computers. Every bit of $h_j(x)$ is just a scaler product over the integers modulo 2. Simple masking would immediately provide a good random hashing for all ranges whose size is a power of 2.

3. The deterministic lower bound

Motivated by [Mehlhorn and Schmidt 82], we investigate the VLSI complexity of a problem R, which has a very efficient Las Vegas solution. For $n \times n$ matrices z, we use the following notation.

Notation
$z_i = (z_{i0}, \ldots, z_{i\,n-1})$ and $\text{shift}(z_i, s) = (z_{i\,n-s}, \ldots, z_{i\,n-1}, z_{i0}, \ldots, z_{i\,n-s-1})$ for $0 \le i, s \le n-1$.

Definition
$R(x_0, \ldots, x_{n-1}, y_0, \ldots, y_{n-1}, s, l)$ holds for $x_i, y_i \in \{0,1\}^n$ and $s, l \in \{0, \ldots, n-1\}$ if
$$\exists i \quad \text{shift}(x_i, s) = y_{i+l \bmod n}$$

Here, s stands for small (within rows) and l for large (moving rows) cyclic shifts.

We use the result, that in the worst case n^2 bits have to be exchanged between the x_i's and the y_i's by every deterministic algorithm deciding R.

Theorem 1 ([Mehlhorn and Schmidt 82])
$R(x_0, \ldots, x_{n-1}, y_0, \ldots, y_{n-1}, 0, 0)$ has deterministic two-way communication complexity n^2 between $(x_0, \ldots, x_{n-1})$ and $(y_0, \ldots, y_{n-1})$.

Theorem 2
$AT_{\text{det}}^2(R) = \Omega(N^2)$, where $N = 2(n^2 + \lceil \log_2 n \rceil)$ is the length of the input.

Proof idea
Let A be the area and T the time of a where-oblivious deterministic chip deciding R. There is a cut of length $O(\sqrt{A})$ such that for some inputs, much information has to cross the cut. In fact, the shifts s and l, as well as some x_j's and y_i's, and some bits of other x_i's and y_i's can be fixed, so that a smaller version of the problem of Theorem 1 needs to be solved with the x_i's and y_i's separated by the cut.

4. A very efficient Las Vegas chip

We describe a Las Vegas algorithm to decide

$$R(x_0, \ldots, x_{n-1}, y_0, \ldots, y_{n-1}, s, l).$$

The principal data structures are two meshes of trees to store the quadratic matrices x and y. This means that we use a standard matrix representation together with a balanced binary tree over every row and over every column. Furthermore, we have a network being able to do any cyclic shift of n bit vectors in time $O(\log n)$. This network can be similar to a FFT graph. Both structures can be implemented in area $A = O(n^2 \, poly(\log n)) = O(N \, poly(\log N))$.

Definition
$shift(r, s)$ is the matrix obtained from r by shifting every row cyclically by a s positions to the right.

Algorithm

1. Choose a random $\log n \times n$ matrix r over $\mathbb{Z}_2$.

2. Compute the matrix products (hashing) $u = r\, x^T$ and $v = \text{shift}(r,s)\, y^T$ (where $x_{ji} \sup T = x_{ij}$).

   Let $u_0,\ldots,u_{n-1}$ and $v_0,\ldots,v_{n-1}$ be the columns of u and shift$(v,-l)$ respectively.

3. Test $u_j \overset{?}{=} v_j$ for $0 \leq j \leq n-1$.

4. For all j with $u_j = v_j$ do

5. if shift$(x_j,s) = y_{j+l}$ then accept and stop.

6. Reject.

Step 4 is done sequentially. But the expected number of repetitions is $O(1)$, in fact it is less than 2. Steps 1,2,3 and each call of step 5 are easily parallelizable. The best way is to work with n-bit vectors at a time, pipelining $\log n$ such operations.

Assume the matrices x and y are stored in the standard way

$$
x = \begin{bmatrix} x_{00} & \cdots & x_{0\,n-1} \\ \vdots & & \vdots \\ x_{n-1\,0} & \cdots & x_{n-1\,n-1} \end{bmatrix}
\qquad
y = \begin{bmatrix} y_{00} & \cdots & y_{0\,n-1} \\ \vdots & & \vdots \\ y_{n-1\,0} & \cdots & y_{n-1\,n-1} \end{bmatrix}
$$

Using all column trees, a row shift(r_i,s) of shift(r,s) is distributed simultaneously to all rows of y. After a pairwise multiplication (AND-operation), addition modulo 2 is done in the row trees. The resulting column vector is sent to the shifting network, shifted by $-l$, and compared componentwise with the outcome of the same procedure (without shifting) from the matrix x, where r_i (instead of shift(r_i,s)) has been fed in. In step 5, one row vector at a time is extracted form x via the column trees and sent to the shifting network.

Theorem 3
$AT^2_{Las\ Vegas}(R) = O(N\ poly\,(\log N))$.

Proof
The algorithm described above is obviously correct. To prove an $O(\log N)$ bound on the expected running time, we show hashing to be so effective that almost all unequal pairs are discovered with high probability in step 3.

Unequality is discovered independently by every bit of the hashed code with probability exactly $1/2$. Using $\log n$ bits, the probability not to detect any fixed unequal pair, is $1/2^{\log n} = 1/n$. Hence, with probability at most $1/n$, step 5 (time $O(\log N)$ has to be done with this unequal pair. Result:

Expected time for one non-matching pair: $O(\log N / n)$.

Expected time for all non-matching pairs: $O(\log N)$.

Expected time for the whole algorithm: $O(\log N)$.

5. Practicability of Las Vegas chips

Las Vegas chips are always correct, and therefore practically useful. But how can they be built? We propose just to toss coins when the chip is designed. So, actually a deterministic chip is built with probabilistic methods. But we do not loose the main advantage of randomized algorithms over probabilistic analysis of deterministic algorithms. We need no a priori assumptions about the input distribution. As long as the input is chosen independently of the chip design, the expected efficiency is good.

One could even go a step further. In our example R, we could modify the algorithm so that even the probability of an input needing more than time $c \log N$ (for some $c > 0$) would be very low, namely less than $2^{-max(100, \log N)}$. Therefore, one could be "practically sure" that this does not occur during the lifetime of a chip.

The modified algorithm would just choose a random $max(100, \log n) + \log n \times n$ matrix in step 1. The probability of not detecting an unequal pair in step 3 would then be bounded by

$$n \cdot 2^{-max(100, \log n) + \log n} \leq \min(2^{-100}, 1/n).$$

The bound $1/n$ for this probability assures an expected running time $O(\log N)$. There is no harm in even transmitting the whole matrix in step 4 of the algorithm with probability $1/n$. At the same time the bound 2^{-100} makes sure that in practice, we can assume that the algorithm always detects every unequal pair in the hashing steps 2 and 3.

A more sophisticated algorithm could even do with less time by inserting another non-random hashing step after each unsuccessful execution of step 5. This could avoid the expensive case that an undetected difference vector occurs several times.

Open problem

Are there relations or functions, more interesting than R, for which our method produces efficient las Vegas chips?

References

J. L. Carter and M.N. Wegman, Universal Classes of Hash Functions, JCSS 18 (1979) 143-154.

M. Fürer, The Power of Randomness for Communication Complexity, 19th STOC 1987, 178-181.

K. Mehlhorn and E. M. Schmidt, Las Vegas is better than Determinism in VLSI, 14th STOC 1982, 330-337.

Converting Affine Recurrence Equations to Quasi-Uniform Recurrence Equations

Yoav Yaacoby [1] *and Peter R. Cappello* [2] *
University of California, Santa Barbara

[1]Electrical and Computer Engineering Dept. [2]Computer Science Dept.

Abstract. *Most work on the problem of synthesizing a systolic array from a system of recurrence equations is restricted to systems of uniform recurrence equations. In this paper, this restriction is relaxed to include systems of affine recurrence equations. A system of uniform recurrence equations typically can be embedded in spacetime so that the distance between a variable and a dependent variable does not depend on the problem size. Systems of affine recurrence equations which are not uniform, do not enjoy this property. A method is presented for converting a system of affine recurrence equations to an equivalent system of recurrence equations that is uniform, except for points near the boundaries of its index sets. A characterization of those systems of affine recurrence equations that can be so converted is given, along with an algorithm that decides if a system is amenable to such a conversion, and a procedure that converts those affine systems which can be converted.*

1 Introduction

A system of uniform recurrence equations, as defined by Karp, Miller, and Winograd [6], maps especially well onto a systolic/wavefront array. Many researchers have either linearly mapped systems of uniform recurrence equations into spacetime, or translated them to systolic/wavefront arrays [7,1,5,3,9]. A system of uniform recurrence equations typically can be mapped linearly into spacetime so that interprocessor communication requires only fixed memory and fixed-length interconnections. This property, in general, is not enjoyed by a system of *affine* recurrence equations (a more general class, first investigated by Delosme and Ipsen [4]). Choffrut and Culik [2] treat a related systolic array problem, using 'folds' to eliminate reflections and rotations that are 2-dimensional. Using this idea, we formulate a 'generalized fold', and consider the following problem: Which systems of affine recurrence

*This material is based upon work supported by the Office of Naval Research under contract nos. N00014-84-K-0664 and N00014-85-K-0553.

equations can be converted, by a generalized fold, to an equivalent system that is uniform, except for points near the 'folds'? These latter systems are called systems of *quasi-uniform* recurrence equations. We provide: 1) a characterization of those systems of affine recurrence equations that can be so converted; 2) an algorithm for deciding if a system of affine recurrence equations can be so converted; 3) a procedure for converting a system of affine recurrence equations to an equivalent system of quasi-uniform recurrence equations.

Where linear embeddings fail, a generalized fold may succeed in enabling a VLSI array implementation.

The complete version of this paper can be found in [10].

2 Definitions

The following definitions are related to a system of recurrence equations (SRE).
Example 1: This SRE factors a symmetric Toeplitz matrix and its inverse into LDL^T:

$$1 \le i \le n, \qquad a_1(i,0) \equiv t_{i-1} \tag{1}$$

$$2 \le i \le n, \qquad a_2(i,0) \equiv t_{i-1} \tag{2}$$

$$1 \le j \le n-1, \qquad a_3(j,j) = a_2(j+1,j-1)/a_1(j,j-1) \tag{3}$$

$$j+1 \le i \le n, \qquad a_3(i,j) = a_3(i-1,j) \tag{4}$$

$$j+1 \le i \le n, \qquad a_1(i,j) = a_1(i-1,j-1) - a_3(i-1,j)a_2(i,j-1) \tag{5}$$

$$j+2 \le i \le n, \qquad a_2(i,j) = -a_3(i-1,j)a_1(i-1,j-1) + a_2(i,j-1) \tag{6}$$

$$a_2(1,0) = 1 \tag{7}$$

$$1 \le j \le n-1,$$

$$1 \le i \le j-1, \qquad a_3(i,j) = a_3(i+1,j) \tag{8}$$

$$1 \le i \le j+1, \qquad a_2(i,j) = -a_3(i+1,j)a_2(j+2-i,j-1) + a_2(i,j-1) \tag{9}$$

Index set: The set of points where an array is computed or used.

Domain of computation: The set of points C_i where an array a_i is computed
   (e.g., $C_2 = \{(i,j) | 1 \le j \le n-1, \ 1 \le i \le n\}$ in Ex. 1).

Dependence map: A function δ_{ij} from the domain of computation of array a_j to
   the index set of a_i, on which the computation of a_j depends (e.g., $\delta_{32}(p) = p + (1,0)^T$ in Ex. 1(9))

Reduced dependence graph (RDG): A directed multigraph with a node for each array in the SRE, and an arc from a_i to a_j associated with every dependence
   map δ_{ji} in the SRE (see [9]). If the RDG is strongly connected, the SRE
   is said to be strongly connected. Since every SRE can be decomposed into
   strongly connected SREs, it is assumed that SREs are strongly connected.
   For example, the RDG of Ex. 1 shown in Fig. 1 is strongly connected.

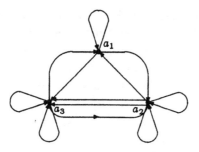

Figure 1: The RDG for the system of recurrence equations of Ex. 1.

Affine dependence: A dependence map of the form: $\delta_{ij}(p) = D_{ij}p + d_{ij}$ where $D_{ij} \in \mathbf{Z}^{n \times n}$, and $d_{ij} \in \mathbf{Z}^n$ (e.g., $\delta_{22}(p) = \begin{pmatrix} -1 & 1 \\ 0 & 1 \end{pmatrix} p + \begin{pmatrix} 2 \\ -1 \end{pmatrix}$ in Ex. 1(9)).

In the remainder of this paper, we assume that D_{ij} is nonsingular and integer.

Uniform dependence: An affine dependence of the form: $\delta_{ij}(p) = p + d_{ij}$ (i.e., $D_{ij} = I$).

A system of affine [uniform] recurrence equations (SARE [SURE]): An SRE where the dependence maps are affine [uniform], and every array is computed in one recurrence equation for its entire domain of computation (e.g., Eq. (9) in Ex. 1 by itself is an SARE).

Domain size parameters: Those parameters in the SRE which instantiate the domains of computation (e.g., n in Ex. 1).

Cycle dependence map: A composition of dependence maps associated with the arcs of a directed cycle in the RDG (not necessarily simple). In most cases, we denote by δ_i a cycle dependence map which starts at a_i, and by δ_{ij} a direct dependence map of a_j on a_i (e.g., δ_{21} and δ_{12} in Ex. 1 constitute a cycle dependence map denoted δ_1 – See Fig. 1).

Dependence length of an index point p with respect to a cycle dependence map δ_i, is the number of cycle compositions which can be made starting with $a_i(p)$ until (and including) a cycle on which an index of one of the arrays (not necessarily a_i) is outside its domain of computation. It is denoted by $\gamma_i(p)$. For a set S of index points, the notation $\gamma_i(S)$ denotes the minimum dependence length of points in S. If $H \subset \mathbf{R}^n$, then $\gamma_i(H) = \min_{p \in H \cap L_i^n} \{\gamma_i(p)\}$, where L_i^n is the set of lattice points in $\mathbf{R}^n$ on which array a_i is defined.

n-dimensional system of recurrence equations: A strongly connected SARE, satisfying: For every cycle dependence map δ_i, $\forall k \in \mathbf{N}$, there exist domain size parameters such that the domain of computation C_i contains H, a k^n hypercube[1], and $\gamma_i(H) > 1$ (e.g., Ex. 1(9) is 2-dimensional).

[1]This means that every lattice point in the hypercube H, which is in $\mathbf{R}^n$, is in C_i.

Partitioned Linear Transformation (PLT): Let V be an n-dimensional vector space. Let $P_0, P_1, \ldots, P_{m-1}$ be m convex polytopes, which partition V. Let $T = \{T_0, T_1, \ldots, T_{m-1}\}$ be a set of nonsingular $n \times n$ matrices. A *Partitioned Linear Transformation* $\mathbf{T} : V \to V$ with respect to $\{P_i\}$ and T is such that $\mathbf{T}(x) = T_i(x)$ if and only if $x \in P_i$. The PLT may be referred to as an m-fold PLT. In Fig. 2, $Q \subset \mathbf{Z}^2 \subset \mathbf{R}^2$. Q is partitioned into Q_0, Q_1, Q_2, Q_3 by its intersection with four polytopes. A PLT is applied to Q such that

$$T_0 = I, \; T_1 = \begin{bmatrix} 0 & 1 \\ 1 & 0 \end{bmatrix}, \; T_2 = \begin{bmatrix} 0 & 1 \\ -1 & 0 \end{bmatrix}, \; T_3 = \begin{bmatrix} -1 & 0 \\ 0 & 1 \end{bmatrix}.$$

The resulting set is also shown in Fig. 2. Since there are four subsets, this is a 4-fold PLT.

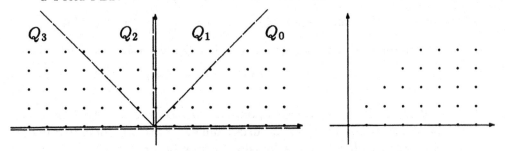

Figure 2: The set Q before and after the PLT.

Uniform subdomain: The largest subset $U \subseteq \bigcup_i C_i$ (where C_i is the domain of computation of the array a_i), such that the SRE defined on this subset is uniform.
Nonuniform subdomain: All index points in $\bigcup C_i - U$.

System of quasi-uniform recurrence equations (SQURE): An SRE that satisfies

1. The SRE of the uniform subdomain is n-dimensional, for some n;

2. $\exists b \in \mathbf{R}$ so that for every value of the domain size parameters, if x is in the nonuniform subdomain, $\max(\min_{x' \in L_i^n - C_i} \|x' - x\|) < b$, where L_i^n is the lattice on which C is defined.

PLT convertible: An SRE that can be converted to an equivalent SQURE (by equivalent we mean that the I/O relationship is the same) by 1) applying an m-fold PLT to the arrays' index sets, and 2) renaming each array such that a_i defined on the index points which are in the polytope P_j of the PLT, is called $a_{ij}, 0 \leq j < m$. Moreover, m does not depend on the domain size parameters.

Convertible: An SRE is *convertible* if it can be transformed, by affine transformations on the index sets of the arrays, into an SARE that is PLT convertible (different transformations may be applied to different index sets).

3 Characterizing and Deciding Convertibility

Let A be a (strongly connected) SARE. Affine transformations can be applied to the index sets of A's arrays such that the resulting RDG has a spanning tree whose arcs have uniform dependence maps associated with them. Such a transformation has been suggested in [4]. A similar procedure is as follows. Choose a spanning tree of the RDG. Start from the root of the tree, and proceed down level by level, applying the affine transformation δ_{ij}^{-1} on the index set of the array a_i (i.e., $p \to D_{ij}^{-1}p - D_{ij}^{-1}d_{ij}$), where a_j is its parent. All the appropriate dependence maps are updated after each transformation. This procedure is called a *tree conversion*.

Theorem 3.1 [10] *An n-dimensional SARE is convertible if and only if after a tree conversion, the linear parts of the dependence maps generate a finite group.*

In fact, for every finite group G of integral $n \times n$ matrices, $|G| \le (2n)!$ [8, pp. 52]. Moreover, if the SARE is convertible with an m-fold PLT, then $\forall D \in G$, $D^k = I$, for some $k \le m$.

The condition in Thm. 3.1 is decided by Alg. Decide: it returns true if and only if the SARE is convertible. The SARE is assumed to be n-dimensional.

ALGORITHM DECIDE

Extract the RDG from the system of recurrence equations;
Perform a tree conversion on the RDG;
$N \leftarrow \{D | \delta(x) = Dx + d$ *is a dependence map* $\}$;
Generate the group $G =< N >$, such that after each generated member $D \in G$ do
    if   (the current size $|G| > (2n)!$) then return(false);
return (true);

The linear parts of cycle dependence maps in the SARE of Ex. 1(9) are: $T_0 = I$, $T_1 = \begin{bmatrix} -1 & 1 \\ 0 & 1 \end{bmatrix}$. Since they form a group G, $|G| = 2$, the SARE is convertible: Alg. Decide returns *true*.

We have assumed that the SARE is strongly connected. If not, we can modify the definition of n-dimensionality so that the same results follow. In the definition of dependence length and an n-dimensional SARE, the word 'cycle' is replaced by 'semicycle'[2]. Also, in the definition of 'tree conversion', the tree has to be a spanning tree in the underlying graph of the RDG.

4 Conversion of an SARE to an SQURE

Let SARE S have dependence maps whose linear parts generate a finite group $G = \{D_i\}_{i=0}^{|G|-1}$, where $D_0 = I$. The procedure below constructs a specific PLT for S (see [11] for proof).

PROCEDURE PLT

1. *Construct [11] a point $p \in \mathbf{R}^n$ such that $D^T p \neq p, \forall D \in G - I$.*

2. *Construct matrix A_0, whose $\text{row}_i = p^T(D_i - I), \forall D_i \in G - I$.*

3. *$\forall D_i \in G$, form the matrix $A_i = A_0 \cdot D_i^{-1}$.*

4. *For $0 \leq i < |G|$, define $F_i = \{x | A_i x \leq 0\}$.*

5. *Construct [11] $q \in F_0^i$ [3] such that for $0 \leq i < |G|; 1 \leq j < |G|, [A_i]_j \cdot q \neq 0$.*

6. *Construct the linear constraints for P_i from those of F_i as follows. If $[A_i]_j \cdot q < 0$ then use $[A_i]_j \cdot x \leq 0$; else use $[A_i]_j \cdot x < 0$. These polytopes are the P_i.*

7. *Corresponding to part P_i, is the linear transformation $T_i = D_i^{-1}$.*

All index points are mapped to P_0 by the PLT above. Since D_i is integer and a root of I (because G is finite), D_i^{-1} is integer. Index points thus are mapped to integer points.

Given a convertible SARE S, the procedure below converts S into an SQURE.

PROCEDURE CONVERT

1. Perform a tree conversion on S.

2. Invoke Proc. PLT with $G = \{D_i\}_{i=0}^{|G|-1}$, the group generated by the linear parts of the direct dependence maps. It produces a set P of polytopes $\{P_i\}_{i=0}^{|G|-1}$, and a set of linear transformations $\{T_i = D_i^{-1}\}_{i=0}^{|G|-1}$ which define a PLT.

3. Partition each domain of computation C_i as follows:

$$\forall 0 \leq j \leq |G| - 1, \quad C_{ij} = C_i \bigcap P_j.$$

Each equation is split into up to $|G|$ equations (some C_{ij}s may be empty).

4. For each array a_i, partition $\mathbf{R}^n$ into $P^{ki} = \{\delta_{ki}^{-1}(P_j)\}_{j=0}^{|G|-1}$, for all direct dependences δ_{ki} [10].

5. For each array a_i, partition $\mathbf{R}^n$ by super-imposing all the partitions P^{ki} computed above.

6. Each C_{ij} is further partitioned into $C_{ij}^{(0)}, C_{ij}^{(1)}, \ldots, C_{ij}^{(h)}$ (the corresponding equations also are split) by restricting to C_{ij} the partition of $\mathbf{R}^n$ for a_i in step 5. This partition satisfies:

$$\forall 0 \leq t \leq h, \ \forall k, \ \delta_{ki}(C_{ij}^{(t)}) \subset P_{s(i,j,k,t)} \ \text{for some } 0 \leq s(i,j,k,t) < |G|.$$

We denote by $C_{ij}^{(0)}$ the domain of computation that satisfies: $\forall k, \ D_{ki}(C_{ij}^{(0)}) \subset P_{s(i,j,k,0)}$. This step ensures that each $C_{ij}^{(t)}$ is mapped by any dependence map of a_i on some array a_k to only one part in partition P.

---

[3] F_0^i is the interior

7. For each equation that computes a_i in the domain of computation $C_{ij}^{(t)}$ do

    (a) update the domain of computation as follows: $C_{ij}^{(t)} \to D_j^{-1} C_{ij}^{(t)}$,

    (b) rename the array on the left side of the equation: $a_i \to a_{ij}$,

    (c) for each term $a_k(\delta_{ki}(p))$ on the right side of the equation, given that $\delta_{ki}(C_{ij}^{(t)}) \subset P_l$ (i.e., $s(i,j,k,t) = l$), do

        i. rename the array: $a_k \to a_{kl}$,

        ii. update and rename the dependence map[4]:

$$\delta_{ki} \to \delta_{kl,ij}, \quad \text{where } D_{kl,ij} = D_l^{-1} D_{ki} D_j, \quad d_{kl,ij} = D_l^{-1} d_{ki}.$$

        The matrices D_l, D_j are members of the group G, while D_{ki}, d_{ki} are the original linear part and translation parts of δ_{ki}. (The above follows from the choice of $T_i = D_i^{-1}$ in Proc. PLT.)
According to the construction of the PLT, for $C_{ij}^{(0)}$, we get $D_{ki} \to I$.

Example 2: Consider the following SARE given in Ex. 1 Eq. (9):

$$1 \le j \le n-1,$$
$$1 \le i \le j+1, \qquad a_2(i,j) = -a_3(i+1,j)a_2(j+2-i,j-1) + a_2(i,j-1) \qquad (9)$$

According to Proc. Convert, we perform the following steps:

1. In this SARE, a tree conversion is not needed.

2. The group G includes the following matrices: $D_0 = I,\quad D_1 = \begin{bmatrix} -1 & 1 \\ 0 & 1 \end{bmatrix}$. The PLT for this case, according to Proc. PLT, is

$$P_0 = \{(i,j)|2i \ge j\}; \quad T_0 = I; \quad P_1 = \{(i,j)|2i < j\}; \quad T_1 = \begin{bmatrix} -1 & 1 \\ 0 & 1 \end{bmatrix}$$

3. In this SARE $C_2 = \{(i,j)|1 \le j \le n-1,\ 1 \le i \le j+1\}$. We partition C_2 according to the PLT, obtaining:

$$C_{20} = C_2 \cap P_0 = \{(i,j)|1 \le j \le n-1,\ j/2 \le i \le j+1\}$$
$$C_{21} = C_2 \cap P_1 = \{(i,j)|1 \le j \le n-1,\ 1 \le i < j/2\}.$$

4. The given SARE has the following dependence maps:

$$\delta_{32}(p) = p + \begin{pmatrix} 1 \\ 0 \end{pmatrix}; \quad \delta_{22}^1(p) = \begin{bmatrix} -1 & 1 \\ 0 & 1 \end{bmatrix} p + \begin{pmatrix} 2 \\ -1 \end{pmatrix}; \quad \delta_{22}^2(p) = p + \begin{pmatrix} 0 \\ -1 \end{pmatrix}$$

---

[4] Superscripts may be used to distinguish names in different equations, or different terms of the same equation.

Their inverses are:

$$\delta_{32}^{-1}(p) = p + \begin{pmatrix} -1 \\ 0 \end{pmatrix}; \quad (\delta_{22}^{1})^{-1}(p) = \begin{bmatrix} -1 & 1 \\ 0 & 1 \end{bmatrix} p + \begin{pmatrix} 3 \\ 1 \end{pmatrix}; \quad (\delta_{22}^{2})^{-1}(p) = p + \begin{pmatrix} 0 \\ 1 \end{pmatrix}.$$

For array a_2, we partition $\mathbf{R}^2$ into:

$$
\begin{aligned}
P^{32} &= \{\delta_{32}^{-1}(P_0) = \{(i,j)\,|\,j/2 - 1 \le i\}, \\
&\qquad \delta_{32}^{-1}(P_1) = \{(i,j)\,|\,i < j/2 - 1\}\} \\
P^{22^1} &= \{(\delta_{22}^{1})^{-1}(P_0) = \{(i,j)\,|\,i \le (j+5)/2\}, \\
&\qquad (\delta_{22}^{1})^{-1}(P_1) = \{(i,j)\,|\,(j+5)/2 < i\}\} \\
P^{22^2} &= \{(\delta_{22}^{2})^{-1}(P_0) = \{(i,j)\,|\,(j-1)/2 \le i\}, \\
&\qquad (\delta_{22}^{2})^{-1}(P_1) = \{(i,j)\,|\,i < (j-1)/2\}\}
\end{aligned}
$$

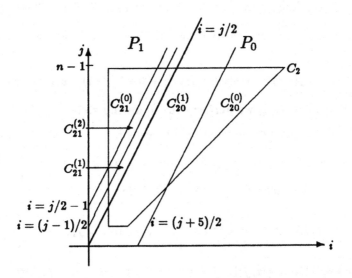

Figure 3: Partitioning the domain of computation C_2.

5. For array a_2, $\mathbf{R}^2$ is partitioned into 4 parts (see Fig. 3):

$$\{(i,j)\,|\,(j+5)/2 < i\}; \qquad\qquad \{(i,j)\,|\,(j-1)/2 \le i \le (j+5)/2\};$$
$$\{(i,j)\,|\,j/2 - 1 \le i < (j-1)/2\}; \quad \{(i,j)\,|\,i < j/2 - 1\}.$$

6. C_{20} is partitioned into (see Fig. 3)

$$
\begin{aligned}
C_{20}^{(0)} &= C_{20} \bigcap (\delta_{22}^{1})^{-1}(P_1) = \{(i,j)\,|\,1 \le j \le n-1, \ (j+5)/2 < i \le j+1\} \\
C_{20}^{(1)} &= C_{20} \bigcap (\delta_{22}^{1})^{-1}(P_0) \bigcap (\delta_{22}^{2})^{-1}(P_0) = \\
&\qquad \{(i,j)\,|\,1 \le j \le n-1, \ j/2 \le i \le \min((j+5)/2, \ j+1)\},
\end{aligned}
$$

and C_{21} is partitioned into

$$
\begin{aligned}
C_{21}^{(0)} &= C_{21} \bigcap \delta_{32}^{-1}(P_1) = \{(i,j)|1 \le j \le n-1,\ 1 \le i < j/2 - 1\} \\
C_{21}^{(1)} &= C_{21} \bigcap (\delta_{22}^1)^{-1}(P_0) \bigcap (\delta_{22}^2)^{-1}(P_0) = \\
&\quad \{(i,j)|1 \le j \le n-1,\ \max(1, (j-1)/2) \le i < j/2\} \\
C_{21}^{(2)} &= C_{21} \bigcap (\delta_{22}^2)^{-1}(P_1) \bigcap \delta_{32}^{-1}(P_0) = \\
&\quad \{(i,j)|1 \le j \le n-1,\ \max(1, j/2 - 1) \le i < (j-1)/2\}.
\end{aligned}
$$

The domains of computation are mapped as follows by the dependence maps

$$
\begin{aligned}
&\delta_{32}(C_{20}^{(0)}) \subset P_0 \quad \delta_{22}^1(C_{20}^{(0)}) \subset P_1 \quad \delta_{22}^2(C_{20}^{(0)}) \subset P_0 \\
&\delta_{32}(C_{20}^{(1)}) \subset P_0 \quad \delta_{22}^1(C_{20}^{(1)}) \subset P_0 \quad \delta_{22}^2(C_{20}^{(1)}) \subset P_0 \\
&\delta_{32}(C_{21}^{(0)}) \subset P_1 \quad \delta_{22}^1(C_{21}^{(0)}) \subset P_0 \quad \delta_{22}^2(C_{21}^{(0)}) \subset P_1 \\
&\delta_{32}(C_{21}^{(1)}) \subset P_0 \quad \delta_{22}^1(C_{21}^{(1)}) \subset P_0 \quad \delta_{22}^2(C_{21}^{(1)}) \subset P_0 \\
&\delta_{32}(C_{21}^{(2)}) \subset P_0 \quad \delta_{22}^1(C_{21}^{(2)}) \subset P_0 \quad \delta_{22}^2(C_{21}^{(2)}) \subset P_1.
\end{aligned}
$$

7. We now update the equation for each of the above domains of computation.

(a) For a_2 computed in $C_{20}^{(0)}$ we have $C_{20}^{(0)} \to D_0^{-1}C_{20}^{(0)} = C_{20}^{(0)}$.

(b) The array a_2 on the left hand side is renamed a_{20}.

(c) On the right hand side we:

    i. rename the arrays as follows: $a_3 \to a_{30};\ a_2^1 \to a_{21};\ a_2^2 \to a_{20}$.

    ii. update and rename the dependence maps as follows:

$$
\delta_{32} \to \delta_{30,20};\quad D_{30,20} = D_0^{-1}D_{32}D_0 = I;\quad d_{30,20} = D_0^{-1}d_{32} = \begin{pmatrix} 1 \\ 0 \end{pmatrix}
$$

$$
\delta_{22}^1 \to \delta_{21,20};\quad D_{21,20} = D_1^{-1}D_{22}^1 D_0 = I;\quad d_{21,20}^1 = D_1^{-1}d_{22}^1 = \begin{pmatrix} -3 \\ -1 \end{pmatrix}
$$

$$
\delta_{22}^2 \to \delta_{20,20};\quad D_{20,20} = D_0^{-1}D_{22}^2 D_0 = I;\quad d_{20,20}^2 = D_0^{-1}d_{22}^2 = \begin{pmatrix} 0 \\ -1 \end{pmatrix}.
$$

The resulting recurrence equation is:

$$
1 \le j \le n-1,\ (j+5)/2 < i \le j+1,
$$
$$
a_{20}(i,j) = -a_{30}(i+1,j)a_{21}(i-3,j-1) + a_{20}(i,j-1) \tag{9b}
$$

This procedure is applied to the other domains of computation, resulting in the following equations:

$$
1 \le j \le n-1,\ j/2 \le i \le \min((j+5)/2,\ j+1),
$$
$$
a_{20}(i,j) = -a_{30}(i+1,j)a_{20}(j-i+2,j-1) + a_{20}(i,j-1) \tag{9a}
$$

$$1 \leq j \leq n - 1, \quad j/2 + 1 < i \leq j - 1,$$
$$a_{21}(i,j) = -a_{31}(i-1,j)a_{20}(i+2,j-1) + a_{21}(i-1,j-1) \tag{9c}$$
$$1 \leq j \leq n - 1, \quad j/2 < i \leq \min(j-1, (j+1)/2),$$
$$a_{21}(i,j) = -a_{30}(j-i+1,j)a_{20}(i+2,j-1) + a_{20}(j-i,j-1) \tag{9d}$$
$$1 \leq j \leq n - 1, \quad (j+1)/2 < i \leq \min(j-1, j/2+1),$$
$$a_{21}(i,j) = -a_{30}(j-i+1,j)a_{20}(i+2,j-1) + a_{21}(i-1,j-1) \tag{9e}$$

Eqs. (9b), (9c) are in the uniform subdomain. All other equations are in the nonuniform subdomain. Further simplification can be achieved by taking the specific structure of this SRE into account. For more detail see [10].

References

[1] P. R. Cappello and K. Steiglitz. Unifying VLSI array design with linear transformations of space-time. In F. P. Preparata, editor, *Advances in Computing Research, vol. 2, VLSI Theory*, pages 23–65, JAI Press, Inc., 1984.

[2] C. Choffrut and K. Culik II. Folding of the plane and the design of systolic arrays. *Information Processing Letters*, 17:149–153, 1983.

[3] J.-M. Delosme and I. C. F. Ipsen. An illustration of a methodology for the construction of efficient systolic architectures in VLSI. In *Proc. 2nd Int. Symp. on VLSI Technology, Systems and Applications*, pages 268–273, Taipei, 1985.

[4] J-M Delosme and I. C. F. Ipsen. *Systolic Array Synthesis: Computability and Time Cones*. Technical Report Yale/DCS/RR-474, Yale, May 1986.

[5] J. A. B. Fortes and D. I. Moldovan. Parallelism detection and algorithm transformation techniques useful for VLSI architecture design. *J. Parallel Distrib. Comput*, 2:277–301, Aug. 1985.

[6] R. M. Karp, R. E. Miller, and S. Winograd. The organization of computations for uniform recurrence equations. *J. ACM*, 14:563–590, 1967.

[7] D. I. Moldovan. On the design of algorithms for VLSI systolic arrays. *Proc. IEEE*, 71(1):113–120, Jan. 1983.

[8] M. Newman. Matrix representation of groups. In *Applied Mathematics Series - 60*, Institute for Basic Standards, National Bureau of Standards, Washington D.C. 20234, July 1968.

[9] S. K. Rao. *Regular Iterative Algorithms and Their Implementation on Processor Arrays*. PhD thesis, Stanford University, October 1985.

[10] Y. Yaacoby and P. R. Cappello. *Converting Affine Recurrence Equations to Quasi-Uniform Recurrence Equations*. Technical Report 18, Dept. Computer Science, UCSB, Santa Barbara, CA 93106, Feb. 1988.

[11] Y. Yaacoby, P. R. Cappello, D. Witt, and K. C. Millett. *Computing a Fundamental Region for a Finite Matrix Group Acting on a Euclidean Space*. Technical Report 4, Dept. Computer Science, UCSB, Santa Barbara, CA 93106, Feb. 1988.

BETTER COMPUTING ON THE ANONYMOUS RING
(Extended Abstract)

Hagit Attiya and Marc Snir[*]

1. INTRODUCTION

A number of processors (n) are arranged in a ring configuration, in which each processor is connected by communication channels to its two neighbors. The processors are indistinguishable from each other, and all execute the same algorithm *(anonymous ring)*. In the *asynchronous* model of computation message transfer time is arbitrary (but always finite). In the *synchronous* model of computation message transfer time is fixed, and all processors are synchronized.

In [ASW] it is shown that deterministic algorithms for many problems in the asynchronous anonymous model require at least $\Omega(n^2)$ messages to be sent in the worst case; synchronous algorithms require $\Omega(n \log n)$ messages in the worst case. Two examples are the problem of computing the XOR of binary input values, and the problem of orienting a ring. Syrotiuk and Pachl showed that these problems can be solved asynchronously with $O(n \sqrt{n})$ messages on the average. Here we show that their result can be further improved. We prove the following results:

The most general problem, of collecting the input values, can be solved in the asynchronous model by a deterministic algorithm using $O(n \log n)$ messages on the average; it can be solved by a probabilistic algorithm that uses $O(n \log n)$ expected number of messages, on any input, with one random bit at each processor. A matching $\Omega(n \log n)$ lower bound on the average complexity is shown for "nonlocal" problems, where the answer is not determined by a short substring of the inputs (in particular, for XOR and orientation). The lower bound holds for bidirectional rings and for nonuniform algorithms that depend depend on the ring size. The same $\Omega(n \log n)$ lower bound holds for asynchronous probabilistic algorithms.

The input collection problem is solved on synchronous rings by an algorithm that uses $O(n)$ messages, on the average. This algorithm can be used to elect a leader in a labeled ring with $O(n)$ messages on the average. A probabilistic algorithm solves the input collection problem with $O(n)$ expected number of messages on any input, using one random bit per processor.

These results provide an example where probabilistic methods provably reduce complexity; a surprisingly small amount of randomization is sufficient to achieve this result.

The worst case performance of probabilistic algorithms is not always equal to the average case performance of deterministic algorithms: We show that the AND function can be computed by a deterministic asynchronous algorithm with $O(n)$ messages, on the average, whereas any asynchronous probabilistic algorithm that computes AND requires $\Omega(n \log n)$ messages, in the worst case.

Finally we examine Monte-Carlo asynchronous algorithms. If a probability p of error is tolerated, than the input collection problem can be solved with $O(n(1 + \log\log(1/p)))$

---

[*] *Hagit Attiya*, Dept. of Computer Science Tel-Aviv University, Tel-Aviv 69978, ISRAEL; *Marc Snir*, IBM T. J. Watson Research Center, Yorktown Heights, NY 10598, USA.

expected number of messages. This implies that a leader can be elected in a labeled asynchronous ring with $O(n)$ messages, and small, constant error probability p. The algorithm is nonuniform, and depends on n, the ring size. Interestingly, a lower bound of $\Omega((1-p)n \log n)$ is given [P] for uniform leader election algorithms (algorithms that work on rings of unknown size).

A Monte-Carlo distributed algorithm may fail by deadlocking, or it may fail by arriving at a wrong answer. If deadlock is prohibited, then we show that any asynchronous algorithm that computes AND with probability of error at most p uses $\Omega(n(\log n - \log\log(1/(1-p))))$ messages, in the worst case. Thus, the reduction in message complexity for Monte-Carlo distributed algorithms is almost entirely due to the acceptance of some probability of deadlock.

2. DEFINITIONS AND PRELIMINARY RESULTS

Consider a system of n processors, numbered $1, \cdots, n$, on a ring (the numbering is not available to the processors). Processor i has links to its two neighbors, $left(i)$ and $right(i)$. We denote by O_i the $orientation$ of processor i: $O_i = 1$ if $right(i) = i + 1$, $O_i = 0$ otherwise. We consider computation problems where inputs are Boolean and a function of the inputs, such as XOR or AND, is computed at each processor. A special case is the problem of $orientation$: Each processor i computes a binary output f_i, such that $f_i = f_j$ iff $O_i = O_j$. It is impossible to solve certain problems (for example XOR and orientation) if the number of processors on the ring is unknown [ASW]. Hence we assume that an algorithm is designed for a specific ring size n; we denote such algorithm by the subscript n.

The $complexity$ of a synchronous algorithm A_n on input I, $C(A_n, I)$ is the number of messages sent in a computation on input I. The $(worst\ case)\ complexity$ of an algorithm A_n, $C_{max}(A_n)$, is maximum of $C(A_n, I)$ over all inputs I. The $average\ complexity$, $C_{aver}(A_n)$ is the average of $C(A_n, I)$ over all inputs. Asynchronous algorithms are non-deterministic; the computation may depend on the order messages are forwarded. We represent this by a $scheduler$. After each transition the scheduler selects the next message to be received. The $complexity$ of an asynchronous algorithm A_n on input I, $C(A_n, I)$ is the number of messages sent in a computation of the algorithm on input I, against a worst scheduler. $C_{max}(A_n)$, and $C_{aver}(A_n)$ are defined as above.

$Deterministic$ algorithms are modeled by deterministic automata; a $probabilistic$ algorithm is modeled by a probabilistic automaton, with probabilistic transitions. A probabilistic algorithm solves a problem $with\ error\ p$ if for any input there is a probability $\geq 1-p$ that all processors halt with a correct answer to the problem. In particular, an $errorless\ probabilistic\ algorithm$ always delivers the right answer. The complexity of an asynchronous probabilistic algorithm A_n on input I, $\overline{C}(A_n, I)$, is the expected number of messages sent against a worst scheduler. The worst case complexity of a probabilistic algorithm, $\overline{C}_{max}(A_n)$, and the average case complexity, $\overline{C}_{aver}(A_n)$, are defined accordingly. The definition for synchronous algorithms is obvious.

Define the $k - neighborhood$ of processor i to be the concatenation of the input values and orientations of the processors at most k apart from processor i, relative to the orientation of i. The $\lfloor n/2 \rfloor$-neighborhood of a processor contains information on the entire ring configuration, relative to the location and orientation of the processor. The $input\ collection$ problem consists of computing for each processor its $\lfloor n/2 \rfloor$-neighborhood. An algorithm that solves the input collection problem can be used to solve any problem that can be computed on a ring.

3. UPPER BOUNDS

3.1. Asynchronous Input Collection Algorithm

For simplicity of description and analysis we first assume the ring to be unidirectional. The input collection problem can be solved with $O(n)$ messages once a leader has been elected on the ring: The leader initiates a message that circles the ring, first collecting all inputs, next distributing them to all processors. It is not always possible to elect a unique a leader on an anonymous ring [An]. However, the algorithm is still correct if several leaders are elected; each of the elected leaders will distribute the inputs independently. We shall exhibit a leader election algorithm that ends by electing a constant number of leaders, on the average. The leader election algorithm resembles the algorithm of [CR]; a processor with maximum label is elected leader. In this algorithm, each processor creates a message that travels around the ring, carrying its originator's id, until it "meets" a processor labeled with a larger id. A message carrying the id k travels distance k on the average. Altogether, $O(n \log n)$ messages are sent on the average. In our model processors are identical; before the leader election algorithm can be run, id's must be computed. We shall label each processor by the number of consecutive ones to its left. Thus, the algorithm consists of three conceptual phases:

(1) Labeling.
(2) Leader election.
(3) Input collection and distribution.

The actual algorithm given below combines phases two and three together: Inputs are collected by the messages used for leader election. The algorithm is described below; $s \cdot t$ is the string obtained by concatenating s and t; $shift(s)$ is the function that shifts the string s cyclically one position to the right.

Algorithm for processor i

{1st phase - label creation}

```
LABEL := 0;
send INPUT to right;
repeat forever
  receive L from left;
  LABEL := LABEL + L ;
  if LABEL = n then break;
  if INPUT = 1 then send L to right;
  if L = 0 then break
end;
```

{2nd phase - leader election and input collection}

```
send (LABEL ,INPUT ) to right;
repeat forever
  receive (L ,SEG ) from left;
  if | SEG | = n
    then begin
      send (L ,shift(SEG )) to right;
      break
    end;
  if L ≥ LABEL then
    send (L ,SEG ·INPUT ) to right
    {else message is not forwarded}
end.
```

If the ring is bidirectional and unoriented, one runs two versions of the algorithm in parallel, one in each direction. The message complexity of the algorithm at most doubles. If the problem has no binary inputs, such as for orientation, then one can compute an "input bit" by comparing the orientation of a processor relative to the orientation of its left neighbor. If each initial orientation is equally probable, then each string of n

zeroes and ones is equally likely to obtain.

Lemma 3.1: *Let W_k be the waiting time until k consecutive successes occur in a sequence of Bernoulli trials, with success probability $p = 0.5$. Then*

$$E(W_k) = 2^{k+1} - 2.$$

Proof: see [Fe, XIII.7, eq. (7.6)]. $\square$

A message initiated in the first phase is forwarded until it encounters a zero on the ring, or until it has done a full circle, if there are no zeroes on the ring. Thus, the expected distance traversed by such message is bounded by

$$E(W_1 \mid W_1 < n) \cdot \text{Prob}[W_1 < n] + n\,\text{Prob}[W_1 \geq n] < 2 + n\,2^{-(n-1)} = O(1).$$

Since exactly n messages are initiated in the first phase, it follows that the expected number of messages transmitted in this phase is $O(n)$.

Let $LABEL_i$ be the label created for processor i, and let X_i be the number of times the message sent by processor i is forwarded in the 2nd phase. Assume that $LABEL_i = k$. If $k+1$ consecutive ones occur at locations $j-k-2, \cdots, j-1$ then $LABEL_j > k$ and processor j does not forward the message initiated by processor i. No message is forwarded more than $2n-1$ times. It follows, by Le. 3.1, that for $1 \leq k \leq \log n$

$$E(X_i \mid LABEL_i = k-1)$$

$$< E(W_k \mid W_k < n-k)) \cdot \text{Prob}[W_k < n-k] + (2n-1) \cdot \text{Prob}[W_k \geq n-k]$$

$$\leq E(W_k) + (2n-1)E(W_k)/(n-k) \leq 2^{k+1} - 2 + (2n-1)(2^{k+1} - 2)/(n-k) \leq 2^{k+3}$$

Summing up over all the values of $LABEL_i$ we obtain that

$$E(X_i) < \sum_{k=1}^{\log n} E(X_i \mid LABEL_i = k-1) \cdot \text{Prob}[LABEL_i = k-1] + (2n-1) \cdot \text{Prob}[LABEL_i \geq \log n]$$

$$\leq \sum_{k=1}^{\log n} 2^{k+3} 2^{-k} + (2n-1)2^{1-\log n} < 8\log n + 4.$$

Since exactly n messages are initiated at this phase, it follows that the average number of messages transmitted is $O(n \log n)$.

We sum up the results of this section in the following theorem.

Theorem 3.2: *For each n here exists a deterministic input collection algorithm IC_n, that works on the asynchronous anonymous ring, and has average complexity $C_{\text{aver}}(IC_n) = O(n \log n)$.*

3.2. AND and Other Problems

The last theorem shows that any computable function, can be computed with $O(n \log n)$ messages, in the average. In §4 we show this result is optimal for problems such as XOR or orientation. For some other problems one can do better.

Theorem 3.3: *For each n there is an asynchronous algorithm AND_n that computes the AND of n inputs on an anonymous ring of length n, with an average number of messages $C_{\text{aver}}(AND_n) = O(n)$.*

Proof: Each processor starts by sending to its right and left a message with its input value, and count one. Afterwards, it forwards the messages it receives, incrementing their count. A processor halts with output zero after it has sent a zero message; it halts with output one if it receives back a one message with count n (i.e. a one message that made a full circle). It is easy to check the algorithm computes AND correctly and that the expected distance traversed by a message is constant. $\square$

A similar algorithm can be used to compute any function which value is determined by a small prefix. Let $f : \Sigma^n \to \Sigma$ be a shift invariant function. Let s be a string of length k that determines the value of $f : f(s \cdot t_1) = f(s \cdot t_2)$, for any strings t_1, t_2 of length $n-k$. We have

Theorem 3.4: *The function f can be computed asynchronously on an oriented ring with $O(nk\,2^k)$ messages, on the average.*

Proof: See [AS]. $\square$

A similar result holds for nonoriented rings (we require then that f be invariant under shifts and reversals). The AND algorithm is a particular case, for $k=1$.

3.3. Asynchronous Probabilistic Algorithms

The input collection algorithm can be easily modified to yield a probabilistic algorithm, that solves the input collection problem in $O(n \log n)$ expected messages, on any input: Select a random bit at each processor, and use this bit to build labels. The expected number of messages sent by this algorithm does not depend on the input, and equals to the average number of messages sent by the input collection algorithm of §3.1. We obtain:

Theorem 3.5: *For each n there exists an errorless probabilistic input collection algorithm PIC_n that uses on any input an expected number of messages $\overline{C}_{\max}(\mathrm{PIC}_n) = O(n \log n)$. This algorithm uses a unique random bit at each processor.*

Thus, any solvable problem can be solved on an asynchronous anonymous ring with $O(n \log n)$ expected messages, using a unique random bit at each processor. In §4 we show this result is optimal for problems such as AND, XOR and orientation. On the other hand, if a positive error probability is tolerated, then the expected number of messages can be reduced to $O(n)$.

Theorem 3.6: *Let p be a fixed positive number. There exist an asynchronous probabilistic input collection algorithm EPIC_n with error probability $\leq p$ and expected number of messages $C_{\max}(\mathrm{EPIC}_n) = O(n(1 + \log\log(1/p))$ on any input.*

Proof: See [AS]. $\square$

3.4. Synchronous Algorithms - Input Collection

Consider the asynchronous input collection algorithm, with the basic three phases: (1) labeling; (2) leader election; and (3) input collection and distribution. The 1st phase requires a linear number of messages, on the average. A more accurate analysis of the leader election process shows that a constant number of leaders are elected on the average; hence, the 3rd phase takes, too, a linear number of messages, on the average. The 2nd phase can be avoided altogether in the synchronous model: We divide the input collection and distribution phase into $n+1$ subphases; at subphase i, which takes $2n$ cycles, only processors with label $n+1-i$ collect and distribute inputs. The algorithm stops as soon as a subphase with active processors occur. Thus, only leaders collect and distribute inputs.

We describe below the algorithm for unidirectional, oriented rings. The algorithm is extended to nonoriented rings, and used to solve the orientation problem, as in the asynchronous case. The total number of cycles used by it is at most $2n^2$. We show that the expected number of messages sent is linear.

Algorithm for processor i

| { First Phase } | { Second Phase } |

```
{ First Phase }

LABEL := 0;
if INPUT = 1 then send message to right;
for j := 1 to n  do
   if received message from left then
   begin
   LABEL := LABEL + 1;
   if INPUT = 1 then send message to right
   end;
```

```
{ Second Phase }

for i := n  downto 0 do
   begin
   if LABEL = i then send INPUT  to right;
   for j := 1 to 2n −1 do
      if received message M then
         if | M | < n then send INPUT&M to right
         else { | M | = n }
         begin
         send shift (M ) to right;
         halt
         end
   end.
```

The first phase is essentially identical to the first phase of the asynchronous input collection algorithm; the expected number of messages sent is $O(n)$. A message is forwarded in the second phase at most $2n−1$ times. If the ring has only ones then n messages are initiated; this happens with probability 2^{-n}. Otherwise, define a *run* on the ring to be a segment of zero or more consecutive ones bordered on its left and its right by zeroes. The number of runs on the ring equals the number of zero inputs. Let *MaxRun* be the number of maximum length runs on the ring; this is the number of messages initiated in the second phase of the algorithm. We terminate the proof by showing that $E(MaxRun) = O(1)$. In order to do so we need the following auxiliary result, which is proven by an argument similar to that used by Rabin [Ra].

Lemma 3.7: *Consider r independent sequences of Bernoulli trials, with probability $p = 0.5$ of success. Let M_r be the number of sequences where the waiting time for the first success is maximum. Then $E(M_r) \le 2$.*

Order the runs on the ring. The last lemma shows that the expected number of maximal length runs among the first r runs on the ring is constant, for any fixed number r. We conclude the proof by showing that with probability $\ge 1 - 1/n$ all the maximal length runs on the ring are among the first $r = n/2 - \sqrt{n} \log n$ runs.

Theorem 3.8: $E(MaxRun) = O(1)$.

Proof Outline: We may assume that the inputs at processors $1, 2, \cdots, n$ are generated by a sequence of n Bernoulli trials, with $p = 0.5$. Let Y be the number of maximum length runs in this sequence, ignoring wrap-around. Clearly, $MaxRun \le Y + 1$. Let F be the number of failures in the sequence. Define the following events:

$A_1 = [\, n/2 - \sqrt{n} \log n < F < n/2 + \sqrt{n} \log n \,]$;
$A_2 = [\,$ there is a run of length $\ge 0.8 \log n$ among the first $n/2 - \sqrt{n} \log n$ runs $]$;
$A_3 = [\,$ there is no run of length $> 0.6 \log n$ among the next $2\sqrt{n} \log n$ runs $]$.

Let $A = A_1 \bigcup A_2 \bigcup A_3$. Then, since $Y \le n$,

$$E(Y) \le E(Y \mid A)\mathrm{Prob}[A] + n(1 - \mathrm{Prob}[A])\,.$$

If A occurs then Y is the number of maximum length runs within the first

$n/2 - \sqrt{n} \log n$ runs. Since each run is a recurrent event, it follows that

$$E(Y \mid A) \text{Prob}[A] \leq E(M_{n/2 - \sqrt{n} \log n}) \leq 2.$$

We conclude the proof by showing that

$$\text{Prob}[A_i] \geq 1 - 1/n, \text{ for } i = 1,2,3.$$

The bound on the probability of A_1 follows from Chernoff's bounds on the tail of a Bernoulli distribution [Ch, Rag]; the bounds on the probabilities of A_2 and A_3 follow from a straightforward analysis. □

We have proven

Theorem 3.9: *For each* n *there exist a deterministic synchronous input collection algorithm* SIC_n *that has average message complexity* $\mathbf{C}_{\text{aver}}(SIC_n) = O(n)$.

The last algorithm can be used to elect a leader in a synchronous labeled ring with $O(n)$ messages on the average, using only comparisons. Assume the ring is oriented. Then each processor computes a bit by comparing its id to the id of its left neighbor. The last algorithm is then run, using these bits as inputs. In addition to collecting the "input" bits, one also collects the processor id's. The processor with the largest id is then elected as leader. This algorithm can be extended to nonoriented rings.

We can modify the deterministic input collection algorithm to obtain a probabilistic input collection algorithm that has average message complexity $O(n)$: Select a random bit at each processor, and use these bits to create labels in the first phase. Run the second phase as before. We obtain

Theorem 3.10: *For each* n *there exist an errorless probabilistic synchronous input collection algorithm* PSIC_n *that uses a unique random bit per processor, and has expected message complexity on any input* $\overline{\mathbf{C}}_{\max}(\text{PSIC}_n) = O(n)$.

4. LOWER BOUNDS

4.1. Deterministic Asynchronous Algorithms

Lower bounds for asynchronous computations are proven using as adversary a suitable scheduler. We use a simple *synchronizing scheduler*, that keeps the computation as symmetric as possible. This scheduler delivers messages in *cycles*. All processors start the execution at cycle one; all messages sent at cycle i are received at cycle $i + 1$.

Lemma 4.1: *Under the synchronizing scheduler, the state of a processor after* i *cycles depends only on its* i *-neighborhood.*

Definition: A function $f : \Sigma^* \to \Sigma$ is *nonlocal* if there exist a constant $0 < c < 1$ such that for any string s of length $|s| \leq cn$ there exist two strings t_1 and t_2 such that $|s \cdot t_1| = |s \cdot t_2| = n$ and $f(s \cdot t_1) \neq f(s \cdot t_2)$.

The function f is nonlocal if its value can not be determined from the value of a small prefix. The XOR function is nonlocal. More generally, we say that a computation problem is *nonlocal* if there is a constant $0 < c < 0.5$ such that on any initial input, the output at a processor can not be determined from the value of its cn-neighborhood. We have

Lemma 4.2: *The orientation problem is nonlocal.*

The proof of the following lemma is immediate from the definition of the synchronizing adversary and Le. 4.1.

Lemma 4.3: *Consider a computation of an algorithm under the synchronizing scheduler. If no message is sent in that computation at cycle* i, *then no transition occurs, and no*

message is sent at any cycle j, for $j > i$.

Corollary 4.4: *Let A_n be an asynchronous algorithm that solves a nonlocal problem on rings of size n. Consider a computation of A_n under the synchronizing scheduler. Then a message is sent by some processor at each cycle i, for $i = 1, \cdots, cn$.*

Let A_n be an algorithm for rings of size n. Let $S_k(A_n)$ be the set of k-neighborhoods that cause a message to be generated by the processor with that neighborhood at the k-th cycle, when the algorithm A_n is executed with the synchronizing scheduler. Our proof uses a counting argument similar to that used by Bodleander [B] to show that the sets $S_k(A_n)$ are large. Since rings are bidirectional, a more delicate counting argument is needed. Let $N(s)$ be the set of strings that appear cyclically in string s (i.e. equal to a prefix of some cyclic shift of s). We have

Lemma 4.5: *Let A_n be an asynchronous deterministic algorithm that solves a nonlocal problem on rings of size n. Let s be the configuration of k successive processors on a ring, where $k \mid n$ and $k < cn$; let r be an integer such that $2r + 1 \leq k$. Then $N(s) \cap S_r(A_n) \neq \emptyset$.*

Proof: Denote $l = n/k$ and look at the configuration $C = s^l$. By Corollary 4.4 a message is sent by some processor at each cycle r, $r = 1, \cdots, k/2$ on the computation of A on C under the synchronizing adversary. The r-neighborhood of this processor is the required string. $\square$

Corollary 4.6: *Let A_n, r and k be as in the previous lemma. Then*

$$| S_r(A_n) | \geq 2^{2r+1}/k.$$

Theorem 4.7: *Let A_n be an asynchronous deterministic algorithm that solves a nonlocal problem on rings of size n. Let $d_1 < d_2 < \cdots < d_r$ be the sequence of divisors of n, ordered in increasing order. Then*

$$C_{aver}(A_n) \geq 0.5n \sum_{d_i \leq cn} (1 - d_{i-1}/d_i).$$

Proof: Let s be a random input of length n. Let t be the r-neighborhood of a fixed processor in s, where $d_{i-1} < 2r + 1 \leq d_i \leq cn$. According to Cor. 4.6,

$$\text{Prob}[t \in S_r(A_n)] \geq 1/d_i.$$

Thus, the expected number of messages sent at cycle r on input s is $\geq n/d_i$, for $r \leq d_i \leq cn$. The total expected number of messages is obtained by summing:

$$C_{aver}(A_n) \geq \sum_{d_i \leq cn} \sum_{d_{i-1} < k \leq d_i,\, k \text{ odd}} n/d_i \geq 0.5n \sum_{d_i \leq cn} (1 - d_{i-1}/d_i).$$

$\square$

Corollary 4.8: *Let $n = 2^k$, then for any asynchronous deterministic algorithm A_n solving a nonlocal problem on rings of size n, $C_{aver}(A_n) = \Omega(n \log n)$.*

Since XOR and orientation are nonlocal it follows:

Corollary 4.9: *Let $n = 2^k$, then for any asynchronous deterministic algorithm A_n computing XOR or orientation on rings of size n, $C_{aver}(A_n) = \Omega(n \log n)$.*

4.2. Probabilistic Algorithms

Let A_n be an errorless probabilistic algorithm with input alphabet Σ. Assume there is a fixed upper bound on the number of messages sent by A_n (the bound may depend on n). We can replace an execution of A_n by a computation whereby each processor first chooses independently a random binary string of fixed length q, next runs a deterministic algorithm A_n^d (with input set $\Sigma \times \{0,1\}^q$). The resulting algorithm is errorless and the average complexity of A_n can be made arbitrarily close to the average complexity of A_n^d. If A_n solves a nonlocal problem, then A_n^d solves, too, a nonlocal problem. Thus, Cor. 4.9 implies the following result.

Theorem 4.10: *Let A_n be an errorless probabilistic asynchronous algorithm that solves a nonlocal problem on an anonymous ring of size $n = 2^k$. Then $\overline{C}_{aver}(A_n) = \Omega(n \log n)$.*

In particular, any errorless probabilistic algorithm that computes XOR or orients a ring has average expected complexity $\Omega(n \log n)$. This implies an $\Omega(n \log n)$ lower bound for the expected number of messages sent on the worst input, for any nonlocal problem.

The last $\Omega(n \log n)$ lower bound on average complexity does not apply to the AND function. Indeed, we have shown in Th. 3.4 that AND can be computed with a linear average number of messages. Nevertheless, we can prove an $\Omega(n \log n)$ lower bound on the expected number of messages sent on the worst input.

Theorem 4.11: *Let A_n be an errorless probabilistic asynchronous algorithm that computes the AND of n inputs. Then the expected number of messages transferred in the computation with input $1, \cdots, 1$ is $\Omega(n \log n)$.*

Proof: Let A_n^d be the associated deterministic algorithm, and consider the computations of this algorithm on inputs of the form $1s_1, \cdots, 1s_n$ under the synchronizing scheduler. The average number of messages sent on these inputs by A_n^d is equal to the expected number of messages sent by A_n on input $1, \cdots, 1$. Since $AND(1, \cdots 1, 1) \neq AND(1, \cdots, 1, 0)$ A_n must run for at least $n/2$ cycles on the input $1, \cdots, 1$. Thus, a computation of A_n^d on any input of the form $1s_1, \cdots, 1s_n$ runs for at least $n/2$ cycles. The lower bound follows, using the same arguments as in Le. 4.5, Cor. 4.6, and Th. 4.7. □

The complexity of AND can be reduced to $O(n)$ if a fixed positive probability for error is allowed. The simple algorithm given in Th. 3.6 either gives the right answer, or deadlocks. If we insist that the algorithm never deadlocks, so that the only failure mode is a wrong answer, then the situation is different; almost nothing is saved by allowing errors. We have

Theorem 4.12: *Let A_n be a probabilistic asynchronous algorithm that computes the AND of $n = 2^s$ inputs with error p, and never deadlocks. Then*

$$\overline{C}_{aver} = \Omega(n(\log n - \log\log(1/(1-p))))$$

Proof: See [AS]. □

The last bound implies that a probabilistic asynchronous algorithm that computes AND and never deadlocks requires $\Omega(n \log n)$ messages, as long as the probability of success is significantly larger than 2^{-n}. Using this result, we can show that Th. 4.11 is valid even for unbounded computations; the same holds true for XOR or orientation. Interestingly, it is possible to compute AND with $O(n)$ messages on a ring of odd length n, with a probability of success $n\, 2^{-n}$, and no deadlock; thus, the last result is optimal.

5. CONCLUDING REMARKS

We have shown in Th. 3.4 that whenever there exists a string of length less than $\log\log n - \log\log\log n$ that determines the value of a function f, then f can be computed in less than $n \log n$ messages, on the average. On the other hand, when the shortest such string has length cn, then the $n \log n$ lower bound applies. This leaves an open gap between $\log\log n$ and cn.

The input collection algorithm uses messages containing up to n bits. It is easy to modify the algorithm so that no more than $O(n^2)$ bits are transferred on the average. Simple information transfer arguments show that this is optimal. The number of bit transfers can be reduced for XOR and orientation to $O(n \log^2 n)$. Similarly, input collection can be done probabilistically with $O(n^2)$ bit complexity, and XOR and orientation with $O(n \log^2 n)$ bit complexity, in the worst case.

Finally, note that the synchronizing scheduler we use to prove lower bounds for asynchronous rings is very simple, and input independent; it merely mimics a synchronous computation (compare with the complex malicious scheduler used in [DG]). This scheduler keeps the computation as symmetric as possible. Here, as in [ASW], the lower bounds reflect the cost of breaking symmetry.

ACKNOWLEDGEMENTS

We would like to thank Michael Benor, Rafi Hassin and Prabhakar Raghavan for their useful suggestions.

REFERENCES:

[An] D. Angluin, "Local and Global Properties in Networks of Processors," *12th Ann. ACM Symp. on Theory of Computing*, (1980) 82-93.

[AS] H. Attiya and M. Snir "Better Computing on the Anonymous Ring," Research Report, IBM T. J. Watson Res. Center, April 1988.

[ASW] H. Attiya, M. Snir and M. Warmuth, "Computing on the Anonymous Ring," *J. of the ACM*, to appear.

[B] H. Bodleander, "Distributed Algorithms: Structure and Complexity," Ph.D. Thesis, University of Utrecht, 1986.

[Ch] H. Chernoff, "A Measure of Asymptotic Efficiency for Tests of a Hypothesis Based on the Sum of Observations," *Annals of Math. Stat.*, **23** (1983) 223-234.

[CR] E. Chang and R. Roberts, "An Improved Algorithm for Decentralized Extrema-Finding in Circular Configurations," *Comm. of the ACM*, **22** (1979) 281-283.

[DG] P. Duris and Z. Galil, "Two Lower Bounds in Asynchronous Distributed Computation," proceedings, *28th Ann. IEEE Symp. on Foundations of Computer Science*, (1987) 326-330.

[Fe] W. Feller, *An Introduction to Probability Theory and its Applications*, Third Edition, Vol. 1. John Wiley & Sons Inc., New York, 1968.

[P] J. Pachl, "A Lower Bound for Probabilistic Distributed Algorithm," *J. of Algorithms*, **8** (1987) 53-65.

[PKR] J. Pachl, E. Korach and D. Rotem, " A New Technique for Proving Lower Bounds for Distributed Maximum-Finding Algorithms," *J. ACM*, **31** (1984) 905-918.

[Ra] M. Rabin, "N-Process Synchronization by $4 \cdot \mathrm{Log}_2 N$-Valued Shared Variable," *J. of Computer and System Sciences*, **25** (1982) 66-75.

[Rag] P. Raghavan, "Randomized Rounding and Discrete Ham-Sandwich Theorems: Provably Good Algorithms for Routing and packing Problems," Report UCB/CSD 87/312, Comp. Sc. Div., Univ. of California Berkeley, July 1986.

[SP] V. Syrotiuk and J. Pachl, "Average Complexity of a Distributed Orientation Algorithm," Research Report CS-87-23, Univ. of Waterloo, March 1987.

Network Complexity of Sorting and Graph Problems and Simulating CRCW PRAMS by Interconnection Networks (Preliminary Version)

Alok Aggarwal

IBM Research Division

T. J. Watson Research Center

Yorktown Heights, New York.

Ming-Deh A. Huang[1]

Department of Computer Science

University of Southern California

Los Angeles, California.

ABSTRACT: A sequential random access machine can permute n data items in n steps. However, Gottlieb and Kruskal have shown that any bounded degree machine with P processors requires $\Omega((n/P) \log P)$ time to permute n data items and this result makes the issue of optimal speedup interesting for interconnection networks. In this paper, we consider the issue of optimal speedup for sorting and graph problems and provide the following results: (1) If a network with P processors can permute P elements in $O(\log P)$ time, then n data items can be sorted on this network in $\Theta((n/P) \log n)$ time when $n \geq P^{1+\epsilon}$. An important consequence of this result is that a single step of any Concurrent Read Concurrent Write PRAM (CRCW-PRAM) algorithm that uses n processors and $O(n)$ memory space, can be simulated optimally (when $n \geq P^{1+\epsilon}$) by any network with P processors that can permute P data items in $O(\log P)$ time. (2) The connected components, biconnected components, and the minimum spanning forest can be determined in optimal time for any network that has P processors as long as $P \leq n^2 / \log^2 n$ and as long as this network can perform a restrictive set of permutations of P items in $O(\log P)$ time. Our paradigm for solving graph problems is quite general and it can be extended to optimally compute the median of n numbers on interconnection networks.

1. INTRODUCTION

Researchers have studied synchronous parallel computation by proposing two classes of models -- the shared memory models and the fixed connection network models. In a shared memory model such as a Parallel Random Access Machine (PRAM), trading time for processors is always possible, i. e., if a problem can be solved in T time steps with P processors then the same problem can be solved in $O(mT)$ time with P/m processors for $m \leq P$. On the other hand, Gottlieb and Kruskal [GK84] have shown that it is not always possible to trade time for the number of processors in the network model because any permutation algorithm for n data items on a bounded de-

---

[1] This work was done when the author was visiting IBM Thomas J. Watson Research Center in 1984-85.

gree, P-processor interconnection network requires $\Omega((n/P)\log P)$ time. Thus, in contrast to the idealized PRAM model, network model is a more restrictive model of parallel computation and in order to understand the complexity issues for the network model, it seems necessary to understand the following constraints that are inherent in this model:

(a) *Granularity:* Under the network model, instead of sharing a global memory, each processor has a local memory. Conflicts occur when several processors want to access variables in the same local memory, even though these variables may be distinct. Granularity especially becomes an issue when the number of processors is less than the input size.

(b) *Communication:* Because of the bounded degree of processors in the network, a processor may require $\Omega(\log P)$ time to access a data element that is not present in its memory even if there are no conflicts while accessing the variable. Consequently, for uniform families of networks with bounded degree nodes, the communication constraint cannot be overlooked.

In this paper, we consider the issue of optimal speedup for sorting and for solving graph problems. In section 2, we show that if a network with P processors can permute P elements in $O(\log P)$ time (possibly, by using an offline algorithm) then n data items can be sorted on this network in $\Theta((n/P)\log n)$ time when $n \geq P^{1+\epsilon}$. However, for any arbitrary network that permutes P elements in $O(\log P)$ time, the leading constant in the time bound of our algorithm is somewhat large. In view of this, we also show that n words can be sorted on P-processor Cube-Connected Cycles (and other similar networks) with optimal speedup such that the leading constant in the time bound is smaller. An important consequence of the sorting result, provided in section 3, states that a single step of any Concurrent Read Concurrent Write PRAM (CRCW-PRAM) algorithm that uses n processors and $O(n)$ memory space, can be simulated optimally by P-processor Cube-Connected Cycles, Shuffle-Exchange Network, or related networks when $n \geq P^{1+\epsilon}$. In section 3, we compare our simulation algorithm with similar algorithms proposed by other researchers.

In [GK84], Gottlieb and Kruskal establish that it is not always advantageous to simulate a PRAM algorithm, stepwise, by an interconnection network since such a simulation will always yield an algorithm that is away from the optimal processor-time product by an $\Omega(\log P)$ factor. In view of this, section 4 presents a systematic approach for solving graph problems on interconnection networks when an undirected graph of n nodes is presented by its adjacency matrix. In particular, we show that connected components, biconnected components, and the minimum spanning forest can be determined in optimal time for any network that has P processors as long as $P \leq n^2 / \log^2 n$ and as long as this network can perform a restrictive set of permutations of P items in $O(\log P)$ time. Our paradigm for solving graph problems is quite general and for networks that can permute P items in $O(\log P)$ time, this approach can be extended to compute the median of n numbers with $P \leq n / \log^2 n$ processors in $O(n/P)$ time. Finally, section 5 discusses some open problems and areas of future research.

An important distinction between the approach that we consider in this paper and the previous approaches is that our approach does not depend upon any particular choice of network family but is universally applicable to any bounded degree network family that uses an offline algorithm to permute P elements in $f(P)$ time.

2. OPTIMAL ALGORITHM FOR SORTING n ELEMENTS ON NETWORKS WITH P PROCESSORS

The optimal sorting algorithm given in this section is first described for Cube-Connected Cycles and then extended to any network family in which a P-processor network can permute P data items (possibly by using an offline algorithm) in $O(\log P)$ time.

Suppose P elements are distributed on Cube-Connected Cycles with P processors such that each processor has one element. Then, Preparata and Vuillemin [PV81] have shown that ASCEND and DESCEND paradigms can be used, *offline*, to obtain any permutation of these P elements in $c \log P$ time, for a suitable positive constant, c. (For the definitions of ASCEND and DESCEND, see [PV81]). Since Cube-Connected Cycles have a recursive structure, Preparata and Vuillemin also show that P elements can be permuted in $c \log(P/2)$ time (using ASCEND or DESCEND paradigms, offline) if this permutation is composed of two permutations where the first permutation is on the first $P/2$ elements and the second permutation is on the last $P/2$ elements. Preparata and Vuillemin exploited this observation to sort n words on Cube-Connected Cycles (CCC) with n processors in $O(\log^2 n)$ time. In Theorem 2.1, we use the same observation to show that n words can be sorted in $O((n/P) \log n)$ time on Cube-Connected Cycles with P processors when $n \geq P^{1+\epsilon}$. Since the processor-time product for this sorting algorithm is $O(n \log n)$, this algorithm is clearly optimal within a constant factor.

Theorem 2.1: For any positive fixed ϵ and for $n \geq P^{1+\epsilon}$, n words can be sorted on Cube-Connected Cycles with P processors in $O((n/P) \log n)$ time.

Proof Idea: We use Columnsort [Lei85] to first show that $2q^3$ elements can be sorted in $O(q^2 \log q)$ time on Cube-Connected Cycles with q processors. Since Cube-Connected Cycles have a recursive structure, Theorem 2.1 will be then established by using this algorithm recursively.

Consider a rectangular matrix of size $2q^2 \times q$ where each entry in the matrix is a distinct word from the $2q^3$ words that have to be sorted. The Columnsort algorithm sorts this matrix in 4 stages where each stage consists of two steps: the first step sorts each column, and the second step permutes the entries of the matrix by a fixed pre-specified amount (i. e., by an amount that does not depend upon any particular data instance). Now, each of the q processors of the Cube-Connected Cycles can simulate the first step of every stage in $O(q^2 \log q)$ time by simply sorting $2q^2$ elements that are present in its memory. Also, since the second step requires permuting the elements by some fixed amount, the *ASCEND* paradigm can be invoked offline and this permutation can be accomplished in $O(q^2 \log q)$ time. Consequently, a rectangular matrix of size $2q^2 \times q$ can be sorted on Cube-Connected Cycles with q processors in $O(q^2 \log q)$ time.

Now, to sort n words on Cube-Connected Cycles with P processors, if $P < \lfloor (n/2)^{1/3} \rfloor$ then we can use the above simulation directly. Otherwise, the first step of the 4 stages of Columnsort can be executed by recursively sorting $\lceil (n/2)^{1/3} \rceil$ sets of at most $\lceil 2^{1/3} n^{2/3} \rceil$ words where $P/\lceil (n/2)^{1/3} \rceil$ processors are assigned to each set. Also, the second step for each of the 4 stages is executed by invoking the *AS-CEND* paradigm. To estimate computation time, let $T(n, P)$ denote the time to sort n words using P processors. Then, it can be readily seen that

$$T(n, P) \leq 4 \times T(\lceil 2^{1/3} n^{2/3} \rceil, P/\lceil (n/2)^{1/3} \rceil) + O((n/P) \log n)$$

when $P \geq \lceil (n/2)^{1/3} \rceil$ and $T(n, P) = O((n/P) \log P)$ when $P < \lceil (n/2)^{1/3} \rceil$. On solving these two inequalities, we obtain $T(n, P) = O((1 + (1/\varepsilon))^\alpha \times (n/P) \log n)$ where $\alpha = (3 - \log 3)/(\log 3 - 1)$. ∎

For a Shuffle-Exchange Network with P processors, it is easy to provide an offline permutation algorithm that follows the ASCEND paradigm and that permutes P elements in $c \log P$ time, for some positive constant c. Consequently, Corollary 2.2 extends Theorem 2.1 for a Shuffle-Exchange network but its proof is omitted.

Corollary 2.2: For any positive fixed ε and for $n \geq P^{1+\varepsilon}$, n words can be sorted on Shuffle-Exchange Network with P processors in $O((n/P) \log n)$ time. ∎

Finally, it may be worthwhile to point out that for both the Shuffle-Exchange Network and the Cube-Connected Cycles, we used the *ASCEND* paradigm to permute elements and the worst-case time for the resulting algorithm is at most $O((1 + (1/\varepsilon))^{\alpha} \times (n/P) \log n)$ where $\varepsilon < 2$ and $\alpha = (3 - \log 3)/(\log 3 - 1)$. A closer look at the proof of Theorem 2.1 shows that to any network with P processors that can permute P elements in $O(\log P)$ time, can in fact sort n words in $O((1 + (1/\varepsilon))^{2/(\log 3-1)} \times (n/P) \log P)$ time irrespective of whether it can execute an algorithm that obeys the ASCEND paradigm. Consequently, the algorithm given in Theorem 2.1 remains optimal (within a constant factor) for any such network but the associated constant factor becomes larger.

Corollary 2.3: Suppose a network with P processors can permute P elements (possibly, by using an offline algorithm) in $O(\log P)$ time where each processor contains one element and the permutation is a one to one permutation. Then, for any positive fixed ε, and for $n \geq P^{1+\varepsilon}$, n words can be sorted on this network with P processors in $\Theta(n \log n/P)$ time. ∎

Before concluding this section, we should note that the sorting network having P processors that has been proposed by Ajtai, Komlos, and Szemeredi [AKS83] can sort n data items in $O((n/P) \log n)$ time for all values of P between 1 and n. However, the leading constant in the time bound for the sorting algorithm on that network is extremely large and that algorithm does not work optimally for other networks.

3. ON-LINE SIMULATION OF CRCW-PRAM ALGORITHMS

Theorem 3.1: Let t, s, and n denote, respectively, the time, memory space, and the number of processors required by a Concurrent Read Concurrent Write PRAM (CRCW-PRAM) algorithm. Then, this algorithm can be simulated in time $O((tn/P) \log n)$ on Cube-Connected Cycles (or Shuffle-Exchange Network) with P processors when $s = O(n)$ and $n \geq P^{1+\varepsilon}$ for any positive constant ε.

Proof Idea: The basic idea is to simulate a step of the CRCW-PRAM by $O((n/P) \log P)$ steps of the network. For this, we use a well known simulation method; variants of this simulation method have been independently proposed by Nassimi and Sahni [NS81], Vishkin [Vi83], and Awerbuch, Israeli, and Shiloach [AIS83]. ∎

On incorporating Corollary 2.3, Theorem 3.1 can be generalized to any network with P processors that uses an offline algorithm for permuting P data items in $O(\log P)$ steps. However, here also the leading constant that depends upon $(1 + (1/\varepsilon))$ is correspondingly larger.

Corollary 3.2: Let t, s, and n denote, respectively, the time, memory space, and the number of processors required by a CRCW-PRAM algorithm. And, let N denote any network with P processors that can permute P data items (possibly, by using an offline algorithm) in $O(\log P)$ time. Then, the CRCW-PRAM algorithm can be simulated by N in time $O((tn/P) \log n)$ when $s = O(n)$ and $n \geq P^{1+\epsilon}$ for any positive constant, ϵ. ∎

Before providing an application of Theorem 3.1, it seems worthwhile to compare our simulation algorithm with similar algorithms proposed by other researchers; such a comparison is made below.

Although, Upfal [Up84] investigated randomized algorithms and Mehlhorn and Vishkin [MV83] carried out a comprehensive study of randomized and deterministic simulations of a step of a PRAM on a fast network, the most efficient method for simulating a PRAM step has been proposed by Upfal and Wigderson [UW84]. They show that for any pair (n,s) where s is polynomial in n, there exists a data organization scheme deriving from a concentrator like bipartite graph which simulates one step of any PRAM algorithm that uses n processors and s memory space, by $O(\log n(\log \log n)^2)$ steps of a Shuffle-Exchange network. However, there are several difficulties with their approach:

(a) No explicit construction of the desired bipartite graph is known.
(b) In theory as well as in practice, one would like the simulation method to be uniformly applicable to any family of networks that is parameterized by the network size. One cannot afford to replicate data and the preprocessing cost of explicitly constructing and storing the bipartite graph given in [UW84] seems excessive.
(c) It is away from the optimal simulation time by an $O((\log \log n)^2)$ factor.

In contrast to [UW84], our algorithm provides a uniform and a universal scheme for simulating CRCW algorithms on families of bounded degree networks. The simulation time of this algorithm is optimal and it does not store a bipartite graph nor does it keep many copies of any data element. However, it works optimally only when $s = O(n)$ and $n \geq P^{1+\epsilon}$ where P is the size of the network. In this context, it may also be pointed out that Awerbuch, Israeli, and Shiloach [AIS83] simulate one step of a CRCW-PRAM algorithm if the number of processors is $n^{1+\epsilon}$ and if the processors are allowed to keep $O(n^\epsilon)$ copies of each data item. Again, the processor-time product of their algorithm is not optimal and the simulation algorithm given by [AIS83] keeps several copies of every data item.

Since a P-processor Cube-Connected Cycles can sort n words in $O((n/P) \log^2 n)$ time, the stepwise simulation given by Awerbuch, Israeli, and Shiloach [AIS83] implies that a single step of CRCW-PRAM algorithm that uses n processors and $O(n)$ memory space, can be executed in $O((n/P) \log^2 n)$ time on the Cube-Connected Cycles. This algorithm coupled with the algorithms given by Shiloach and Vishkin [SV82] and Tarjan and Vishkin [TV84] imply that the connected and the biconnected components of a graph with n nodes and m edges can be computed in $O((m/P) \log^3 n)$ time on Cube-Connected Cycles having P processors. Similarly, the tree functions given in [TV84] can be computed on P-processor Cube-Connected Cycles or a P-processor Shuffle-Exchange Network in $O((n/P) \log^3 n)$ time. These simple observations improve upon the result by Gopalakrishnan et al. [GKR85,GKR85a] by an $O(\log n)$ factor but by using Theorem 3.1, we show how to do even better when $n \geq P^{1+\varepsilon}$.

Corollary 3.3: The connected components and the biconnected components of an undirected graph G with n vertices and m edges can be obtained in $O((m/P) \log^2 n)$ time on Cube-Connected Cycles, a Shuffle-Exchange Network, or any network with P processors as long as this network can permute P words in $O(\log P)$ time and as long as $m \geq P^{1+\varepsilon}$ where ε is a fixed positive constant.

Proof: Shiloach and Vishkin [SV82] have provided a parallel algorithm that uses $O(\log n)$ time, $O(m)$ memory space, $n + m$ processors, and that computes the connected components of a graph with n nodes and m edges. Similarly, Tarjan and Vishkin [TV84] have provided a CRCW-PRAM algorithm that takes $O(\log n)$ time and $O(m)$ memory space using $n + m$ processors and that computes the biconnected components of a graph with n nodes and m edges. Consequently, we get the desired result by incorporating Theorem 3.1 and Corollary 3.2. ∎

4. OPTIMAL PROCESSOR-TIME PRODUCT FOR GRAPH PROBLEMS

The simulated CRCW algorithms are often non-optimal because of the extraneous factors that arise due to excessive routing of data. As shown by Gottlieb and Kruskal [GK84], there are problems for which the best achievable speedup on a P-processor network is $O(P/\log P)$. Consequently, unlike the CRCW-PRAM, where processors can always be traded for time, a loss of $\log P$ factor in the processor-time product is imminent if we step-wise simulate an arbitrary CRCW-PRAM algorithm. However, we now demonstrate a paradigm for solving many graph problems that achieves an optimal processor-time product for almost all values of P.

Theorem 4.1: Let $[A_P]$ denote a family of bounded degree networks that is characterized by a permuting function f so that the time to permute P data items on a network in this family having P processors in which one data item is present in each processor, is $f(P)$. If $f(P) = O(\log^k P)$ for any constant $k \geq 1$ then any network in this family that has

$n^2/f(n)$ log n processors can compute the connected components, biconnected components, and the minimum spanning forest, of an n-node graph in $O(f(n) \log n)$ time.

Proof Idea: We provide a sketch of the proof only; the technical details will appear in the final version of the paper.

Suppose we want to find the connected components of an n-node graph that is given by an adjacency matrix of size n by n and suppose a network N belongs to this family and has $n^2/f(n) \log n$ processors. Let τ be any spanning tree of N. Since the network has bounded degree, by removing some edges we can partition τ into $(n/f(n) \log n)^2$ subtrees of processors such that each subtree contains $\Theta(f(n) \log n)$ processors. Note that the i-th subtree in τ corresponds to a group of processors in N, call this group g_i and denote the j-th processor of the i-th group by $P_{i,j}$. (To simplify the presentation, assume that each group has exactly $f(n) \log n$ processors.) Partition the adjacency matrix of size n by n into $(n/f(n) \log n)^2$ submatrices of size $f(n) \log n$ by $f(n) \log n$. For $0 \leq i \leq (n/f(n) \log n)^2 - 1$, index these submatrices, S_i, in row-major order, and, partition each S_i further into $f(n) \log n$ submatrices of size $(f(n) \log n)^{1/2}$ by $(f(n) \log n)^{1/2}$. For $0 \leq j \leq f(n) \log n - 1$, index the submatrices of each S_i in row-major order and call these $R_{i,j}$. See figure. Now, feed $R_{i,j}$ to processor $P_{i,j}$ such that every group, g_i, receives an entire submatrix of size $f(n) \log n$ by $f(n) \log n$ and perform the following steps:

(a) For each processor, find the connected components of its matrix that is of size $(f(n) \log n)^{1/2}$ by $(f(n) \log n)^{1/2}$ in $O(f(n) \log n)$ time. Note that after completion of this step, each processor has at most $(f(n) \log n)^{1/2}$ edges of the graph.

(b) Since the submatrix associated with each processor was "compressed" in the previous step, each group has $O((f(n) \log n)^{3/2})$ edges at the beginning of this step. If $i = k \times (n/f(n) \log n) + r$, then group g_i computes the connected components of its adjacency matrix that contains the edges between the vertices of the following two sets: $[v_{((k-1)f(n) \log n)+1}, \cdots, v_{(kf(n) \log n)}]$ and $[v_{((r-1)f(n) \log n)+1}, \cdots, v_{(rf(n) \log n)}]$. Using a simple technical lemma, it can be shown that the connected components of each group can be found in $O(f(n) \log n)$ time. Thus, at the end of this step, there are at most $f(n) \log n$ edges for each group and we can redistribute the edges such that each processor of the group has a constant number of edges.

(c) At the beginning of this step, all the edges of a set of $n/f(n) \log n$ contiguous set of vertices belong to the processors of group g_i, for any i. Now, the ideas of *supervertices* [HCS79] and *broadcast* and *merge* [Hu85] can be invoked to obtain the connected components of the given graph. This is the most complex part of the algorithm. Basically, it consists of $\log n$ iterations where each iteration consists of a constant number of repetitions of two substeps. The first substep requires the permutation of data from the processors of one group to another group, and the second substep requires the permutation within each group. In the final version of the paper, we will show that both substeps of an iteration can be performed in $O(f(n))$ time so that the entire computation for this step can be performed in $O(f(n) \log n)$ time. ∎

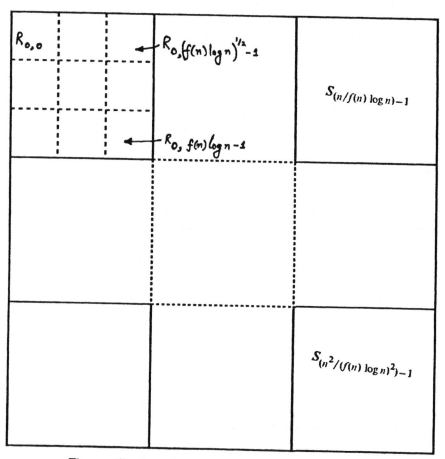

Figure: All submatrices are indexed in row-major order.

For P-processor networks that can permute data in $O(\log P)$ steps, Theorem 2.1 achieves the optimal processor-time product when $T = O(\log^2 n)$. Since this is the best known time bound even for the CREW-PRAM, one cannot hope to achieve a better time bound for the network model unless one achieves a better time bound for CREW-PRAMs. Also, since the permuting function for shuffle-exchange networks and

Cube-Connected Cycles is logarithmic in the number of processors, we obtain optimal processor-time product for solving graph problems on these networks. Theorem 2.1 will be extended to network families with arbitrary permuting functions (such as d-dimensional pyramids and meshes [Ag84]) in the final version of the paper.

Unlike the sorting problem where the number of words to be sorted remains the same throughout the computation, the amount of data is significantly reduced for most graph problems. From the proof of Theorem 2.1, it can be seen that our paradigm for solving graph problems uses this fact crucially and, in fact, we can use this property to achieve optimal speedup for finding the median of a set of n words on an arbitrary network. We again state this result for networks that can permute data in polylogarithmic time; we will generalize it to arbitrary networks in the final version of this paper:

Theorem 4.2: Let $f(P)$ denote the permuting function for a family $[A_P]$ such that $f(P) = O(\log^k P)$ for any constant $k \geq 1$. Then, a network in this family that has $O(n/f(n) \log n)$ processors can compute the median of n elements in $O(f(n) \log n)$ time.

5. DISCUSSION

In this paper, we demonstrated optimal algorithms for sorting n numbers and for solving some graph problems on arbitrary networks with P processors. However, several problems remain open and some of these are discussed below:

(a) Our algorithm for sorting n words works for a class of networks rather than some specific networks but the leading constant in the time bound for this algorithm is polynomial in $(1 + (1/\varepsilon))$. It would be worthwhile to reduce the leading constant in the time bound. Also, it seems really challenging to provide an algorithm that works optimally on Cube-Connected Cycles (or a Shuffle-Exchange network) even when $n = o(P^{1+\varepsilon})$ for every $\varepsilon > 0$.

(b) An important direction of research would be to extend our approach for solving graph problems to other problems and, if possible, to a class of problems. We believe that problems such as that of determining the connected components in a picture with black and white pixels can be solved efficiently using this approach. In general, this approach should be amenable to problems in which input data has to be substantially reduced to obtain the output.

(c) The approach that we used for solving graph problems reduced the input data and this approach has been used in the past. See [AH86,Hu85,RS85,AS83], for example. Nevertheless, it seems that this paper is the first to present a set of unified results for a class of networks rather than a particular network. It improves upon

Awerbuch and Shiloach's result [AS83] in that it uses only $n^2/\log^2 n$ processors (instead of n^2 processors) for finding a Minimum Spanning Forest on a Shuffle-Exchange network. However, even this approach does not do well for sparse graphs. In particular, it remains open whether the connected components of a graph with n nodes and m edges can be determined, say, in $O(\log^2 n)$ time using a Shuffle-Exchange network with only $m/\log^2 n$ processors.

(d) Tight bounds on area-time complexity of graph problems also remains open. For finding the connected components of an n-node graph in Thompson's VLSI model [Th80], Ja'Ja [Ja81] has shown a lower bound of $AT^2 = \Omega(n^2 \log^2 n)$. By adapting our connectivity algorithm for a mesh of trees and by replacing every leaf of this network by an $O(\log n)$ bit register, connected components can be found in $O(\log^2 n)$ time and $O(n^2 \log^2 n)$ area. This yields $AT^2 = O(n^2 \log^6 n)$ and although this improves the AT^2 result due to Hambrusch [Ha83] by a $\log^2 n$ factor, this upper bound is still away from the lower bound by a $\log^4 n$ factor.

(e) In [GK84], Gottlieb and Kruskal establish a lower bound on time for *online* simulation of a CRCW-PRAM algorithm by an interconnection network. However, the only known lower bound for simulating a CRCW-PRAM algorithm, *offline,* that uses n processors and $O(n)$ memory space, by a bounded degree network that uses P processors, is $\Omega((n/P) + \log P)$ whereas the best upper bound for offline simulation is the same as that for the online simulation.

ACKNOWLEDGEMENTS: The authors thank Janos Simon and Clyder Kruskal for pointing [AIS83] and [GK84].

REFERENCES
[Ag84] A. Aggarwal, "A Comparitive Study of X-Tree, Pyramids, and Related Machines" Proc. of 25th Ann. Conference on Foundations of Computer Science, pp. 89-99, 1984.
[AH85] M. J. Atallah and S. E. Hambrusch, "Solving Tree Problems on a Mesh Connected Processor Array," Proc. of the 26th Ann. Conference on the Foundations of Computer Science, pp. 222-231, 1985.
[AKS83] M. Ajtai, J. Komlos, and E. Szemeredi, "An O(n log n) Sorting Network," Proc. of 15th Ann. Symposium on Theory of Computing, pp. 1-9, 1983.
[AIS83] B. Awerbuch, A. Israeli, and Y. Shiloach, "Efficient Simulation of PRAM by an Ultracomputer," Technical Report 120, Israel Scientific Center, 1983.
[AS83] B. Awerbuch and Y. Shiloach, "New Connectivity and MSF Algorithms for PRAMs and Ultracomputer," Technical Report 122, Israel Scientific Center, 1983.
[GK84] A. Gottlieb and C. P. Kruskal, "Complexity Results for Permuting Data and Other Computations on Parallel Processors," Journal of the ACM, Vol. 13, No. 2, pp. 193-209, 1984.

Note added: Tom Leighton has brought to the attention of the authors that a similar result on sorting provided in this paper has been obtained independently by Cypher and Sanz which also appears in this proceedings.

[GKR85] P. S. Gopalakrishnan, L. N. Kanal, and I. V. Ramakrishnan, "Finidng Connected Components on SIMD Computers," Tech. Report, Univ. of Maryland, 1985.

[GKR85a] P. S. Gopalakrishnan, L. N. Kanal, and I. V. Ramakrishnan, "Computing Tree Functions on SIMD Computers," Technical Report, Univ. of Maryland, 1985.

[Ha82] S. E. Hambrusch, "The Complexity of Graph Problems in VLSI," Ph. D. Dissertation, Penn. State University, 1982.

[Hu85] M.-D. A. Huang, "Solving Some Graph Problems with Optimal or Near-Optimal Speedup on Mesh-of-Trees Networks," Proc. of the 26th Ann. Conference on the Foundations of Computer Science, pp. 232-240, 1985.

[HCS79] D. S. Hirschberg, A. K. Chandra, and D. V. Sarwate, "Computing Connected Components on Parallel Computers," Comm. of ACM, pp. 461-464, 1979.

[Ja81] J. Ja'Ja, "The VLSI Complexity of Graph Problems," Technical Report, Penn. State University, 1981.

[Lei85] F. T. Leighton, "Tight Bounds on the Complexity of Parallel Sorting," IEEE Trans. on Computers, Vol. C-34, No. 4, pp. 344-354, 1985.

[MV83] K. Mehlhorn and U. Vishkin, "Randomized and Deterministic Simulations of PRAMs by Parallel Machines with Restricted Granularity of Parallel Memories," Proc. of the 9th Workshop on Graph Theoretic Concepts in Computer Science, Fachbereich Mathematik, Universitat Osnabruck, June 1983.

[NMB83] D. Nath, S. N. Maheshwari, and P. C. P Bhatt, " Efficient VLSI Networks for Parallel Processing Based on Orthogonal Trees," IEEE Trans. on Computers, Vol. C-32, No. 6, pp. 569-581, 1983.

[NS81] D. Nassimi and S. Sahni, "Data Broadcasting in SIMD Computers," IEEE Trans. on Computers, Vol. C-30, No. 2, pp. 101-107, 1980.

[PV81] F. P. Preparata and J. Vuillemin, "The Cube-Connected Cycles: A Versatile Network for Parallel Computation," Comm. of ACM, Vol. 24, No. 5, pp. 300-309, 1981.

[RS85] J. Reif and Q. Stout, *Personal Communication,* 1985.

[SV82] Y. Shiloach and U. Vishkin, "An O(log n) Parallel Connectivity Algorithm," Journal of Algorithms, Vol. 3, pp. 128-146, 1982.

[Th80] C. D. Thompson, "A Complexity Theory for VLSI," Ph. D. Dissertation, Carnegie-Mellon University, 1980.

[TV84] R. E. Tarjan and U. Vishkin, "Finding Biconnected Components and Computing Tree Functions in Logarithmic Time," Proc. of the 25th Ann. Conference on Foundations of Computer Science, pp. 12-20, 1984.

[Up84] E. Upfal, "A Probabilistic Relation Between Desirable and Feasible Models of Parallel Computation," Proc. of the 16th Ann. Symposium on Theory of Computing, pp. 258-265, 1984.

[UW84] E. Upfal and A. Wigderson, "How to Share Memory in a Distributed System," Proc. of 25th Ann. Conference on the Foundations of Computer Science, pp. 171-180, 1984.

[Vi83] U. Vishkin, "Implementation of Simultaneous Memory Address Access in Models That Forbid It," Journal of Algorithms, Vol. 4, pp. 45-50, 1983.

ANALYSIS OF A DISTRIBUTED SCHEDULER
FOR COMMUNICATION NETWORKS

Yossi Malka, Shlomo Moran and Shmuel Zaks*

Department of Computer Science
Technion, Haifa, Israel

ABSTRACT

Let $G = (V, E)$ be a given network, and d a positive integer. A *d-scheduling* of G is an infinite sequence of rounds $[r_1, r_2, \cdots]$,such that for each i, r_i is a non-empty subset of V, and the distance between any two nodes in r_i is greater than d. A *d-scheduler* is a protocol that determines a d-scheduling of G. Of special interest is the case $d = 1$, which corresponds to a proper communication schedule of the half-duplex model, where information can move in either direction of a communication line, but not simultaneously. This case corresponds also to a proper scheduling of processes in a resource sharing system, where an edge represents a resource shared by two processes, and every process needs all its resources to operate. Another application of this scheduler is a collision-free protocol for radio-networks (this corresponds to $d = 2$).

In this paper a simple d-scheduler is presented and analyzed. We first show that the resulting scheduling is periodic and fair. Then we give a complete characterization of this scheduling for trees and cycles. We study the period length of that scheduling, and the main result is a worst-case exponential lower bound for this length. We also study other issues concerning these schedulings; optimal rate schedulings are given for some classes of graphs, although the problem of finding optimal schedulings for any d is NP-complete.

1. INTRODUCTION

A distributed system consists of processors with distinct identities, connected by a communication network. Each processor has some communication lines connecting it to some other processors. The underlying structure of the network is an undirected <u>connected</u> graph $G = (V, E)$, with a set of nodes $V = \{1, 2, ..., n\}$, $n > 1$ (representing the processors of the network), such that $(i, j) \in E$ iff the processors i and j are connected by a communication line (see [E] for basic Graph Theory terminology). The processors' only means of communication is messages, that are transmitted through the communication lines.

The communication lines are usually one of the following: *simplex* (one-way communication), *half–duplex* (two-way communication, but not in both directions simultaneously), or *full–duplex* (two-way communication, with no restriction). The half-duplex model is widely used in both theory and practice. For example, fiber optics form such a communication media. In theory this model is studied also for its simplicity comparing to the regular two-way model ([vLT]).

A *d–scheduling* of a network is a sequence of rounds, each consisting of some of the nodes of the network, s.t. the distance between any two nodes participating in the same round is greater than d. For example, synchronizer A in [A] is a 1-scheduling protocol. The rate of such a scheduling is the average fraction of nodes participating in a round. In

---

* The research of this author was supported by the Fund for Research in Electronics, Computers and Communications administered by the Israeli Academy of Sciences and Humanities

[BG] a 1-scheduling protocol is presented, and an algorithm to compute the rate of the resulting 1-scheduling is given. 2-scheduling protocols are presented in [CK] and in [CS].

In this paper we analyze a simple scheduling protocol, that converges to a periodic and strongly fair scheduling, regardless of the network topology or the initial configuration. In the case where the network is a tree, the resulting 1-scheduling is guaranteed always to converge to an optimal scheduling. A complete characterization for this 1-scheduling is also given for the case where the network is a cycle. An optimal d-scheduling is also achieved, by a simple efficient algorithm, for the case where the network is a tree. Optimality is shown to be achievable efficiently for a class of graphs that contains perfect graphs, though the problem of providing an optimal scheduling is NP-complete (see [GJ]). Our main result concerns bounds on the period length of the scheduling; in particular, we show that the period length can be exponential. This result was overlooked in [B], where it is argued that the period length is at most n.

The properties of our scheduling scheme makes the special case $d=1$ suitable to be used as a simple synchronizer for the half-duplex model (for synchronizers in different models see [A], [SvL], [CCGZ]), or as a scheduler of a resource sharing system(as in [BG]). Another special case, $d=2$, can be used in radio networks, as a collision-free multi-hop channel access protocol (in this case nodes cannot use the channel simultaneously if they are neighbors or have a common neighbor (see [T])).

The rest of this paper is organized as follows. In Section 2 we present the d-scheduler. Section 3 describes its basic properties, and Section 4 deals with its behavior on cycles. In Section 5 we state additional properties of the scheduler, and prove that the period length of the resulting scheduling can be $e^{\Omega(\sqrt{n})}$. Section 6 deals with optimal rate schedulings; we first prove that finding optimal schedulings of our d-scheduler is NP-complete, and then we show that optimality is achieved efficiently for certain classes of graphs. In this paper we skip some of the lemmas and most of the proofs. A complete discussion can be found in [MMZ].

2. BASIC DEFINITIONS AND THE d-COMPARISON-SCHEDULING PROTOCOL

A *scheduling* of a network G is a mechanism that assigns to each node a permission to use the communication lines, without violating given restrictions. More precisely, let $G = (V, E)$ be a given network. A subset U of V is d-sparse if for each $u, v \in U$, the distance between u and v is greater than d. A d-scheduling of G consists of an infinite sequence of *rounds* $r = [r_1, r_2, \cdots]$ where each r_i is a d-sparse subset of V. Let $x_i = |r_i|$; *rate* (r), the rate of a d-scheduling r, is:

$$rate(r) = \limsup_{i \to \infty} \frac{x_1 + x_2 + \cdots + x_i}{i \, n}$$

(intuitively, *rate* (r) is the expected fraction of processors in a round in r). A scheduling r is said to be *optimal* (for a given G and d) if *rate* (r) is the largest possible.

For a given d-scheduling r, a node v and an integer i, $r_v(i)$ is the number of rounds in $r_1, ..., r_i$ that contain v. r is *fair* if $r_v(i)$ is an unbounded function for every $v \in V$, and is *strongly fair* if there exists a constant k, such that for every integer i and for every pair of nodes u, v, $|r_v(i) - r_u(i)| \le k$. r is *eventually periodic* if for some integer p and for all sufficiently large i, $r_i = r_{i+p}$. r is *periodic* if this holds for all $i \ge 1$. If r is eventually periodic, then the *period* of r is $r_i, ..., r_{i+p-1}$, where p is the smallest possible, and is termed the *period length* of r. A scheduling is m-*fair* if it is eventually periodic and each of the processors appears exactly m times within a period. Note that an eventually periodic scheduling is strongly fair iff it is m-fair for some m. A scheduling is *simple* if it is 1-fair. By the definition of the rate of a d-scheduling, that the rate of an eventually periodic scheduling $r = [r_1,...,r_{i-1},(r_i,...,r_{i+p-1})^*]$ (* denotes an infinite repetition of the sequence in the parentheses) is $\frac{1}{np} \sum_{k=0}^{p-1} x_{i+k}$, and that the rate of an $m-fair$ eventually periodic d-scheduling with a period of length p is m/p. Finally, a d-*scheduling protocol* is a protocol that determines a d-scheduling.

The design of optimal and fair scheduling protocols appears to be a hard problem (see, e.g., [EGMT]). Moreover, even in special cases where the design of such protocols is easy (e.g., for tree or cycle networks), the distributed implementation of such protocols is not obvious. We present a simple d-scheduling protocol, which is relatively easy to implement.

The d-comparison scheduling protocol

Initialization: assign distinct labels from 1 to n to the processors.

(1) A processor participates in a given round iff no processor at distance $\le d$ from it has a smaller label.

(2) After a processor is scheduled for a certain round it increases its label by n.

Increasing the labels indefinitely does not cause any problem, since the algorithm can be implemented using a bounded range of labels.

Example: For the following graph, where processor i is labeled by i,

the resulting 1-comparison scheduling is $\frac{1}{3}, 2, \frac{1}{4}, \frac{2}{3}, \frac{1}{4}\frac{2}{3} \cdots$ (every column represents a round and consists of the nodes participating in that round). We denote this by $\frac{1}{3}, 2, \begin{bmatrix} 1 & 2 \\ 4 & 3 \end{bmatrix}^*$. The 2-comparison scheduling for the above graph is $\begin{bmatrix} 1, 2, 3, 4 \end{bmatrix}^*$.

These two schedulings are simple.

Note that the behavior of a d-comparison scheduling might depend on the initial labeling of the processors in the network. We denote by $DS = (G, f)$ a distributed network where the underlying graph is the undirected graph $G = (V, E)$, with a set of nodes $V = \{1, 2,..., n\}$ $n > 1$, and f is a labeling function $f : V \rightarrow \{1, 2,..., n\}$ that assigns a label $f(v)$ to a node v. Unless otherwise stated, we assume that f is a permutation of the elements $1, 2,..., n$. $f^k(i)$ is the label of processor i before round k. Thus $f^1(i) = f(i)$.

Let $\delta_G(u, v)$ denote the distance between u and v in G. For a given graph $G = (V, E)$ and an integer $k \ge 1$, define the $k-power$ *graph* of G to be $G^k = (V, E^k)$, where $E^k = \{(u, v) | u, v \in V$ and $1 \le \delta_G(u, v) \le k\}$. For a given network $DS = (G, f)$, and a given d, we define the *priority graph* $PG_t^d = (V, E_t^d)$, $t \ge 1$, at the beginning of round t of the d-comparison scheduling, as the directed graph obtained from G^d by directing the edges from processors with higher labels to those with smaller ones, where processor v is labeled with $f^t(v)$. We can view a d-comparison scheduling as a sequence of priority graphs. When $d = 1$, PG_t will be written instead of PG_t^1. One can easily verify that PG_t^d is acyclic, and that the processors that are scheduled for round t are the sinks in PG_t^d (a sink in a directed graph is a node with no out-going edges).

3. BASIC PROPERTIES OF THE d-COMPARISON SCHEDULING

In this section we state the basic properties of the d-comparison scheduling, and prove that it is eventually periodic, and strongly fair. A characterization of 1-comparison schedulings for trees is also given.

The following Lemmas 1-5 describe basic properties of the d-comparison scheduling.

Lemma 1: For every network $DS = (G, f)$ and

every integer $d \geq 1$, the d-comparison scheduling is infinite (i.e., no deadlock can occur). ∎

Lemma 2: For every network $DS = (G, f)$, every integer $d \geq 1$, and every pair of distinct nodes $u, v \in V$ s.t. $\delta_G(u, v) \leq d$, between every two rounds that contain u in the d-comparison scheduling there exists exactly one round that contains v. ∎

Lemma 3: For every $DS = (G, f)$, every $d \geq 1$, and every pair of nodes $u, v \in V$ s.t. $\delta_{G^d}(u, v) = k$, it holds that $|r_u(i) - r_v(i)| \leq k$ for every $i \geq 1$. ∎

From Lemma 1 and Lemma 3 we get:

Lemma 4: For every network $DS = (G, f)$ and every integer $d \geq 1$, every processor appears in an infinite number of rounds in the d-comparison scheduling (i.e., no starvation can occur). ∎

Lemma 5: For every $DS = (G, f)$, every $d \geq 1$, and every $i \geq 1$, PG_{i+1}^d is determined from PG_i^d by reversing all edges adjacent to sinks. ∎

Lemma 5 implies that for every $DS = (G, f)$ the d-comparison scheduling is the same as the 1-comparison scheduling for $DS' = (G^d, f)$.

The basic property of the protocol is:

Theorem 1: For every network $DS = (G, f)$ and every integer $d \geq 1$, the d-comparison scheduling is eventually periodic and every processor appears in the same number of rounds within the period. ∎

It follows from Theorem 1 that the d-comparison scheduling is strongly fair.
Note: Not every such scheduling is simple; for this it suffices to note that for the graph

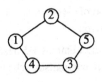

the   1-comparison   scheduling   is

$$\begin{bmatrix} 1 & 2 & 1 & 2 & 4 \\ 3 & 4 & 5 & 3 & 5 \end{bmatrix}^*,$$ in which every processor is scheduled twice within a period.

A complete characterization of the 1-comparison scheduling for a trees follows:

Theorem 2: For every network $DS = (G, f)$, where $G = (V, E)$ is a tree, every 1-comparison scheduling is simple, and its rate is $\frac{1}{2}$. ∎

Note: This implies that, given a tree network, with any 1-sparse subset of processors that start the protocol, the 1-comparison scheduling always converges to an optimal simple 1-scheduling.

4. CYCLES

1-comparison schedulings for a cycle play an important role in our discussion. We start with the following lemma:

Lemma 6: Let $DS = (G, f)$ be a distributed network, where $G = (V, E)$ is a cycle, and let $r = [r_1, ..., r_{i-1}, (r_i, ..., r_{i+p-1})^*]$ be a 1-comparison scheduling for the network. Then:

1. $x_k \leq x_{k+1}$, for every $k \geq 1$ (recall that $x_k = |r_k|$).
2. $x_i = x_{i+1} = \cdots = x_{i+p-1}$ (that is, the number of processors that are scheduled for each round within the period is constant). ∎

For a cycle network we define a *rotating scheduling* to be a scheduling that starts with any 1-sparse subset of processors, and r_{i+1} is constructed by rotating the set of processors in r_i (the sinks in PG_i) one position on the cycle in a given direction, for every $i \geq 1$. For example, the scheduling in the example preceding Theorem 2 is rotating.

For a network $DS = (G, f)$, its d-comparison scheduling $r = [r_1, ..., r_{i-1}, (r_i, ..., r_{i+p-1})^*]$, and the corresponding priority graphs $PG_1^d, ..., (PG_i^d, ..., PG_{i+p-1}^d)^*$, we say that a priority graph is *in* the period if it is one of $PG_i^d, ..., PG_{i+p-1}^d$.

The following theorem provides a complete characterization of the 1-comparison scheduling in cycle networks:

Theorem 3: For every network $DS = (G, f)$, where $G = (V, E)$ is a cycle, the periodic part of the 1-comparison scheduling is a rotating scheduling. ■

The following corollaries apply to 1-comparison schedulings for cycles with n processors:

Corollary 1: The period length is a divisor of n. ■

Corollary 2: a. A priority graph of a cycle is in a period iff all edges not adjacent to sinks are oriented in the same direction. b. Between consecutive priority graphs in the period the sinks rotate in a direction opposite to that of these edges. ■

For any priority graph PG in the period of a 1-comparison scheduling of a cycle, define D_v to be the set of all distances along the direction opposite to the sink's rotation, from a sink v to all the sinks in PG (including itself, in which case the distance is taken as n). Define $D = \bigcap_{v \text{ is a sink}} D_v$. Note that $D \neq \varnothing$ because $n \in D_v$ for every sink v.

Corollary 3: Let PG_i be a priority graph in the period of a 1-comparison scheduling of a cycle. The period length is equal to the minimum element in D. ■

A *segment* of a priority graph is a sequence of consecutive nodes and edges, starting with a node, and ending with an edge. Two segments of a priority graph of a cycle are said to be *equivalent* if they are of the same length, and each pair of respective edges are both oriented clockwise, or both oriented counterclockwise.

Corollary 4: A priority graph of a cycle is in a period of length p of a 1-comparison scheduling iff p is the smallest integer s.t. the priority graph can be partitioned into n/p equivalent segments of length p. ■

Corollary 5: For every cycle and every integer k, $k \leq \lfloor n/2 \rfloor$, there exists a 1-comparison scheduling with rate k/n. ■

5. WORST CASE LOWER BOUND ON THE PERIOD LENGTH

In this section we state some additional properties of 1-comparison schedulings and use them to prove that their period length can be exponential. This contradicts [B], where it was claimed that the period length is at most equal to the number of nodes in the graph. Some of the definitions are taken from [BG].

5.1. INDEPENDENT AND CRITICAL CYCLES

We restrict the discussion to 1-comparison schedulings, even though some results are applicable to d-comparison schedulings. By Lemma 5 we can use the 1-comparison scheduler to produce d-comparison schedulings, $d \geq 1$, for every graph G, by executing it on G^d. Thus, results concerning the rate of a 1-comparison scheduling of G^d apply also to the d-comparison scheduling of G. This relation enables us to state some interesting results for d-comparison schedulings.

An *orientation* of a graph $G = (V, E)$ is a directed graph θ whose underlying graph is G. Let $sinks(\theta)$ be the set of all sinks in θ. In this paper we use only acyclic orientations. Let $\Theta(G)$ be the set of all acyclic orientations of a given graph G.

From every orientation θ of a graph G, we can derive a labeling function that has this orientation as its priority graph. This, together with Lemma 5, enables us to view a 1-comparison scheduling for a graph G as a sequence of orientations $\theta_1, \theta_2, \ldots$, all in $\Theta(G)$, s.t. $r_i = sinks(\theta_i)$, for every $i \geq 1$, and θ_{i+1} is obtained from θ_i by reversing all edges adjacent to sinks. This scheduling will be denoted as $r(\theta_1)$. We view the period of $r(\theta)$ as a series of orientations $\alpha_0, \alpha_1, \ldots, \alpha_{p-1}$, where p is the period length. Each of these orientations will be referred to as orientations of the period. An orientation θ is *periodic* if $r(\theta)$ is periodic.

For $\theta \in \Theta(G)$ let $\gamma(\theta)$ denote the rate of $r(\theta)$, and let $\gamma^*(G) = \max_{\theta \in \Theta(G)} \gamma(\theta)$. Let K be the set of all simple cycles in G. For a cycle $C \in K$ given as $v_1, \ldots, v_{|C|}$, we define the clockwise direction to be the direction when traversing the cycle from v_1 to $v_{|C|}$. Let $e_{cw}(C, \theta)$ $(e_{ccw}(C, \theta))$ be the number of edges oriented clockwise (counter-clockwise) in θ. Define

$$\rho(C, \theta) = \frac{m}{p}, \quad \text{where} \quad p = |C| \quad \text{and}$$

$m = \min\{e_{cw}(C, \theta), e_{ccw}(C, \theta)\}$. Note that the ratio ρ remains constant throughout the 1-comparison scheduling (reversing all edges adjacent to sinks does not change ρ).

The following is Corollary 17 in [BG].

Lemma BG: If G is not a tree, then $\gamma(\theta) = \min_K \rho(C, \theta)$, for all $\theta \in \Theta(G)$. ∎

In view of this lemma, we define a *critical* cycle in an orientation $\theta \in \Theta(G)$, to be a simple cycle C such that $\rho(C, \theta) = \gamma(\theta)$. The above result will be useful in showing that the period length can be exponential, and that the problem of finding an optimal scheduling is NP-complete.

Let θ/H, where $\theta \in \Theta(G)$ and H is a subgraph of G, be the restriction of the orientation θ to H, and let the set of nodes in H be V_H. Let also $r(\theta)/V_H$ be the scheduling obtained from $r(\theta)$ by deleting all nodes not in V_H. It is possible that $r(\theta)/C$ has empty rounds, but this will not be the case when this operation is used. A cycle C in G is said to be *independent in* $r(\theta)$, if $r(\theta/C) = r(\theta)/V_C$. That is, if we orient C's edges as in θ, the 1-comparison scheduling we get when C operates by itself, is the same as the scheduling obtained from $r(\theta)$ by deleting all nodes not in C. C is said to be *independent in* $r(\theta)$'s period if $r(\alpha_0/C) = r(\alpha_0)/V_C$

(recall that α_0 is the first orientation in the period of $r(\theta)$). Note that a cycle can be independent in a period of $r(\theta)$, but not independent in $r(\theta)$ as a whole.

Lemma 7: Let C be simple cycle in a graph G. For all $\theta \in \Theta(G)$, C is critical in θ iff C is independent in $r(\theta)$'s period. ∎

For the following let $CRIT(\theta)$ be the set of all critical cycles in a periodic orientation θ. Define also $l(\theta)$ to be the least common multiple of all period lengths of $r(\theta/C)$, where $C \in CRIT(\theta)$.

Corollary 6: For a periodic orientation $\theta \in \Theta(G)$, the period length of $r(\theta)$ is at least $l(\theta)$. Furthermore, if every *node* lies on a critical cycle then the period length is $l(\theta)$. ∎

Before turning to prove the lower bound, we state the following lemma, that gives a family of graphs and orientations that have 1-comparison schedulings with short periods.

Lemma 8: If every *edge* in a periodic orientation $\theta \in \Theta(G)$ lies on a critical cycle, then the period length of $r(\theta)$ is the minimum over all period lengths of schedulings $r(\theta/C)$, where $C \in CRIT(\theta)$. ∎

5.2. THE LOWER BOUND

We now turn to construct a directed graph $\vec{G}_t$, for every $t > 1$, whose 1-comparison scheduling has a period length which is exponential in the number of nodes in the graph. These graphs consist of cycles $\vec{C}_i$, that have $4i$ nodes and exactly i sinks, for every $i \geq 1$. Whenever $i \not\equiv 0 \pmod{3}$, $\vec{C}_i$'s 1-comparison scheduling has a period length equal to the size of $\vec{C}_i$.

We start by constructing $\vec{C}_i$. Choose any node v_i as a sink. Place other sinks, in the clockwise direction, the first at distance 5 from v_i, the second at distance 5 from the first, and a third one at distance 2 from the second. Continue placing sinks on the cycle·using these distance differences 5,5,2,5,5,2... . Direct edges not adjacent to sinks clockwise. In this manner one sink is placed on every segment of four nodes on the cycle, starting at v_i.

The counter-clockwise neighbor of v_i will be denoted by v'_i. Figure 1 contains $\vec{C}_1, \vec{C}_3, \vec{C}_4$. Note that the constructed cycle $\vec{C}_i$ is a periodic orientation (Corollary 2). Therefore by Lemma 6, exactly i nodes participate in every round of $r(\vec{C}_i)$, and hence by definition its rate is $1/4$.

Lemma 9: The period length of the 1-comparison scheduling for $\vec{C}_i$ is $4i$ when $i \not\equiv 0 \,(mod\ 3)$, and is 12 otherwise. ∎

The graphs $\vec{G}_t$, $t > 1$, are constructed using the cycles $\vec{C}_i$'s. For a given integer $t \geq 1$, let $q_1, q_2, \ldots, q_k$ be all the prime numbers not greater than t. For each j, $1 \leq j \leq k$, select the greatest integer e_j s.t. $q_j^{e_j} \leq t$, and let $a_j = q_j^{e_j}$.

The graph $\vec{G}_t$ consists of the cycle $\vec{C}_1$ and all cycles $\vec{C}_{a_j}$, $1 \leq j \leq k$, with additional edges $v'_{a_j} \rightarrow v_1, 1 \leq j \leq k$. $\vec{G}_4$ is shown in Figure 1. Note that $\vec{G}_t$ has no other cycles except for the $\vec{C}_i$'s. Note also that all cycles in $\vec{G}_t$ have their ratio ρ equal $1/4$, and therefore by Lemma BG we have that the rate of $r(\vec{G}_t)$ is $1/4$.

The following theorem gives the lower bound, which is the main result of this paper.

Theorem 4: For every $t > 1$, $r(\vec{G}_t)$ is periodic, and its period length is $e^{\Omega(\sqrt{n})}$, where n is the number of nodes in $\vec{G}_t$.

Sketch of proof: We show that the orientation $\vec{G}_t$ is periodic, hence each of the $\vec{C}_i$'s is independent in $r(\vec{G}_t)$.

Consider $r(\vec{G}_t)$'s period length p. By Lemma 9 the period length of every scheduling $r(\vec{C}_{a_j})$ is $4a_j$ when $j \neq 2$, and 12 otherwise (a_2 is a power of 3). Since $\vec{G}_t$ is periodic, and since every node is on a critical cycle, we have by Corollary 6 that $p = lcm\,(4, 4a_1, 12, 4a_3$

$\ldots, 4a_k)$, thus $p = 4 \cdot 3 \prod_{j \neq 2} a_j \geq \dfrac{12}{t} \prod_{j=1}^{k} a_j$. Since $(\prod_{j=1}^{k} a_j)(\prod_{j=1}^{k} q_j) > t^k$, and $a_j \geq q_j$ for all $1 \leq j \leq k$, we conclude that $\prod_{j=1}^{k} a_j > t^{k/2}$. There are k prime numbers smaller than t, and therefore by the Prime Number Theorem we have that $k \approx t/\ln t$ ([Nak]). Hence, $t^{k/2} \approx e^{t/2} = e^{\Omega(\sqrt{n})}$, which completes the proof. ∎

6. THE COMPLEXITY OF FINDING AN OPTIMAL RATE d-COMPARISON SCHEDULING

In this section we prove that finding an optimal d-comparison scheduling is NP-complete. Then we discuss cases where optimality can be reached efficiently.

6.1. NP-COMPLETENESS RESULT

For a graph G and a labeling function f let $rate\,(f, d)$ be the rate of the d-comparison scheduling of $DS = (G, f)$.

Π: OPTIMUM RATE d-COMPARISON SCHEDULING
INSTANCE: Graph $G = (V, E)$ and integers $p \geq 2, m \geq 1$.
QUESTION: Is there a labeling function f s.t. $rate\,(f, d) > m/p$?

It has been shown in [BG] that for $d = 1$, Π is NP-complete. We generalize this result for all $d \geq 1$. The proof of membership in NP is different and simpler than the one in [BG], and uses the property of cycles that are independent in a scheduling. The reduction we use is a generalization of the one in [BG]. We first need the following two lemmas.

Lemma 10: For an orientation $\theta \in \Theta(G)$, let v be a node on a critical cycle C of size p, with $\rho(C, \theta) = m/p$. Then $r_v(kp) \leq km + p/2$, for every integer $k \geq 1$. ∎

Lemma 11: For an orientation $\theta \in \Theta(G)$, if a cycle C is independent in the first kp rounds of $r(\theta)$, where p is the size of C and k is the smallest integer s.t. $kp \geq 2n^3$, and θ/C is a periodic orientation, then C is independent in $r(\theta)$'s period $(n = |V|)$. ∎

Lemma 12: $\Pi \in NP$.

Sketch of proof: We first argue that it suffices to show that given an orientation θ of G^d, one can compute $\gamma(\theta)$ non-deterimistically, in polynomial time.

For the graph G^d guess an orientation θ and a cycle C, that is independent in $r(\theta)$'s period, s.t. θ/C is a periodic orientation. By Lemma 7 C is critical, that is, $\rho(C,\theta) = \gamma(\theta)$. Computing $\rho(C,\theta)$ takes $O(|C|) \leq O(n)$ time. By Lemma 11 it suffices to produce the first $O(n^3)$ rounds of $r(\theta)$ to verify that C is independent in $r(\theta)$'s period. This can be done in $O(n^3|E|)$ time (every round takes at most $O(|E|)$ time). Thus we can compute $\gamma(\theta)$ non-deterministically in $O(n^3|E|)$ time. ∎

Define a *distance-d colouring* of a graph $G = (V,E)$ to be a function $g : V \rightarrow \{1,...,k\}$, such that $g(u) \neq g(v)$, for every pair of distinct nodes u,v where $\delta_G(u,v) \leq d$. We consider the problem:

Π': DISTANCE-d CHROMATIC NUMBER
INSTANCE: Graph $G = (V,E)$, positive integer $k \leq |V|$.
QUESTION: Is there a distance-d colouring of G's nodes with k colours or less ?

Π' is NP-complete ([J]). We prove Π's completeness by reducing Π' to Π.

In the following we make use of another notion of 1-comparison schedulings. It is possible to have a 1-comparison scheduling, $\theta_1, \theta_2,...$, where $r_i \subseteq sinks(\theta_i)$. We call such a scheduling *non-greedy* in contrast to the usual 1-comparison scheduling which is greedy. It has been shown in [BG] that the rate obtained by a greedy 1-comparison scheduling is not smaller than the rate obtained by any non-greedy one, starting with the same orientation. Furthermore, given a colouring of G's nodes

using k colours $c_0,...,c_{k-1}$, we can derive a periodic non-greedy 1-comparison scheduling by letting every node with colour c_i participate in round i $0 \leq i \leq k-1$, and this scheduling's rate is $1/k$. The greedy 1-comparison scheduling starting from an orientation θ derived by directing every edge towards the node with the lower colour, has a rate $1/k$.

Theorem 5: Π is NP-complete.

Sketch of proof: We have already shown in Lemma 12 that Π is in NP. We now prove the completeness of Π.

Let $G' = (V',E')$ and an integer k be an instance of Π'. An instance $G = (V,E)$ of Π is obtained by adding a new node x, and connecting it to every node in G' by a path of $\lfloor d/2 \rfloor$ new nodes. That is, for every $v \in V'$, add the path $x - y_{v,1} - ... - y_{v,\lfloor d/2 \rfloor} - v$. If d is even, connect all nodes $y_{v,1}, v \in V'$ to form a clique. To complete the reduction, set $m = 1$ and $p = k + \lfloor d/2 \rfloor \cdot |V'| + 2$.

It can be easily verified that: 1. if $\delta_{G'}(u,v) > d$ then $\delta_G(u,v) > d$; 2. in every distance-d colouring of G, every node in $V - V'$ is assigned a colour not assigned to any other node in G. It is shown that: G' has a distance-d colouring with k colours or less iff G has a labeling f s.t. $rate(f,d) > 1/(k + \lfloor d/2 \rfloor \cdot |V'| + 2)$. ∎

6.2. FINDING AN OPTIMAL RATE d-COMPARISON SCHEDULING EFFICIENTLY

In some cases it is still possible to find an optimal d-comparison scheduling efficiently in spite of the intractability of the general problem. In this section we discuss such cases and give a characterization of the optimal scheduling achieved.

A *k-tuple colouring* of a graph G, $k \geq 1$, is a function that assigns k distinct colours to each node, such that no two neighbors have a colour in common. $\chi^k(G)$ denotes the *k–chromatic number* of G ($\chi(G)$ is G's chromatic number), which is the minimum number of colours needed to give a k-tuple colouring for G. The *multichromatic number* of G, $\chi^*(G)$, is defined as:

$$\chi^*(G) = \min_{k \geq 1} \frac{\chi^k(G)}{k}$$

([CJ] have shown that the use of min instead of inf is valid). It is shown in [BG] that

$$\frac{1}{\chi(G)} \leq \gamma^*(G) \leq \frac{1}{\chi^*(G)} \qquad (*)$$

for every graph G.

Let $\chi^{(d)}(G)$ be the smallest integer k such that there exists a distance-d colouring of G using k colours. An optimal distance-d colouring for trees is given by Algorithm A in [MMZ], for all $d > 1$. It is not hard to see that $\chi^{(d)}(G) = \chi(G^d)$.

For a given graph G, let $\omega(G)$ denote the maximum size of a clique in G. G is said to be *perfect* if for every subgraph H, $\omega(H) = \chi(H)$. G is said to be *semi–perfect* if $\omega(G) = \chi(G)$.

Theorem 6: For every graph G, and every integer $d \geq 1$ such that G^d is semi-perfect, there exists a simple d-comparison scheduling with rate $1/\chi(G^d)$ which is optimal. ∎

By the proof of Theorem 6, it suffices to give an optimal distance-d colouring for G to get an optimal rate d-comparison scheduling when G^d is semi-perfect. In [MMZ] we show how this can be exploited in order to find optimal d-comparison schedulings for trees, for any $d \geq 1$.

Unfortunately recognizing a semi-perfect graph is NP-complete, and even knowing that a given graph is semi-perfect, it is still NP-hard to find an optimal colouring ([L]). Nevertheless, we can do much better with perfect graphs. It has been shown in [GLS] that an optimal colouring for a perfect graph can be found in polynomial time (the complexity of testing perfectness is unknown). The algorithm

given there is complicated and seems impractical for our purposes (see [L] for an overview of this subject). For some classes of perfect graphs (i.e., triangulated graphs, comparability graphs, interval graphs, permutation graphs) there exist simple algorithms for testing whether a graph is in such a sub-class, and to find optimal colourings efficiently ([G], [BC]). We can use these algorithms to test whether the graph for which we want a scheduling, is perfect, and then find an optimal colouring which in turn gives an optimal rate and simple scheduling. Note the importance of the simplicity of the period, as by Theorem 4 the period length might be exponential.

Another property of perfect graphs that may be useful, is the ability to compose two such graphs to produce a larger perfect graph(see [Bi] and [H]), and thus "tailor" a perfect graph for as many nodes as required.

Acknowledgement: We wish to thank I. Cidon and M. Sidi for introducing a 2-scheduling protocol that motivated this research.

REFERENCES

[A] B. Awerbuch, *Complexity of network synchronization*, **Journal of the ACM** 32, 1985, pp. 804-823.

[B] V.C. Barbosa, *Concurrency in systems with neighborhood constraints*, Ph.D. Dissertation, Computer Sci. Dept., Univ. of California, Los Angeles, 1986.

[BC] C. Berge and V.Chvátal, **Topics on Perfect Graphs, Annals of Discrete Math.**, 21, North Holland, Amsterdam, 1984.

[Bi] R.E. Bixby, *A composition for perfect graphs*, in [BC] pp. 221-224.

[BG] V.C. Barbosa, E. Gafni, *Concurrency in systems with neighborhood constraints*, **Proc. of Distributed Computing Systems**, 1987, pp. 448-455.

[CCGZ] C.T. Chou, I. Cidon, I. Gopal and S. Zaks, *Synchronizing asynchronous bounded delay networks*, RC 12274, IBM T. J. Watson Research Center, Yorktown Heights, N.Y., U.S.A., Oct. 1986.

[CJ] F.H. Clarke and R.E. Jamison, *Multicolourings, measures, and games on graphs*, **Discrete Math.** 14 (1976), pp. 241-245.

[CK] I. Chlamtac, and S.Kutten, *A spatial reuse TDMA/FDMA for mobile multi-hop radio networks*, **INFOCOM** Conf. Proc., Mar. 1985, pp. 385-393.

[CS] I. Cidon and M. Sidi, *A distributed assignment algorithm for multi-hop packet-radio networks*, IBM Communications/Computer Science RC 12563 (#56508), 1987.

[E] S. Even, **Graph Algorithms**, Computer Science Press, 1979.

[EGMT] S. Even, O. Goldreich, S. Moran and P. Tong, *On the NP-completeness of certain network testing problems*, **Networks** 14 (1984), pp. 1-24.

[G] M. Golumbic, **Algorithmic Graph Theory and Perfect Graphs**, Academic Press, New york, 1980.

[GJ] M.R. Garey and D.S. Johnson, **Computers and Interactability : A Guide to the Theory of NP-Completeness**, Freeman, San Francisco, CA, 1979.

[GLS] M.Grotschel, L. Lovász and A. Schrijver, *The ellipsoid method and its consequences in Combinatorial Theory*, **Combinatorica** 1(2), 1981, pp. 169-197.

[H] W.L. Hsu, *Decompositions of perfect graphs*, **J. Comb. Theory**(B) 43, 1987, pp. 70-94.

[J] D.S. Johnson, *The NP-Completeness column: an ongoing guide*, **J. of Algorithms**, 1982, pp. 184.

[L] L. Lovász, **An Algorithmic Theory of Numbers, Graphs, and Convexity**, Society for Industrial and Applied Math., Philadelphia PA, 1986, pp. 75-82.

[MMZ] J. Malka, S. Moran and S. Zaks, *Analysis of a distributed scheduler for communication networks*, Technical Report 495, Computer Sci. Dept., Technion, ISRAEL, Feb. 1988.

[NaK] W. Narkiewitz and S. Kanematsu, **Number Theory**, World Scientific, 1983.

[SvL] A. A. Schoone and J. van Leeuwen, *Simulation of parallel algorithms on a distributed network*, Technical Report RUU-CS-86-1, Dept. of Computer Sci., Univ. of Utrecht, The Netherlands, Jan. 1986.

[T] A. S. Tanenbaum, **Computer Networks**, Prentice-Hall, 1981.

[vLT] J. van Leeuwen and R. B. Tan, *An improved upper bound for distributed election in bidirectional rings of processor*, Technical Report RUU-CS-85-23, Dept. of Computer Sci., Univ. of Utrecht, The Netherlands, Aug. 1985.

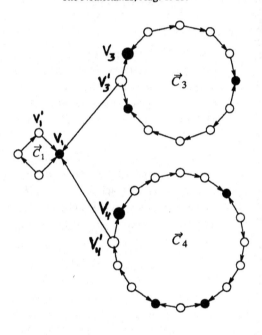

Figure 1: $\vec{G}_4$, $n=32$, with period length 48.

Weighted Distributed Match-Making
(Preliminary Version)

Evangelos Kranakis

Centrum voor Wiskunde en Informatica Kruislaan 413, 1098 SJ Amsterdam,
The Netherlands

Paul M. B. Vitányi

Centrum voor Wiskunde en Informatica and Universiteit van Amsterdam,
Kruislaan 413, 1098 SJ Amsterdam, The Netherlands

ABSTRACT

In many distributed computing environments, processes are concurrently executed by nodes in a store-and-forward network. Distributed control issues as diverse as name-server, mutual exclusion and replicated data management, involve making matches between processes. The generic paradigm is a formal problem called "distributed match-making". The applications require solutions to weighted versions of the problem. We define new multi-dimensional and weighted versions, and the relations between the two, and develop a very general method to prove lower bounds on the complexity as a trade-off between number of messages and "distributedness". The resulting lower bounds are tight in all cases we have examined.

1. Introduction

A distributed system consists of computers (*nodes*) connected by a communication network. Each node can communicate with each other node through the network. There is no other communication between nodes. Distributed computation entails the concurrent execution of more than one process, each process being identified with the execution of a program on a computing node. Communication networks come in two types: broadcast networks and store-and-forward networks. In a broadcast network a message by the sender is broadcasted and received by all nodes, including the addressee. In such networks the communication medium is usually suited for this, like ether for radio. An example is Ethernet. Here we are interested in the latter type, store-and-forward networks, where a message is routed from node to node to its destination. Such networks occur in the form of wide area networks like Arpa net, but also as the communication network of a single multi-computer. The necessary coordination of the separate processes in various ways constitutes distributed control. The situation gets more complicated by assuming that processes can migrate from host to host, e.g., to balance the load in the system.

We focus on a common aspect of seemingly unrelated issues in this area, such as name server, mutual exclusion and replicated data management. Namely, processes residing in different nodes need to find each other, without knowing the host addresses of each other in advance. E.g., in a name-server a client process wants to know the host address of a server process providing a particular service; in distributed mutual exclusion a process that wants to enter the critical section needs to know whether some other process wants to do so as well (see [7] for a general overview). This aspect is formalized in [4] as the paradigm "Distributed Match-Making." Roughly speaking, the problem consists in associating with each node v in the network two sets of network nodes, $P(v)$ and $Q(v)$, such that the intersection $P(v) \cap Q(v')$ for each ordered node pair (v,v') is nonempty. We want to minimize the average of $|P(v)| + |Q(v')|$, the average taken over all pairs (v,v'). This average is related to the amount of *communication* (number of messages) involved in implementations of the distributed control issues mentioned. In the application to name-server: v is a server that posts its whereabouts in all nodes $P(v)$, and v' is a client that looks for a particular service (as provided by v) in all nodes in the query set $Q(v')$. Nodes in $P(v) \cap Q(v')$ can establish contact between v and v' by e.g. sending a message to v with the address of v'. In distributed mutual exclusion the interpretation is about the same, except that there is no difference between client and server, i.e. $P(v) = Q(v)$, see e.g. [3], [4]. For application to replicated data management see [4], final version. We make the simplifying assumption that the involved processes do not migrate during execution of a match-making instance.

Previously, for instance in name servers in distributed operating systems, only ad hoc solutions were proposed, e.g, [5] and references in [4]. Lack of any theoretical foundation necessarily entailed that comparisons of the relative merit of different solutions could only be conducted on a haphazard basis. The question about how to distribute the name-server in a distributed operating system that is currently being implemented [6], prompted our initial investigation in distributed match-making [4]. Our analysis leads to a natural quantification of the distributedness of a match-making algorithm, and trade-offs between number of messages and distributedness. Thus, the complexity results hold for the full range from centralized via hierarchical to totally distributed algorithms for match-making. As pointed out in [4], in many applications we are actually interested in *weighted* versions, i.e., we want to minimize the average of $|P(v')| + \alpha(v',v)|Q(v)|$. It turns out that to do so we have to look at *multi-dimensional* versions first. We develop a very general argument to obtain lower bounds on both versions that include as special case the ones in [4]. The structure of the paper is as follows. First we formally define the multidimensional version and the weighted version of the problem. In the next section we derive the lower bound trade-off (Theorem 1) on the multidimensional case. We then show that the lower bound is tight for the binary n-cube topology and projective n-space topology, by exhibiting distributed algorithms that match the lower bound. In the final section, we derive the promised lower bound on the weighted version of distributed match-making (Theorem 2). This development enhances applicability of the theory of distributed match-making in practical situations.

1.1. Formal Framework

To simplify notation from now on let the set N of network nodes be equal to $\{1,...,n\}$. Let $\mathbf{P} = (P_a,...,P_s)$ be a communication strategy in a given network as follows. (For convenience, with some abuse of notation, we use letters a through s to denote both node variables and the numbers 1 through s.) For each $j = a,...,s$, $P_j: N \rightarrow 2^N$ is a total function, and for each s-tuple $(a',b',...,s')$ of nodes $P_a(a') \cap P_b(b') \cap ... \cap P_s(s') \neq \varnothing$. For any s-tuple $(a',b',...,s')$ of nodes let $m[\mathbf{P}](a',...,s') = |P_a(a')| + ... + |P_s(s')|$ be the number of messages required for the match-making instance $(a,...,s)$ following strategy $\mathbf{P}$. The average number $M[\mathbf{P}]$ of point-to-point messages necessary for match-making is (deleting here and elsewhere $[\mathbf{P}]$ because $\mathbf{P}$ is understood):

$$M = n^{-s}\sum m(a',...,s'), \tag{1}$$

with the sum taken over $(a',...,s') \in N^s$. Let us interpret the case $s=2$ in terms of the name-server, in order to give the intuitive background for considering weighted versions. Since a server i posts its whereabouts at all the nodes in $P(i)$, by sending messages to all these nodes, and a client j queries each node in $Q(j)$, we have $\mathbf{P}=(P,Q)$. The number $m(i,j)$ of point-to-point messages in the match-making instance (i,j) must be at least $|P(i)| + |Q(j)|$. Another more general situation arises when the average call for a service i by a client j occurs $\alpha(i,j)$-times more often than the average posting of a service available at i. Here one wants to minimize (1), with $m(i,j) = |P(i)| + \alpha(i,j)|Q(j)|$. A similar case arises when in the match-making instance (i,j) the server i is allowed to post $p(i,j)$-many times to the nodes in $P(i)$ and the customer j is allowed to query $q(i,j)$-many times the nodes in $Q(j)$ in order to increase reliability of the network. In this case the number $m(i,j)$ of point-to-point messages is equal to $p(i,j)|P(i)| + q(i,j)|Q(j)|$.

In contrast to the *post-query* case ($s=2$), which is best visualized in two dimensions, the more general case ($s>2$) is best visualized in s dimensions. (Each axis is marked with a node from $1,...,n$ and at the vertex $(a,...,s)$ a point of the intersection $\cap P_r$ is located.) To obtain lower bounds on the complexity of the weighted versions and the versions with retransmission, it turns out that it is advantageous to analyse the general s-dimensional case first.

2. The s-Dimensional Lower Bounds

In this section the main lower bound results are derived. In order to be able to prove the most general results possible it will be necessary to formulate the required concepts with a higher level of abstraction than in the introduction. The motivation however is derived from the previous section, and the results are necessary to resolve weighted match-making in the next section.

Let $N, N_a,...,N_s$ be nonempty sets, and $n = |N|$, $n_a = |N_a|$,..., $n_s = |N_s|$. For convenience we set $N = \{1,...,n\}$. It is important to note that, in this general setting, N_a, ..., N_s are *arbitrary* finite sets (of integers), in particular, they can have more elements than N. Consider a *strategy* $\mathbf{P} = \{P_a(a'),...,P_s(s') : a' \in N_a,...,s' \in N_s\}$, with total mappings $P_r : N_r \to 2^N$, and $p_r(x) = |P_r(x)|$, for $r \in \{a,...,s\}$. Let K_i be the set of s-tuples $(a',...,s')$ such that $i \in P_a(a') \cap \cdots \cap P_s(s')$ and let $k_i = |K_i|$. (It is clear that if each of these intersections is nonempty then $k_1 + \cdots + k_n \geqslant n_a \cdots n_s$, and equality holds if all intersections are singleton sets.) For the given strategy $\mathbf{P}$ define the *product* Π and the *sum* M associated with $\mathbf{P}$ by the following formulas:

$$\Pi = (n_a \cdots n_s)^{-1} \sum p_a(a') \cdots p_s(s') ,$$

$$M = (n_a \cdots n_s)^{-1} \sum [p_a(a') + \cdots + p_s(s')] ,$$

with the sums taken over $(a',...,s') \in N_a \times ... \times N_s$. Further, for $r \in \{a,...,s\}$ define

$$M_r = n_r^{-1} \sum p_r(r')$$

(with summation over $r' \in N_r$), so that

$$\Pi = M_a \cdots M_s \text{ and } M = M_a + \cdots + M_s . \tag{2}$$

The main result of the section is the following

Theorem 1. *For any strategy* $\mathbf{P}$ *the following inequalities hold:*

$$\Pi \geqslant (n_a \cdots n_s)^{-1} \left[\sum_{i \in N} k_i^{1/s} \right]^s \text{ and } M \geqslant s(n_a \cdots n_s)^{-1/s} \left[\sum_{i \in N} k_i^{1/s} \right].$$

Remark If $n_a = \cdots = n_s = n$ then

$$\Pi \geqslant \left[n^{-1} \sum_{i=1}^n k_i^{1/s} \right]^s \text{ and } M \geqslant sn^{-1} \left[\sum_{i=1}^n k_i^{1/s} \right].$$

Additionally considering the symmetric case where all k_i's are equal, viz., $k_i = n^{s-1}$, $i = 1,...,n$. Then Theorem 1 specializes to the important "truly distributed" case: $\Pi \geqslant n^{s-1}$ and $M \geqslant sn^{(s-1)/s}$. We will find matching upper bounds below.

Remark. M equals the right-hand side of the inequality in which it occurs, exactly when $M_a = \cdots = M_s$, i.e. the strategy $\mathbf{P}$ is optimal exactly when the average number of messages is equally balanced in all directions.

Proof: The following inequality, also known as *inequality of the arithmetic and geometric means,* holds for s-many nonnegative real numbers $\alpha, \ldots, \sigma$,

$$\alpha + \cdots + \sigma \geqslant s(\alpha \cdots \sigma)^{1/s}. \tag{3}$$

In fact, equality holds exactly when all the summands are equal [2]. Thus, the inequality in the Theorem concerning the sum M follows immediately from the inequality concerning product Π, identities (2), and inequality (3). It is only left to prove the inequality concerning Π. For each $r \in \{a,...,s\}$ and each $i \in N$, define the set $H_{r,i} \subseteq N_r$ such that $r' \in H_{r,i}$ iff for some s-tuple $(a',..,r',.,s')$ holds

$$i \in P_a(a') \cap ... \cap P_r(r') \cap ... \cap P_s(s').$$

Set $h_{r,i} = |H_{r,i}|$. Clearly, for all $i = 1,...,n$,

$$h_{a,i} \cdots h_{s,i} = |H_{a,i} \times \cdots \times H_{s,i}|$$

$$\geqslant |\{(a', \ldots, s') : i \in P_a(a') \cap \cdots \cap P_s(s')\}| = k_i . \tag{4}$$

Now, for all $r \in \{a,...,s\}$,

$$\sum_{i \in N} h_{r,i} \leqslant \sum_{i \in N} |\{r' : i \in P_r(r')\}|$$

$$= \sum_{i \in N} \sum_{r' \in N_r} |\{(i,r') : i \in P_r(r')\}|$$

$$= \sum_{r' \in N_r} |\{i : i \in P_r(r')\}|$$

$$= \sum_{r' \in N_r} p_r(r') = n_r M_r . \tag{5}$$

To obtain the lower bound on Π, we now proceed as follows.

$$\Pi = M_1 \cdots M_s \quad \text{(by (2))}$$

$$\geqslant (n_a \cdots n_s)^{-1} \left[\sum_{\alpha \in N} h_{a,\alpha} \right] \cdots \left[\sum_{\sigma \in N} h_{s,\sigma} \right] \quad \text{(by (5))}$$

$$= (n_a \cdots n_s)^{-1} \sum_{\alpha,...,\sigma \in N} h_{a,\alpha} \cdots h_{s,\sigma}$$

Set $S(\alpha,...,\rho,\sigma) = h_{a,\alpha} \cdots h_{r,\rho} h_{s,\sigma}$. By cyclically rotating the indices $\alpha, \ldots, \rho, \sigma$ of $S(\alpha,...,\rho,\sigma)$ one obtains the following s-many summands:

$$a_1 = S(\alpha,...,\rho,\sigma) = h_{a,\alpha} \cdots h_{s,\sigma}$$

$$a_2 = S(\beta,...,\sigma,\alpha) = h_{a,\beta} \cdots h_{s,\alpha}$$

$$\cdots \qquad \cdots \tag{6}$$

$$a_s = S(\sigma,...,\pi,\rho) = h_{a,\sigma} \cdots h_{s,\rho}$$

Using inequalities (3) and (4) and *regrouping* terms in the resulting product $a_1 \cdots a_s$ it is easy to see that $a_1 + \cdots + a_s \geqslant s(a_1 \cdots a_s)^{1/s} \geqslant s(k_\alpha \cdots k_\sigma)^{1/s}$. After adding the s-many summands of (6), each one summed with respect to $\alpha,...,\sigma$, dividing again by s to eliminate s-multiple copies, and taking into account the last inequality, we obtain:

$$\sum_{\alpha,...,\sigma \in N} h_{a,\alpha} \cdots h_{s,\sigma} \geqslant \sum_{\alpha,...,\sigma \in N} (k_\alpha \cdots k_\sigma)^{1/s} = \left[\sum_{i \in N} k_i^{1/s} \right]^s$$

This completes the proof of the lower bound of Π, and hence the proof of the theorem is complete. $\square$

Corollary. Both propositions 1 and 2 of [4] are immediate consequences of Theorem 1.

3. Optimality

We show that Theorem 1 is optimal in some special cases (which are of sufficient generality), by exhibiting matching strategies.

(**Multidimensional Cube Network**) Let the number of nodes be $n = 2^d$ and suppose that s is a divisor of d. Addresses of nodes consist of d bits, like $u_1 u_2 \cdots u_d$. Nodes are connected by an edge exactly when they differ by a single bit. Let $\mathbf{P} = (P_1, ..., P_s)$ be a strategy, and, for each $r \in \{1, ..., s\}$, let $P_r(u_1 \cdots u_d)$ be the set

$$\{x_1 \cdots x_{(r-1)d/s} u_{(r-1)d/s+1} \cdots u_{rd/s} x_{rd/s+1} \cdots x_d : x_i \in \{0,1\}\}.$$

Clearly, each of the above sets has size $2^{(s-1)d/s}$ and $k_i = 2^{(s-1)d} = n^{s-1}$. Thus, one easily obtains that $M \leqslant sn^{(s-1)/s}$, i.e. the average number of point-to-point message transmissions is at most $sn^{(s-1)/s}$. In view of Theorem 1 this strategy is also optimal.

(**Multidimensional Projective Plane**). Consider generalized mutual exclusion in a distributed setting, where $s-1$ processors are allowed to be in the critical section simultaneously, but not s or more processors. For background and nondistributed solutions we refer to [1]. In [3], Maekawa considers the distributed version of mutual exclusion for $s=2$, the commonly studied variant. In our terminology, for mutual exclusion with $s=2$ we can set $P_1(i)=P_2(i)$, which is some sort of symmetry condition. Each instance of mutual exclusion contains a match-making instance [4]. For the truly distributed case, with $k_1=...=k_n=n$ and $s=2$ we find that on the average each match-making instance takes at least $2\sqrt{n}$ messages [4]. Maekawa obtains a similar lower bound, and exhibits an algorithm that achieves $5\sqrt{n}$ [3]. Theorem 1 gives a lower bound of $sn^{(s-1)/s}$ for the generalized version. We exhibit an algorithm that achieves this. The s-dimensional projective plane $PG(s,k)$ has $k^s + k^{s-1} + \cdots + 1 = n$ nodes, each node is incident to $k^{s-1} + k^{s-2} + \cdots + 1$ hyperplanes, and each hyperplane contains $k^{s-1} + k^{s-2} + \cdots + 1$ nodes. Each s-element set of hyperplanes intersects in precisely one node. Let $\mathbf{P}=(P_1, ..., P_s)$ be a symmetric strategy with each query set $S(i) = P_1(i) = \cdots = P_s(i)$ of a node i consists of the set of $k^{s-1} + k^{s-2} + \cdots + 1$ nodes incident to a hyperplane containing node i. It does not matter which hyperplane we pick, because any s hyperplanes intersect in a single node. The average cost M of point-to-point messages associated with a particular mutual exclusion instance is therefore (generalizing Maekawa's method for $s=2$ [3]) $O(s(k^{s-1} + k^{s-2} + \cdots + 1)) \approx O(sn^{(s-1)/s})$. In view of Theorem 1 this strategy is also optimal.

4. Weighted Distributed Match-Making

We can now examine weighted distributed match-making. This can be formulated as communication strategies with multiple transmissions allowed. We use Theorem 1 to derive significant lower bounds on the average number of message transmissions in distributed networks when multiple transmissions are allowed. Consider a strategy $\mathbf{P}=(P_a,...,P_s)$, with all parameters as above, and define a weighted version of m. I.e., define the number of messages for the match-making instance $S=(a',...,s')$ as $m[\mathbf{P}](S) = l_a(S)p_a(a') + \cdots + l_s(S)p_s(s')$, where each $l_a(S),...,l_s(S)$ is a positive integer. Then, with S as above, define $N_{r,r'}$, for all $r \in \{a,..,s\}$ and $r' \in N_r$ so that it satisfies:

$$(n_a \cdots n_s) M[\mathbf{P}] = \sum_{S \in N_a \times ... \times N_s} l_a(S)p_a(a') + \cdots + l_s(S)p_s(s')$$

$$= \sum_{a' \in N_a} \left[\sum_{S \in \mathbf{S}_a} l_a(S) \right] p_a(a') + \cdots + \sum_{s' \in N_s} \left[\sum_{S \in \mathbf{S}_s} l_s(S) \right] p_s(s')$$

$$(\text{with } \mathbf{S}_a = \{a'\} \times N_b \times ... \times N_s, \ldots, \mathbf{S}_s = N_a \times ... \times N_r \times \{s'\})$$

$$= \sum_{a' \in N_a} N_{a,a'}p_a(a') + \cdots + \sum_{s' \in N_s} N_{s,s'}p_s(s'), \tag{7}$$

where $N_{a,a'} = \Sigma_{S \in \mathbf{S}_a} l_a(S)$, etc. Define $N'_r = \sum_{r' \in N_r} N_{r,r'}$. Consider the following related strategy $\mathbf{Q}$ for the set of nodes N. $\mathbf{Q} = \{Q_a(a'),..., Q_s(s'): a' \in N'_a,...,s' \in N'_s\}$, such that, for each $r \in \{a,...,s\}$ and each $y \in N_r$ there are $N_{r,y}$ distinct x's, with $Q_r(x)=P_r(y)$. I.e., $\mathbf{Q}$ is formed from the strategy $\mathbf{P}$ by repeating each set $P_r(y)$, $N_{r,y}$-times. Let $q_r(x) = |Q_r(x)|$. Note that we have chosen the definitions such that

$$\sum_{r' \in N'_r} q_r(r') = \sum_{r' \in N_r} N_{r,r'}p_r(r'),$$

for all r from a through s. Then we can relate $M[\mathbf{P}]$ with $\Pi[\mathbf{Q}]$:

$$M[\mathbf{P}] = (n_a \cdots n_s)^{-1} \left[\sum_{a' \in N'_a} q_a(a') + \cdots + \sum_{s' \in N'_s} q_s(s') \right] \quad (\text{by (7)})$$

$$\geqslant s(n_a \cdots n_s)^{-1} \left[\left[\sum_{a' \in N'_a} q_a(a') \right] \cdots \left[\sum_{s' \in N'_s} q_s(s') \right] \right]^{1/s} \quad (\text{by (3)})$$

$$= s(n_a \cdots n_s)^{-1}(N'_a \cdots N'_s)^{1/s}\Pi[\mathbf{Q}]^{1/s} \quad (\text{by definition})$$

$$\geqslant s(n_a \cdots n_s)^{-1} \sum_{i \in N} k_i[\mathbf{Q}]^{1/s} \quad (\text{by Theorem 1})$$

It remains to compare the quantities $k_i[\mathbf{P}]$, $k_i[\mathbf{Q}]$. This can be done by comparing the sizes of the sets $K_i[\mathbf{P}]$, $K_i[\mathbf{Q}]$. Now, for each s-tuple $(a',...,s')$ such that

$i \in P_a(a') \cap ... \cap P_s(s')$ there are at least $N_{a,a'} \cdots N_{s,s'}$ s-tuples $(a'', ..., s'')$ such that $i \in Q_a(a'') \cap ... \cap Q_s(s'')$. Namely, there are $N_{r,r'}$ copies of $P_r(r')$, for r, r' from a, a' through s, s', in $\mathbf{Q}$. Therefore, each $(a', ..., s') \in K_i[\mathbf{P}]$ corresponds to a disjoint subset of at least $N_{a,a'} \cdots N_{s,s'}$-many s-tuples in the set $K_i[\mathbf{Q}]$'s. Hence, it has been proved that

$$M[\mathbf{P}] \geqslant s(n_1 \cdots n_s)^{-1} \sum_{i \in N} \left[\sum \{ N_{a,a'}...N_{s,s'} : (a', ..., s') \in K_i[\mathbf{P}] \} \right]^{1/s} \tag{8}$$

In particular, with some computation we can specialize the general result (8) to:

Theorem 2. *For any strategy* $\mathbf{P}$, *if there are positive integers* $\lambda_a, ..., \lambda_s$ *such that for all* $(a', ..., s')$ *holds* $m(a', ..., s') = \lambda_a p_a(a') + \cdots + \lambda_s p_s(s')$, *then*

$$M \geqslant \frac{s(\lambda_1 \cdots \lambda_s)^{1/s}}{n} \sum_{i=1}^{n} k_i^{1/s} \tag{9}$$

Moreover, the quantity M *equals the right-hand side of the inequality above, exactly when* $\lambda_a M_a = \cdots = \lambda_s M_s$. $\square$

Corollary. Routine calculation shows that Theorem 2 also holds for *rational* λ's. (Hint: for $\lambda_r = p_r / q_r$ apply Theorem 2 for $\mu_r = c\lambda_r$ with $c = q_a...q_s$ ($r \in \{a, ..., s\}$). This gives an inequality for cM. Substituting the λ's for the μ's, we can cancel c on both sides of the inequality.)

References

[1] Fischer, M.J., Lynch, N.A., Burns, J.E., and Borodin, A., *Distributed FIFO allocation of identical resources using small shared space*, Massachusetts Institute of Technology, Cambridge, Mass., Report MIT/LCS/TM-290, October 1985.

[2] Hardy, G. H., Littlewood, J. E. and Polyá, G., *Inequalities*, Cambridge University Press, 1934.

[3] Maekawa, M., *A $\sqrt{N}$ Algorithm for Mutual Exclusion in Decentralized Systems*, ACM Transactions on Computer Systems 3 (1985), pp. 145-159.

[4a] Mullender, S. J. and Vitányi, P. M. B., *Distributed Match-Making for Processes in Computer Networks, Preliminary Version*, Proceedings of the 4th annual ACM Symposium on Principles of Distributed Computing, 1985, pp. 261-271.

[4b] Mullender, S. J. and Vitányi, P. M. B., *Distributed Match-Making*, Algorithmica, (1988).

[5] Powell, M.L. and Miller, B. P., *Process Migration in DEMOS/MP*, Proceedings of the 9th ACM Symposium on Operating Systems Principles, 1983, pp. 110-119.

[6] Tanenbaum, A.S. and S.J. Mullender, *The Design of a Capability Based Distributed Operating System*, The Computer Journal, 29(1986).

[7] Tanenbaum, A.S., *Computer Networks*, Prentice-Hall, 1981.

A Tradeoff Between Information and Communication in Broadcast Protocols

Baruch Awerbuch [1]   Oded Goldreich [2]   David Peleg [3]   Ronen Vainish [2]

Abstract

This paper concerns the message complexity of broadcast in arbitrary point-to-point communication networks. *Broadcast* is a task initiated by a *single* processor that wishes to convey a message to all processors in the network. We assume the widely accepted model of communication networks, in which each processor initially knows the identity of its neighbors, but does not know the entire network topology. Although it seems obvious that the number of messages required for broadcast in this model equals the number of links, no proof of this basic fact has been given before.

We show that the message complexity of broadcast depends on the exact complexity measure. If messages of unbounded length are counted at unit cost, then broadcast requires $\Theta(|V|)$ messages, where V is the set of processors in the network. We prove that if one counts messages of *bounded length* then broadcast requires $\Theta(|E|)$ messages, where E is the set of edges in the network.

Assuming an intermediate model in which each node knows the topology of the network in radius $\rho \geq 0$ from itself, we prove tight upper and lower bounds of $\Theta(\min\{|E|, |V|^{1+\frac{\Theta(1)}{\rho}}\})$ on the number of messages of bounded length required for broadcast. Both the upper and the lower bounds hold for both synchronous and asynchronous network models.

The same results hold for the construction of spanning trees, and various other global tasks.

1   Introduction

Broadcast [DM] is one of the most fundamental tasks in distributed computing. It is initiated by a single processor, called the *source*, wishing to distribute a message (*the initial message*) to all processors in the network.

[1]Dept. of Math., MIT, Cambridge, MA 02139, baruch@theory.lcs.mit.edu Partially supported by Air Force contract TNDGAFOSR-86-0078, ARO contract DAAL03-86-K-0171 and NSF contract CCR8611442.

[2]Dept. of Computer Science, Technion, Haifa, 32000, Israel. Partially supported by the New York Metropolitan Research Fund.

[3]Dept. of Computer Science, Stanford University, Stanford CA 94305. Partially supported by ONR contract N00014-85-C-0731.

We consider the standard model of distributed computing, which is a point-to-point communication network. The network is modeled by an undirected graph $G(V, E)$ whose nodes represent processors and whose edges represent bidirectional communication links (cf. [A1, Bu, FL, GHS]). Communication itself is either synchronous or asynchronous — all our results hold for both cases. An *elementary* message may contain only a constant number of bits and a constant number of node identities. Longer messages must be chopped into elementary messages prior to transmission. The *communication complexity* of an algorithm is the total number of (elementary) messages sent in a worst-case execution.

The two most basic and well-known algorithms for broadcast in a point-to-point communication network are *tree broadcast* and *flooding*. The tree broadcast algorithm requires the existence of a spanning tree that is known to all processors. Given such a tree, broadcast can be performed with only $|V| - 1$ messages. In case such a tree is not available, it has to be constructed first. Note, however, that the problem of constructing a spanning tree from a single initiator is equivalent in terms of communication complexity to the problem of broadcasting a single message. This follows from the fact that any broadcast algorithm can also be used to build a tree in the network; the parent of a node in that tree is the neighbor from which the first message is received.

In contrast, the flooding algorithm makes no initial assumptions. This algorithm achieves its task by simply "flooding" the links of the network, i.e., forwarding the message over *all* links. Clearly, this requires $\Theta(|E|)$ messages.

When discussing the applicability of these (and other) broadcast algorithms to a communication network, a central issue is the amount of knowledge available at the nodes regarding the topology of the network. There are two common models, representing the two possible extreme situations. In the first model (which we denote KT_∞ for reasons which will become clear later) one assumes that every node has full knowledge of the network topology. In this model, it is obvious that broadcast can be performed with the minimal number of messages, i.e., the communication complexity of the problem is $\Theta(|V|)$. This is because each node can use its knowledge in order to locally construct the (same) spanning tree without sending any message. Then, the tree broadcast algorithm can be applied.

The standard model for a communication network, which we denote KT_1, assumes very little knowledge. That is, initially each processor knows only its own identity and the identity of its neighbors, but nothing else. In this model, a well-known "folk theorem" asserts that "flooding" is the best that can be done, i.e., that $\Theta(|E|)$ is a tight bound for the communication complexity of the problem. However, to the best of our knowledge, no proof of this lower bound (or for that matter, of any lower bound higher than $\Omega(|V|)$) was given before. At first glance the claim seems obvious. Indeed, the claim *is* obvious if we consider the even more extreme model KT_0, based the (unnatural) assumption that a node does not know even the identities of its neighbors. The intuition behind the $\Omega(|E|)$ lower bound for KT_0 is that in this case "every edge must be traversed at least once". However, slightly shifting from this extreme model towards the more common (and more reasonable) KT_1 model, this intuition fails, as is implied by the following algorithm.

Consider a "traveler" which performs a Depth-First Search (DFS) traversal (cf. [E]) on the communication graph. Observe that by carrying the list of nodes visited so far, the traveler may avoid traversing non-tree edges (or "backward" edges) since at any point during the search the traveler knows which nodes have already been visited. Thus, the traveler will not traverse every graph edge, but only $n - 1$ tree edges (each being traversed exactly twice).

While this algorithm indicates that there is no need to traverse each graph edge, it does not disprove the above "folk theorem". Indeed, observe that the total number of *elementary* messages sent is not $2|V|$, but rather $O(|V|^2)$, as the lists carried by the traveler may contain up to $O(|V|)$ node identities; thus the traversal of an edge may require $O(|V|)$ elementary messages.

In this paper, we (finally) prove the above "folk theorem" for the standard KT_1 model. More precisely, we show that in a communication network where each node knows only its neighbors, the number of elementary messages required for broadcast is $\Omega(|V|)$.

Once we establish this gap between the two extreme models, it becomes interesting to look at intermediate points, in which processors are allowed only partial knowledge of the topology, and investigate the implications of such knowledge with regard to the communication complexity of the broadcast operation. These intermediate points attempt to capture common situations in which nodes know more about their near-by vicinity than about other regions of the network. We formalize such situations by introducing a (mainly theoretical) hierarchy of models KT_ρ defined as follows: In the model KT_ρ (for an integer $\rho \geq 0$), every node knows the topology of a subgraph of radius ρ around it. Hence the models KT_0 and KT_1 described earlier correspond to the lowest two levels of this hierarchy, while KT_∞ corresponds to the highest levels, i.e., the models KT_ρ with ρ being the diameter of the network or larger.

For this hierarchy of models, we prove a general tradeoff result. For every $\rho \geq 0$, the number of elementary messages required for broadcast in the model KT_ρ is $\Theta(\min\{|E|, |V|^{1 + \frac{\Theta(1)}{\rho}}\})$. To be more precise, we can prove the following.

Theorem 1: *There exists a constant $c > 0$ such that for every integer $\rho \geq 0$ and for any graph $G(V, E)$, broadcast can be performed in the model KT_ρ, using at most $O(\min\{|E|, |V|^{1 + \frac{c}{\rho}}\})$ messages. This upper bound holds even if the network is asynchronous.*

Theorem 2: *There exists a constant $c' > 0$ such that for every integer $\rho \geq 0$ there exists a family of graphs F_ρ, where each graph has m edges and n nodes, with*

$m = \Omega(n^{1+\frac{c'}{\rho}})$, such that any protocol that works correctly on all graphs in F_ρ in the model KT_ρ sends at least $\Omega(m)$ messages over a constant fraction of the graphs from F_ρ. This lower bound holds even if the network is synchronous, all the nodes start the protocol at the same round, and the size of the network is known.

Our results suggest that there exists an inherent tradeoff between the information that the nodes have about the communication graph, and the number of messages needed to perform the broadcast. The more knowledgeable nodes are about the network, the cheaper it is to perform broadcast.

One of the novelties of our work is that it applies to a network of arbitrary topology, and takes *full* advantage of the fact that network's topology is initially *unknown*. For example, as a corollary we get also that constructing a spanning tree in a network whose topology is *unknown* is harder than constructing a spanning tree in a network whose topology is *known*.

Our result enables one to prove an $\Omega(|E| + |V|\log|V|)$ lower bound on the the communication complexity of any spanning tree construction algorithm, thus implying optimality of the algorithm of [GHS].

Some of the above results have been reported in an earlier version of this paper [AGV]. Results somewhat weaker than [AGV] have been independently obtained by [RK]. (The lower bound of [RK] does not hold if the size of the network is known.)

The rest of the paper is organized as follows. In Section 2 we define the model used for the main result and state the problem. In Section 3 we present the upper bound. In Section 4 we outline the lower bound proof.

2 The Model

Our communication model consists of a point-to-point communication network, described by a simple undirected graph $G(V, E)$, where the nodes represent network processors and the edges represent bidirectional communication channels operating between them.

Whenever convenient, we will assume that $V = \{1, 2, ..., |V|\}$. Initially, (unique) ID's are assigned to the processors (nodes) of the graph G. These ID's are taken from an ordered set of integers $S = \{s_1, s_2, ...\}$ where $s_i < s_{i+1}$ for every $i \geq 1$. Thus a system configuration consists of a graph G and an ID-assignment, which is a one-to-one mapping $\phi : V \to S$.

One can distinguish between *synchronous* and *asynchronous* network models, as in [A1]. For the lower bound, we assume here that communication is synchronous; i.e., communication takes place in "rounds", where processors transmit only in the very beginning of a round and all messages are received by the end of the round.

Clearly, the lower bound holds also if communication is asynchronous. For the upper bound, we assume that the network is asynchronous. Our results hold assuming either synchronous or asynchronous communication.

A *protocol* is a local program executed by all the nodes in the network. A node starts executing a protocol either by means of a special *wake-up* signal, or as a result of receiving a message of the protocol. The set of nodes which can possibly receive a wake-up signal is called the *initiators* of the protocol. A protocol achieving a given task should work on every network G, and every assignment ϕ of ID's to the processors of G.

The local program at a node has local *input* and local *output* variables. Our hierarchy of models KT_ρ (for $\rho \geq 0$) is characterized by the local inputs regarding the topology. In the model KT_ρ, the input to the local program at a node v contains the subgraph of G induced by all nodes at radius ρ from v, where each node is labeled by a distinct ID.

We assume that all messages sent by the protocol contain only a bounded number of ID's. Let us denote this bound by B. (Alternatively, we may allow longer messages, but charge them by the number of processor ID's they contain.) Our complexity measure will be the number of messages containing at most constant number B of node ID's, sent in the worst-case execution of the protocol on the network $G(V, E)$.

Finally let us give a precise statement of the problem of *broadcast from a single source*. One of the nodes is marked as *source*, and it has a certain *value*. The fact that a node is a source, and the value which needs to be broadcast is kept at a special input tape. This value should be disseminated from the source node to all nodes in the network, which will write it on their output tape.

3 The Upper Bound

In this section we prove Theorem 1, that is, we show that for any integer $\rho \geq 0$ and for any graph $G(V, E)$, in the model KT_ρ broadcast can be performed with at most $O(\min\{|E|, |V|^{1+\frac{c}{\rho}}\})$ messages for some constant $c > 0$. This upper bound holds even if the network is asynchronous.

The key observation behind the algorithm is that if a node knows the structure of the subgraph induced by the vertices in radius ρ around it, then it can "see" all *short* cycles (defined here as cycles of length $2\rho + 1$ or less) going through it. This enables us to "open" all short cycles locally, by "deleting" the heaviest edge (the one with the highest weight) in each such cycle. More precisely, assume some (locally computable) assignment of distinct weights to the edges. Define a subgraph $\bar{G}$ of G by marking the heaviest edge in every short cycle "unusable" and taking $\bar{G}$ to consist of precisely all "usable" edges. Observe that given the partial topological knowledge of the nodes, such edge deletions can be performed locally by the nodes incident to the edges, without sending a single message.

Let $\bar{G}(V, \bar{E})$ denote the graph resulting from the deletions above. Let $g(G)$ the *girth* of a graph G, i.e. the length of a smallest cycle in G. (A single edge is not considered a cycle of length 2, so $g(G) \geq 3$ for every G.)

Lemma 3.1: *The graph $\bar{G}$ is connected.* ∎

Another immediate consequence of the above marking process is that all short cycles (of length less than or equal to $2\rho + 1$) are opened, and hence we have

Lemma 3.2: *The girth of $\bar{G}$ satisfies $g(\bar{G}) \geq 2\rho + 2$.* ∎

We need the following proposition.

Proposition 3.3 [A1, PS]: *There exists a constant $c' > 0$ such that for any graph $G(V, E)$ and for any $k \geq 1$ there exists a subgraph $G'(V, E')$ such that*

1. $|E'| \leq O(|V|^{1 + \frac{c'}{k}})$.

2. *For every edge $(u, v) \in E$, the distance between u and v in G' is at most k. (I.e., G' is a k-spanner of G [PS].)*

Corollary 3.4: *There exists a constant $c' > 0$ such that any graph $G(V, E)$ with girth $g(G) \geq 3$ has at most $|E| = O(|V|^{1 + \frac{c'}{g(G) - 2}})$ edges.* ∎

It follows from Lemma 3.2 and Cor. 3.4 that $|\bar{E}| = O(|V|^{1 + \frac{c}{\rho}})$ for $c = c'/2$. We can now perform broadcast on $\bar{G}$ using by the standard flooding algorithm described earlier, which requires $O(|\bar{E}|) = O(|V|^{1 + \frac{c}{\rho}})$ messages. This completes the proof of our first Theorem.

4   The Lower Bound

Our lower bound proof proceeds in several stages. In the first stage we prove the claim only for $\rho = 1$, and only in a restricted model of *comparison protocols*, with the only initiator being the source node. The proof is then extended to the general model, which allows arbitrary values of ρ, arbitrary computations at nodes, and arbitrary set of initiators. For lack of space we omit these latter stages from this abstract.

At this point, we restrict the local computations of the program which involve processors' ID's to comparing two ID's. Formally, this can be achieved by asserting that the program has local variables of two types: variables of type *identity* (*ID-typed*) and *ordinary* variables. Initially, the ID-typed variables are empty, while the ordinary variables may contain some constants (e.g. 0 and 1). The local computations of the program are of two corresponding types:

1. Comparing two ID-typed variables and storing the result of the comparison in an ordinary variable.

2. Performing an arbitrary computation on ordinary variables and storing the result in another ordinary variable.

Messages consist of the values of some of the variables of the local program. Without loss of generality we may further assume that all "send" instructions are of the form

if $vo_1 = vo_2$ then send the message $(id_1, id_2, ..., id_B; \overline{vo})$ to processor id_{B+1},

where id_i (resp., vo_i) is the value of the i-th ID-typed (resp., ordinary) variable and $\overline{vo}$ is the sequence of values of all ordinary variables of the sending processor.

The reason for restricting the permissible operations of local programs on ID's to comparisons is that this makes it easy to prove the existence of "equivalent" ID's. Loosely speaking, these are ID's whose substitution in the processors of the network essentially "preserves" the execution.

Definition 4.1: Two *ID's* $x, y \in S$ are *adjacent* if they appear consecutively in S. Two *ID-assignments* ϕ_0, ϕ_1 are *adjacent* if for every node $v \in V$, the ID's $\phi_0(v)$ and $\phi_1(v)$ are adjacent.

Clearly, if $|S| \geq 2|V|$ then there exist two adjacent assignments ϕ_0, ϕ_1 with disjoint range. (For example, let $\phi_i(j) = s_{2j+i}$.)

In the following definition we assume, without loss of generality, that each processor can send at most one message to each of its neighbors in each round.

Definition 4.2: An *execution* $EX(\Pi, G, \phi)$ of a protocol Π on a graph $G(V, E)$ with an ID-assignment ϕ is the sequence of messages sent during the corresponding run, where messages sent during the run appear in the sequence in lexicographic order of triples (ROUND, SENDER, RECEIVER). For simplicity, we append to each message in the sequence a *header* consisting of the corresponding (ROUND, SENDER, RECEIVER) triple.

Definition 4.3: Given a protocol Π, a graph $G(V, E)$ and two ID-assignments ϕ_0 and ϕ_1, we say that the executions $EX_0 = EX(\Pi, G, \phi_0)$ and $EX_1 = EX(\Pi, G, \phi_1)$ are *essentially the same* if when substituting in Ex_0, for every $v \in V$, the ID value $\phi_0(v)$ by $\phi_1(v)$, we get the execution EX_1. Similarly, we say that the values of an ID-typed variable id_j are essentially the same at a certain point in these two runs if it contains, respectively, $\phi_0(v)$ and $\phi_1(v)$ for some $v \in V$.

Lemma 4.1: *Let ϕ_0 and ϕ_1 be two adjacent ID-assignments. Then for any pro-*

tocol Π *and any graph* $G(V, E)$, *the executions* $EX(\Pi, G, \phi_0)$ *and* $EX(\Pi, G, \phi_1)$ *are essentially the same.* ∎

Corollary 4.2: *Let* ϕ_0 *and* ϕ_1 *be adjacent executions with disjoint ranges. For any n-bit string* α, *let* ϕ_α *be an ID-assignment such that for every* $i \in V$ $\phi_\alpha(i) = \phi_{\alpha_i}(i)$, *where* α_i *is the i-th bit of* α. *Then for any protocol* Π *and any graph* $G(V, E)$, *all the executions* $EX(\Pi, G, \phi_\alpha)$ *(for every such* α) *are essentially the same.* ∎

We are now ready to introduce the family of networks to be used in our lower bound proof.

Definition 4.4: For any given graph $G(V, E)$ and for every edge $e = (u, v)$ in E, we define the graph $G^{(e)}(V^{(e)}, E^{(e)})$ so that

$$V^{(e)} = V \bigcup \{u', v'\}, \quad \text{where} \quad u', v' \notin V$$

$$E^{(e)} = (E - \{e\}) \bigcup \{(u, v'), (v', u'), (u', v)\}$$

Further, for any given graph $G(V, E)$ define the family of graphs

$$C_G = \{G\} \bigcup \{G^{(e)} : e \in E\}.$$

Note that for every edge e, the auxiliary graph $G^{(e)}$ is a copy of G, except that the edge $e = (u, v)$ is replaced by the path $u - v' - u' - v$. In our lower bound argument we concentrate on the graph G, switching whenever required to one of the auxiliary graphs $G^{(e)}$, and relying on the observation that the protocol is also correct when run on $G^{(e)} \in C_G$. This is the underlying principle of the following main Lemma, which asserts that neighbors must "hear" of one another during any execution of a broadcast protocol.

Lemma 4.4: *Let* $G(V, E)$ *be an arbitrary graph, let* S *be a set of ID's* $(|S| \geq 2|V|)$, *and let* Π *be a protocol which achieves broadcast on each graph in* C_G. *Then there exists an ID-assignment* $\phi : V \to S$ *such that during the execution* $EX(\Pi, G, \phi)$, *for every edge* $(u, v) \in E$, *at least one of the following three events takes place:*

(i) *A message is sent on* (u, v).

(ii) *Processor* u *either sends or receives a message containing* $\phi(v)$.

(iii) *Processor* v *either sends or receives a message containing* $\phi(u)$.

Proof: The intuition behind the proof is that, in case the Lemma does not hold for $e \in E$, no processor in the network can distinguish the case in which it takes part in an execution on G from the case in which it takes part in an execution on $G^{(e)}$. The only potential difference between these executions lies in whether u and

v are neighbors or not, where $e = (u, v)$. But this neighborhood relation can not be tested if no messages bearing the ID of one processor are communicated from/to the other. This intuition needs careful formalization, which is sketched below.

Assume, to the contrary, that the Lemma does not hold with respect to some graph $G(V, E)$, ID set S and protocol Π. Let $\phi_0 : V \to S$ and $\phi_1 : V \to S$ be two adjacent ID-assignments with disjoint ranges. By the assumption, there exists an edge $e = (u, v) \in E$ such that during the execution $EX(\Pi, G, \phi_0)$ none of the above events ((i), (ii) or (iii)) takes place. Define the ID-assignment $\psi^{(u)}$ as

$$\psi^{(u)}(j) = \begin{cases} \phi_0(j), & j \in V - \{u\}, \\ \phi_1(j), & j = u, \end{cases}$$

and define $\psi^{(v)}$ similarly. Define

$$\begin{aligned} EX_1 &= EX(\Pi, G, \psi^{(u)}), \\ EX_2 &= EX(\Pi, G, \psi^{(v)}). \end{aligned}$$

By Corollary 4.2, these executions and $EX(\Pi, G, \phi_0)$ are essentially the same.

Finally define the ID-assignment $\psi : V^{(e)} \to S$, where

$$\psi(j) = \begin{cases} \phi_0(j), & j \in V, \\ \phi_1(u), & j = u', \\ \phi_1(v), & j = v', \end{cases}$$

and denote

$$EX_3 = EX(\Pi, G^{(e)}, \psi).$$

Recall our assumption that, except for the initial transmission by the source, a processor sends a message only after receiving one. Thus, unless processors u' and v' receive messages from v or u, they do not send any messages during EX_3. Next, we claim that the executions EX_1, EX_2 and EX_3 are essentially the same. (Proof omitted.) Consequently, we conclude that u (resp., v) does not send a message to v' (resp., u') in EX_3 (i.e., the execution of Π on $G^{(e)}$ with ID's ψ). This contradicts the correctness of Π, and the Lemma follows. ∎

Lemma 4.4 provides us with an accounting method for charging messages sent during the execution of any broadcast protocol to the links of the network. We stress that this does *not* imply that messages *must actually be sent over every link*.

Theorem 4.5: *Let $G(V, E)$ be an arbitrary graph, and let Π be a protocol achieving broadcast on every network of the family C_G. Then the message complexity of Π on G is $\Omega(|E|)$.*

Proof: Let ϕ be an ID-assignment such as in Lemma 4.4. We employ the following *charging rule* to messages sent during the execution $EX(\Pi, G, \phi)$. For every message containing $\phi(w)$ that is sent from the processor u to the processor v, we charge

(1) the edge (u, v),

(2) the pair (possibly edge) (u, w), and

(3) the pair (possibly edge) (w, v).

Claim 4.5.1: *The number of links that get charged for a single message sent during the execution $EX(\Pi, G, \phi)$ is at most $2B + 1$.* ∎

Claim 4.5.2: *Each link is charged at least once.* ∎

Let C denote the total charge placed by the above rules on the execution $EX(\Pi, G, \phi)$, and let M denote the total number of messages sent during that execution. Combining the above claims, we get $|E| \leq C \leq (2B + 1)M$. Recalling that B is a constant, the Theorem follows. ∎

In the full paper we will extend the result of Theorem 4.5 by getting rid of the simplifying technical restrictions imposed on the comparison model. In particular, we will show how to handle the case where multiple initiators are allowed (or in fact, where all processors wake up at round 0 of the run), processors know the size of the network and no restrictions are placed on the local programs. We will also show the extension to general $\rho \geq 0$.

ACKNOWLEDGEMENTS

We wish to thank Avi Wigderson for raising the question related to knowledge of the size of the network. Also, we would like to thank Rüdiger Reischuk for helpful comments on proof of the lower bound for the case of multiple initiators. Thanks are also due to Yishay Mansour for collaborating with us on the proof of the upper bound.

References

[A1]   B. Awerbuch, "Complexity of Network Synchronization", *Jour. of ACM*, Vol. 32, No. 4, 1985, pp. 804-823.

[AGV]   B. Awerbuch, O. Goldreich and R. Vainish, "On the Message Complexity of Broadcast: Basic Lower Bound", *Technical memo*, MIT/LCS/TM-325, April 1987.

[Bo]   B. Bollobas, *Extremal Graph Theory*, Academic Press, 1978.

[Bu]   J.E. Burns, "A Formal Model for Message Passing Systems", TR-91, Indiana University, (1980).

[DM]   Y.K. Dalal and R. Metcalfe, "Reserve Path Forwarding of Broadcast Packets", *Comm. ACM*, Vol. 21, No. 12, pp. 1040-1048, 1978.

[E]   S. Even, *Graph Algorithms*, Computer Science Press, 1979.

[FL]   G.R. Frederickson and N.A. Lynch, "The Impact of Synchronous Communication on the Problem of Electing a Leader in a Ring", *Proc. 16th ACM Symp. on Theory of Computing*, 1984, pp. 493-503.

[GHS]   R.G. Gallager, P.A. Humblet and P.M. Spira, "A Distributed Algorithm for Minimum Weight Spanning Tree", *ACM Trans. on Program. Lang. and Systems*, Vol. 5, 1983, pp. 66-77.

[GRS]   R.L. Graham, B.L. Rothschild and J.H. Spencer, *Ramsey Theory*, John Wiley & Sons, 1980.

[PS]   D. Peleg and A. Schäffer, Graph Spanners, Manuscript, Sept. 1987.

[RK]   R. Reischuk and M. Koshors "Lower bound for Synchronous Systems and the Advantage of Local Information" *Proc. 2nd International Workshop on Distributed Algorithms*, Amsterdam, June 1987.

Families of Consensus Algorithms

Amotz Bar-Noy [*]   Danny Dolev [†]

Abstract

Three main parameters characterize the efficiency of algorithms that solve the Consensus Problem. The ratio between the total number of processors and the maximum number of faulty processors (n and t, respectively), the number of rounds, and the size of any single message. Lower bounds exist for each one of the three. In this paper we present two families of algorithms, each achieving the lower bound for one parameter and a trade-off between the other two. The first family includes algorithms where, given an integer k, the algorithm always requires the minimal possible number of rounds ($t + 1$), with $n = k(3t + 1)$ processors and messages of size at most $t^{O(t/k)}$. To the second family belong algorithms in which all messages are of one bit size, the number of processors is $t^{O((k+1)/k)}$, and the number of rounds is $t + t^{O((k-1)/k)}$. These two families are based on a very simple algorithm with $(2t + 1)(t + 1)$ processors using the minimal number of rounds and the minimal message size (one bit).

---

[*]Stanford University. Part of the work was done while the author was in the Hebrew University of Jerusalem. Supported in part by a Weizmann fellowship, and by contract ONR N0014-85-C-0731.

[†]IBM Almaden Research Center and Hebrew University Jerusalem, Israel.

1 Introduction

In a distributed systems it is essential to agree on common values. This paper presents two families of algorithms to solve Distributed Consensus, a problem considered to be at the very base of fault tolerant distributed computing. The problem's importance follows from the need to overcome the uncertainty that faults introduce. Three main parameters characterize the efficiency of algorithms that solve the Consensus Problem. The ratio between the total number of processors and the maximum number of faulty processors, the number of rounds, and the maximum size of any single message. The families of algorithms presented in this paper have the property that one of the parameters reaches its optimal value, and a trade-off between the other two is provided. These families are a further step in the way to finding a complete characterization of the possible solutions to the Consensus Problem.

1.1 The Consensus Problem

Given a distributed system of n processors, $p_1, \ldots, p_n$, each having an initial binary value v_i ($v_i \in \{0, 1\}$), where up to t of the processors might be *faulty*. One looks for a finite time algorithm at the completion of which the following two conditions hold:

- *validity* – All the non-faulty (*correct*) processors agree (decide) on the same binary value.

- *consistency* – If the initial value was the same for all the correct processors, then this value will be the final *decision* value.

This problem is called the *Consensus Problem*. It is a variant of the known Byzantine General Problem [PSL], in which the processors are required to agree on a binary value of one of them called the general. These two questions were asked in various models. In this paper we concentrate on a basic and simple model:

The system is completely synchronized; the communication is via a complete and reliable network, (i.e., each processor can send a message to every other processor and messages arrive unaltered and on time); the processors are deterministic, and no randomized operations are allowed; and the faulty processors can even be malicious (they might collude in order to prevent reaching the agreement).

Assume the existence of synchronous *rounds* of communication such that in each round a processor performs three operations: it sends messages, receives messages, and performs internal computation. In the last round of the algorithm, after the internal computation is completed, the processor must *decide*.

Three complexity measures determine the efficiency of a solution: The ratio between n and t, the maximal number of rounds required in the worst case, denoted by r, and the maximum size in bits of a single message, denoted by m. Traditionally researchers use the total number of messages (or bits) sent during the algorithm, instead of the third parameter. The results in this paper suggest that in order to explore the correct trade-off among the parameters, the bound on a message size may be a better measure.

1.2   Known Results

There are three known lower bounds for these complexity measures: $n \geq 3t + 1$ [PSL], $r \geq t + 1$ [DS,DLM] and $m \geq 1$ (obvious). The known upper bounds optimize two of the parameters but, "pay" in the third one:

- The *Full Information* algorithm uses $n \geq 3t + 1$ processors and $r = t + 1$ rounds, but there are messages of size $m = n^t = t^{O(t)}$ [PSL,BDDS].

- The *Linear Size Message* algorithm, where $m = O(n) = O(t)$ (almost optimal), uses minimal number $n \geq 4t + 1$, of processors, but takes $r = 2t$ rounds [BDDS].

- The *Square* algorithm with $m \leq n$ and $r = t + 1$, but $n \geq 2t^2 - 3t + 2$ [DRS].

Recently a family of algorithms with an additional parameter d was presented [Co,BDDS]. Every algorithm in this family uses only $n = 4t + 1$ processors, takes $r = t + \lceil t/d \rceil$ rounds, and the message size is at most $t^{O(d)}$. This family becomes the Full Information algorithm when $d = t$, and the Linear Size Message algorithm when $d = 1$.

1.3 Our Results

The basic algorithm, to be presented in the sequel, requires $n = (2t + 1)(t + 1)$ processors, $r = t + 1$ rounds and uses messages of 1-bit size. In this algorithm each processor sends a binary message exactly once, hence its name, the *Beep Once* algorithm. From this algorithm we derive the following two families of algorithms, both use a parameter k:

- The *Disjoint Sets* algorithms: For every k, $1 \leq k \leq t + 1$, there exists an algorithm with $n = k(3t + 1)$ processors and message size $m = t^{O((t+1)/k)}$. All the algorithms require exactly $r = t + 1$ rounds.

- The *Beep* algorithms: For every k, $1 \leq k < \infty$, there exists an algorithm with single bit messages. It uses $n = 4t^{\frac{k+1}{k}} + t^{\frac{k-1}{k}} \leq 5t^{1+\frac{1}{k}}$ processors, and the number of rounds is, $r = \sum_{i=0}^{k} t^{\frac{i}{k}} + t^{\frac{k-1}{k}} \leq t + 3t^{1-\frac{1}{k}}$. Asymptotically, the result is,

$$\forall_{1 \leq k < \infty, 0 < \epsilon \leq 1} \exists T \; s.t. (t \geq T \wedge n \geq 5t^{1+\frac{1}{k}}) \Longrightarrow r \leq (1 + \epsilon)t.$$

Observe that almost every algorithm can be simulated by a single-bit algorithm by encoding messages using bits sent at specific times; i.e., using many more rounds. Our algorithms require sending only single bits without using extra rounds.

The first family matches the lower bound on the number of rounds and gives a trade-off between the number of processors and the maximum size of a single message. In the second family the algorithms are Beep algorithms, i.e. a processor can send at most one bit to any other processor in a round. Here the trade-off is between the number of processors and the number of rounds.

One can see that in all of the algorithms mentioned in this section the following formula holds:

$$(1 + \log_t m) \cdot (r - t) \cdot \left(\frac{n}{t}\right) = O(t).$$

Conjecture: This bound is tight. That is, for every algorithm this product is greater than $\Omega(t)$.

In Section 2 we present the Beep Once algorithm, and the two families of algorithms based on it are presented in Sections 3 and 4.

2   The Beep Once algorithm

This section describes the basic algorithm. In this algorithm the number of processors is $n = (2t + 1)(t + 1)$. It requires $r = t + 1$ rounds and uses 1-bit messages.

The algorithm: Partition the n processors into $t + 1$ disjoint sets, each of cardinality $2t + 1$. Denote these sets by $S_1, S_2, \ldots, S_{t+1}$. There are $t + 1$ rounds in the algorithm. In round number i only processors from set S_i send messages. In the algorithm, whenever a message that is supposed to be sent does not arrive, the receiver assumes a default value of 0 is received.

- $i = 1$: Every processor in S_1 sends its initial value to every processor in S_2.

- $1 < i < t + 1$: Each processor $p \in S_i$ received $2t + 1$ bits in round $i - 1$. Processor p sends the majority value of all these $2t + 1$ values to all the processors in S_{i+1}.

- $i = t + 1$: The processors in S_{t+1} send the majority value of the $2t + 1$ values they have received from S_t to all the n processors.

- **Decision:** The decision value for every processor is the majority value of the $2t + 1$ bits it receives from the processors in S_{t+1}.

Proof of correctness: In each set S_i, there are $2t + 1$ processors, at least $t + 1$ of them are correct. If these $t + 1$ processors send to S_{i+1} (or to everybody in case $i = t+1$) the same value, this will be the majority value of all the correct processors in the set S_{i+1} (or the decision value).

In such a case, the correct processors in S_{i+1} also forward the same value. This process continues until decision is made on this value. In particular, the consistency condition holds, because all the correct processors have the same initial value (which will be the decision value).

There are $t+1$ disjoint sets and at most t faulty processors, impling the existence of a set S_i that does not contain any faulty processors. Therefore, all the correct processors from S_{i+1} (or all processors when $i = t + 1$) receive the same $2t + 1$ bits and compute the same majority value. By the previous argument this will be the decision value and the validity condition holds. ∎

3   The Disjoint Sets algorithms

A generalization of the previous algorithm to create a family of algorithms is presented in this section. For any given k, $1 \le k \le t + 1$, the algorithm requires $k(3t + 1)$ processors, uses messages of size $t^{O((t+1)/k)} = t^{O(d)}$, where $d = \lceil (t + 1)/k \rceil$, and lasts $t + 1$ rounds.

These algorithms employ as a sub-algorithm the Full Information algorithm (the version appearing in [BDDS]). This algorithm has $3t + 1$ processors, runs $t + 1$ rounds and requires messages of size $t^{O(t)}$. In each round every processor sends to every other processor all the information it has received. After $t + 1$ rounds the processors are able to reach the desired decision. The Full Information algorithm has the following two properties:

- The *preservation* property: When all the correct processors start with the same initial value, then for every d, $1 \le d \le t + 1$, if the processors make their decision after d rounds, then all the correct ones reach the same decision value.

- The *listening* property: In case there are more than $3t + 1$ processors, the algorithm can run with $3t + 1$ active processors and every other processor can reach the desired decision using only the messages of the $t + 1$-st round.

The Disjoint Sets algorithm for $1 \le k \le t + 1$: The processors are partitioned into k disjoint sets each of cardinality $3t + 1$. Denote these sets by $S_1, S_2, \ldots, S_k$. Choose $d_1, d_2, \ldots, d_k$ to be either d or $d - 1$, such that $\sum_{i=1}^{k} d_i = t + 1$. (Recall that $d = \lceil (t + 1)/k \rceil$.)

- $i = 1$: The processors from S_1 run the Full Information algorithm, using their initial values, for d_1 rounds. In the last round they send their messages to all the processors in S_2.

- $1 < i \le k$: Each processor $p \in S_i$, takes as an initial value the decision value of the algorithm that has just finished in S_{i-1} (using the listening property). The processors run the Full Information algorithm on these values for d_i rounds, and at the last round they send their messages to S_{i+1} (when $i = k$, to all the processors).

- **Decision:** The decision value for every processor is the outcome of the Full Information algorithm that processors in S_k have run.

Proof of correctness: The proof is similar to that of the Beep Once algorithm. The preservation property of the Full Information algorithm implies that if all the correct processors in S_i start with the same value, then this will be the initial value for $S_{i+1}, \ldots, S_k$ and the decision value. In particular, if all the correct processors have the same initial value this will be the decision value; and the consistency condition holds.

There are k disjoint sets and at most t faulty processors. Therefore, there exists i, such that S_i contains at most $d_i - 1$ faulty processors. Otherwise, there would be at least $\sum_{i=1}^{k} d_i = t+1$ faulty processors, which is impossible. The correctness of the full information algorithm implies that at the end of this stage i, all the processors reach the same decision value. As mentioned above, this will be the decision value of all the correct processors. Thus, the validity condition holds.

It is clear that the algorithm takes $t+1$ rounds. Since $d_i \leq d$, for every i, the full information algorithm implies that the bound on the message size is $t^{O(d)} = t^{O((t+1)/k)}$. ∎

4 The Beep Family of algorithms

Another generalization of the Beep Once algorithm is presented in this section. Again we use a parameter k, $1 \leq k < \infty$. For simplicity assume that $t = s^k$, for some integer s. This time the number of processors is:

$$n = 4t^{1+\frac{1}{k}} + t^{1-\frac{1}{k}} \leq 5t^{1+\frac{1}{k}}.$$

In each algorithm the number of rounds is:

$$r = \sum_{i=0}^{k} t^{\frac{i}{k}} + t^{\frac{k-1}{k}} \leq t + 3t^{1-\frac{1}{k}}.$$

Every processor is allowed to send only single bit messages. Thus, it can either *Beep* or be silent (1 or 0).

We construct recursively this family of algorithms. Each instance of it is denoted by $A(n,t,k)$, where

$$1 \leq k < \infty, \quad 2 \leq t, \quad n = 4t^{1+\frac{1}{k}} + t^{1-\frac{1}{k}}.$$

Case $k = 1$: $n > (2t+1)(t+1)$ and apply the Beep Once algorithm.

Case $k > 1$: partition the n processors into $t^{1/k}$ disjoint sets, $S_1, S_2, \ldots, S_\ell$, $\ell = t^{\frac{1}{k}}$. In each set the number of processors is, $n' = \frac{n}{\ell} = 4t + t^{\frac{k-2}{k}}$. In the sequel, let $t' = t^{\frac{k-1}{k}}$.

The algorithm uses an initial round followed by ℓ stages of rounds.

- In the initial round every processor sends its initial value to the processors of the set S_1.

- In stage number i, $1 \leq i \leq \ell$, the processors of S_i apply the algorithm $A(n', t', k-1)$. The initial value for this sub-algorithm is the majority value of the messages received from the set S_{i-1} (in case $i = 1$ these are the values of the initial round). At the end of the sub-algorithm the processors send a bit value to all processors in S_{i+1} (or everybody in case $i = m$) according to the following rules:

  - The *forward* rule: If the majority of the values received from the set S_{i-1} (or during the initial round, in case $i = 1$) was *supported* by $n' - t$ messages, then they send this value.

  - The *calculate* rule: Otherwise, they send the outcome of the sub-algorithm.

- **Decision:** Each processor decides on the majority value from the n' bits it received from the set S_ℓ.

Proof of correctness: First note that the processors can indeed apply the sub-algorithm because,

$$n' = 4t + t^{\frac{k-2}{k}} = 4(t^{\frac{k-1}{k}})^{\frac{k}{k-1}} + (t^{\frac{k-1}{k}})^{\frac{k-2}{k-1}} = 4t'^{\frac{(k-1)+1}{k-1}} + t'^{\frac{(k-1)-1}{k-1}}.$$

The proof is by induction on k. When $k = 1$ it follows from the correctness of the Beep Once algorithm. Assume that the algorithm is correct for $1 \leq k - 1$ and let us prove it for k.

If all the correct processors start with the same initial value, then in each set S_i the processors apply the forward rule when they send messages to S_{i+1}. In each set there are at least $n' - t > n'/2$ correct processors, thus this value is carried from set to set until all the processors decide on it and the consistency condition holds.

A pigeon hole argument implies the existence of a set S_i with at most t' faulty processors. Otherwise, there would be $\ell t' > t$ faulty processors. If, in this set S_i, there is a processor that received a support of size at least $n' - t$ to a value (0 or 1), then every other correct processor in this set received for this value a majority of at least $n' - 2t > n'/2$. The correctness of $A(n', t', k-1)$ implies that all the correct processors send the same value to the set S_{i+1} (or to everybody in case $i = \ell$) and this value is preserved until it becomes the decision value. Otherwise, all of them use $A(n', t', k-1)$ to decide and will end up having the same value. From the same argument as before it follows that all the correct processors send the same value and this value is forwarded until decision is reached and the validity condition holds.

Note that in the sub-algorithms $A(n', t', k-1)$ the initial round is not needed, because the sub-algorithms can use the initial round of the main algorithm. ∎

Number of rounds: Denote by $r(n, t, k)$ the number of rounds in the algorithm $A(n, t, k)$. We prove by induction that

$$r(n, t, k) = \sum_{i=0}^{k} t^{\frac{i}{k}} + t^{\frac{k-1}{k}}.$$

When $r = 1$, then $r(n, t, 1) = t + 1$. In order to simplify the proof define $r(n, t, 1) = t + 2$ (one can always add a dummy round). This is the desired value, because

$$r(n, t, 1) = t^{\frac{0}{1}} + t^{\frac{1}{1}} + t^{\frac{0}{1}} = t + 2.$$

When $k > 1$, an equivalent to that construction of the algorithm implies,

$$r(n, t, k) = 1 + t^{\frac{1}{k}} r(n', t', k-1) =$$

$$1 + t^{\frac{1}{k}} \left(\sum_{i=0}^{k-1} t'^{\frac{i}{k-1}} + t'^{\frac{k-2}{k-1}} \right) = 1 + t^{\frac{1}{k}} \left(\sum_{i=0}^{k-1} (t^{\frac{k-1}{k}})^{\frac{i}{k-1}} + (t^{\frac{k-1}{k}})^{\frac{k-2}{k-1}} \right) =$$

$$1 + t^{\frac{1}{k}} \left(\sum_{i=0}^{k-1} t^{\frac{i}{k}} + t^{\frac{k-2}{k}} \right) = t^{\frac{0}{k}} + \sum_{i=1}^{k} t^{\frac{i}{k}} + t^{\frac{k-1}{k}} = \sum_{i=0}^{k} t^{\frac{i}{k}} + t^{\frac{k-1}{k}}.$$

∎

Acknowledgment: The authors would like to thank Ray Strong for his part in developing the Beep Once Algorithm.

References

[BDDS] A. Bar-Noy, D. Dolev, C. Dwork, and H. R. Strong, "Shifting Gears: Changing Algorithms on the Fly to Expedite Byzantine Agreement," *proceedings, the 6th Annual ACM SIGACT-SIGOPS Symposium on Principles of Distributed Computing*, Aug. 87, pp. 42-51.

[Co] B. A. Coan, "A Communication-Efficient Canonical Form for Fault-Tolerant Distributed Protocols," *Proceedings, the 5th Annual ACM SIGACT-SIGOPS Symposium on Principles of Distributed Computing*, Aug. 1986.

[DLM] R. DeMillo, N. A. Lynch, and M. Merritt, "Cryptographic Protocols," *Proceedings, the 14th Annual ACM Symposium on Theory of Computing*, pp. 383-400, 1982.

[DRS] D. Dolev, R. Reischuk, and H. R. Strong, "Early Stopping in Byzantine Agreement," *IBM Research Report RJ5406 (55357)*, 1986.

[DS] D. Dolev, and H. R. Strong, "Authenticated Algorithms for Byzantine Agreement," *Siam Journal on Computing*, Vol. 12, pp. 656-666, 1983.

[PSL] M. Pease, R. Shostak, and L. Lamport, "Reaching Agreement in the Presence of Faults," *JACM*, Vol. 27, 1980.

Uniform Self-Stabilizing Rings[1]

James E. Burns

Georgia Institute of Technology

Jan Pachl

University of Waterloo

Abstract. *A self-stabilizing system has the property that it eventually reaches a legitimate configuration when started in any arbitrary configuration. Dijkstra originally introduced the self-stabilization problem and gave several solutions for a ring of processors [Dij74]. His solutions, and others that have appeared, use a distinguished processor in the ring, which can help to drive the system toward stability. Dijkstra observed that a distinguished processor is essential if the number of processors in the ring is composite [Dij82]. We show that there is a self-stabilizing system with no distinguished processor if the size of the ring is prime. Our basic protocol uses $\Theta(n^2)$ states in each processor, where n is the size of the ring. We also give a refined protocol which uses only $\Theta(n^2 / \ln n)$ states.*

1 Overview

A distributed self-stabilizing system is a network of finite-state machines which, when started in an arbitrary initial configuration, always converges to a "legitimate" configuration. (A formal definition is given below.)

Since the introduction of the problem by Dijkstra [Dij74], there has been considerable interest in problems of this type [BGW,HK84,Kru79,Lam84,Tch81]. One characteristic of all proposed solutions is that there is at least one distinguished processor. In [Dij82], Dijkstra commented on the fact that a solution with no distinguished processor is impossible in rings of composite size. In this paper we show that the problem does have a solution with no distinguished processor if the size of the ring is prime.

We present two solutions. In the first solution, the number of states in every processor is $(n-1)(n-2)$ for a ring of (prime) size n. The second solution is obtained by compressing the number of states to $(n-1)\,\pi(\lceil n/3 \rceil)$, where $\pi(x)$ is the number of primes smaller than or equal to x. Thus, by the prime number theorem, the number of states of each processor in the second solution is $O(n^2 / \ln n)$.

---

[1]This work was supported in part by the Natural Sciences and Engineering Research Council of Canada grant A0952. Authors' current addresses: James E. Burns, School of Information and Computer Science, Georgia Institute of Technology, Atlanta, GA 30332-0280. Jan Pachl, IBM Research Laboratory, Saumerstrasse 4, 8803 Ruschlikon, Switzerland.

2   Self-stabilizing Rings

For an integer $n \geq 2$, an **n-processor ring** is a pair (Γ, Δ), where $\Gamma = \Sigma_0 \times \cdots \times \Sigma_{n-1}$ is a set of **configurations** with finite state sets $\Sigma_0, \ldots, \Sigma_{n-1}$, and $\Delta = \delta_0, \ldots, \delta_{n-1}$ is a sequence of **transition relations**. Each δ_i is a relation from $\Sigma_{i-1} \times \Sigma_i \times \Sigma_{i+1}$ to Σ_i (here and in the sequel, all arithmetic on indices is modulo n). We refer to Σ_i and δ_i as the state set and the transition relation, respectively, of **processor** i or P_i. Thus each P_i is a (possibly nondeterministic) finite state machine whose transitions depend on the states of its neighbors in the ring.

For $a \in \Sigma_{i-1}$, $b, d \in \Sigma_i$ and $c \in \Sigma_{i+1}$, we write

$$a \ \underline{b} \ c \to d \qquad \text{in } P_i$$

if $d \in \delta_i(a, b, c)$.

For $0 \leq i \leq n - 1$, we write $\gamma \overset{i}{\to} \gamma'$ if $\gamma, \gamma' \in \Gamma$, $\gamma = (a_0, a_1, \ldots, a_{n-1})$, $\gamma' = (a'_0, a'_1, \ldots, a'_{n-1})$, $a_j = a'_j$ for $j \neq i$, and $a_{i-1} \ \underline{a_i} \ a_{i+1} \to a'_i$ in P_i. In this case we also say that P_i is **enabled** in γ, and that P_i **moves** from γ to γ'. The notation $\gamma \to \gamma'$ means that $\gamma \overset{i}{\to} \gamma'$ for some i. A **computation** (of a given ring) is a finite or infinite sequence $\gamma_0 \gamma_1 \ldots$ such that $\gamma_{j-1} \to \gamma_j$ for all j greater than zero and less than the length of the sequence. We write $\gamma \Rightarrow \gamma'$ if there is a finite computation $\gamma_0 \gamma_1 \ldots \gamma_k$ such that $\gamma_0 = \gamma$ and $\gamma_k = \gamma'$.

Definition 2.1 *A ring (Γ, Δ) is **self-stabilizing** if and only if there is a set $\Lambda \subseteq \Gamma$, called the **legitimate configurations** of S, such that the following five conditions are satisfied.*

1. *[No deadlock] For every $\gamma \in \Gamma$ there is a $\gamma' \in \Gamma$ such that $\gamma \to \gamma'$.*

2. *[Closure] For every $\lambda \in \Lambda$, every λ' such that $\lambda \to \lambda'$ is in Λ.*

3. *[No livelock] Every infinite computation of S contains a configuration in Λ.*

4. *[Mutual exclusion] For every $\lambda \in \Lambda$, exactly one processor is enabled.*

5. *[Fairness] For every processor i, every infinite computation consisting of configurations in Λ contains an infinite number of moves by P_i.*

A more restrictive definition would require that legitimate configurations form a cycle. Our solution presented in Section 3 satisfies this stronger requirement.

The ring is **uniform** if $\Sigma_0 = \Sigma_1 = \cdots = \Sigma_{n-1}$ and $\delta_0 = \delta_1 = \cdots = \delta_{n-1}$. It was pointed out by Dijkstra [Dij82] that there is no uniform n-processor self-stabilizing ring for a composite size n. (Other similar results about impossibility of solving certain problems in distributed systems in the presence of symmetry were established by Angluin [Ang80]). On the other hand, it is easy to construct a uniform 2-processor self-stabilizing ring (with 3 states in each processor). Seger [Seg86] found a uniform 3-processor self-stabilizing ring (with 2 states in each processor).

In the next section we show that there is a uniform n-processor self-stabilizing ring for every prime $n \geq 3$.

The ring is **unidirectional** if $\delta_i(a,b,c)$ is independent of c for each i and all $a \in \Sigma_{i-1}$, $b \in \Sigma_i$. Since our solutions are unidirectional and uniform, we write transition rules in the form $a \ \underline{b} \rightarrow d$ instead of $a \ \underline{b} \ c \rightarrow d$, and omit the qualification "in P_i".

3   The First Protocol

Let the size of the ring be $n \geq 3$. The states of our processors are composed of two parts, the label and the tag. Labels range over $\{0, 1, \ldots, n-2\}$, and tags over $\{0\} \cup \{2, 3, \ldots, n-2\}$. The composite state will be written in the form *label.tag*.

The protocol is defined by the following two rules, in which the expressions $a+1$ and $b-a$ are computed modulo $n-1$.

A. If $b \neq a+1$ and ($b \neq 0$ or $t \neq b - a$ or $t < u$ or $t = 0$) then

$$a.t \quad \underline{b.u} \rightarrow (a+1).(b-a)$$

B. If $t \neq u$ and $a+1 \neq 0$ then

$$a.t \quad \underline{(a+1).u} \rightarrow (a+1).t$$

Let Λ be the set of all cyclic permutations of configurations of the following form (underlining indicates the enabled state):

$$0.0 \quad 1.0 \quad \ldots \quad (i-1).0 \quad i.0 \quad \underline{i.0} \quad (i+1).0 \quad \ldots \quad (n-2).0$$

for $i = 0, 1, \ldots, n-2$.

We shall prove the following theorem:

Theorem 3.1 *If n is prime, then Λ is a set of legitimate configurations such that the ring defined by Rules A and B is self-stabilizing.*

Thus, we need to show that all five conditions of Definition 2.1 hold. Three of these are easy to show.

Lemma 3.2 *Conditions 2, 4, and 5 of Definition 2.1 hold.*

Proof: Rule B does not apply to configurations in Λ. Rule A applies to exactly one processor in every configuration in Λ, so Condition 4 (mutual exclusion) holds. Let processor i be the processor enabled in configuration $\lambda \in \Lambda$. Applying Rule A to λ gives a new configuration λ' such that $\lambda' \in \Lambda$ and processor $(i+1) \pmod n$ is enabled at λ'. Thus, Conditions 2 (closure) and 5 (fairness) both hold. □

The remaining two conditions in Definition 2.1 will now be established in a series of lemmas.

In the sequel, all arithmetic on processor indices is modulo n, and all arithmetic on state labels is modulo $n - 1$. (Sometimes the modulus operator is included for emphasis.)

Definition 3.1 *Let $\gamma \in \Gamma$ be a configuration. Two consecutive processors, say P_i and P_{i+1}, with states a.t and b.u in γ form a **gap** of γ, which we designate by $(a.t, b.u)$, if $b \neq a + 1 \pmod{n-1}$. The **gap size** $g(P_i, \gamma)$ is defined to be $b - a \pmod{n-1}$. A **segment** of γ is a maximal cyclically contiguous sequence of processors $s = (P_i, P_{i+1}, \ldots, P_{j-1}, P_j)$ which contains no gaps; the gap size of s, denoted $g(s, \gamma)$, is $g(P_j, \gamma)$. We refer to P_j as the **right end** of s and P_i as the **left end** of s. A segment is **zero-based** if the label of its left end is 0.*

For example, a configuration for seven processors with state sequence (for processors P_0 through P_6):

$$1.3 \quad 2.3 \quad 3.1 \quad 5.2 \quad 0.2 \quad 1.1 \quad 0.4$$

has two segments: (P_6, P_0, P_1, P_2) (which is zero-based) and (P_3, P_4, P_5), and, hence, two gaps: $(3.1, 5.2)$ and $(1.1, 0.4)$, of sizes 2 and 5, respectively.

The following lemma is the only place where we need the fact that n is prime.

Lemma 3.3 *If n is prime, then some processor is enabled for every configuration in Γ (so Condition 1 holds).*

Proof: Let γ be any configuration in Γ and suppose that no procesor is enabled at γ. Recall that Rule A applies if there are two consecutive states $a.t$ and $b.u$ such that $b \neq a + 1$ and ($b \neq 0$ or $t \neq b - a$ or $t < u$ or $t = 0$). Then γ must be composed only of zero-based segments, since otherwise Rule A would apply to some processor. There are two cases:

Case 1: There is only one segment in γ, and hence only one gap, say $(a.t, b.u)$. Then $b = 0$ because the segment is zero-based, and $a = 0$ because the length of the segment is n. Thus $b - a = 0$ and Rule A applies for any value of t, which is a contradiction.

Case 2: There is more than one segment in γ. Thus the length of every segment is at most $n - 1$, and every segment contains exactly one label 0 (in its left-end processor). Since Rule B does not apply, all tags within each segment are equal. Since Rule A does not apply, the tag of the right-end processor of each segment is equal to the gap size of the segment. Consequently, all gaps have the same size (otherwise we would have $t < u$ for some gap $(a.t, b.u)$ and Rule A would apply). But since all the segments are zero-based and all the gaps have the same size, it follows that all the segments have the same length. Since n is a prime, all the segment lengths are 1. Thus all the labels in the ring are 0, and all the gap sizes are 0. That means that Rule A applies, which is the desired contradiction. □

The protocol works because segments continually move around the ring. To show that all segments but one are eventually consumed by this process, we first show that the continual movement of segments is guaranteed.

Lemma 3.4 *In every infinite computation, every processor executes a move by Rule A infinitely many times.*

Proof: If any processor i makes an infinite number of moves by Rule A, then so must processor $i - 1$ (because processor i can only move by Rule A if the label of processor $i - 1$ has changed since i's last move). Thus by induction, if any processor moves an infinite number of times by Rule A then they all do, so it is sufficient to show that there must be *some* processor which makes an infinite number of moves by Rule A.

Suppose to the contrary that there is an infinite computation of S with only a finite number of moves using Rule A. Then, there is a configuration γ in the computation after which the labels of all the processors (and hence all the segments) are fixed and all future moves are by Rule B. However, the left end of any segment cannot make any move by Rule B, and if P_i is not the left end of its segment then P_i can make at most one more move than P_{i-1} after γ is reached. Thus no P_i can make infinitely many moves by Rule B after γ, which contradicts the assumption that the computation is infinite. □

Up to now, we have only been concerned with segments as parts of a single configuration, but now we must deal with them as dynamic objects which change during a computation. Since transitions can destroy segments, things can get rather complicated. Fortunately, we will only have to deal with parts of computations where segments persist. This is a consequence of the following easy lemma.

Lemma 3.5 *If $\gamma, \gamma' \in \Gamma$, $\gamma \Rightarrow \gamma'$, then the number of segments in γ' is at most the number of seqments in γ.*

Proof: Rule B does not affect segments or gaps. It is easy to see that Rule A cannot increase the number of segments, so the result follows. □

By the lemma, any infinite computation reaches a point after which the number of segments is constant. A computation in which the number of segments is constant is **quiet**. The following definition captures the idea that in a quiet computation segments and gaps preserve their "identity".

Definition 3.2 *Let $C = \gamma_0\gamma_1 \cdots$ be a quiet computation. A **dynamic gap** of C is a function z from $\{0, 1, \ldots\}$ into processors $\{P_0, \ldots, P_{n-1}\}$ satisfying the following properties:*

1. *$z(0)$ is the right end of a segment in γ_0.*

2. *Let $0 < j < length(C)$ and i be such that $\gamma_{j-1} \xrightarrow{i} \gamma_j$. If $\gamma_{j-1} \xrightarrow{i} \gamma_j$ is a move by Rule A and $z(j-1) = P_{i-1}$ then $z(j) = P_i$, otherwise $z(j) = z(j-1)$.*

Let z be a dynamic gap of a quiet computation $C = \gamma_0\gamma_1 \cdots$. For $0 \leq j < length(C)$, processor $z(j)$ is the right end of a segment of γ_j, which we denote $s_z(j)$. Then s_z is a function, called a **dynamic segment**, from $\{0, 1, \ldots\}$ into segments. Recall that, by Definition 3.1, we have $g(z(j), \gamma_j) = g(s_z(j), \gamma_j)$ for $0 \leq j < length(C)$.

Lemma 3.6 *Let $C = \gamma_0\gamma_1 \cdots$ be a quiet computation and z a dynamic gap of C. Then $g(z(j-1), \gamma_{j-1}) = g(z(j), \gamma_j)$ for $0 < j < length(C)$.*

Proof: Rule B does not affect gaps or segments. Rule A can only cause a change of gap size if a gap is destroyed during the move, but this cannot happen in a quiet computation, so the gap size of every dynamic gap in C is constant during C. ☐

In light of the lemma, for a quiet computation C with dynamic gap z, define $g(z)$ to be the common value of $g(z(j), \gamma_j)$ over $0 \leq j < length(C)$.

Lemma 3.7 *Let $C = \gamma_0\gamma_1 \cdots$ be an infinite quiet computation, and let z be a dynamic gap of C. Then*

- *there is a $j \geq 0$ such that the state of processor $z(j)$ in γ_j is $0.g(z)$, and*

- *there is a $k \geq 0$ such that the state of processor $z(k) + 1$ in γ_k is $0.u$ for some u.*

Proof: These facts follow from Lemma 3.4 using easy inductions on j and k. ☐

The next lemma specifies how tags can be arranged during a quiet computation. This result will be useful in proving that the number of gaps is eventually reduced to one. We say that a segment is **well-formed** if all its tags are equal to its gap size.

Lemma 3.8 *Let $C = \gamma_0\gamma_1 \cdots$ be an infinite quiet computation. Then there is a $j_0 \geq 0$ such that for all $j \geq j_0$, all segments of γ_j are well-formed.*

Proof: Let z be any dynamic gap of C. By Lemma 3.7, there is $j_1 \geq 0$ such that the state of processor $z(j_1)$ in γ_{j_1} is $0.g(z)$. By Lemma 3.4, there is $j_2 > j_1$ such that processor $z(j_1)$ is the left end of the segment $s_z(j_2)$ of γ_{j_2}. Choose the smallest such $j_2 > j$; then processor $z(j_1)$ is in the segment $s_z(j)$ of γ_j for $j_1 \leq j \leq j_2$.

Induction on j shows that for $j_1 \leq j \leq j_2$ the tags of the processors

$$z(j_1), z(j_1) + 1, \ldots, z(j)$$

in γ_j are equal to $g(z)$. Therefore the segment $s_z(j_2)$ of γ_{j_2} is well-formed, and it follows that the segment $s_z(j)$ of γ_j is well-formed for all $j \geq j_2$. Since there are only finitely many dynamic segments in C, the lemma follows. □

We are finally ready to prove the crucial lemma.

Lemma 3.9 *There is a point in every infinite computation (regardless of the initial configuration) after which every configuration has exactly one segment.*

Proof: Let $C = \gamma_0\gamma_1 \cdots$ be any infinite computation and (by Lemmas 3.5 and 3.8) choose j_0 such that $C' = \gamma_{j_0}\gamma_{j_0+1} \cdots$ is quiet, and all the segments in each γ_j, $j \geq j_0$, are well-formed. It follows that no move in C' is by Rule B.

We first show that for all dynamic gaps z of C' we have $g(z) = 0$. If not, then there are two dynamic gaps z and z' such that s_z immediately precedes $s_{z'}$ in the ring, $g(z) \geq g(z')$, and $g(z) \neq 0$. By Lemma 3.7, there exists $j_1 \geq j_0$ such that the state of the processor $i = z(j_1) + 1$ in γ_{j_1} is $0.g(z')$. The state of processor $z(j_1)$ in γ_{j_1} is $a.g(z)$ for some a. It follows that processor i cannot move by Rule A in γ_{j_1}. In fact, processor i can subsequently move only if processor $i - 1 = z(j_1)$ has moved first. But such a move of processor $i - 1$ would decrease the number of segments. Since such a move exists by Lemma 3.4, this contradicts the assumption that C' is quiet.

Thus all the gap sizes in every γ_j, $j \geq j_0$ are equal to 0. This implies that either there is a single segment, as desired, or there are n segments of length one. But the latter case is impossible, for a single move by Rule A would decrease the number of segments, violating the assumption that C' is quiet. Since there cannot be n segments in a quiet computation, it follows that there is exactly one segment. □

This allows the proof of the theorem to be easily completed.

Proof: [of Theorem 3.1] By Lemma 3.9, in any infinite computation after finitely many moves there is only one well-formed segment of length n. The size of the only gap in the ring is 0, hence the tags of the single segment are all 0. But this describes an element of Λ. Thus we have proved Condition 3 (no livelock) in Definition 2.1. The remaining conditions were shown to hold by Lemmas 3.2 and 3.3. □

4 Reducing the number of states

We have shown that self-stabilization can be achieved for a uniform system on a ring of size n, if n is prime. In our solution, the state space of each processor has cardinality $(n - 1)(n - 2) = \Theta(n^2)$. We can improve this with a slightly more complicated protocol.

Our revised protocol uses the original Rule B and replaces Rule A by the following rule, in which $a + 1$ and $b - a$ are again computed modulo $n - 1$.

Rule A′. If $b \neq a + 1$ and ($b \neq 0$ or $t \neq f(b - a)$ or $t < u$ or $t = 0$) then

$$a.t \quad \underline{b.u} \rightarrow (a + 1).f(b - a)$$

where f is a function from $\{0\} \cup \{2, 3, \ldots, n - 2\}$ to non-negative integers. Our task is to choose f (which may depend on n) so as to minimize the number of tag values while maintaining correctness. Labels range over $\{0, 1, \ldots, n - 2\}$, tags over the range of f. The set Λ is the same as in Section 3.

Lemma 4.1 *If f is any function such that $f(0) = 0$ and $f(k) > 0$ for $k > 0$, then the protocol defined by Rule A′ and Rule B satisfies conditions 2, 3, 4, and 5 of Definition 2.1.*

Proof: Rule A′ is equivalent to Rule A on legitimate configurations, so Lemma 3.2 (and hence Conditions 2, 4, and 5) holds for the revised protocol. Regardless of what f is, an infinite quiet computation must again eventually have all of its tags equal to 0. By the restriction on f, this can only happen when there is a single segment which constitutes a legitimate configuration. This implies that Condition 3 holds. □

Now we construct f for which the remaining condition, no deadlock, holds as well. For any integer $x \geq 2$, let $p(x)$ be the smallest prime divisor of x. Assume $n > 6$, and define $h(0) = 0$ and $h(k) = \min(p(n - k), \lceil n/3 \rceil)$ for $2 \leq k \leq n - 2$.

Lemma 4.2 *If $f = h$, then the protocol defined by Rule A′ and Rule B satisfies Condition 1 (no deadlock) for prime $n > 6$.*

Proof: Suppose there is a deadlocked configuration, γ. As in the proof of Lemma 3.3, it follows that all segments of γ are zero-based, and their lengths have a common divisor. That contradicts the primality of n.

If γ has only one segment then we obtain a contradiction as in Case 1 in the proof of Lemma 3.3. Assume that there are at least two segments in γ. Then all tags in γ have the same value $t \neq 0$, and for every gap size $g(P_i, \gamma)$ we have $t = h(g(P_i, \gamma))$ (otherwise Rule A′ or Rule B would apply). Since $t \neq 0$, no gap size is 0. Let s be a segment of γ, and let $a.t$ be the state of its right end, P_i. Then $n - g(P_i, \gamma) = n - ((0 - a) \pmod{n - 1}) = a + 1 = length(s)$. Thus we get

$$t = h(g(P_i, \gamma)) = \min(p(length(s)), \lceil n/3 \rceil).$$

If $t < \lceil n/3 \rceil$ then $p(length(s)) = t \geq 2$ for every segment s of γ; hence the length of every segment is divisible by t, which contradicts the primality of n.

If $t \geq \lceil n/3 \rceil$ then $length(s) \geq p(length(s)) \geq n/3$ for every segment s of γ; hence γ has exactly two segments and (since n is odd), $length(s)$ is even for a segment s. But then $p(length(s)) = 2 < \lceil n/3 \rceil$, which is a contradiction. □

From these two lemmas, we have the following theorem.

Theorem 4.3 *For any ring of prime size n there is a uniform self-stabilizing system that uses $O(n^2/\ln n)$ states per processor.*

Proof: If $f = h$, the number of tags used by the revised protocol is $\pi(\lceil n/3 \rceil)$, where $\pi(x)$ is the number of primes smaller than or equal to x. By the prime number theorem ([Gro84], p. 169), $\pi(\lceil n/3 \rceil) \approx \frac{n}{3}/\ln\frac{n}{3}$. Thus the theorem follows from the lemmas for $n > 6$. We have already noted that there exist solutions for $n = 2, 3, 5$. □

We would like to show that our revised protocol is optimal in terms of shared space. However, this appears to be a difficult problem. The strongest result we know of is by Carl-Johan Seger [Seg87]. Seger has shown that any uniform unidirectional self-stabilizing procotol for a ring of n processors must use at least $n - 1$ states per processor.

References

[Ang80]   D. Angluin. Local and global properties in networks of processors. *Proc. 12th Symp. on Theory of Computing (Los Angeles, CA, April 1980)*, ACM, 1980, pp. 82–93.

[BGW]   G.M. Brown, M.G. Gouda, and C.-L. Wu. Token systems that self-stabilize. Manuscript.

[Dij74]   E.W. Dijkstra. Self-stabilizing systems in spite of distributed control. *Communications of the ACM 17*, 11 (1974) 643–644.

[Dij82]   E.W. Dijkstra. Self-stabilization in spite of distributed control (EWD391). Reprinted in *Selected Writing on Computing: A Personal Perspective*, Springer-Verlag, Berlin, 1982, 41–46.

[Dij86]   E.W. Dijkstra. A belated proof of self-stabilization. *Distributed Comput-ing 1*, 1 (1986) 5–6.

[Gro84]   E. Grosswald. *Topics from the Theory of Numbers*. Birkhäuser Boston, 1984.

[HK84]   R.W. Haddad and D.E. Knuth. A programming and problem-solving seminar. Stanford University Technical Report STAN-CS-85-1055, June 1985.

[Kru79]   H.S.M. Kruijer. Self-stabilization (in spite of distributed control) in tree-structured systems. *Information Processing Letters 8*, 2 (1979) 91–95.

[Lam84]   L. Lamport. Solved problems, unsolved problems, and non-problems in concurrency. Invited address at the 2nd Symp. on Principles of Dis-tributed Computing (Montreal, August 17-19, 1983), included in *Proc. of the 3rd Symp. on Principles of Distributed Computing*, ACM, 1984, pp. 1–11.

[Seg86]   C.-J. Seger. A note for the graduate course CS 760. University of Water-loo, October 2, 1986.

[Seg87]   C.-J. Seger. Private communication.

[Tch81]   M. Tchuente. Sur l'auto-stabilization dans un réseau d'ordinateurs. *RAIRO Inf. Theor. 15*, 1981, 47–66.

The Complexity of Selection Resolution, Conflict Resolution and Maximum Finding on Multiple Access Channels

CHARLES U. MARTEL
Division of Computer Science
University of California at Davis
Davis, California, 95616

THOMAS P. VAYDA
Department of Computer Science
California State University at Chico
Chico, California, 95929

1. Introduction

In this paper we discuss the complexity of several fundamental problems for a *multiple access broadcast network* (MABN). An MABN is a computer network in which there is a single shared communication channel and every network node can receive all messages sent over the channel. There are many MABNs currently in use; the Ethernet is one ubiquitous example [11,17]. We use a commonly accepted model for studying this type of environment: the *time slotted broadcast channel* [6]. In this model the network nodes *contend* for access to the channel and can only transmit during discrete synchronized time intervals called *slots*. During each slot, either zero, one, or more than one node can transmit a message. In the third case no valid information can be received other than the feedback that a *collision* has occurred. We study two channel feedback models: in the *normal* channel model the only feedback a node gets from a collision is that two or more nodes tried to access the channel; in the *perfect feedback* channel model the feedback is an integer indicating the number of nodes which tried to transmit.

Efficient use of an MABN requires distributed algorithms which coordinate access to the channel while using a small number of slots. We present lower bounds and fast algorithms for three fundamental problems relevant to the efficient use of MABNs: Selection Resolution, Conflict Resolution, and Maximum Finding. In each of these problems there is a set C of two or more *contending* nodes that require access to the channel. The *selection resolution* problem requires that *one* node in C be selected for transmission, while the *conflict resolution* problem requires that *every* element of C gets to transmit. In the *maximum finding* problem each node in C has a real value and the problem is to quickly allow the node of maximum value to transmit. For each of the three problems we give new lower bounds. In addition, we *resolve* the complexity of selection resolution and maximum finding by presenting new algorithms which are within a small constant of the corresponding lower bound.

2. Main Results and Their Significance

Good selection and conflict resolution algorithms are essential for the effective use of multiple access broadcast networks. Additionally, selection and conflict resolution as well as maximum finding are the fundamental operations required for the implementation of many distributed MABN algorithms [3,8,14,20]. Since MABNs are an important class of networks, these problems have been closely studied from around 1976, when MABNs first emerged, to the present time [5,6,7,9,17,21]. Our results provide significant improvements over prior work in this area and settle the *exact* complexity of selection resolution and maximum finding. They also help to answer two long standing questions: "Is the tree algorithm optimal?" and "Are unrestricted MABN algorithms more powerful than divide and conquer ones?" [2,3,6,7,9,12]

In our discussions of MABNs we use n to denote the total number of nodes on the network, c the number of contenders (i.e. $c = |C|$) and assume that $c \geq 2$. The algorithms always know the value of n, but the value of c may or may not be known.

2.1. Selection Resolution

Our main result is a new lower bound for deterministic selection resolution which is essentially *tight*: at least $\lceil \log(n-c)-0.322... \rceil$ slots are required by any deterministic algorithm. This bound holds even if c is known in advance, and even in the perfect feedback channel model. We show that this bound is near optimal when c is known by presenting an algorithm which uses at most $\lceil \log(n-c) \rceil +1$ slots. This essentially resolves the complexity of deterministic selection resolution for both the normal and perfect feedback channel models.

2.2. Conflict Resolution

We prove new lower bounds for deterministic conflict resolution algorithms, both for general algorithms and for the restricted class of divide and conquer algorithms. In the unrestricted case we show that even if c is known in advance the number of slots needed is at least: $(7/4)c+\lfloor \log(4n/7c) \rfloor$ if $c < (4/7)n$; and n if $c \geq (4/7)n$. This bound is *tight* for the upper range of c. For divide and conquer algorithms we obtain a stronger bound, which is tight for $c \geq n/2$.

2.3. Maximum Finding

For the maximum finding problem we consider probabilistic algorithms. In this setting we establish the complexity of maximum finding by showing that $\Theta(\log c)$ expected slots are necessary and sufficient to find the maximum of c elements. Our main contribution is a new $O(\log c)$ algorithm which improves on Dechter and Kleinrock's best previous upper bound of $O((\log\log c)\log c)$ expected slots [3]. The lower bound of $\Omega(\log c)$ expected slots can be obtained from results by Fich, Ragde and Wigderson on CRCW PRAMS with constant memory [4,18].

An important current topic is the relationship between different parallel models of computation. Probabilistic selection resolution algorithms which run in time T, allow a CRCW PRAM with arbitrary or priority write to simulate a step of a *random* CRCW PRAM (where a random element gets to write in case of a conflict) in $O(T)$ time. Our maximum finding algorithm introduces a new technique which solves a sequence of decreasing size selection resolution problems in amortized $O(1)$ expected slots per selection. This allows for a fast simulation of a sequence of write steps on a random CRCW PRAM when the write conflicts decrease in size.

3. Model Description and Notation

We assume that a set of nodes share a single MABN. We denote this set of nodes by N and its cardinality by n. As described in the introduction, we use the time slotted broadcast channel model and both the normal and perfect feedback channel models. The slotted time broadcast network model has been widely used to study the selection resolution, the conflict resolution and the maximum finding problems [5,6,7,9,20,21]. It has frequently been shown that results for the slotted broadcast model will usually apply to actual multiple-access channels.

Many analyses of a multiple-access channel assume that there is a global *channel access algorithm* which works as follows [2,6,7,10]. Initially access to the channel is unrestricted. As soon as a collision occurs a *restricted access algorithm* is invoked (usually a selection or conflict resolution algorithm). Until this restricted access algorithm is completed, no messages other than those associated with the restricted access algorithm are transmitted, and only those nodes involved in the initial collision transmit during this algorithm's execution. Since all further discussion will be about such restricted access algorithms, we will simply refer to them as algorithms. This paper deals with the design and analysis of such algorithms, but we allow a somewhat more general definition of the setting in which they can be executed.

We assume that an algorithm is invoked when a specified signal (such as a collision or a valid message) is sent to all the network nodes. The signal activates a set of nodes C, called *contenders*, which will participate in the algorithm (often, C is the set of nodes which just participated in a collision). Initially, the only information a node has is n and whether or not it is in C (in some cases we will also assume that the nodes know $c = |C|$).

The algorithms we consider are implemented as distributed programs executed independently and concurrently by each of the contending nodes. Usually each node will execute an identical copy of the program, but this is not assumed when proving lower bounds. In each time slot, the program a node is executing will determine whether or not that node is to transmit a message. In a *deterministic* algorithm the decision is based solely on the nodes' identity and the history of channel responses which have occurred since the invocation of the algorithm whereas in a *probabilistic* algorithm a locally generated random variable can also be used in making the decision. In designing deterministic algorithms we assume that the nodes are numbered from 1 to n and that each node knows its own number. Without this assumption it is *impossible* to implement deterministic algorithms for some of the problems under consideration. Usually, the probabilistic algorithms will have much better expected performance than the deterministic algorithms, but have unbounded worst case performance [21]. Deterministic algorithms are important since bounds can be proven for their worst case performance. Worst case bounds are crucial in many settings, such as control of real time systems.

Following Greenberg [6], we will describe an algorithm's execution as a sequence of *queries* $Q_1, Q_2, Q_3, \ldots$ where $Q_t \subset N$ for $t = 1, 2, \ldots$ and Q_t denotes the set of nodes enabled to transmit during the t-th slot. During slot t the nodes in $Q_t \cap C$ transmit. We also consider *divide and conquer* algorithms. Intuitively, a divide and conquer algorithm always uses queries that are subsets either of a previous query or of the complement of a previous query. More formally, we maintain a family F of sets of nodes called *atoms* such that each query must be a subset of an atom. Initially F consists of the single element N, the set of all nodes. After the first query Q_1, $F = \{Q_1, N - Q_1\}$. In general the i-th query Q_i must be a subset of some atom A. After this query, $F = (F - A) \cup Q_i \cup (A - Q_i)$. (Thus A is split into two sets: Q_i and its complement).

4. Selection Resolution

In this section we prove a new lower bound for deterministic selection resolution when c is known in advance and then describe an algorithm which nearly achieves the lower bound. We begin by reviewing the celebrated *tree algorithm*, which can be used both for selection and conflict resolution [2,3,7,10]. The tree algorithm for selection begins by dividing the nodes into two (nearly) equal sized sets Q and $\bar{Q}$, then enables the contenders in Q to transmit. If the result is a good transmission then we are done, on a collision the procedure is called recursively with Q and on a null it is called with $\bar{Q}$. It is easy to show that this algorithm uses at most $\lceil \log n \rceil$ slots.

Our lower bound holds for the perfect feedback channel model, therefore it also applies to the normal channel model. We start by giving a high level description of the adversary strategy used. Initially we only know that c of the n nodes are contenders. During each slot, the algorithm selects some subset Q of N and all of the nodes in $T = C \cap Q$ transmit in the slot. The adversary chooses T depending on the size of Q and C according to the following rules.

- If Q is "small" or $c \leq 3$, the adversary decides that T is empty, so no node transmits and all contenders are in $N - Q$.

- If Q is "large", the adversary declares that all the contenders are in Q; so, a collision occurs.

- If Q is neither "small" nor "large" and $c \geq 4$, then k^* contenders are put into Q and the remaining $c - k^*$ contenders are put into $N - Q$ and a collision occurs. The value of k^* will be a function of n and c and obeys $2 \leq k^* \leq c - 2$. Intuitively, k^* is chosen so that the two resulting subproblems (on the sets Q and $N - Q$) are as close to equally hard as possible.

In all three cases, the result of a query is two new subproblems which are just like the original problem: a set of nodes of known size which contains a known number of contenders. The proof hinges on two main insights; there exists an optimal algorithm with a simple form and that the difficulty of a problem instance can be characterized by a simple function.

Lemma 4.1: There exists an *optimal* algorithm for the perfect feedback channel which is divide and conquer and furthermore whose queries satisfy $Q_i \supset Q_{i+1}$ or $\bar{Q}_i \supset Q_{i+1}$ for $i \geq 1$.

Proof Sketch: Any algorithm can be represented by a decision tree, in which each node represents a query, each arc a possible response and another query follows each possible response, except for leaf queries that produce a good slot.

A fairly straightforward modification of the decision tree yields an algorithm whose worst case performance has not been degraded and meets the conditions of the lemma. The most subtle part of the proof is showing the correctness of this modified algorithm. Applying this construction to an optimal algorithm produces a new optimal algorithm which satisfies the statement of the lemma. For a detailed proof see [19]. □

This result shows that there is an optimal algorithm in which each query is a subset of a previous query or its complement and thus allows us to focus our attention on the easier of the two subproblems created by each query.

The main insight needed to complete the proof is a simple measure of the difficulty of a problem instance. We call such a measure a *potential function*. Specifically, if we have a problem with n total nodes, and $c > 1$ contenders, the potential of this problem is $n - c$. The potential is zero if $c = 1$, and ∞ if $c = 0$. The potential function allows us to fill in the details of the adversary strategy. Given a set N of n nodes containing c contenders and a query $Q \subset N$ of size q

(1) if $q \leq \lceil (n-c)/2 \rceil$ or $2 \leq c \leq 3$ then return a null and put all contenders into $N - Q$,

(2) else if $\lceil (n-c)/2 \rceil < q \leq n/2$ and $4 \leq c$ then return a collision, put k^* contenders into Q and $c - k^*$ contenders into $N - Q$. where $k^* = MAX \{2, \lceil q + (c-n)/2 \rceil \}$,

(3) else if $q > n/2$ then exploit "symmetry" as follows: treat $N - Q$ as the query, apply one of the above rules, and return the obvious "complement" of the response.

The next lemma shows that this strategy prevents the potential from falling too quickly:

Lemma 4.2: If $1 \leq p_0 < \infty$ is the potential of the current problem and p_1 is the minimum potential of the two new subproblems created by a divide and conquer query, then $p_1 \geq (1/2) p_0 - 2$.

Proof Sketch: The proof follows from a case analysis of the adversary strategy. Depending on the size of the query and the number of contenders, there are twelve cases to consider. Symmetry reduces this to the six cases produced by using rules (1) and (2). The cases represent each of the possible subproblems produced by the query, which in turn depend on the rule the adversary uses to respond to the query. The adversary can use rule (1) for two reasons and these produce the first two cases. When the adversary uses rule (2), two new subproblems are created by each of the two possible choices for k^*.

We will prove the result for one of the six cases; the others can be shown in a similar manner. Suppose that adversary using rule (2), responds with a collision and sets $k^* = \lceil q + (c-n)/2 \rceil$. This results in two new subproblems of sizes (q, k^*) and $(n-q, c-k^*)$. Now we consider the second of these subproblems. Here $p_0 = n - c$ and $p_1 = (n-q) - (c - k^*)$ thus we get

$$p_1 = n - q - c + k^* = n - q - c + \lceil q + (c-n)/2 \rceil \geq \frac{2n - 2c - n + c}{2} = \frac{n-c}{2} > \frac{p_0}{2} - 2.$$

We note that a stronger inequality holds in all but two of the six cases. □

Theorem 4.3 Any deterministic selection resolution algorithm for the perfect channel model requires at least $\lceil \log(n-c) - 0.322... \rceil$ slots for $1 \leq n - c \leq n - 2$.

Proof Sketch: Let p_t represent the minimum potential at the start of slot t. An optimal algorithm can be assumed to have the form given in Lemma 4.1, thus by Lemma 4.2, p_t satisfies the following recurrence relation: $p_1 = n - c$ and $p_t \geq (1/2) p_{t-1} - 2$ for $t > 1$.
This relation is easily bounded as follows: $p_{t+1} \geq \frac{1}{2^t} p_1 - 2 \sum_{i=0}^{t-1} \frac{1}{2^i} > \frac{1}{2^t} p_1 - 4$ for $t \geq 1$.

Thus $p_{t+1} > 1$ whenever $t \leq \log(n-c) - \log(5)$. As long as the potential is greater than or equal to 1, at least two more slots are required to obtain a good transmission in the worst case. Adding on these two more slots, shows that in the worst case, at least a total of

$\log(n-c) + 2 - \log(5) = \log(n-c) - \log(5/4)$ slots are required. Since the number of slots must be an integer, taking the ceiling of this quantity completes the proof. □

This beats the previous best lower bound of $\log(n/c)$ for this problem when $c > 3$ [3,9]. The bound also shows that the well known tree algorithm, whose worst case performance is $\lceil \log n \rceil$, is within one slot of optimality for $2 \le c \le n/2$. This helps to explain why the tree algorithm has essentially withstood all attempts at improvement since its discovery over 10 years ago.

We now show that a simple modification the tree algorithm is always within two slots of the lower bound. If $n - c = 1$ then it is easy to show that 2 slots are both necessary and sufficient. When $n - c > 1$, if $c > n/2$, start with a query Q of size $n-c$ (else use $|Q| = n/2$). If the result is null, then all of the remaining nodes are candidates and we are done with one more slot; if one node transmits, then we are done. If there is a collision, we apply the tree algorithm to the set Q and get a transmission in at most $\lceil \log(n-c) \rceil$ additional slots. Thus we have shown that

Theorem 4.4 If $c < n - 1$ is known, selection resolution requires at most $1 + \lceil \log(n-c) \rceil$ slots.

5. Conflict Resolution

In this section we present new lower bounds for conflict resolution both for general algorithms and for divide and conquer ones. Since the best deterministic algorithms known for conflict resolution are divide and conquer, they are an interesting class of algorithms to investigate [6,7,12]. In this section, space does not permit a complete description of the proofs; for details see [19].

Theorem 5.1 Any deterministic divide and conquer collision resolution algorithm, in the worst case, uses at least $2c + \lfloor \log(n/(c-1)) \rfloor - 3$ slots when $c < n/2 + 1$ and n slots when $c \ge n/2 + 1$. This holds even if the algorithm is initially told c, the number of contending nodes.

NOTE: The second part of this result provides a *tight* bound for the upper range of c since a trivial polling algorithm uses exactly n slots. Greenberg also investigated divide and conquer algorithms, and showed that at least $c + (1/8)c \log(n/c)$ slots are needed [6]. It is easy to show that our bound is better when $c > n/256$.

Proof Sketch: We first use an adversary strategy to show that if $n = 2c - 2$, then at least n slots are required. We then show that if c is small compared to n, the adversary can force the algorithm to waste $\lfloor \log(n/2(c-1)) \rfloor$ slots in order to produce a subproblem with a total of $2c - 2$ nodes and containing all c of the original contenders. For problems with higher densities of contenders it is also easy to adapt the adversary strategy to retain the bound of n slots.

We now describe the general approach that the adversary uses for the case where $n = 2c - 2$. As the algorithm runs, certain nodes will be revealed either as *known* contenders or *known* noncontenders (e.g. if a query Q returns null, all nodes in Q are known noncontenders). The remaining nodes will be considered *unknown*, even though there may be restrictions on combinations of these nodes based on feedback from previous queries (i.e. if query Q returns collision, any individual node in Q may or may not be a contender, but at least two nodes in Q must be contenders).

To simplify the bookkeeping, occasionally the adversary reveals some extra information; information that could *not* be deduced from the channel feedback. The divide and conquer assumption allows the adversary to form disjoint sets of nodes which can never be "overlapped" by future queries. A slot in which no node transmits (i.e. a null or collision slot) is called a *wasted* slot. The adversary uses several techniques to force the algorithm to waste many slots.

The details of the adversary's strategy are described by a decision table that specifies a response to all possible queries. A typical response might inform the algorithm that a query of size four contains *exactly* two contenders; note that this could not be obtained from the normal channel feedback. A case analysis of each response shows that at least one wasted slot is associated with each response that reveals a contender. More specifically, we show that if at the start, at least half the nodes are contenders, then the strategy maintains the following invariants:

(1) among the unknown nodes at least half are always contenders,

(2) *only* queries consisting of exactly one known contender can result in a good transmission,

(3) for each revealed contender, at least one noncontender is also revealed and,

(4) each revealed noncontender eventually causes at least one wasted slot.

The invariants (3) and (4) assure that at least one wasted slot can be associated with each non-contender. Since each contender must use one slot to transmit, the bound of n slots follows. $\Box$

For unrestricted algorithms we have been able to show that

Theorem 5.2 Any deterministic conflict resolution algorithm uses at least $7/4\,c + \lfloor \log(4n/7c) \rfloor$ slots when $c < \dfrac{4}{7}n$ and n slots when $c \geq \dfrac{4}{7}n$. This holds even if c is known initially.

Proof Sketch: This can be proved using a technique similar to that used in the proof of Theorem 5.1. However, the proof requires a more complicated adversary strategy since now the adversary must deal with queries that can overlap several sets which have been used in earlier queries. $\Box$

The best previous lower bound for unrestricted algorithms is due to Greenberg and Winograd:

$$\frac{c\,[\log(\sqrt{n/c})+1]}{\log(c)-1}$$

for $8 \leq c \leq 2n/3$, c *even* and c *divides* $2n$ (for other values of c it is close to this but messier to describe) [9]. It is easy to verify that our bound is better when $c \geq \sqrt{n}/2$.

A fascinating, well studied and long standing question is: "Are unrestricted collision resolution algorithms more effective than divide and conquer ones?" [6,7,9,12]. Our results provide a partial answer: Theorems 5.1 and 5.2 together show that the answer is **no** when $c \geq 4/7n$! Insights gained while working on this problem supports the following:

CONJECTURE: The answer is no for all $c \geq n^{\varepsilon}$, $\varepsilon > 0$ and perhaps even for all $c \geq 2$.

6. Maximum Finding

In this section we resolve the expected time complexity of *probabilistic* (and thus also deterministic) maximum finding algorithms. An input for this problem consists of a set C of nodes each of which has an arbitrary real value. We assume that c, the size of C, and the distribution of the nodes' values are initially not known by the algorithm. We first describe an O(log$c$) expected time probabilistic algorithm and then show its optimality by proving that at least $\Omega(\log c)$ expected slots are required. The algorithm has a very simple high level structure.

Repeat until C is empty.
    1. Randomly select a node in C to transmit its value x.
    2. Update C by deleting all nodes whose value is $\leq x$.

This approach was suggested in [3,14] and they showed that the expected number of iterations of this loop is O(log$c$). If c is known, a random contender can be selected in expected constant time by having each contender transmit with probability $1/c$, and repeating this until exactly one node transmits [17,21] (see procedures *Test*, and *Resolve* in Fig 1). However, in the above loop, even if the size of C is known initially, it will *not* be known after the first iteration. Thus generally, the selection of a random node must be obtained from a set of *unknown* size.

Probabilistic selection resolution algorithms can applied to get a random node to transmit quickly. This yields an O((loglog$c$)log$c$) expected time algorithm if the random transmission step uses Willard's *optimal* O(loglog$c$) selection resolution algorithm [21]. Our innovation is to show that the **sequence** of selection resolution problems can be solved in expected O(1) slots per transmission by proving that the sequence only requires a *total* expected time of O(log$c$) slots.

Suppose that $C_1, C_2, \ldots, C_k$ are sets of sizes $c_1 \geq c_2 \geq \ldots \geq c_k$. We will now describe in detail a fast algorithm which does a sequence of k selection resolutions on the sets C_i, $i=1,2,\ldots,k$ in expected time O($k+\log(c_1)$).

The algorithm starts by finding an estimate $\hat{c}_1$ for c_1, using a probabilistic doubling search (see procedure *Move_Up* in Fig. 1). If the search does not result in a transmission, then all nodes in C_1 execute $Test(\hat{c}_1)$, described in Fig. 1, until a successful transmission occurs.

For each subsequent set C_i, we form an estimate $\hat{c}_i$ as follows. As a starting point, we use the previous estimate $\hat{c}_{i-1}$ and all nodes in C_i execute $Test(\hat{c}_{i-1})$ (see Fig. 1). Three cases can result:

(1) If the result is a good transmission, we are done and $\hat{c}_i$ is set equal to $\hat{c}_{i-1}$.

(2) If the result is a null, we assume that the current estimate is too high, and halve our estimate (so $\hat{c}_i=\hat{c}_{i-1}/2$) and all nodes in C_i execute $Test(\hat{c}_i)$. If the test results in a null we continue to halve our estimate; if the test results in a good transmission then we are done with this set, and if the test results in a collision, the estimate is frozen at the previous value, and used repeatedly until a successful transmission occurs (see procedure *Move_Down* in Fig. 1).

(3) If the result is a collision, then we assume that the estimate is too low, and successively double our estimate until a null or a successful transmission occurs. If a null follows the doubling of a previous estimate, the current estimate is again frozen, and used repeatedly until a successful transmission occurs (see procedure *Move_Up* in Fig. 1).

A complete description of the algorithm is given in Fig. 1, which is given at the end of the paper. In order to analyze the running time for this algorithm we will prove several lemmas.

Lemma 6.1 The total expected number of slots used when the estimate has been frozen is $O(k)$.

Proof: The only way the estimate $\hat{c}_i$ gets frozen is when a null was obtained in response to $Test(\hat{c}_i)$, and a collision to $Test(\hat{c}_i/2)$. Willard showed that an estimate with this property uses expected $O(1)$ slots to obtain a successful transmission [21]. (See procedure *Resolve*.) □

A slot in which the estimate has not yet been frozen will be called an *estimation slot*. In order to count the total number of estimation slots we will partition them into four types:

● *stop* slot: either a successful transmission occurs or the current estimate is frozen.
● *up* slot: the current estimate is already too high ($\hat{c}_i \geq c_i$), yet a collision causes it to be increased.
● *down* slot: the current estimate is too low ($\hat{c}_i \leq c_i$), yet a null causes it to be decreased.
● *normal* slot: the current estimate is too high and a null result causes the estimate to decrease, or the current estimate is too low and a collision causes the estimate to increase.

Lemma 6.2 The total expected number of stop, up and down slots is $O(k)$.

Proof: There are a total of k stop slots since the algorithm assures that exactly one stop slot associated with each set C_i. (Slots used after an estimate has been frozen are covered by Lemma 6.1.)

For any set C_i, the number of up slots is bounded by the expected number of successive times that $Test(\hat{c}_i)$ results in a collision when $\hat{c}_i \geq c_i$. It is shown in [21] that

$$P[Test(\hat{c}_i) = \text{collision}] < \frac{c_i^2}{2\hat{c}_i^2} \leq \frac{1}{2},$$

since $\hat{c}_i \geq c_i$. Thus the expected number of up slots associated with C_i is bounded by two, so the total expected number of up slots is bounded by $2k$.

Similarly, the expected number of down slots associated with a set C_i, is bounded by the expected number of successive times that $Test(\hat{c}_i)$ results in a null when $\hat{c}_i \leq c_i$. Now

$$P[Test(\hat{c}_i) = \text{Null}] = \left(1 - \frac{1}{\hat{c}_i}\right)^{c_i} < \left(1 - \frac{1}{c_i}\right)^{c_i} < \frac{1}{e},$$

since $\hat{c}_i \leq c_i$. Thus the expected number of down slots associated with each set C_i is bounded by $\dfrac{1}{1 - 1/e} = \dfrac{e}{e-1} \cong 1.58$, and thus the total expected number of down slots is bounded by $1.58k$. □

Lemma 6.3 The total number of expected normal slots is $O(k + \log c_1)$.

Proof: If there were no up or down slots executed then there would be at most $\log c_1$ normal slots to increase the initial estimate up to c_1 and then at most $\log c_1 - \log c_k$ normal slots to decrease the estimate down to c_k. Unfortunately, up or down slots may move our estimate too high or too low, and then normal slots will have to be used to move the estimate back to its proper value.

However, any normal slot that is used to correct the effect of some previous up or down slot will be charged to that (up or down) slot. Thus the total number of expected normal slots is at most $\log c_1 + \log c_1 - \log c_k +$ (expected number of up and down slots). Using Lemma 6.2 to bound the last term yields the desired result. □

Theorem 6.4 The total expected time used by the multiple select algorithm to complete a sequence of selection resolutions on sets of sizes $c_1 \geq c_2 \geq \ldots \geq c_k$ is $O(k + \log c_1)$.

Proof: Follows from Lemmas 6.1-6.3 since they account for all slots used by the algorithm. □

The maximum finding algorithm creates a nested sequence of sets of contenders that clearly satisfy the conditions of Theorem 6.4 with $k = O(\log c)$, the expected number of transmissions, and $c_1 = c$. Thus we can use the *Multiple_Select* procedure (see Fig. 1) to get the required sequence of transmissions for step 1. In addition, we must detect when we have finally found the maximum and that C is now empty. An efficient method uses the fact that while trying to estimate the size of an empty set, the procedure *Move_Down* will repeatedly halve the estimate of the size of the current set. Eventually the estimate will drop below one, at which time $Test(1)$ will be a executed. A null response to the $Test(1)$ indicates that we are done. Thus we have shown that

Theorem 6.5 Our maximum finding algorithm has an expected running time of $O(\log c)$ slots.

Our analysis of the *Multiple_Select* procedure was geared for showing the asymptotic time bound; we did not attempt to prove a tight bound for the constant. However, simulations of this algorithm suggest that the expected number of slots to find the maximum is approximately $3\log c$.

We note that there are variations on this algorithm which also achieve $O(\log c)$ expected running time. Karl Abrahamson independently discovered a very similar algorithm in which the estimate is never frozen; it is always doubled after a collision and halved after a null [1].

We conclude by showing that our maximum finding algorithm is asymptotically optimal.

Theorem 6.6 The complexity of finding the maximum of c real numbers on a MABN is $\Theta(\log c)$ expected slots.

Proof: Theorem 6.5 gives the upper bound. For the lower bound we use two facts:

(1) $\Omega(\log c)$ expected slots are required by a randomized maximum finding algorithm on a CRCW PRAM with arbitrary write and using constant space [18].

(2) A CRCW PRAM with arbitrary write and constant space can simulate each step of a MABN in $O(1)$ time [4].

Thus a maximum finding algorithm for a MABN which used fewer than $\Theta(\log c)$ expected slots would imply a maximum finding algorithm for a CRCW PRAM with arbitrary write and using constant space, which is faster than its lower bound. □

We note that this lower bound shows that this version of the maximum finding problem is harder than maximum finding when the nodes' values are drawn from a known probability distribution. In the latter version, the maximum can be found in $O(\log\log c)$ expected slots [16,19].

7. Applications and Conclusion

Good selection and conflict resolution algorithms are essential both for making effective use of MABNs and for implementing many distributed MABN algorithms [3]. Lower bounds are particularly important in this area, since they demonstrate the maximum effectiveness of MABNs. Our lower bounds for the perfect feedback channel are especially significant since they demonstrate that improving the channel feedback will not lead to much better performance.

Maximum finding is a fundamental operation for resource allocation problems such as load balancing and also for many distributed algorithms such as merging, sorting, and weighted graph algorithms [3,13,16,20]. Thus, an *optimal* maximum finding algorithm is clearly significant.

An important current topic is the relationship between different parallel models of computation. We will now show that Theorem 6.4 can be used for a fast simulation of a special type of CRCW PRAM. Define a *random concurrent write* to be such that if n processors try and write to the same location in a single step, each one is equally likely to succeed. A direct consequence of Theorem 6.4 is that a CRCW PRAM which writes the value of an arbitrary processor when two or more processors try to write to the same location can do a sequence of k random writes with write conflicts of sizes $c_1 \geq c_2 \geq \ldots \geq c_k$ in $O(k + \log c_1)$ time. This follows from the fact that such a PRAM can simulate a step of a MABN in $O(1)$ time [4], and the fact that our collision resolution procedure selects a random processor to write out of each set C_i. This situation arises in a number of settings and in particular the simulation results in a fast parallel version of quicksort on a CRCW PRAM with arbitrary write [15].

8. Acknowledgement

We would like to thank Faith Fich and Prabhakar Ragde for valuable discussions on the relationship of this work to CRCW PRAMS and for improving the lower bound proof of theorem 6.6.

Procedure *Multiple_Select*; { This procedure does a sequence of selection resolutions on sets $C_1, \ldots, C_k$ of sizes $c_1 \geq c_2 \geq \ldots \geq c_k$ in $O(k + \log c)$ expected slots. }

Function *Test* (c);
 Each node in the current set transmits with probability $1/c$.
 Null, Good, or Collision is returned depending on the number of nodes which transmit.

Procedure *Resolve* (c);   { Repeatedly call Test(c) until a good transmission occurs. }
 repeat
  result $\leftarrow$ *Test* (c)
 until result = Good;

Procedure *Move_Up* (VAR $\hat{c}$);   { Increase the estimate until a Null or Good results. After
 repeat   a Null, the current estimate is used to get a transmission. }
  $\hat{c} \leftarrow 2*\hat{c}$; result $\leftarrow$ *Test* $(\hat{c})$;
 until result <> Collision;
 If result = Null **then** *Resolve* $(\hat{c})$;

Procedure *Move_Down* (Var $\hat{c}$); { Decrease the estimate until a Collision or Good results.
 repeat   After a Collision the estimate is used to get a transmission. }
  $\hat{c} \leftarrow \hat{c}/2$; result $\leftarrow$ *Test* $(\hat{c})$;
 until result <> Null;
 $\hat{c} \leftarrow \hat{c}*2$;   { Set the estimate to the last Null value. }
 If result = Collision **then** *Resolve* $(\hat{c})$;

{ Main Body of Procedure *Multiple_Select* }
 $\hat{c}_1 \leftarrow 1$;   { Start by estimating c_1 then select from C_1 using *Move_Up* }
 Move_Up $(\hat{c}_1)$;
 for $i \leftarrow 2$ **to** k **do**   { Estimate then select from each of the other sets C_i in turn. }
  $\hat{c}_i \leftarrow \hat{c}_{i-1}$;   { Start by using the old estimate. }
  result $\leftarrow$ *Test* $(\hat{c}_i)$;
  if result = Null **then** *Move_Down* $(\hat{c}_i)$ **else** **if** result = Collision **then** *Move_Up* $(\hat{c}_i)$;

Figure 1. The Multiple Select Algorithm.

9. References

[1]   Abrahamson, K., Personal Communication, Department of Computer Science, Washington State University, September, 1987.

[2]   Capetanakis, J. I., Tree Algorithms for Packet Broadcast Channels, *IEEE Transactions on Information Theory* , Vol. **OT-25**, No. 5, (September 1979) pp. 505-515.

[3]   Dechter, R. and Kleinrock, L., Broadcast Communications and Distributed Algorithms. *IEEE Transactions on Computers*, Vol. **C-35**, No. 3, (March 1986) pp. 210-219.

[4]   Fich, F.E., Ragde, P.L., and Wigderson, A., Relations Between Concurrent-Write Models of Parallel Computation, To appear in *SIAM Journal on Computing*.

[5]   Greenberg, A., On the Time Complexity of Broadcast Communication Schemes. *Proceedings of the 14th Symposium on the Theory of Computing,* (1984) pp. 354-364.

[6]   Greenberg, A. G., Efficient Algorithms for Multiple Access Channels, T.R. No. #83-08-01 Department of Computer Science, University of Washington, Seattle, August 1983.

[7]   Greenberg, A. G., Flajolet, P. and Ladner, R. E., Estimating the Multiplicities of Conflicts to Speed Their Resolution in Multiple Access Channels, *Journal of the Association for Computing Machinery* , Vol. **34**, No. 2, (April 1987) pp. 289-325.

[8]   Greenberg, A. G., Lubachevsky, B. D. and Odlyzko, A. M., Simple, Efficient Asynchronous Parallel Algorithms for Maximization. *Proceedings of the Fourth Annual ACM Symposium on Principles of Distributed Computing*, (August 1985) pp. 300-308.

[9]   Greenberg, A. and Winograd, S., A lower bound on time needed in the worst case to resolve conflicts deterministically in multiple access channels. *Journal of the ACM*, Vol. **32**, No. 3, (July 1985) pp. 589-596.

[10] Hayes, J. F., An Adaptive Technique for Local Distribution. *IEEE Transactions on Communications* , Vol. **COM-26**, No. 8, (August 1978) pp. 1178-1186.

[11] IEEE Project 802, *Draft IEEE Standard 802.3 CSMA/CD Access Method and Physical Layer Specifications*. IEEE Computer Society, Revision D, December 1982

[12] Komlos, J. and Greenberg, A. G. An Asymptotically Fast Non-adaptive Algorithm for Conflict Resolution in Multiple Access Channels, *IEEE Trans. Inf. Theory* (March 1985)

[13] Levitan, S., Algorithms for Broadcast Protocol Multiprocessors. *Proceedings of the 3rd International Conference on Distributed Computer Systems*, (1982) pp. 666-671.

[14] Levitan, S. P. and Foster, C., Finding an Extremum in a Network. *Proceedings Ninth International Symposium of Computer Architecture*, (April 1982) pp. 321-325.

[15] Martel, C., Gusfield, D. A Fast Parallel Quicksort Algorithm, Preprint, U.C. Davis, 1988

[16] Martel, C. and Vayda, T. Load Estimation, Problem Size Reduction and Maximum Finding in Multiple Access Broadcast Networks. Proc. of 25th Annual Allerton Conf., Oct. 1987.

[17] Metcalfe, R. and Boggs, D., Ethernet: Distributed Packet Switching for Local Computer Networks. *Communications of the ACM*, Vol. **19**, No. 7, (July 1976) pp. 395-404.

[18] Ragde, P. Lower Bounds for Parallel Computation, Ph.D. Thesis, U.C. Berkeley 1986.

[19] Vayda, T. P., On the Complexity of Parallel Algorithms for Multiple Access Networks, Ph.D. Thesis, Division of Computer Science, U.C. Davis, Davis Ca. 1988.

[20] Wah, B. and Juang, J., Resource Scheduling for Local Computer Systems with a Multiaccess Network. *IEEE Transactions on Computers*, Vol. **C-34**, No. 12, (December 1985)

[21] Willard, D. E., Log-Logarithmic Selection Resolution Protocols in a Multiple Access Channel. *SIAM Journal of Computing*, Vol. **15**, No. 2, (May 1986) pp. 468-477.

Optimal Routing Algorithms for
Mesh-Connected Processor Arrays

—Extended Abstract—

Danny Krizanc[†]   Sanguthevar Rajasekaran[‡]   Thanasis Tsantilas[§]

Aiken Computation Laboratory
Harvard University
Cambridge, Mass., 02138
U.S.A.

Abstract

We show that there is a randomized oblivious algorithm for routing any (partial) permutation on an $n \times n$ grid in $2n + O(\log n)$ parallel communication steps. The queues will not grow larger than $\Theta(\log n)$ with high probability. We then modify this to obtain a (non-oblivious) algorithm with the same running time such that the size of the queues is bounded by a constant with high probability. For permutations with locality, where each packet has to travel distance at most L in either the horizontal or the vertical direction, a generalization of the algorithm routes in time $3L + o(L)$, while the queue size remains bounded by $\Theta(\log n)$ with high probability. Finally, we show that for a general class of oblivious deterministic routing strategies, $\Omega(n^2)$ time is required if we want to have constant size queues.

1. Introduction

One of the central questions in parallel computation, is whether we can simulate efficiently an idealistic parallel computer by a realistic one. We view a realistic computer as some (large) number of processors, each with its own local memory, connected via a sparse network of communication links. Sorting and routing turn out to be key problems in this simulation (see [9,13] for a general discussion).

In this paper we consider the problem of routing n^2 packets through an $n \times n$ mesh connected processor array. The mesh-connected processor array (also referred to as the grid) consists of n^2 processors. The interconnections between them are defined by a two dimensional $n \times n$ grid with no wraparound connections. That is, each processor is connected to its four grid neighbours to the left, to the right, above and below, provided they exist. Edges are assumed to be bidirectional and at most one packet of information can travel through an edge along each direction in unit time.

Mesh-connected processor arrays can be categorized into two groups, viz., SIMD (Single Instruction Multiple Data), and MIMD (Multiple Instruction Multiple Data) depending on the instruction streams allowed for the processors. An instruction refers to an arithmetic operation, or a communication step with a neighbour in a specified

---

[†]Supported in part by NSF Grant NSF-DCR-86-00379 and by an NSERC Postgraduate Scholarship
[‡]Supported in part by NSF Grant NSF-DCR-85-03251 and ONR contract N00014-80-C-0647
[§]Supported in part by NSF Grant NSF-DCR-86-00379

direction. The model we use in this paper is the MIMD mesh-connected processor array where each processor can communicate with all its neighbours in a single step (this is the same model used by Valiant and Brebner [14] and by Schnorr and Shamir [8] in their lower bound proof).

Due to its large diameter, this model does not give rise to efficient simulations. (See however [7] where we define a class of mesh-like networks which have n^2 nodes but less than $2n$ diameter. Optimal routing algorithms for this class are presented in that paper.) However, the mesh is an important model of parallel computation for several reasons. First, the simplicity of the interconnection pattern as well as its regularity make it ideally suited for highly dense VLSI implementation. Second, it is a reasonable interconnection pattern if we assume that the transmission times are proportional to the length of the communication links. Further, a large variety of efficient algorithms for fundamental problems have been developed during the past years to run on the mesh. See [1,10] for examples of existing mesh-like prototype architectures.

The algorithms we present are capable of realizing the following communication requests: (a) *permutations;* initially there is one packet at each processor, each with distinct destination where it is to be routed, and (b) *partial permutations;* initially there is at most one packet at each processor, each with distinct destination. At every step of the routing algorithm, each packet is at some node. During routing steps, packets are transmitted through an edge. Every node has a queue associated with each outgoing communication link, and the queueing discipline determines how the packets are distributed in the queues. One of the main objectives of the routing strategy is to keep the size of the queues small, in our case $O(\log n)$ or constant with high probability, otherwise, it is easy to achieve a $2n - 2$ running time (see Fact 2, section 2).

We present two routing algorithms which are based on the two stage strategy introduced by Valiant [12]. During the first stage, for each packet a processor is chosen at random according to some probability distribution, and the packet is routed to that node. Then, in the second stage the packet is routed to its destination. Randomization is used in an effective way in order to avoid large bottlenecks. The algorithms are *distributed*, in the sense that no global information about the permutation is needed (cf. [5,4,6] for studies in the non-distributed context). The first of them is also *oblivious*, that is, the path taken by each packet depends only on its source and destination, and does not use information from other packets. The presence of other packets affects only the rate at which the path is traversed.

Our main results are the following:

- In section 3.2 (Theorem 1), we give a time-optimal randomized oblivious algorithm for realizing a partial permutation. Its running time is $2n + O(\log n)$ with very high probability. Further, the size of the queues will not grow more than $\Theta(\log n)$ with high probability. Notice that $2n - 2$, the diameter of the mesh, is the trivial distance lower bound, hence the optimality of our algorithm. The best previously known upper bound was $3n + o(n)$ and due to Valiant and Brebner [14].

- The previous algorithm can be modified, as pointed out to us by Tom Leighton, so that the size of the queues is reduced to constant. In section 3.3 (Theorem 2), we give a randomized (non-oblivious) algorithm for realizing a partial permutation whose running time is $2n + O(\log n)$ with high probability, and the size of the queues will not grow more than a constant with high probability. Hence this algorithm is time optimal as well as queue size optimal.

- Let L be the maximum distance a packet has to travel along either the horizontal or the vertical direction. Assume that the maximum horizontal as well as the maximum vertical source-destination distances are globally known. This situation arises naturally when specific known algorithms are implemented on the mesh. In section 4 (Theorem 3), we give an algorithm similar to the above that preserves *locality*, that is, it routes a partial permutation in $3L + o(L)$ steps with high probability. [N.B. $3L$ is an overestimation since when $L = n$ the algorithm is the same as the $2n + o(n)$ one.] This can be adapted to an algorithm that preserves *individual locality*, namely, each packet is routed to its destination in time proportional to its origin-destination distance. We also describe a second adaptation that allows routing in $O(L)$ steps in the case that L is not known to the processors.

- In section 5 (Theorem 4) we show that for a general class of *oblivious deterministic* routing strategies, $\Omega(n^2)$ time is required, if we want to have constant size queues.

Recently, Kunde [3] presented a deterministic *non-oblivious* algorithm (based on sorting) whose running time is $2n + O(n/f(n))$ where $f(n)$ is a bound on the size of the queues. If $f(n) = \Theta(\log n)$, then the running time of the algorithm is $2n + O(n/\log n)$. Using the same queue size, the randomized algorithm that we present in section 3.2 achieves running time $2n + O(\log n)$ and is oblivious. The algorithm presented in section 3.3 is the only one known requiring only constant queue size and running in optimal time. It is an open problem whether one can achieve optimal running time and constant size queues deterministically.

2. Some Definitions and Preliminary Facts

The problem of routing on a parallel machine is defined as follows. Each processor in the machine has a packet of information. Each packet has a destination address written on it that specifies the processor to which it is to be sent. The task of the parallel machine is to send the packets to their correct destinations simultaneously so that at most one packet passes down any wire at any time and all of them arrive at their destinations quickly.

By *routing function* we mean a rule that specifies for each packet a directed path it should take and associates a *priority number* with each packet. Specification of priorities of packets is also called *queueing discipline*. These priorities serve the purpose of resolving contentions for the same edge by more than one packets. The problem of routing is then the selection of a routing function. Our algorithms apply to communication requests that form *partial permutations* (as defined in the introduction).

We use two different queueing disciplines:

Q: Any queueing discipline with the property that "packets that have already started the current phase have higher precedence over those who have not". This implies that when a packet starts to move, it will never be delayed again.

Q': Packets that have farther to go have higher priority.

First we present a result about routing on a one-dimensional mesh. See [7] for more results concerning one-dimensional problems.

Fact 1: Consider an n-node chain graph, i.e., a graph with node-set $\{1, 2, \ldots, n\}$ and edge-set $\{\{i, i+1\} \mid i = 1, \ldots, n-1\}$. Suppose that n packets are distributed arbitrarily in the nodes of the graph, each having a distinct destination. Further, let L be the maximum over all origin-destination distances. Then, using queueing discipline Q', a packet starting at node i will reach its destination within $\max\{L, b(i)\}$ steps at most (edges are bidirectional), where $b(i)$ denotes the distance between i and the boundary in the direction the packet is moving (i.e., $n - i$ or $i - 1$ depending on whether the packet is moving from left to right or from right to left).

Proof: Consider a packet q at node i and destined for j. Assume (w.l.o.g.) it is moving from left to right. q can only be delayed by the packets with destinations $> j$ and which are to the left of their destinations. Let $k_1, k_2, \ldots, k_n$ be the number of such packets (at the beginning) at nodes $1, 2, \ldots, n$ respectively. (Notice that $\sum_{l=1}^{n} k_l \leq n - j$).

Let m be such that $k_m > 1$ and $k_{m'} \leq 1$ for $m < m' \leq n$. Call the sequence $k_{m+1}, \ldots, k_n$ the *free sequence*. Realize that a packet in the free sequence will neither delay nor be delayed by any other packet in the future. It is easy to see that at every time step, at least one new packet joins the free sequence. Thus, after $n - j$ steps, all the packets that can possibly delay q would have joined the free sequence. q needs only an additional $j - i$ steps, at the most, to reach its destination. The case the packet moves from right to left is similar.

Fact 2: There is a *naive* deterministic algorithm that routes each packet to its destination in time $2n - 2$: For a packet at (i, j) destined at (r, s), first send the packet to (i, s) along the row, and then to its destination along the column. Use queueing discipline Q' to resolve collisions. However, queues of size n might be necessary, which makes the use of the algorithm impractical.

We shall need the following bound for the probabilistic analysis of our algorithms (see [14] and references therein):

Fact 3: Let $S_{N,p}$ be a random variable having binomial distribution with parameters N and p (i.e., $S_{N,p}$ is the sum of N independent Bernoulli random variables each with mean p). Then, for $m \geq Np$

$$P(S_{N,p} \geq m) \stackrel{\text{def}}{=} B(m; N, p) \leq \exp[m \ln(Np) - m \ln m + m - Np].$$

3. A Time Optimal Routing Algorithm

Valiant and Brebner [14] gave a randomized routing algorithm for routing a (partial) permutation on an $n \times n$ MIMD mesh that runs in time $3n + O(n^{3/4})$ with very high probability. The size of the queues does not exceed $\Theta(\log n)$ with high probability. In this section we present a time optimal randomized oblivious algorithm that runs in time $2n + O(\log n)$ with high probability. The maximum queue length is again $\Theta(\log n)$ with high probability. Then we show how this algorithm can be modified in order to get a (non-oblivious) randomized algorithm with constant queue size, while the running time remains the same. Before describing our optimal algorithm, we show how to extend Valiant and Brebner's algorithm in order to achieve optimal running time.

3.1 A Simple $2n + o(n)$ Algorithm

The routing algorithm is a modification of the one proposed by Valiant and Brebner [14], and operates in three phases:

PHASE I Divide the rows into $1/\epsilon$ strips of ϵn rows each ($\epsilon \geq 1/\log n$). For a packet at (i, j) destined at (r, s), pick a processor (k, j) at random in the same column and strip as (i, j). Send the packet to (k, j) along the column.

PHASE II Send the packet to (k, s) along the row.

PHASE III Send the packet to (r, s) along the column.

We use queueing disciplines $\mathcal{Q}$ and $\mathcal{Q}'$ to resolve contentions during phases II and III respectively. In [14] $\mathcal{Q}$ is used for both phases.

Queue size analysis: First, we note that the size of the queues at each phase, will not grow more than the maximum of the sizes of the queues at the beginning and at the end of that phase. It can be shown that the average queue size at the end of phases I and II is 1 and $1/\epsilon$ respectively (whence the restriction that $\epsilon \geq 1/\log n$ if we want to have an $O(\log n)$ bound on the size of the queues). Application of fact 3 shows that the probability that any one processor will ever have to store more than $\Theta(\log n)$ packets is smaller than inverse polynomial.

Routing time analysis: Phase I requires at most ϵn steps, since there are no delays involved. This is because initially we have one packet at each node so packets will move continually until they reach their destination.

For phase II we use queueing discipline $\mathcal{Q}$. Consider a packet q at (k, t) moving to the right (the same analysis holds for packets moving to the left). As shown in [14] the delay of this packet will be at most the total number of packets starting the phase at $\{(k, l) \mid l \leq t\}$. There are $t\epsilon n$ packets that could be in this set of nodes, each with probability $1/\epsilon n$. So, the probability that at least m packets will delay q is at most $B\left(m; t\epsilon n, \frac{1}{\epsilon n}\right)$. Suppose that $m = t + g$. Since the length of the route is at most $n - t$ this bounds the probability that the packet will take more than $n + g$ steps to finish phase II. Taking $g = n^\delta, 1/2 < \delta < 1$, and using Fact 3, we can show that

$$B\left(t + n^\delta; t\epsilon n, \frac{1}{\epsilon n}\right) \leq \exp\left(-C' n^{2\delta - 1}\right)$$

for an appropriately chosen $C' > 0$.

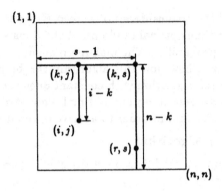

FIG. 1. Normal path length of a packet: $(i - k) + (s - 1) + (n - k)$

Thus the probability that at least one packet will take time more than $n + n^\delta$ is at most $n^2 \exp(-C'n^{2\delta-1}) < \exp(-Cn^{2\delta-1})$, $C > 0$.

For phase III we use queueing discipline $\mathcal{Q}'$. Fact 1, shows that any packet will reach its destination within n steps.

We therefore conclude (choosing, say, $\delta = 3/4$) that there is a randomized oblivious algorithm for routing on an $n \times n$ mesh with the property that any partial permutation can be realized within $2n + \epsilon n + n^{3/4}$ ($\epsilon \geq 1/\log n$) steps with probability at least $1 - \exp(-C\sqrt{n})$ for some positive constant C. Further, the probability that the size of the queues will grow larger than $M \log n$ is less than n^{-D} for some constant $D > 0$ which depends on M, and large enough M. Choosing $\epsilon = 1/\log n$ we obtain a $2n + (n/\log n) + n^{3/4}$ algorithm which is time-optimal.

3.2 Our Optimal Routing Algorithm

In the $2n + \epsilon n$ algorithm described above, we consider the effect on the running time when the three phases are coalesced. That is, a packet that finishes phase I before ϵn steps can start phase II without waiting for the other packets to finish phase I and so on. We expect some packets to finish faster now.

The only possible conflict between phases is between phases I and III. A packet that is doing its phase I and a packet that is doing its phase III might contend for the same edge. Always, under such cases, we give precedence to the packet that is doing its phase I. If a packet q doing its phase III contends for an edge with a packet that is doing its phase I, it means that q has completed its phases I and II within ϵn steps. Since after ϵn steps from the start of the algorithm every packet will be doing either its phase II or phase III, q will reach its destination within $(1 + \epsilon)n$ steps. Packets like q are not interesting to our analysis. Thus we will not mention them hereafter in any analysis. A packet doing its phase II can not be delayed by packets doing their phase I or phase III and vice versa.

In summary, no packet suffers additional delays due to overlap of phases. It is also easy to see that the maximum queue length at any node does not increase due to overlap.

We shall now modify the algorithm of section 3.1 to obtain an algorithm whose running time is $2n + O(\log n)$ with high probability. Consider a packet q initially at

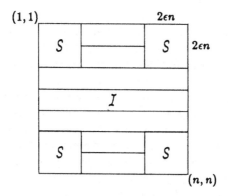

FIG. 2. Superior and inferior packets

node (i, j) with (r, s) as its destination. Assume without loss of generality, (r, s) is below (i, j) and to the right (the other cases can be argued along the same lines). If in phase I q chooses a position (k, j) (in the strip (i, j) is in), then q will finish phase I in $|i - k|$ steps. Also q will take at the most $(s - 1) + n^\delta$ steps to finish phase II (with high probability) and $(n - k)$ steps to finish phase III (see figure 1 and the time analysis in the previous section). Therefore q takes $|i - k| + (s - 1) + (n - k)$ (plus lower order terms) total steps to complete all the three phases. Call this sum the *normal path length* of q. We require the normal path length of every packet to be $\leq 2n$. For this purpose we divide the grid into two regions S and I, and call the packets having their origin in these two regions *superior packets* and *inferior packets* respectively (see figure 2). Our routing algorithm will route the inferior packets as usual using the $(2 + \epsilon)n$ algorithm given above. Superior packets are given special treatment. Coalescence of phases is assumed for both the types of packets. Details of the algorithm follow.

Algorithm for inferior packets: Inferior packets use the usual algorithm of section 3.1. Every column is divided into strips of length ϵn. In phase I, a packet at (i, j) with a destination (r, s) is sent to a random position (k, j) in its strip along the column j. In phase II, the packet is sent to (k, s) along row k and finally in phase III, the packet is sent to (r, s) along the column s.

Algorithm for superior packets: Each node (i, j) in the upper two S squares chooses a random k in $\{2\epsilon n + 1, 2\epsilon n + 2, \ldots, 2\epsilon n + (1/4)n\}$ and sends its packet, q, to (k, j) along column j. If (r, s) is the destination of q, q is then sent to (k, s) along row k and finally along column s to (r, s). (ϵ is chosen to be less than $1/8$, so that superior packets in the upper (lower) half of the grid remain in the upper (lower) half after the randomization phase.) A symmetric algorithm is used for the packets in the lower two S squares.

Queueing Disciplines: (1) In phases I and III no distinction is made between superior and inferior packets. (2) If a packet doing its phase I and a packet doing its phase III contend for the same edge, the packet doing its phase I takes precedence. (3) Queueing discipline Q' is used for phase III. (4) In phase II, superior packets take precedence over inferior packets. Among superior (inferior) packets queueing discipline Q is used.

Routing Time Analysis:

(a) Superior packets: A superior packet does not suffer any delay in phase I.

A superior packet q starting its phase II at (k, t) can be delayed by at the most the number of superior packets that will ever start their phase II from a node to the left of (k, t) in row k (assuming w.l.o.g., q is moving from left to right). Following an analysis similar to the one of section 3.1, we see that there are $2\epsilon nt$ packets that could delay q, each with probability $4/n$ (i.e., the expected delay is $8\epsilon t < t$ if $\epsilon < 1/8$). So the probability that at least m packets will delay (k, t) is at most $B\left(m; 2\epsilon nt, \frac{4}{n}\right)$. Let $t > \alpha \log n$, where α is an appropriately chosen small constant. Then, by choosing $m = t + \alpha \log n$, we can show that q will complete phase II in at most $n + \alpha \log n$ steps with probability at least $1 - n^{-\beta}$ where β depends on α. If $t < \alpha \log n$ then by choosing $m = \alpha' \log n$ we can show that the probability that the delay is larger than $\alpha' \log n$ is smaller than inverse polynomial. Therefore, superior packets will complete phase II in $n + O(\log n)$ steps with high probability.

A superior packet that moves in the direction of its destination in phase I spends at the most n steps in total in phases I and II. A packet that was moving in a direction opposite to its destination in phase I will complete phases I and II in a total of $2((1/4)n + 2\epsilon n)$ steps. Since ϵ is chosen to be less than $1/8$, this total is at most n. And hence all the superior packets will complete all the three phases in $2n + O(\log n)$ steps with high probability.

(b) Inferior packets: Consider an inferior packet q that starts phase II at row k. We have the following case analysis according to whether the packets start phase II in the two upper and bottom strips of size ϵn, or not.

case 1: $k \leq 2\epsilon n$ or $k > n - 2\epsilon n$.

Suppose q starts phase II at row k, $k \leq 2\epsilon n$, and w.l.o.g. it moves from left to right. We observe that at the end of phase I, the S-regions are empty, hence no packet will start phase II at an S region. Suppose q starts the phase at column $2\epsilon n + t$. Similarly to the analysis in section 3.1, the only packets that could delay q are the ones that start the phase at columns $2\epsilon n + 1, \ldots 2\epsilon n + t$. So with very high probability the delay will be at most $t + n^\delta$. Since q has to travel a distance of at most $n - t - 2\epsilon t$, it will complete phase II in at most $n - 2\epsilon n + n^\delta$ steps. q spends at most ϵn steps in phase I (since it suffers no delay during that phase) and at most n steps to finish phase III (from Fact 1). Thus q takes in total at most $2n - \epsilon n + n^\delta$ steps to complete all three phases, which is at most $2n$ for small enough ϵ.

case 2: $2\epsilon n \leq k \leq n - 2\epsilon n$.

q spends at most ϵn steps in phase I and at most $n - 2\epsilon n$ steps in phase III, since, $n - 2\epsilon n$ is the maximum distance from row k to any boundary. We want to analyse the delay experienced by q during phase II. This delay is due to the presence of both superior and inferior packets. Suppose the packet starts the phase at column t. The delay due to inferior packets is, as already known, $t + n^\delta$ with very high probability. We estimate the delay due to superior packets. The number of superior packets that could ever do their phase II at a given row k is $8\epsilon^2 n^2$, each with probability $4/n$. That is, the expected number of superior packets that will start phase II at row k is $32\epsilon^2 n$. Using Fact 2, we can show that the probability that the delay is greater than $\alpha\epsilon^2 n$ ($\alpha > 32$) is smaller than inverse polynomial. So the total running time of the packet in this case is $\epsilon n + n - 2\epsilon n + n + N^\delta + \alpha\epsilon^2 n \leq 2n$ for small enough ϵ.

Thus all the inferior packets will complete all the three phases in at most $2n$ steps with high probability.

It is easy to see that the queue length at any node will not exceed $\Theta(\log n)$ at any time with high probability. Consider any node (i,j) in the mesh. During phase I, the only packets that can add to the queue size of (i,j) are those that start their phase II from (i,j). The number of such packets is $O(\log n)$ with high probability. During phase II, the only packets that can contribute to the queue size of (i,j) are those that start their phase III form (i,j) and they are $O(\log n)$ w.h.p. in number. During phase III, the queue size of (i,j) can increase by at the most one (viz., by the packet whose destination is (i,j)). And hence the queue size of every node is $O(\log n)$ w.h.p.

The results are summarized in the following theorem.

Theorem 1: *There is a randomized oblivious algorithm for routing on an $n \times n$ mesh with the property that any partial permutation can be realized within $2n + K \log n$ steps with probability at least $1 - n^{-C}$ for some constant $C > 1$ which depends on K. Further, the probability that the size of the queues will grow larger than $M \log n$ is less than n^{-D} for some constant $D > 0$ which depends on M, and large enough M.*

3.3 A Time-Optimal Algorithm With Constant Size Queues

We shall now modify the previous algorithm in order to reduce the size of the queues to constant. In the $2n + O(\log n)$ algorithm of the above section we chose ϵ to be a constant. One can see that the expected queue size at any node at the end of phase I and II is a constant. Use of fact 3 showed that with high probability the size of the queues at any node will not grow larger than $\Theta(\log n)$.

The idea is, instead of considering individual processors to "divide" each column into consecutive blocks of $\log n$ nodes each. Then the expected number of packets per block is $\Theta(\log n)$, and using fact 3 we get that with high probability we have $\Theta(\log n)$ packets per block, or, $\Theta(1)$ packets per processor.

We use therefore the same algorithm as in section 3.2. However, we do not store packets at their target row and column at the end of phases I and II respectively, rather we redistribute packets within each block so that to obtain a constant number of packets per node. Redistributing packets introduces at most $O(\log n)$ delay, and is the reason for the algorithm being non-oblivious. The details of the implementation as well as the analysis of the algorithm will appear in the final version of the paper. The results are summarized in the following theorem.

Theorem 2: *There is a randomized (non-oblivious) algorithm for routing on an $n \times n$ mesh with the property that any partial permutation can be realized within $2n + K \log n$ steps with probability at least $1 - n^{-C}$ for some constant $C > 1$ which depends on K.*

Further, the probability that the size of the queues will grow more than a constant M is less than n^{-D} for some constant $D > 0$ which depends on M, and large enough M.

4. Routing With Locality

In the previous algorithm, packets that have to travel a small distance *e.g.*, $\sqrt{n}$, might require time proportional to n to reach their destination. Here we describe an algorithm similar to the one of the previous section, with the additional feature that it preserves *locality*.

Let $L^{(h)}$ and $L^{(v)}$ denote the maximum among all origin-destination distances along the horizontal and vertical axis respectively, and let $L = \max\{L^{(h)}, L^{(v)}\}$. In other words, L is the maximum distance a packet has to travel in either the horizontal or the vertical direction. We assume that both $L^{(h)}$ and $L^{(v)}$ are known to the processors. We describe an algorithm that routes each packet to its destination in $O(L)$ steps.

If $L = O(\log n)$, we use the naive algorithm described in fact 2 (a generalization of this fact shows that $2L$ steps suffice). If $L = \Omega(\log n)$, we use an algorithm similar to the one presented in section 3.1. The only distinction is that in phase I the rows are divided into strips consisting of $L^{(v)}/\log L^{(v)}$ rows each (or, in general, any function $f(L^{(v)}) = o(L^{(v)})$ as long as $L^{(v)}/f(L^{(v)}) = O(\log n)$).

Using similar techniques, which are omitted here, we can prove the following theorem.

Theorem 3: *Let $L^{(h)}$ and $L^{(v)}$ denote the maximum among all origin-destination distances along the horizontal and vertical axis respectively, and let $L = \max\{L^{(h)}, L^{(v)}\}$. Then, provided that $L^{(h)}$ and $L^{(v)}$ are globally known, there is a randomized oblivious algorithm with the property that any partial permutation can be realized within $3L + o(L)$ steps with probability at least $1 - n^{-\Omega(1)}$. The queue size will not grow more than $\Theta(\log n)$ with probability at least $1 - n^{-\Theta(1)}$. [N.B. 3L is an overestimation since when $L = n$ the algorithm is the the same as the $2n + o(n)$ one.]*

Remark: We can adapt this algorithm to obtain a routing scheme that preserves *individual locality*, namely, if packet π has to travel distance l_π, it is routed to its destination in $O(l_\pi)$ steps with high probability. This scheme is roughly a repeated application of the above algorithm; at stage k, packets that travel distance l, $2^{k-1} < l < 2^k$, are routed to their destination. The details are left for the final version.

Finally, we mention that we can obtain an $O(L)$ algorithm even if $L^{(h)}, L^{(v)}, L$ are unknown. During phase I, packet π that travels distance l_π, randomizes within a strip consisting of l_π rows. Phases II and III remain the same.

5. A Lower Bound

In this section we prove a general lower bound for the time required to route on any network with a limited queue size by a particular class of routing strategies. We then apply this result to the case of an $n \times n$ mesh with $o(\log n/\log\log n)$ queue size. For our purposes, a network is a digraph $G = (V, E)$ where the nodes may be thought of as processors and the edges as directed communication links between the processors. If

$(u, v) \in E$ then processor u can send one packet to processor v in a single communication (time) step. A processor with queue size k can store up to k packets between communication steps.

A *pure source-oblivious* routing strategy has the following properties:

1. For every origin-destination pair, (s, t), the strategy provides a route (possibly randomly chosen) from s to t for a packet to follow independent of the routes chosen for the other packets (i.e., the strategy is *oblivious*).

2. The next edge in the route of a packet depends only on the packet's present location, its final destination and (possibly) some random bits (i.e., the strategy is *source-oblivious*).

3. The queueing discipline (i.e., the method of resolving contentions for the same edge by more than one packet) has the minimal property that if at a particular time there are one or more packets in queue for a particular edge then one of them is chosen to cross the edge in the next communication step (i.e., the strategy is *pure*).

As a consequence of the definition a pure source-oblivious strategy for a network G with queue size k must insure for all permutations and for all nodes of G, if at a particular time a node has j packets in its queue, i of which are scheduled to leave during the next communication step, and there are l packets which are going to arrive at the node during the next communication step then $j - i + l \leq k$ (i.e. nodes never contain more packets than their queue size).

We prove the following

Theorem 4: *Let $G = (V, E)$ be a n node network having maximum degree d and queue size k. The time required in the worst case by any deterministic pure source-oblivious routing strategy for G is $\Omega(\frac{n}{d^4 k (8k)^{5k}})$.*

The $n \times n$ mesh has n^2 nodes and degree 4 so we have:

Corollary: *Any deterministic pure source-oblivious routing strategy for the $n \times n$ mesh with $o(\log n / \log \log n)$ queue size requires $\Omega(n^{1+\epsilon})$ time, for some $\epsilon > 0$, in the worst case. In particular, if the size of the queues is bounded by a constant, the running time is $\Omega(n^2)$.*

This result suggests that to achieve $O(n)$ routing on a mesh with a small queue size one must sacrifice either determinism or obliviousness or both though it doesn't exclude the possibility of a deterministic oblivious strategy. The algorithm in Section 3.2 is a randomized pure oblivious strategy requiring $\Theta(\log n)$ size queues. The constant queue size algorithm presented in Section 3.3 is both randomized and non-oblivious and the deterministic family of algorithms presented by Kunde [3] are non-oblivious.

Acknowledgment We would like to thank John Reif and Les Valiant for their constant encouragement and support. We would also like to thank Tom Leighton for pointing out that we can reduce the size of the queues to constant and for allowing us to include his ideas here. Umesh Vazirani suggested to explore the locality issue. We would also like to thank Benny Chor, Mihály Geréb-Graus, Tom Leighton and Umesh Vazirani for many fruitful discussions.

References

[1] K. E. BATCHER, *Design of a Massively Parallel Processor*, IEEE Trans. Comp., **29** (1980), pp. 836–840.

[2] M. KUNDE, *Optimal Sorting On Multi-Dimensionally Mesh-Connected Computers*, STACS 1987, Lecture Notes in Computer Science 247, pp. 408–419, Springer-Verlag.

[3] M. KUNDE, *Routing and Sorting on Mesh-Connected Arrays*, Aegean Workshop on Computing (AWOC 1988), to appear.

[4] D. NASSIMI AND S. SAHNI, *An Optimal Routing Algorithm for Mesh-Connected Parallel Computers*, J. ACM, **27** (1980) pp. 6–29.

[5] S. E. ORCUTT, *Implementation of Permutation Functions in Illiac IV-Type Computers*, IEEE Trans. Comp., **25** (1976) pp. 929–936.

[6] C. S. RAGHAVENDRA AND V. K. P. KUMAR, *Permutations on Illiac IV-Type Networks*, IEEE Trans. Comp., **35** (1986) pp. 662–669.

[7] S. RAJASEKARAN AND TH. TSANTILAS, *An Optimal Randomized Routing Algorithm for the Mesh and A Class of Efficient Mesh-like Routing Networks*, 7th Conference on Foundations of Software Technology and Theoretical Computer Science, Pune, India, Lecture Notes in Computer Science 287, pp. 226–241, Springer-Verlag.

[8] C. P. SCHNORR AND A. SHAMIR, *An Optimal Sorting Algorithm for Mesh Connected Computers*, In Proc. 18th ACM Symposium on Theory of Computing, 1986, pp. 255–263.

[9] J. T. SCHWARTZ, *Ultracomputers*, ACM Trans. Prog. Lang. Syst., **2** (1980), pp. 484–521.

[10] L. SNYDER, *Supercomputers and VLSI: The Effect of Large-Scale Integration on Computer Architecture*, in Advances in Computers, vol. 23 (1984), M. C. Yovits, ed., Academic Press, pp. 1-33.

[11] C. D. THOMPSON AND H. T. KUNG, *Sorting on a Mesh-Connected Parallel Computer*, Comm. ACM, **20** (1977) pp. 263–270.

[12] L. G. VALIANT, *A Scheme for Fast Parallel Communication*, SIAM J. Comp. **11** (1982), pp. 350–361.

[13] L. G. VALIANT, *Parallel Computation*, In Proc. 7th IBM Symposium on Mathematical Foundations of Computer Science, 1982, pp. 171–189. Also, Tech. Report TR-16-82, Aiken Computation Laboratory, Harvard University, Cambridge, Mass.

[14] L. G. VALIANT AND G. J. BREBNER, *Universal Schemes for Parallel Communication*, In Proc. 13th ACM Symposium on Theory of Computing, 1981, pp. 263–277.

Routing and Sorting on Mesh-Connected Arrays

- Extended Abstract -

Manfred Kunde *
Institut für Informatik
Technische Universität
Arcisstr. 21
D-8000 München 2
W.-Germany

Abstract

The problems of sorting and routing on $n_1 \times \ldots \times n_r$ mesh-connected arrays of processors are studied. A new sorting algorithm for r-dimensional meshes, $r \geq 3$, is presented. On meshes without wrap-around-connections it only needs $2(n_1 + \cdots + n_{r-1}) + n_r$ steps which is asymptotically optimal. For meshes with wrap-arounds the number of steps is asymptotically $n_1 + n_2 + \cdots + n_r$ which is very close to the lower bound of $(n_1 + \cdots + n_{r-1}) + n_r/2$. Furthermore, for two-dimensional meshes a new deterministic routing algorithm is given for $n \times n$ meshes where each processor has a buffer of size $f(n) < n$. It needs $2n + O(n/f(n))$ steps on meshes without wrap-arounds. Hence it is asymptotically optimal and as good as randomized algorithms routing data only with high probability.

1. Introduction

It has turned out that in the performance of parallel computation data movement plays an important role. Especially for VLSI-architectures, where a regular net of simple processing cells and local communication between these cells is required [FK, KL], the number of parallel data transfer during a computation is a fundamental performance measure. In this paper we present routing and sorting algorithms on mesh-connected arrays [KH, NS, TK] where the number of parallel data transfers is asymptotically minimal or nearly minimal.

An $n_1 \times \ldots \times n_r$ mesh-connected array or grid of processors is a set $mesh(n_1, \ldots, n_r)$ of $N = n_1 n_2 \ldots n_r$ identical processors where each processor $P = (p_1, \ldots, p_r)$, $0 \leq p_i \leq n_i - 1$,

---

* This work was supported by the Siemens AG, München

is directly interconnected to all its nearest neighbours only. A processor $Q = (q_1,\ldots,q_r)$ is called nearest neighbour of P if and only if the distance between them is exactly 1.

For a mesh without wrap-around connections the distance is given by $d(P,Q) = |p_1 - q_1| + \cdots + |p_r - q_r|$. If wrap-around connections are allowed we define $d_{wrap}(P,Q) = |(p_1 - q_1) \bmod n_1| + \cdots + |(p_r - q_r) \bmod n_r|$ where $-b/2 \le a \bmod b \le b/2$ is assumed for arbitrary integers a and b. If diagonal connections are also be given, then the distance function can be defined by $d_{diag}(P,Q) = max_{1 \le i \le r} |p_i - q_i|$ in the case without wrap-arounds. Analogously one can describe $d_{wrap,diag}(P,Q)$.

The control structure of the grid of processors is thought to be of the MIMD type (Multiple Instruction Multiple Data). The main assumptions are:

- each processor has its own program memory
- different processors can perform different instructions at the same clock period
- there is a global clock
- each processor can send data only to its nearest neighbours during one clock period
- each processor has only a limited number of registers for data (e.g. the buffer size is constant or $log\ N$).

Grids of processors are of special interest because they have a regular communication structure and for dimension $r \le 3$ physical local communication is possible. In the two-dimensional case the grids are a well-suited model for a VLSI architecture.

Figure 1   Mesh–connected arrays and index functions

a) 3 x 3 mesh

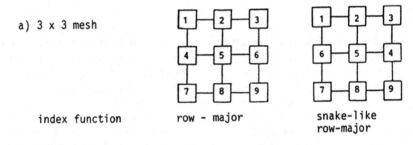

index function   row - major   snake-like
                                              row-major

b) 3 x 3 x 3 mesh-connected cube

with

snake-like indexing

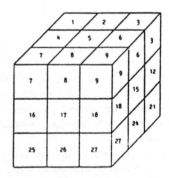

For the sorting problem we assume that N elements from a linearly ordered set are initially loaded in the N processors, each receiving exactly one element. W.l.o.g. we assume that the initial loading of a processor P denoted by $cont_0(P)$ is an integer. The processors are thought to be indexed by a certain one-to-one mapping f from $mesh(n_1, \ldots, n_r)$ onto $\{1, \ldots, N\}$. With respect to this function the sorting problem is to move the i-th smallest element to the processor indexed by i for all $i = 1, \ldots, N$. Hence a sorting problem can be described by the triple $(mesh, cont_0, f)$.

For the routing problem N elements or packets are loaded in the N processors. Each packet has a destination address specifying the processor to which it has to be sent. Different packets have different addresses. The routing problem is to transport each packet to its address. Thus the problem is described by the pair $(mesh, address)$ where address is a one-to-one mapping from mesh onto itself.

Note that the routing problem $(mesh, address)$ can be solved by a sorting algorithm for the problem $(mesh, f \circ address, f)$ for which the initially contents is given by $cont_0'(P) = f(address(P)) = f(Q)$ for all $P \in mesh$, meaning that the contents of P must be transported to $Q = address(P)$.

This fact can be used for a strategy for solving sorting problems $(mesh, cont_0, g)$ with arbitrary index function g, which even may be chaotic. In this case take a well-suited index function f and do the following

1. Solve $(mesh, cont_0, f)$
2. Solve $(mesh, f \circ g^{-1} \circ f, f)$.

After the first step the contents of $P \in mesh$ with $f(P) = i$ is the i-th smallest element which has to be transported (routed) to processor Q with index $g(Q) = i$. Hence the address of the packet in P is $address(P) = Q = g^{-1}(i) = g^{-1}(f(P))$. Therefore solve the routing problem $(mesh, g^{-1} \circ f)$ which can be treated as the sorting problem $(mesh, f \circ g^{-1} \circ f, f)$.

We call this method the standard algorithm for solving $(mesh, cont_0, g)$ by using the (hopefully well-behaved) index function f.

Note that this construction also is valid for other networks than meshes. Indeed, it works for arbitrary graphs.

Models of computation

In order to get lower and upper bounds for sorting and routing on meshes the possible steps for computations must be described very carefully. We will use two models. The first one allows interchanges of data between two directly neighboured processors and data shifts on cycles of processors. (Compare figure 2). The interchange of data may be caused by a comparison or not. Note that after each parallel step consisting of several independent data interchanges and shifts exactly one element (packet) is in each processor.

In the second model all operations of the first model are allowed. The additional feasibilities are that there may be data stored in a processor until a limited buffer is filled. That means on the other hand that there might be processors which contain no data during the computation. But from problem formulation it is clear that at the end of the computation each processor must contain an element (packet).

For 2-dimensional $n \times n$ meshes without wrap-arounds several sorting algorithms have been proposed for the first model or can easily be transformed into algorithms which fit for that model [KH, LSSS, MSS, NS, SI, SS, TK]. The fastest ones need about $3n + O(low\ order\ term)$ steps for snake-like indexing [TK,SS] or blockwise snake-like indexing [MSS]. For such indexings the algorithms are asymptotically optimal as shown in [SS, Ku2].

Figure 2   Models of computation

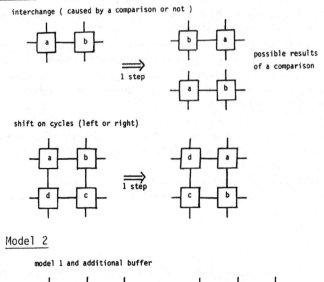

Note that for arbitrary indexing g the standard algorithm shows that at most $6n + O(low\ order\ term)$ steps are then necessary. This is relatively good since there are index functions which need approximately at least $4n$ steps [Ku4].

For wrap-around meshes a lower bound of $3n/2$ can be shown for arbitrary indexings [Ku4]. Algorithms for snake-like indexings needing $5n/2 + low\ order\ terms$ are given in [Ku3] and for blockwise snake-like indexing of $2n$ in [MSS] which is known so far to be the best algorithm. Using the standard algorithm sorting with respect to arbitrary index functions can therefore be done within $4n$ steps.

For r-dimensional meshes $r \geq 3$ there have been several approaches to solve this problem [Ku1, Ku3, NS, S, TK]. The best algorithm for an $n \times \cdots \times n$ cube needs only $(2r - 1)n + low\ order\ term$ steps for sorting with respect to snake-like indexing [Ku3] which is asymptotically optimal for this type of indexing [Ku2]. For meshes with wrap-around-connections the lower bound is $(r - 1/2)n$ and the best algorithm given in [Ku3] asymptotically needs $((3r - 1)/2)n$ steps.

In the second section of this paper we present a sorting algorithm for wrap-around meshes only needing $rn + O(n^{1-1/r})$ steps which is a nearly 50% improvement compared with the approach in [Ku3]. The new algorithm is asymptotically optimal within a factor of at most $1+1/(2r-1)$. If wrap-arounds are not allowed, then the algorithm is asymptotically optimal.

Note that the sorting problem with arbitrary index function can therefore be solved by the standard algorithm with asymptotically $(4r - 2)n$ steps and with $2rn$ steps in the case where wrap-around-connections are given.

All results concerning the standard algorithm could be improved if better routing algorithms would be available. Unfortunately for model 1 the best sorting algorithms are more or less the best routing algorithms. But in the case that a bigger buffer size is given, then we are able to show that for the two-dimensional case better algorithms can be obtained.

Here we touch on another question: can routing be performed in substantially fewer steps than sorting? This is still an open question for both the models proposed in this paper. But we can show in section 3 that routing in model 2 can be done within $2n + O(n/f(n))$ steps in the case where buffer size $f(n)$ is given. Hence this routing algorithm asymptotically matches the distance bound of $2n - 2$ provided that $f(n)$ is a strictly increasing function and not bounded by any constant. If wrap-around-connections are given then the performance can be improved to $n + O(n/f(n))$ steps which again is asymptotically optimal in the above mentioned cases.

For example, for buffer size $f(n) = \log n$ the proposed deterministic algorithm is faster than a randomized algorithm presented in [VB] and as good as that one in [RT], both algorithms reaching their execution times with high probability only.

2. A sorting method for r-dimensional meshes

In this section we will give a relatively simple method for sorting on r-dimensional meshes, $r \geq 3$. Some basic ideas are from an algorithm which sorts on a plane of processors [MSS]. For simplicity we only discuss the case of a 3-dimensional cube that is an $n \times n \times n$ mesh.

A fundamental method for obtaining asymptotically optimal or nearly optimal sorting algorithms is to divide the cube of n^3 processors into small cubes called blocks, sort all these blocks and then perform something like a merge operation to get the whole cube sorted [Ku3]. In our case blocks are given by
$$B(k_1, k_2, k_3) = [k_1 n^{2/3}, (k_1+1)n^{2/3} - 1] \times [k_2 n^{2/3}, (k_2+1)n^{2/3} - 1] \times [k_3 n^{2/3}, (k_3+1)n^{2/3} - 1]$$
for all $k_i, 0 \leq k_i \leq n^{1/3} - 1$, $i = 1, 2, 3$. That is there are n blocks each one containing $(n^{2/3})^3 = n^2$ processors.

For a fixed k_3, $0 \leq k_3 \leq n^{1/3} - 1$, the union of blocks $\bigcup_{k_1=0}^{n^{1/3}-1} \bigcup_{k_2=0}^{n^{1/3}-1} B(k_1, k_2, k_3)$ is called the k_3-th plane of blocks and for fixed k_1 and k_2 the set $\bigcup_{k_3=0}^{n^{1/3}-1} B(k_1, k_2, k_3)$ is called a column of blocks. (Compare Figure 3).

Note that each plane of blocks consists of exactly $n^{2/3}$ blocks. On the other side the $k_3 - th$ plane of blocks can be viewed as the union of $n^{2/3}$ 2-dimensional meshes of type $[0, n - 1] \times [0, n - 1] \times \{i\}$ where $k_3 n^{2/3} \leq i < (k_3 + 1)n^{2/3}$. The third coordinate i is the number of the corresponding plane. We assume that the $n^{2/3}$ blocks in the k_3-th plane of blocks are numbered by the same numbers as the planes. Hence we have a one-to-one correspondence between 3-dimensional blocks and 2-dimensional planes in each plane of blocks. Furthermore we demand that the numbering of blocks is done according to snake-like row-major indexing in the case where k_3 is even and opposite snake-like row-major indexing provided k_3 is odd. That means the blocks in the total cube are numbered in a snake-like manner where the numbering is from 0 to $n - 1$. The block with block number

Figure 3   Partitioning of the mesh into blocks

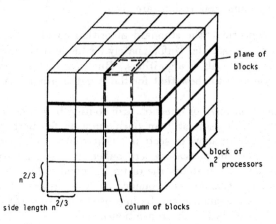

i is denoted by B_i for $i = 0, \ldots, n-1$. We say that the block number of a processor $P = (p_1, p_2, p_3)$ is i, if P is contained in B_i.

A fundamental operation of the following sorting algorithm is to transport the contents of all n^2 processors in a block i to the n^2 processors in a plane numbered by i for all $i = 0, \ldots, n-1$. This operation is called the block-to-plane-transportation (Figure 4). It can be performed by shift operations done in parallel on all planes of blocks.

The precise definition of block-to-plane-transportation is given as follows. For all processors $P = (p_1, p_2, p_3)$ move the contents to processor $Q = btp(P) = (q_1, q_2, q_3)$ with $q_1 = (p_1 + \lfloor (p_3 \bmod n^{2/3})/n^{1/3} \rfloor \cdot n^{2/3}) \bmod n$, $q_2 = (p_2 + (p_3 \bmod n^{1/3}) \cdot n^{2/3}) \bmod n$, and $q_3 = blocknumber((p_1, p_2, p_3))$. The transportation can be performed in three steps following one another.

block-to-plane-transportation (btp)

For all processors $P = (p_1, p_2, p_3)$ in parallel:
1. Move contents of P to processor $((p_1 + \lfloor (p_3 \bmod n^{2/3})/n^{1/3} \rfloor \cdot n^{2/3}) \bmod n, p_2, p_3)$
   (that is a shift in parallel to the first axis)
2. Move contents of P to processor $(p_1, (p_2 + (p_3 \bmod n^{1/3}) \cdot n^{2/3}) \bmod n, p_3)$
   (that is a shift in parallel to the second axis)
3. Move contents of P to processor $(p_1, p_2, blocknumber((p_1 - \lfloor (p_3 \bmod n^{2/3})/n^{1/3} \rfloor \cdot n^{2/3}) \bmod n, (p_2 - (p_3 \bmod n^{1/3}) \cdot n^{2/3}) \bmod n, p_3))$.

Note that by the block-to-plane-transportation not only the contents of a block numbered by i is distributed to a plane i, but also the contents of different planes within the block are transported to different columns of blocks.

For the sorting algorithm we assume that the processors are indexed by the following index function f which might be called block-wise-snake-like indexing as in [MSS]:
$$f(p_1, p_2, p_3) = blocknumber((p_1, p_2, p_3)) \cdot n^2 + g(p_1 \bmod n^{2/3}, p_2 \bmod n^{2/3}, p_3 \bmod n^{2/3})$$
where g is any index function for a cube of processors with sidelength $n^{2/3}$. g may be viewed as an indexing for each single block. Let us call an index function f for an $a \times a \times a$ cube to be balanced iff for all intervalls $[ca+1, (c+1)a]$, $0 \le c \le a^2 - 1$, and all planes of processors $A = [0, a-1] \times [0, a-1] \times \{i\}$, $0 \le i \le a-1$, there is a processor $P \in A$ with $f(P) \in [ca+1, (c+1)a]$.

Figure 4   Block-to-plane-transportation

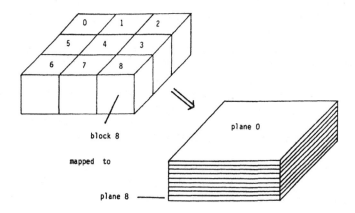

block 8

mapped to

plane 8 ——

Sorting algorithm
1. Sort all blocks with respect to balanced index functions
2. Perform a block-to-plane-transportation
3. Sort all columns of blocks downwards
   (that is with respect to snake-like indexing for $n^{2/3} \times n^{2/3} \times n$ meshes)
4. Perform a plane-to-block-transportation
   (that is an inverse block-to-plane-transportation.)
5. For all $i = 0, \ldots, n - 2$ sort all unions of neighboured blocks $B_i \cup B_{i+1}$, with respect
   to index function $h_i : B_i \cup B_{i+1} \rightarrow \{1, \ldots, 2n^2\}$ with $h_i(P) < h_i(Q)$ iff $f(P) < f(Q)$.

The correctness of the algorithm can be shown by the help of the zero-one principle
[Kn] which tells us that an algorithm sorts an arbitrary given initial loading of arbitrary
integers if and only if all initial loadings only consisting of zeroes and ones are sorted
correctly.

Let us assume that there is an arbitrary initial loading of zeroes and ones. Then in
block i after step 1 all of the planes in the block contain nearly the same number of ones,
since the index function for the block was balanced. More precisely for each block i there
is an integer z_i such that each plane within the block contain z_i or $z_i + 1$ ones. After step 2
the contents of the planes of the blocks are distributed to different columns of blocks.
Hence each column of blocks contain at least $z = \sum_{i=0}^{n-1} z_i$ ones and at most $z + n$ ones.

After step 3 all the ones in each column of blocks have been sorted to the bottom.
That is in each column of blocks at least the $\lfloor z/n^{4/3} \rfloor$ lower planes are totally filled by
ones and at most the $\lceil (z + n)/n^{4/3} \rceil$ planes from below may contain ones. That means
that at most two planes of the total cube may contain both zeros and ones (namely those
planes numbered by $n - 1 - \lfloor z/n^{4/3} \rfloor$ and $n - \lfloor z/n^{4/3} \rfloor$.)

By step 4 the contents of neighboured planes are transported to neighboured blocks.
Hence there are only two neighboured blocks which can contain both zeroes and ones.
After step 5 all of the blocks are sorted.

For the complexity of the algorithm note that step 1 and 5 only need $O(n^{2/3})$ steps.
Step 3 can be shown to be done within $n + O(n^{2/3})$ steps. The block-to-plane-transportation
or its inverse can be done by $2n + O(n^{2/3})$ if the mesh has no wrap-around-connections.

Hence for meshes without wrap-around-connections $5n + O(n^{2/3})$ steps are sufficient. If there are wrap-arounds the block-to-plane-transportation or its inverse only costs $n + O(n^{2/3})$ steps. That is in the wrap-around case we need only $3n + O(n^{2/3})$ steps.

The algorithm can be generalized to arbitrary dimensions. Blocks are then r-dimensional submeshes of sidelength $n^{1-1/r}$ and planes are $(r-1)$-dimensional meshes with sidelength n. Note that $(n^{1-1/r})^r = n^{r-1}$. Step 2 in the generalized version can be described by $(r$ - block$)$-to-$((r-1)$-plane$)$-transportation. In step 4 the inverse operation has to be performed. The result for a r-dimensional cube without wrap-arounds is then $(2r-1)n + O(n^{1-1/r})$ which is as good as the algorithm proposed in [Ku3] and asymptotically optimal for the given index function which can easily be seen by the joker argument [Ku2, Ku4].

In the case of wrap-around-connections it can be shown that $rn + O(n^{1-1/r})$ steps are necessary. The best lower bound known until yet is $(r - 1/2)n$ for this indexing scheme. That means the proposed algorithm differs at most by a factor of $1 + 1/(2r - 1)$ from an optimal algorithm.

It is not hard to see that the method can be generalized to r-dimensional meshes of size $n_1 \times n_2 \times \cdots \times n_r$. Neglecting low order terms the number of steps needed by the algorithm is $2(n_1 + \cdots + n_{r-1}) + n_r$ for meshes without wrap-arounds and $n_1 + \cdots + n_r$ for wrap-around meshes.

3. An optimal deterministic routing algorithm

In this section we present a routing algorithm for two-dimensional meshes. For the sake of brevity we will only discuss the case of $n \times n$ meshes. The model of computation differs from that one used in the last section in so far that each processor is able to store a limited amount of data during the computation. For an $n \times n$ mesh we assume that the size of the additional buffer is $2f(n)$, $1 \leq f(n) < n/2$. Typical buffer sizes are for example $\log n$, $n^{1/4}$ or a constant c.

As in the last section divide the total mesh into blocks of processors. In this case we assume a block to be a $n/f(n) \times n/f(n)$ submesh. That is the side length depends on the given buffer size.

The address of a packet is the name of that processor (r, c), $0 \leq r, c \leq n - 1$, where the packet has to be sent to. We call r the row address and c the column address of the packet. For the first step of the algorithm we assume the addresses to be linearly ordered by their row addresses. That is $(r_1, c_1) \leq (r_2, c_2)$ if and only if $r_1 \leq r_2$.

Routing algorithm
1. Sort all blocks with row-major indexing
2. In each column c, $0 \leq c \leq n-1$: transmit packets with row address r to processor in row r.
3. In each row r, $0 \leq r \leq n-1$: transmit packets with column address c to processor in column c.

In order to proof the correctness of our algorithm let us observe all packets with row address r for an arbitrary r, $0 \le r \le n-1$. Note that the number of such packets is exactly n. Take an arbitrary column of blocks. Assume that these blocks are numbered from 1 to $f(n)$. Let $b(i)$ be the number of packets with row address r in block i, $1 \le i \le f(n)$. After the first step of the routing algorithm the packets in block i are ordered with respect to their row addresses. Hence all packets with row address r have been sent to processors in neighboured rows of processors in block i. Since the side length of a block is $n/f(n)$ that means that in each column of block i there are at most $\lceil b(i)/(n/f(n)) \rceil$ packets with row address r (compare Figure 5).

Therefore in each column of the total block column there are at most

$$\sum_{i=1}^{f(n)} \lceil f(n) \cdot b(i)/n \rceil \le \sum_{i=1}^{f(n)} (b(i) \cdot f(n)/n + 1)$$

$$\le (f(n)/n) \cdot \sum_{i=1}^{f(n)} b(i) + f(n) \le (f(n)/n) \cdot n + f(n) \le 2f(n)$$

packets with row address r.

Figure 5   Columns of blocks after the first step of the routing algorithm

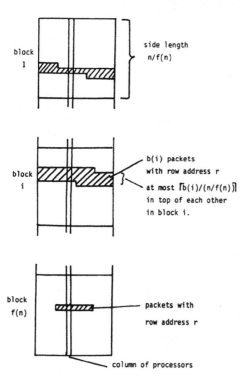

Hence in each column c, $0 \le c \le n - 1$, of the total mesh there are at most $2f(n)$ packets with row address r. These at most $2f(n)$ packets are then transported to processor

(r, c) which can store all these packets because its buffer is large enough. In the last step the packets are then distributed to their destination processor.

The first step of the algorithm costs $O(n/f(n))$ steps, each of the second and the third steps can be performed by n steps [VB, RT]. That means in total $2n + O(n/f(n))$ steps are sufficient for routing on meshes without wrap-around connections. If there are wrap-arounds the last two steps can be performed together in n steps.

It is easily seen that the method can be applied to $a \times b$ meshes. If the buffersize is given by a function $g(a, b)$, the complexity is then $a + b + O((a + b)/g(a, b))$ for meshes without wrap-arounds and $(a+b)/2+O((a+b)/g(a, b))$ in the wrap-around case. Note that for both types of meshes the distance bound asymptotically is matched if the buffersize is given as a strictly increasing unbounded function. That means that in all these cases the method is asymptotically optimal. Hence this deterministic algorithm is asymptotically faster than the randomized algorithm in [VB] and as good as the randomized algorithm in [RT].

Furthermore, it should be pointed out that the routing algorithm only needs $2d_{max} + O(n/f(n))$ steps, if one knows in advance the maximum distance $d_{max} = \{d(P, address(P)) | P \in mesh\}$ for a problem $(mesh, address)$. Hence for local routing problems with small d_{max} the routing algorithm can make use of the locality.

4. Conclusion

In this paper we presented sorting algorithms for r-dimensional $n_1 \times \cdots \times n_r$ meshes without additional buffer for each processor. If no wrap-around connections are given, then the algorithm needs $2(n_1 + \cdots + n_{r-1}) + n_r$ parallel data movements (neglecting low order terms) which is asymptotically optimal for the used type of index functions. For wrap-around meshes the method works with $n_1 + \cdots + n_r$ steps, which again is very fast. However, it is still open, if it is asymptotically optimal.

The proposed sorting algorithms can be used to solve the routing problem on this type of meshes, but the question arises, whether there are algorithms for the easier looking routing problem which are faster than algorithms for the more complex sorting problem.

If for each single processor an additional buffer of size $log(n_1 + n_2)$ or $(n_1 + n_2)^{1/4}$ is given, then we were able to show that for 2-dimensional meshes there are routing algorithms which match the distance bound, and therefore are asymptotically optimal. It remains an open question whether sorting can be solved as fast as routing on meshes with additional buffer.

Concerning the question routing versus sorting the result of this paper is that in the 2-dimensional case routing on meshes with additional buffer can be done faster than sorting on meshes without additional buffer.

References:

[FK]   Foster, M.J., Kung, H.T.: The design of special-purpose VLSI-chips. IEEE Comput. **13**, 26-40 (1980)

[KH]   Kumar, M., Hirschberg, D.S.: An efficient implementation of Batcher's odd-even merge algorithm and its application in parallel sorting schemes. IEEE Trans. Comput. C-**32**, 254-264 (1983)

[Ku1]   Kunde, M.: A general approach to sorting on 3-dimensionally mesh-connected arrays. In: Händlers, W. (eds.) Lect. Notes Comput. Sci., vol. 237, pp. 84-95. Berlin-Heidelberg-New York-Tokyo: Springer 1986

[Ku2]   Kunde, M.: Lower bounds for sorting on mesh-connected architectures. Acta Informatica 24, 121-130 (1987).

[Ku3]   Kunde, M.: Optimal sorting on multi-dimensionally mesh-connected computers. Proceedings of STACS 87. In: Brandenburg, F.J., Vidal-Naquet, G., Wirsing, M. (eds.) Lect. Notes Comp. Sci., vol. 247, pp. 408-419. Berlin-Heidelberg-New York-Tokyo: Springer 1987

[Ku4]   Kunde, M.: Bounds for l-section and related problems on grids of processors; submitted

[KL]   Kung, H.T., Leiserson, C.E.: Systolic arrays for VLSI. Symposium on Sparse Matrix Computation 1978, Proceeding, Duff, I.S., Stewart, C.G. (eds.) 1978

[Kn]   Knuth, D.E.: The art of computer programming , vol. 3: Sorting and Searching, Addison Wesley, Reading, 1973, pp. 224-225.

[LSSS]   Lang, H.-W., Schimmler, M., Schmeck, H., Schröder, H.: Systolic sorting on a mesh-connected network. IEEE Trans. Comput. C-**34**, 652-658 (1985)

[MSS]   Ma, Y., Sen, S., Scherson, I.D.: The distance bound for sorting on mesh-connected processor arrays is tight. Proceedings FOCS 86, pp. 255-263

[NS]   Nassimi, D., Sahni, S.: Bitonic sort on a mesh-connected parallel computer. IEEE Trans. Comput. C-**28**, 2-7 (1979)

[RT]   Rajasekaran, S., Tsantilas, Th.: An optimal randomized routing algorithm for the mesh and a class of efficient mesh-like routing networks. 7th Conference on Foundations of Software Technology and Theoretical Science, Pune, India

[SI]   Sado, K., Igarashi, Z.: A fast parallel pseudo-merge sort algorithm. Gunma University, Technical Report, Japan, 1985

[S]   Schimmler, M.: Fast sorting on a three dimensional cube grid. Technical Report 8604, University of Kiel, Germany, 1986

[SS]   Schnorr, C.P., Shamir, A.: An optimal sorting algorithm for mesh-connected computers, pp. 255-263. Proceedings STOC 1986. Berkley 1986

[TK]   Thompson, C.D., Kung, H.T.: Sorting on a mesh-connected parallel computer. CACM **20**, 263-271 (1977)

[VB]   Valiant, L.G., Brebner, G.J.: Universal schemes for parallel communication. Proceedings STOC 81, pp. 263-277.

Time Lower Bounds for Parallel Sorting on a Mesh-Connected Processor Array

Yijie Han[*]   Yoshihide Igarashi[*†]

[*]Department of Computer Science
University of Kentucky
Lexington, KY 40506

[†]Department of Computer Science
Gunma University
Kiryu, 376 Japan

ABSTRACT

We prove that $(1+\sqrt{6}/2)n$ is a time lower bound independent of indexing schemes for sorting n^2 items on an $n \times n$ mesh-connected processor array. We distinguish between indexing schemes by showing that there exists an indexing scheme which is provably worse than the snake-like row-major indexing for sorting. We also derive lower bounds for various indexing schemes. All these results are obtained by using the chain argument which we provide in this paper.

1. Introduction

Parallel sorting algorithms and their time complexities have been intensively studied[1-19]. Although there exist sorting algorithms of time complexity $O(\log N)$ [1,2,4,8,11], the structure of such algorithms are complicated and their realization is extremely difficult. A mesh-connected processor array is widely accepted as a realistic model of parallel computers. It consists of $N=n^2$ identical processors arranged in a square. A number of parallel sorting algorithms on the mesh-connected model have been reported[7,9,10,12-19]. Some of these algorithms are good in the practical sense and fast for realistic values of n. Schnorr and Shamir [17] and Ma et al.[9] have designed asymptotically fast sorting algorithms on the mesh-connected model. The time complexities of their algorithms are $3n+O(n^{3/4})$ steps and $4n+O(n^{3/4}\log n)$ steps, respectively.

Kunde[6] and Schnorr and Shamir [17] have proved a $3n-2n^{1/2}-3$ lower bound on the number of steps for sorting n^2 items into the snake-like row-major order. They used a technique called the joker zone argument[6]. Since these discoveries, it becomes curious whether the snake-like row-major indexing is the best indexing scheme for sorting[17]. A question whether the distance bound of $2n$ is ultimately achievable by using some kind of super indexing schemes has also been raised[9].

The only known lower bound independent of indexing schemes for sorting on an $n \times n$ mesh-connected processor array was the distance bound of $2n$. No previous results showed that some indexing schemes are better than others.

The aim of this paper is to provide a general technique to derive lower bounds for various indexing schemes and to derive a good lower bound independent of indexing schemes. In this paper we show that $2.22n$ is a lower bound for any indexing scheme on the $n \times n$ mesh-connected processor array. We therefore answered the question posed by Ma *et al.*[9]. We use a tool which we call the chain argument to obtain this lower bound. By using the chain argument we give a constructive proof showing that $4n - 2(2n)^{1/2} - 3$ is a time lower bound for sorting by a certain indexing scheme. Combined with Schnorr and Shamir's result[17], this shows that some indexing schemes are provably worse than the snake-like row-major indexing for sorting. By using the chain argument we also derive lower bounds for various indexing schemes.

The present paper is divided into 6 sections. In section 2 we will define our mesh-connected model, illustrate commonly used indexing schemes and give some terminologies. In section 3, a basic tool, called the chain argument, is described. In section 4 we derive time lower bounds for various indexing schemes. We also show the existence of a poor indexing scheme. In section 5 we derive a good lower bound independent of indexing schemes.

2. Preliminaries

We consider a general model of a synchronous $n \times n$ mesh-connected processor array as given in [17]. The processor array is denoted by $M[1..n, 1..n]$. Each processor at location (i, j), $1 \leq i, j \leq n$, is denoted by $M[i, j]$. A subarray consisting of processors $M[i, j]$ such that $p \leq i \leq q$ and $r \leq j \leq t$ is denoted by $M[p..q, r..t]$. Processor $M[i, j]$ is directly connected to its neighbors $M[i, j-1]$, $M[i-1, j]$, $M[i+1, j]$ and $M[i, j+1]$, provided they exist. This model is shown in Fig. 1. All n^2 processors work in parallel with a single clock, but they may run different programs. As for sorting computation, the initial contents of $M[1..n, 1..n]$ are assumed to be a permutation of n^2 linearly ordered data items, where each processor has exactly one data item. The final contents of $M[1..n, 1..n]$ are the sorted sequence of the items in a specified order. In one step each processor can communicate with all of its directly-connected neighbors. The interchange of data items and the replacement of the data item in a processor with the data item of its directly connected processor can be done in one step. We should note that a data item can only be moved to one of its neighbors in one step. The computing time is defined to be the number of parallel steps of the basic operations to reach the final configuration of the processor array. As pointed out by Schnorr and Shamir[17], such a model is stronger than most models used for deriving upper bounds.

An indexing scheme I on array $M[1..n, 1..n]$ is a one-to-one mapping from $\{1, ..., n\} \times \{1, ..., n\}$ to $\{1, ..., n^2\}$, *i.e.* $I: (i, j) \rightarrow k$. Sorting by an indexing scheme I is to sort the items into the order defined by I.

The row-major indexing and the snake-like row-major indexing are commonly accepted ways to order the processor array. Other indexing schemes are also used[3]. They are diagonal-major indexing, snake-like diagonal-major indexing, file-major indexing, and q-column-block-major indexing. These indexing schemes are shown in Fig. 2. The rules of indexing for these schemes may be clear from the illustrations. We briefly explain how processors are ordered by the file-major indexing and by the q-column-block-major indexing in the following. In the file-major indexing the whole processor array $M[1..n, 1..n]$ is divided into 4 subfiles $S_1 = M[1..n/2, 1..n/2]$, $S_2 = M[1..n/2, n/2+1..n]$, $S_3 = M[n/2+1..n, 1..n/2]$, $S_4 = M[n/2+1..n, n/2+1..n]$. Any processor in S_i proceeds any processor in S_j if $i < j$. The order of the processors in the same subfile is recursively defined in the same way. In the q-

column-block-major indexing the i-th block is $M[1..n, iq+1..(i+1)q]$, $i=0,1,...,n/q-1$, where n is assumed to be a multiple of q. The order of any processor in the i-th block proceeds the order of any processor of the j-th block if $i<j$. The order of processors in the same block is by the row-major indexing.

The distance between $M[i_1, j_1]$ and $M[i_2, j_2]$ is defined as $|i_1-i_2|+|j_1-j_2|$ and denoted by $d((i_1,j_1),(i_2,j_2))$. A zone S of $M[1..n, 1..n]$ is denoted by a subset of $\{1,...,n\}\times\{1,...,n\}$. The area of zone S is the number of processors in S. For convenience $M[1..n, 1..n]$ also denotes the zone of all processors in $M[1..n, 1..n]$. For an indexing scheme I the index of $M[i,j]$ is denoted by $I(i,j)$. For example, if I is the row-major indexing on $M[1..4, 1..4]$, $I(2,4)=8$. For an indexing scheme I, a chain C is a sequence of pairs $((i_1,j_1), (i_2,j_2), ... (i_c,j_c))$, $(i_k,j_k) \in \{1,...,n\}\times\{1,...,n\}$, $1\leq k\leq c$, such that $I(i_1,j_1), I(i_2,j_2), ..., I(i_c,j_c)$ is a consecutive integer sequence in increasing order. In this case the length of C is $c-1$. For the row-major indexing on $M[1..4, 1..4]$, $((2,4), (3,1), (3,2), (3,3))$ is a chain of length 3. It is obvious that no chain in $M[1..n, 1..n]$ could have its head and tail joined to form a circle.

3. The Chain Theorem

We shall frequently consider the computation performed in a duration called a sweep. As shown in Fig. 3(a), a sweep is defined by two lines and a direction. The two lines are parallel to a diagonal of the mesh-connected processor array and the direction is orthogonal to the lines. One of the lines is called the start line and the other is called the stop line. The computation performed during the sweep as specified in Fig. 3(a) is shown in Fig. 3(b). The frontier of the computation is initially at the start line x and propagates along direction Z. When the frontier hits the extreme of the processor array (point A in Fig. 3(a)), it will be bounced back and then propagates backwards until it hits the stop line y. Referring to Fig. 3, the length of the sweep is defined to be the sum of the distance from line x to point A and the distance from point A to line y. Therefore the length of a sweep represents the minimum number of computing steps required to accomplish the sweep. The region enclosed by the start line x and corner D is called the stretching zone. The items in the stretching zone are used to stretch one position into a chain. This is demonstrated in the proof of the chain theorem. The region enclosed by stop line y and corner A is called the residing zone. We shall demonstrate the existence of a chain of a certain length in the residing zone. Notice that if the length of a sweep is no less than $2n-2$ then the area of residing zone is no less than the area of the stretching zone.

Our first theorem, called the chain theorem, relates the computing time of a sorting algorithm, the length of a sweep on mesh M and the length of a chain in the residing zone.

Theorem 1(Chain Theorem): *For any sorting algorithm of time complexity T and a sweep of length $T+1$, there exists a chain C of length S in the residing zone of the sweep, where S is the area of the stretching zone .*

Proof: Let $SORT$ be a sorting algorithm of time complexity T. Referring to Fig. 3(a), let k be the distance between line x and point A. During the first $k-1$ steps of $SORT$ the computation happens at point A is not affected by the initial contents in the stretching zone (in Fig. 3(a) the stretching zone is ΔFED). Let a be the item which is at point A immediately after the $(k-1)$-st step of $SORT$. Item a is independent of the initial contents of any processor in the stretching zone, and its final sorted position must be in the residing zone since the time complexity of $SORT$ is T. By assigning different input values to the processors in the stretching zone, we can force (stretch) a into $S+1$ different sorted positions. These different positions are all within the residing zone and they form a chain of length S. $\square$

The next theorem is immediate form the Chain Theorem and will be often used to show time lower bounds for various indexing schemes in the following sections.

Theorem 2: *For an indexing scheme I and a sweep of length T, if there is no chain of length S in the residing zone, where S is the area of the stretching zone, then there is no algorithm of time complexity $T-1$ for sorting the items on the mesh-connected processor array by indexing scheme I (i.e., T is a time lower bound for sorting by I).* $\square$

4. Lower Bounds for Various Indexing Schemes

By using the chain argument we can derive new lower bounds for various indexing schemes. Theorems 3 and 4 in the following were proved by Kunde[6] and by Schnorr and Shamir[17]. Here we give alternative proofs by using the chain argument.

Theorem 3: *A lower bound for sorting by the row-major indexing is $3n-(2n)^{1/2}-2$ steps.*

Proof: We define the sweep (shown in Fig. 4, where $k=(2n)^{1/2}$) such that the stretching zone and the residing zone of the sweep are $\{m[i,j] \mid d((i,j),(n,1))\leq(2n)^{1/2}-1\}$ and $\{M[i,j] \mid d((i,j),(1,n))\leq n-1\}$, respectively. The length of the sweep is $3n-(2n)^{1/2}-2$. The area of the stretching zone is $k(k+1)/2>n-1$. However, the longest chain in the residing zone has length $n-1$. By Theorem 2, a lower bound for sorting by the row-major indexing is $3n-(2n)^{1/2}-2$ step. $\square$

Theorem 4: *A lower bound for sorting by the snake-like row-major indexing is $3n-2(n-2)^{1/2}-2$ steps.* $\square$

The proof of Theorem 4 is similar to the proof of Theorem 3, and it is omitted here.

We use the chain argument to derive new lower bounds. These lower bounds are listed in Theorems 5, 6 and 7.

Theorem 5: *A lower bound for sorting by the diagonal-major indexing or the snake-like diagonal-major indexing is $3n-4$ steps.*

Proof: We choose the sweep (shown in Fig. 5) such that the stretching zone and the residing zone are $\{M[i,j] \mid i\geq j\}$ and $\{M[i,j] \mid (i,j)\neq(n,1)\}$, respectively. That is, the stretching zone is the lower-left triangle while the residing zone consists of all processors except the one at point C. The length of the longest chain in the residing zone is equal to the area of the stretching zone minus 1. The sweep has length $3n-4$. By Theorem 2, a lower bound for sorting by these indexing schemes is $3n-4$ steps. $\square$

Theorem 6: *A lower bound for sorting by the file-major indexing is $2.75n-4$ steps.*

Proof: Consider the sweep (shown in Fig. 6.) such that the stretching zone and the residing zone are $\{M[i,j] \mid i\leq j+n/4-1\}$ and $\{M[i,j] \mid (i,j)\neq(1,n)\}$, respectively. The length of the sweep is $2.75n-4$. The area of the stretching zone is $(23n^2-12n)/32$. However, the longest chain in the residing zone has length $n^2(1/2+1/8+...)<(23n^2-12n)/32$. Hence, by Theorem 2, a lower bound for sorting by the file-major indexing is $2.75n-4$ steps. $\square$

Theorem 7: *Let n be a multiple of k. A lower bound for sorting by the (n/k)-column-block-major indexing is $3n+2n/k-(2/k)^{1/2}n-4$ steps.*

Proof: Consider the sweep (shown in Fig. 7.) such that the stretching zone and the residing zone are $\{M[i,j] \mid d((i,j),(n,1))\leq(2/k)^{1/2}n-1\}$ and $\{M[i,j] \mid d((i,j),(n,1))\geq(n/k)(k-2)+1\}$, respectively. The length of the sweep is $3n+2n/k-(2/k)^{1/2}n-4$. The length of the longest chain in the residing zone is smaller than the area of the stretching zone. By Theorem 2, a lower bound for the (n/k)-column-block-major indexing is $3n+2n/k-(2/k)^{1/2}n-4$ steps. $\square$

For various values of k we show lower bounds for the (n/k)-column-block-major indexing scheme in Table 1. We also illustrate the relation between lower bounds and values of k in Fig. 8.

By using the chain argument we show that there exists an indexing scheme with a lower bound greater than $3n$.

Theorem 8: *There exists an indexing scheme with lower bound of $4n-2(2n)^{1/2}-3$ steps for sorting.*

Proof: Consider the sweep defined in Fig. 9(a). Notice that the start line is immediate to the lower right of the stop line. Dividing the sorted chain of length n^2-1 evenly into $S+1$ chains as shown in Fig. 9(b). Define indexing scheme I such that the S dividing points are located outside the residing zone (*i.e.*, in the stretching zone) and the others are located in the residing zone. The longest chain in the residing zone has length no greater than $\lceil \dfrac{n^2}{S+1} \rceil$. Let $S=n$. Then the length of the sweep is $4n-2(2n)^{1/2}-3$. By Theorem 2, $4n-2(2n)^{1/2}-3$ is a lower bound for sorting by indexing scheme I. $\square$

Since $3n$ is the known upper bound for sorting by the snake-like row-major indexing[17], we have proved that there exists an indexing scheme which is worse than the snake-like row-major indexing for sorting.

5. A Lower Bound for Any Sorting Order

We apply the chain theorem to derive a lower bound independent of indexing schemes.

Theorem 9: *A lower bound independent of index schemes for sorting is $2.22n$.*

Proof: Let I be an arbitrary but fixed indexing scheme. Referring to Fig. 10, the area of square $EFGH$ is $2k^2$, therefore there is a point (i,j) satisfying:

(a). (i,j) is located outside square $EFGH$;

(b). $n^2/2-k^2 \leq I(i,j) \leq n^2/2+k^2$.

Without loss of generality we may let a be point (i,j) as shown in Fig. 10. Because of property (b), the longest chain in polygon $PQDBA$ has length at most $\max\{I(i,j), n^2-I(i,j)\} \leq n^2/2+k^2$.

Define a sweep such that ST, PQ and Z are the start line, the stop line and the direction of the sweep, respectively. The length of the sweep is $2n+\epsilon$. The area of the stretching zone (polygon $SACDT$) is $n^2-(n-k+\epsilon)^2/2$.

By Theorem 2, a lower bound of $2n+\epsilon$ will be held when $n^2-(n-k+\epsilon)^2/2 \geq n^2/2+k^2$.

Solving the equation $n^2-(n-k+\epsilon)^2/2=n^2/2+k^2$ we obtain

$$\epsilon = \sqrt{n^2-2k^2}-n+k.$$

Taking derivative, we find the maximum value of ϵ to be $(\sqrt{6}/2-1)n \approx 0.22n$. $\square$

Recent results of Han, Igarashi and Truszczynski[5] have raised the lower bound to $2.27n$.

6. Concluding Remarks

Our main results shown in this paper are a lower bound for any sorting order and the existence of a poor indexing scheme. These results were obtained by using the chain argument. Although it has been shown in this paper that it is impossible to sort in $2n$ steps, the question remains open whether there exist sorting algorithms with less than $3n$ time.

Acknowledgement

We wish to thank F.D. Lewis and M. Truszczynski for comments and discussions on this paper.

References

[1]. Ajtai, M., J. Komlós, and E. Szemerédi. An $O(N\log N)$ sorting network. Proc. 15th ACM Symp. on Theory of Computing, 1-9(1983).

[2]. Bilardi, G., and Preparata, F.: A minimum area VLSI architecture for $O(\log N)$ time sorting, Proc. 16th Annual ACM Symp. on Theory of Computing, 67-70(1984).

[3]. Bitton, D., Dewitt, D.T., Hsiao, D.K., and Menon, J.: A taxonomy of parallel sorting, ACM Computing Survey, 16, 287-318(1984).

[4]. Cole, R.: Parallel merge sort. 27th Symp. on Foundations of Comput. Sci., IEEE, 511-516(1986).

[5]. Han, Y., Y. Igarashi, M. Truszczynski.: Indexing schemes and lower bounds for sorting on a mesh-connected computer. manuscript.

[6]. Kunde, M.: Lower bounds for sorting on mesh-connected architectures. Acta Informatica, 24, 121-130(1987).

[7]. Lang, H.-W., Schimmler, M., Schmech, H. and Schröder, H.: Systolic sorting on a mesh-connected network, IEEE Trans. Comput., C-34, 652-658(1985).

[8]. Leighton, T.: Tight bounds on the complexity of parallel sorting. IEEE Trans. Comput., C-34, 344-354, 1985.

[9]. Ma, Y., Sen, S. and Scherson, I.D.: The distance bound for sorting on mesh-connected processor arrays is tight. 27th Symp. on Foundations of Comput. Sci., IEEE, 255-263(1986).

[10]. Nassimi, D., and Sahni, S.: Bitonic sort on a mesh-connected parallel computer, IEEE Trans. Comput., C-27(1979).

[11]. Reif, J.H. and Valiant L.G.: A logarithmic time sort for linear size networks, J. ACM, 34, 60-76(1987).

[12]. Sado, K. and Igarashi, Y.: A divide-and-conquer method of the parallel sort, IECE of Japan, Tech. Commit. of Automata and Languages, AL84-68, 41-50(1985).

[13]. Sado, K. and Igarashi, Y.: A fast pseudo-merge sorting algorithm, IECE of Japan, Tech. Commit. of Automata and Languages, AL84-68, 41-50(1985).

[14]. Sado, K. and Igarashi, Y.: Fast parallel sorting on a mesh-connected processor array, Proc. of Japan-U.S. Joint Seminar, Discrete Algorithms and Complexity Theory(Johnson, D.S. et al. eds), Academic Press, New York, 161-183(1986).

[15]. Sado, K. and Igarashi, Y.: Some parallel sorts on a mesh-connected processor array, J. Parallel and Distributed Computing, Vol. 3, pp. 389-410, 1986.

[16]. Scherson, I.D., Sen, S. and Shamir, A.: Shear sort: A true two-dimensional sorting technique for VLSI networks, Proc. 1986 Int. Conf. on Parallel Processing, 903-908(1986).

[17]. Schnorr, C.P. and Shamir, A.: An optimal sorting algorithm for mesh-connected computers, Proc. 18-th ACM Symp. on Theory of Computing, 255-263(1986).

[18]. Thompson, C.D. and Kung, H.T.: Sorting on a mesh-connected parallel computer, Commun. ACM, 20, 263-271(1977).

[19]. Thompson, C.D.: The VLSI complexity of sorting, IEEE Trans. Comput. C-32, 1171-1184(1983).

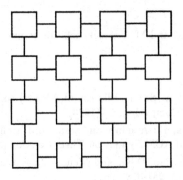

Fig. 1. A mesh-connected processor array.

1	2	3	4
5	6	7	8
9	10	11	12
13	14	15	16

(a) Row-major indexing

1	2	3	4
8	7	6	5
9	10	11	12
16	15	14	13

(b) Snake-like row-major indexing

1	3	6	10
2	5	9	13
4	8	12	15
7	11	14	16

(c) Diagonal-major indexing

1	3	4	10
2	5	9	11
6	8	12	15
7	13	14	16

(d) Snake-like diagonal-major indexing

1	2	5	6
3	4	7	8
9	10	13	14
11	12	15	16

(e) File-major indexing

1	2	9	10
3	4	11	12
5	6	13	14
7	8	15	16

(f) 2-column-block-major indexing

Fig. 2. Indexing schemes.

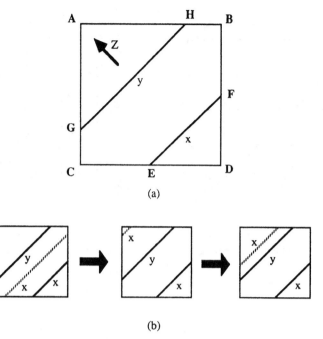

(a)

(b)

Fig. 3. A sweep.

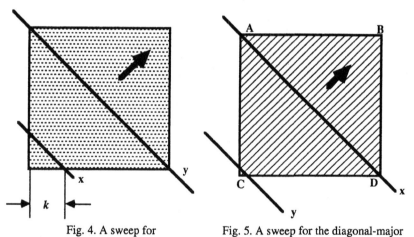

Fig. 4. A sweep for
the row-major indexing.

Fig. 5. A sweep for the diagonal-major
and the snake-like diagonal-major indexings.

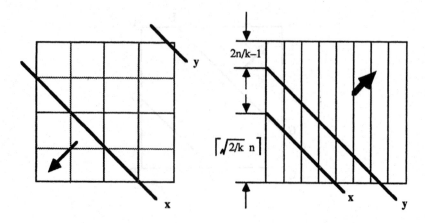

Fig. 6. A sweep for the
subfile-major indexing.

Fig. 7. A sweep for the
(n/k)-column-block-major indexing.

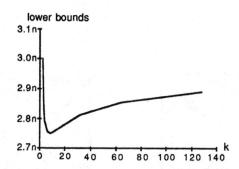

Fig. 8. Lower bounds for the (n/k)-column-block-major indexing.

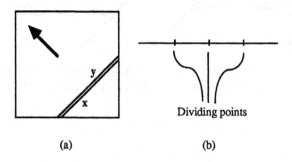

(a)

(b)

Fig. 9. A poor indexing scheme.

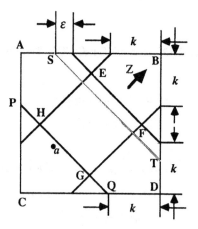

Fig. 10. A sketch for Theorem 9.

k	lower bound	k	lower bound
2	$3n$	10	$2.752n$
3	$2.850n$	16	$2.771n$
4	$2.792n$	32	$2.812n$
6	$2.755n$	64	$2.854n$
8	$2.750n$	128	$2.890n$

Table 1. Lower bounds for the (n/k)-column-block-major indexing.

A CORRECTION NETWORK FOR N-SORTERS

Manfred Schimmler

Christoph Starke

Institut für Informatik und Praktische Mathematik

Christian-Albrechts-Universität Kiel

Olshausenstr. 40

D-2300 Kiel 1

ABSTRACT. A correction network C is introduced that can be added to an arbitrary N input sorting net in order to achieve single fault tolerance. Multiple (m) fault robustness ist attained by adding C^m. For single fault correction C is proved to be assymptotically optimal.

KEYWORDS: Sorting networks, comparators, fault tolerance, reliability, Hamming distance.

1. INTRODUCTION

The high integration density in VLSI technology has created an increasing probability of fabrication faults. Therefore, although it is cheap to produce large quantities of a single chip or wafer, only a small fraction of them can be expected to function correctly. One way of increasing the yield is by designing fault tolerant algorithms.

Another aspect of fault tolerance is to increase the reliability of a chip. A useful technique is to introduce redundant components in order to keep the whole system reliable even in the presence of several operating faults.

In this paper we present a method for adding a fault correcting network to an arbitrary sorting net. It is very simple and needs only small effort in additional hardware and additional delay time. Furthermore, we will show that the number of additional comparators and the number of additional delay stages is asymptotically optimal for single fault correction of arbitrary sorting nets.

The problem has already been investigated in [6]. The authors found a correction mechanism for k faults in an N-network, consisting of only k(2N-3) additional comparators. Unfortunately, the number of additional delay stages is k(2N-3), too. Therefore, even a single fault tolerant N-sorter has time complexity $\Theta(N)$.

In [4] a multiple half-fault tolerant N-sorter is presented. For $N = 2^n$ it consists of one block of the 'balanced sorting network' [2], consisting of log N stages of N/2 comparators. The output of this block is recirculated back as input. If the block is fault free, sorting requires log N passes through the network. If it is not fault free, the number of necessary passes increases with the number of faulty comparators, but the network can still sort. There are n pairs of 'critical' comparators. The network fails only if both comparators of such a pair are faulty.

2. DEFINITIONS

Let R denote the set of real numbers. An *N-comparator* (or comparator if N is understood) is a pair [i:j], with $0 \le i, j \le N-1$, $i \ne j$. Associate with the N-comparator [i:j] the mapping from R^N to R^N defined by

$$\langle x_0, x_1, \ldots, x_{N-1} \rangle \ [i:j] := \langle x'_0, x'_1, \ldots, x'_{N-1} \rangle$$

where $x'_k = x_k$ if $k \notin \{i, j\}$, $x'_i = \min(x_i, x_j)$. We call $[i:j]$ a *standard comparator* if $i < j$.

An *N-comparator stage* (or comparator stage if N is understood) s is a set of r N-comparators $\{[i_0:j_0], [i_1:j_1], \ldots, [i_{r-1}:j_{r-1}]\}$, where $i_0, j_0, i_1, j_1, \ldots, i_{r-1}, j_{r-1}$ are pairwise distinct. The N-comparator stage s defines a mapping from R^N to R^N which is the arbitrary sequential composition of the mappings associated with $[i_0:j_0], [i_1:j_1], \ldots, [i_{r-1}:j_{r-1}]$.

An *N-network* A is a sequence of N-comparator stages $s_1, s_2, \ldots, s_k$. The N-network A defines the mapping from R^N to R^N by successively applying the mappings induced by $s_1, s_2, ..., s_k$. A *standard N-network* is an N-network consisting of standard comparators only.

A vector $x = \langle x_0, x_1, ..., x_{N-1} \rangle \in R^N$ is *sorted*, iff $x_{i-1} \le x_i$ for $1 \le i \le N-1$.

An *N-sorter* A is an N-network that satisfies the condition:

$\forall x \in R^N$: $xA = x' \Rightarrow x'$ is sorted.

A very useful tool in the studies of sorting networks is the

Zero One Principle:

Let A be an N-network. If xA is sorted for every $x \in \{0,1\}^N$ then A is an N-sorter.

Proof: See [3, 5.3.4, Thm. Z].

Using the Zero one Prinziple we can restrict our studies of sorting networks to input vectors $x \in \{0,1\}^N$.

For $x \in \{0,1\}^N$ the sorted version x_s denotes the nondecreasing sequence consisting of the same number of 0's and 1's as x:

$$x_s := xA \text{ for an arbitrary N-sorter A.}$$

Fig. 1 shows for example the 4-sorter $\{[0:1], [3:2]\}, \{[0,2], [1:3]\}, \{[0:1], [2:3]\}$. drawn as defined in [3, p. 222]: Comparators are drawn as vertical rows between horizontal data lines. Elements to be sorted travel through the network from left to right, and whenever they meet a comparator they are compared and possibly interchanged such that the larger element appears at the line of the arrow head and the smaller one at the line of the arrow tail after passing the comparator. They arrive at the right end of the network in nondecreasing order.

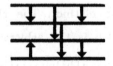

Figure 1:   The bitonic 4-sorter [Ba]

3. FAULT MODEL

A correct comparator performs a comparison-exchange as depicted in Fig. 2a.

We shall consider three different types of functional faults: Fig. 2b shows the *full-fault*, a comparator, producing the maximum of its inputs instead of the minimum and vice versa. The *half-fault* comparator (Fig. 2c) leaves the inputs unchanged, and the *x-fault* (Fig. 2d) comparator exchanges the inputs independant of their values. Each of these faults is assumed to be static, i.e. the comparator behaves always in the same faulty way.

Figure 2:

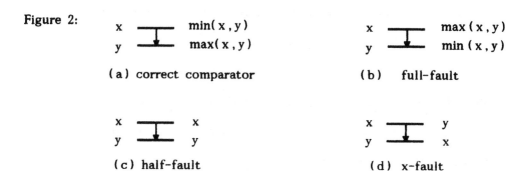

(a) correct comparator (b) full-fault

(c) half-fault (d) x-fault

Of course there are many other possible faults as for example dynamic faults, stuck-at faults, technology dependent faults such as stuck open faults in CMOS etc. We do not want to discuss each of these fault types but we are able to apply our correction method to arbitrary faults, if the considered sorter enables us to bypass faulty comparators.

Our aim is to find a solution for the half-fault correction problem. The reason for choosing this specific fault type is the fact that the mapping of a half-fault comparator is the identity, and thus it behaves like a simple storage cell. Assume now we are able to bypass each of the faulty comparators in a given sorting network by cutting out the comparator's logic and by connecting the input and the output lines, we get a network that behaves exactly like the original sorting net, where every fault (of arbitrary type) is replaced by a half-fault. If we can add now some half-fault correcting network (correcting an appropriate number of half faults), the result is a fault free sorter.

Of course, this construction is particularly useful, if the additional correction net is also allowed to contain faulty comparators which can be bypassed, if necessary. The same construction is not possible with full-faults or x-faults instead of half-faults, because the last stage of the resulting network could be faulty. In this case, a full-fault as well as an x-fault produces unsorted output sequences which means that the network is not a correct sorter.

In the remainder of the paper, a *k-fault N-sorter* denotes a network obtained from an N-sorter by replacing k comparators with half-fault comparators.

4. LOWER BOUNDS

Let C be an N-network that corrects any 1-fault N-sorter A', i.e.

$$x \, A' \, C = x_s \text{ for every } x \in \{0,1\}^N.$$

In this section we will prove two lower bounds for C:

<u>Lemma 1.</u> The number of comparators of C is $\Omega(N)$.

<u>Lemma 2.</u> The number of comparator stages of C is $\Omega(\log N)$.

For the proofs we use the following

Proposition 1:

For any i, j, $0 \le i, j \le N-1$ there is a 1-fault N-sorter A'_{ij} satisfying the following conditions for every $x \in \{0,1\}^N$:

$$(x_s)_k = (x \, A'_{ij})_k \text{ for } k \notin \{i, j\}$$
$$(x_s)_i = (x \, A'_{ij})_j ,$$
$$(x_s)_j = (x \, A'_{ij})_i .$$

In other words: A'_{ij} sorts every x except of the items on positions i and which are exchanged.

Proof: Take an arbitrary N-sorter S. If $i < j$ we define $A_{ij} = S, \{[j:i]\}, \{[i:j]\}$.

A_{ij} is an N-sorter. Let A'_{ij} be A_{ij} with a half fault in the last comparator [i:j]. Obviously, A'_{ij} satisfies the conditions of Proposition 1.

Proof of Lemma 1:

From Proposition 1 we know the N-network A'_{0j} for every j, $1 \le j \le N-1$. An N-network C correcting an arbitrary A'_{0j} must be able to move the wrong item from position 0 to any position between 1 and N-1. Therefore, every position must appear at least once in a comparator of C. Hence, C consists of at least N/2 comparators.

Proof of Lemma 2:

Again we consider the N-networks A'_{0j} for $1 \le j \le N-1$. The correcting N-network C must be able to move the wrong item from position 0 to any position between 1 and N-1. Since every comparator connects only two lines, at least $\log N$ comparator stages are needed to obtain a path from position 0 to any other position.

In Section 6, the N-network C is presented, correcting every 1-fault N-sorter. C meets the lower bounds in the number of comparators and the number of comparator stages.

5. SYMMETRIC NETWORKS

To simplify the proof of the Theorem in Section 7 we need some more notations and a few lemmata:

For $x = \langle x_0, x_1, \ldots, x_{N-1} \rangle \in \{0,1\}^N$ we define the *complement*

$$\bar{x} := \langle \bar{x}_0, \bar{x}_1, \ldots, \bar{x}_{N-1} \rangle, \quad \bar{x}_i := 1 - x_{N-1-i}, \quad i \in \{0,1,\ldots,N-1\}.$$

For a comparator $[i:j]$ the complement is defined by

$$\overline{[i:j]} = [N-1-j:N-1-i].$$

The complement $\bar{s}$ of a comparator stage $s = \{[i_0:j_0], [i_1:j_1], \ldots, [i_{r-1}:j_{r-1}]\}$ is defined by $\bar{s} := \{\overline{[i_0:j_0]}, \overline{[i_1:j_1]}, \ldots, \overline{[i_{r-1}:j_{r-1}]}\}$
and the complement of an N-network $A = s_1, s_2, \ldots, s_k$ is $\bar{A} = \bar{s}_1, \bar{s}_2, \ldots, \bar{s}_k$.

An N-network A is called *symmetric* iff $\bar{A} = A$.

Lemma 3: For $x \in \{0,1\}^N$ and $0 \le i, j \le N-1$ the following equations hold:

(i) $\bar{\bar{x}} = x$

(ii) $\overline{\overline{[i:j]}} = [i:j]$

(iii) $\overline{\bar{x}\,\overline{[i:j]}} = x[i:j]$

Proof: (i) and (ii) is obvious, (iii) is easy and can be found in [5]

Lemma 4: For every N-network A and every $x \in \{0,1\}^N$: $\overline{\bar{x}\,\bar{A}} = xA$

Proof: The proof is a simple induction on the total number m of comparators in A. It can be found in [5], too.

<u>Symmetric Network Lemma</u>: For every symmetric N-network A and every $x \in \{0,1\}^N$, $\quad \overline{\overline{x} A} = x A$

Proof: Definition of symmetric networks and Lemma 4.

6. THE CORRECTION NETWORK

Let n be an integer greater than 1 and $N = 2^n$. In this section we shall define the N-network C that can be used for single half-fault correction. We show that C consists of $2 \log N - 1$ comparator stages, and of $3.5 N - 2 \log N - 5$ comparators.

Definition: Let $N = 2^n$. The *correcting N-network* C is defined by

$$C := s_1, s_2, \ldots, s_{2n-1} \text{ with}$$

$$s_1 := \{ [2i : 2i+1] \mid 0 \le i \le 2^{n-1} - 1 \}$$

$$s_j := \{ [2^j i : 2^j i + 2^{j-1}] \mid 0 \le i \le 2^{n-j} - 1 \} \cup$$
$$\{ [N-1-(2^j i + 2^{j-1}) : N-1-2^j i] \mid 0 \le i \le 2^{n-j} - 1 \} \quad \text{for } 2 \le j \le n-1$$

$$s_j := \{ [2^{2n-j-1} i - 2^{2n-j-2} : 2^{2n-j-1} i] \mid 1 \le i \le 2^{j-n+1} - 1 \} \cup$$
$$\{ [N-1-2^{2n-j-1} i : N-1-(2^{2n-j-1} i - 2^{2n-j-2})] \mid 1 \le i \le 2^{j-n+1} - 1 \}$$
$$\text{for } n \le j \le 2n-3$$

$$s_{2n-2} := \{ [2i-1 : 2i] \mid 1 \le i \le 2^{n-1} - 1 \}$$

$$s_{2n-1} := \{ [2i : 2i+1] \mid 0 \le i \le 2^{n-1} - 1 \}$$

As an example Figure 3 shows the network C for $N = 32$ ($n = 5$).

To assure that C is welldefined we need to prove the following

<u>Proposition 2</u>: For every $r \in \{1, 2, \ldots, 2n-1\}$ and for all comparators $[i_0 : j_0]$ and $[i_1 : j_1]$ occurring in s_r, i_0, i_1, j_0, j_1 are pairwise distinct.

Proof: For stage 1, 2n-2, and 2n-1 the proposition is obviously true. For $2 \le k \le n-1$ the comparators of the kth stage are of the form

$$(*) \quad [(2i) 2^{k-1} : (2i+1) 2^{k-1}] \quad \text{or}$$

$$(**) \quad [N-1-(2i+1) 2^{k-1} : N-1-(2i) 2^{k-1}] \quad \text{for } 0 \le i \le 2^{n-k} - 1$$

For two different comparators of the form (*) the proposition is true. The same holds for two comparators of the form (**). Now let $[i_0 : j_0]$ be an (*) type comparator and $[i_1 : j_1]$ a (**) type one. We have $i_0 \neq j_0$ as well as

$i_1 \neq j_1$. And since i_0 and j_0 are even while i_1 and j_1 are odd, they must be pairwise distinct. The same argument holds for the kth comparator stage where $n \leq k \leq 2n - 2$, which completes the proof.

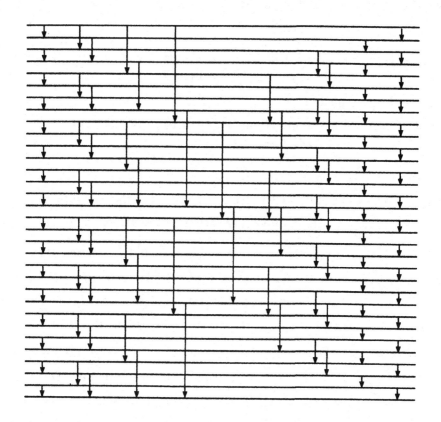

Figure 3:   The Correction N-network C for N = 32

Now we count the total number of comparators in C. The first stage consists of 2^{n-1} comparators. For $2 \leq k \leq n-1$ comparator stage k has $2 * 2^{n-k}$ comparators. For $n \leq k \leq 2n - 3$ stage k consists of $2 * (2^{k-n+1} - 1)$ comparators. Stages $2n - 2$ and $2n - 1$ have $2^{n-1} - 1$ and 2^{n-1} comparators, respectively. Therefore, the total number of comparators in C is

$$m = 2^{n-1} + 2 \sum_{k=2}^{n-1} 2^{n-k} + 2 \sum_{k=n}^{2n-3} (2^{k-n+1} - 1) + 2^{n-1} - 1 + 2^{n-1} = 3.5 * 2^n - 2n - 5.$$

Observe that C is a symmetric standard network. In particular each comparator stage s_j of C is symmetric and so is every subnetwork $s_1, s_2, ..., s_j$, for $1 \leq j \leq 2n - 1$.

7. CORRECTION PROPERTY OF C

The *Hamming Distance* of two n bit vectors x and y is the number of bits in which they differ: $D(x,y) = \sum_{i=1}^{n-1} |x - y|$

Lemma 5: Let $x \in \{0,1\}^N$ and $[i:j]$ a standard comparator in an N-network. Then $D(x[i:j], x_s) \le D(x, x_s)$.

Proof: see [6], Lemma 2.

Lemma 6: Let A' be a k-fault N-sorter. Then for any $x \in \{0,1\}$,
$$D(xA', x) \le 2k .$$

Proof: see [6], Lemma 4.

Theorem: Let $x \in \{0,1\}^N$ with $D(x, x_s) > 0$. For the correcting N-network $C = (s_1, s_2, ..., s_{2n-1})$ the following relation holds: $D(xC, x_s) \le D(x, x_s) - 2$

Proof: Let b be the number of 0's in x, i.e. $x_s = 0^b 1^{N-b}$. Since $D(x, x_s) > 0$, b must be greater than 0 and smaller than N. We call the set of positions $\{0, 1, ..., b-1\}$ the *zero area*, and the set $\{b, b+1, ..., N-1\}$ the *one area*. A *wrong 1* is a 1 positioned in the zero area and a *wrong 0* a 0 in the one area.

If there is any comparator in C that compares a wrong 1 with a wrong 0, these two elements change their positions and the Hamming Distance to the sorted sequence is reduced by two. Since the Hamming Distance cannot be enlarged again by standard comparators (Lemma 5), after passing the network C we would have $D(xC, x_s) \le D(x, x_s) - 2$.

Therefore, it is sufficient to construct a contradiction to the following assumption:

(∗) There is no wrong 1 which is compared with a wrong 0 in C.

Let P_0 be the position of the first wrong 0 in x, i.e.
$$P_0 = \min\{ p \in \{b, b+1, ..., N-1\} \mid x_p = 0 \} .$$

We want to construct the sequence of positions $P_0, P_1, ..., P_{2n-2}$, which we shall show to be the path of this 0 through the network C:

Let $p_{n-1} p_{n-2} ... p_0$ be the binary representation of P_0, i.e. $P_0 = \sum_{i=0}^{n-1} p_i 2^i$.

Since $P_0 \geq b$ we can define the constant K:

$$K := \max\{j \mid 0 \leq j \leq n-1 \text{ and } \sum_{i=j}^{n-1} p_i 2^i \geq b\}.$$

Now we define the sequence of positions P_j:

for $0 \leq j \leq K$ $\qquad\qquad P_j := \sum_{i=j}^{n-1} p_i 2^i$

for $K+1 \leq j \leq 2n-K-2$ $\qquad P_j := P_K$

for $2n-K-1 \leq j \leq 2n-2$ $\qquad d_j := \max\{i \in \{0,1\} \mid P_{j-1} - i*2^{2n-2-j} \geq b\}$

$$P_j := P_{j-1} - d_j * 2^{2n-2-j}.$$

d_j is welldefined because $P_{j-1} \geq b$. We need some observations concerning the P_j:

(1) For $0 \leq j \leq K$ P_j is a multiple of 2^j.

(2) For $0 \leq j \leq K$ $P_j = P_{j-1} - p_{j-1} 2^{j-1}$.

The definition of K implies

$$P_K = \sum_{i=K}^{n-1} p_i 2^i \geq b > \sum_{i=K+1}^{n-1} p_i 2^i$$

Therefore, $0 \leq P_K - b < p_K * 2^K$ which implies $p_K = 1$ and

(3) $0 \leq P_K - b < 2^K$

For $j = 2n - K - 2$ we have $P_j = P_K$. From (1) we know that P_j is a multiple of 2^{2n-2-j}.

For $2n-K-1 \leq j \leq 2n-2$ we have

$$P_j = P_K - \sum_{i=2n-K-1}^{j} d_i 2^{2n-2-i}$$

$$= 2^K \sum_{i=K}^{n-1} p_i 2^{i-K} - 2^{2n-2-j} * \sum_{i=2n-K-1}^{j} d_i 2^{j-i}$$

$$= 2^{2n-2-j} * (2^{K-2n+2+j} * \sum_{i=K}^{n-1} p_i 2^{i-K} - \sum_{i=2n-K-1}^{j} d_i 2^{j-i})$$

Therefore, we know

(4) $\forall j,\ 2n-K-2 \le j \le 2n-2\ \exists$ integer $m_j : P_j = m_j * 2^{2n-2-j}$.

where $1 \le m_j \le 2^{j-n+2}$, because $0 < P_j = m_j * 2^{2n-2-j} < 2^n$.

Now we want to prove that the sequence of positions P_j is the path of the first wrong 0 through the network:

(5) $\forall j,\ 0 \le j \le 2n-2\ :\ x(s_1, s_2, ..., s_j)_{P_j} = 0$.

The proof is a straightforeward induction on j which can be found in [5].

For $2n-K-2 \le j \le 2n-2$ we need an additional statement concerning the distance of P_j from the zero area:

(6) $\forall j,\ 2n-K-2 \le j \le 2n-2\ :\ 0 \le P_j - b < 2^{2n-2-j}$.

(6) can be prooved by a simple induction on j, too.

(6) means in particular for $j = 2n-2$ that $0 \le P_{2n-2} - b < 1$ or in other terms $P_{2n-2} = b$. From (5) we know $x(s_1, s_2, ..., s_{2n-2})_{P_{2n-2}} = 0$, and thus we get

(7) $x(s_1, s_2, ..., s_{2n-2})_b = 0$.

Since the complement $\overline{x}$ consists of N-b 0's and b 1's, we get from equation (7): $\overline{x}(s_1, s_2, ..., s_{2n-2})_{N-b} = 0$, and therefore,

$$\overline{\overline{x}(s_1, s_2, ..., s_{2n-2})_{N-1-(N-b)}} = 1 .$$

Because of the Symmetric Network Lemma this is equivalent to

(8) $x(s_1, s_2, ..., s_{2n-2})_{b-1} = 1$,

since $s_1, s_2, ..., s_{2n-2}$ is a symmetric network.

Now (7) and (8) show that there is a 0 in position b and a 1 in position b-1 after 2n-2 stages. Since in stage 2n-2 there are the comparators [2i-1:2i] they would have been interchanged if b is even. Hence b must be odd. But then in stage s_{2n-1} there is a comparator [b-1:b]. This comparator compares a wrong 1 with a wrong 0, a contradiction to (*).

Corollary: Given an N-sorter S. If in the network SC^k at most k comparators are replaced by half-fault comparators, the resulting network is still an N-sorter.

Proof: Let r be the number of comparators replaced in S. Then at most k-r copies of C are faulty and at least r copies of C are fault free. Lemma 6 ensures that for the network S' obtained from S by replacing the r comparators the following condition holds: $D(xS', x_s) \le 2r$.

Due to the theorem each of the fault free copies of C reduces the Hamming Distance by at least 2. Therefore, after passing the r fault free copies the sequence is sorted.

8. CONCLUSIONS

We have introduced an efficient way to achieve fault tolerance in sorting networks. We have shown the asymptotic optimality of our method for single half-fault correction. It is certainly of further interest to find lower bounds for multiple half-fault correction and correction networks that meet these bounds.

REFERENCES

[1] Batcher, K.E., *Sorting Networks and Their Applications*, AFIPS Spring Joint Computer Conference, Vol 23, 307-314, (1968)

[2] Dowd, M., Perl, Y., Rudolph, L., Saks, M., *The Balanced Sorting Network*, in Proc. ACM Princ. Distrib. Comput., 161-172, (1983)

[3] Knuth, D.E., *The Art of Computer Programming*, Vol 3. Sorting and Searching, Addison Wesley (1973)

[4] Rudolph, L., *A Robust Sorting Network*, IEEE Trans. Comput., C-34, 326-335, (1985)

[5] Schimmler, M., Starke, C., *A Correction Network for N-Sorters*, Technical Report 8714, Institut für Informatik und Praktische Mathematik, Kiel, W.-Germany, (1987)

[6] Yao, A.C., Yao, F.E., *On Fault Tolerant Networks for Sorting*, SIAM J. on Comput., Vol 14, 120-128, (1985)

Cubesort: An Optimal Sorting Algorithm
for Feasible Parallel Computers

R. Cypher
University of Washington, Seattle, WA 98195

J.L.C. Sanz
IBM Almaden Research Center, San Jose, CA 95120

Abstract

This paper studies the problem of sorting N items on a P processor parallel machine, where $N \geq P$. The central result of the paper is a new algorithm, called cubesort, that sorts $N = P^{1+1/k}$ items in $O(k\, P^{1/k} \log P)$ time using a P processor shuffle-exchange. Thus for any positive constant k, cubesort provides an asymptotically optimal speed-up over sequential sorting. Cubesort also sorts $N = P \log P$ items using a P processor shuffle-exchange in $O(\log^3 P/\log\log P)$ time. Both of these results are faster than any previously published algorithms for the given problems. Cubesort also provides asymptotically optimal sorting algorithms for a wide range of parallel computers, including the cube-connected cycles and the hypercube. An important extension of the central result is an algorithm that simulates a single step of a Priority-CRCW PRAM with N processors and N words of memory on a P processor shuffle-exchange machine in $O(k\, P^{1/k} \log P)$ time, where $N = P^{1+1/k}$.

1. Introduction

This paper presents a new parallel algorithm for sorting N items using P processors, where $N \geq P$. This new algorithm can be implemented efficiently on a wide range of parallel computers, including the hypercube, the shuffle-exchange and the cube-connected cycles. In particular, the algorithm runs in $O((N \log N)/P)$ time on any of the above architectures, provided $N = P^{1+1/k}$ for some positive constant k. This is the first sorting algorithm for any of the above architectures that obtains this performance. In addition, the sorting algorithm will be extended to obtain an efficient simulation of a Priority-CRCW PRAM using a hypercube, shuffle-exchange or cube-connected cycles. The remainder of this section reviews models of parallel computers and examines previous work in the field of parallel sorting.

The models of parallel computers that will be used in this paper are the PRAM [5], the hypercube [10], the shuffle-exchange [10] and the cube-connected cycles [11]. These models operate in an SIMD mode, with all of the processors performing the same instruction at any given time. The PRAM is a shared memory model in which all processors can access a common memory in unit time. The Priority-CRCW PRAM allows multiple processors to read from or write to a single memory location simultaneously. In the case of simultaneous writes to a single location, the lowest numbered processor attempting to write to that location succeeds.

The hypercube, the shuffle-exchange and the cube-connected cycles consist of a set of processors, each containing a local memory, that communicate with one another using a fixed interconnection network. In the hypercube, the P processors are numbered 0 .. P-1 and processors i and j are connected if the binary representations of i and j differ in exactly 1 bit position. In the shuffle-exchange, the P processors are numbered 0 .. P-1 and processors i and j are connected if j = Shuffle(i, P), j = Unshuffle(i, P) or j = Exchange(i) where Shuffle(i, P) = 2i mod (P-1), Unshuffle(i, P) = j iff Shuffle(j, P) = i, and Exchange(i) = i+1 - 2(i mod 2). The cube-connected cycles contains P processors, where $P = 2^K$ and $K = R + 2^R$. The processors are numbered with pairs (b, c) where b is a (K-R) bit number and c is an R bit number. Processor (b, c) is connected to processor (d, e) if b = d and c = e+1, if b = d and c = e-1, or if c = e and the binary representations of b and d differ in only the c-th bit position. The shuffle-exchange and the cube-connected cycles are *feasible* models because each processor is connected to only a fixed number of other processors.

One of the earliest results in parallel sorting was obtained by Batcher. In [3], Batcher presented the bitonic sorting algorithm. In [12], Stone showed that the bitonic sort could be implemented on a shuffle-exchange. This yields an $O(\log^2 N)$ time sort for N = P numbers on the shuffle-exchange.

In [4], Baudet and Stevenson show how any parallel algorithm for sorting N items with P = N processors that is based on comparisons and exchanges can be used to obtain an algorithm for sorting N items with P < N processors. By applying their technique to the bitonic sort on the shuffle-exchange, they obtained an $O((N/P) \log (N/P) + (N/P) \log^2 P)$ time sorting algorithm when $P \le N$. Their algorithm provides an optimal speed-up over sequential comparison sorting only when $P = O(2^{\text{sqrt}(\log N)})$.

An algorithm for a special case of the sorting problem was given by Gottlieb and Kruskal [6]. They presented a shuffle-exchange algorithm for the permutation problem, where the N numbers to be sorted are in the range 1 through N and where each number appears exactly once. Their algorithm requires $O(P^{9/2} + (N/P) \log P)$ time and gives optimal speed-up over sequential comparison sorting when $P = O((N \log N)^{2/9})$. In their paper, Gottlieb and Kruskal state that they do not know of an optimal algorithm for the permutation problem when P is not in $O((N \log N)^{2/9})$. The current paper thus improves upon Gottlieb and Kruskal's result in two ways. First, the algorithm presented in this paper solves the general sorting problem rather than the permutation problem. Second, the algorithm presented here gives optimal speed-up when $N = P^{1+1/k}$ for any positive constant k.

A breakthrough in parallel sorting was obtained by Ajtai, Komlos and Szemeredi [2]. They created a network for sorting N items that consists of O(N log N) comparators and has O(log N) depth. This network was used by Leighton to create a feasible parallel machine that sorts in O(log N) time when $P = N$ [8].

Unfortunately, there are two serious difficulties with Leighton's technique. First, the technique performs poorly for $P < 10^{100}$. In contrast, the algorithm presented in this paper has a much smaller constant of proportionality and is much more likely to be useful in practice. Second, Leighton's network is not a standard network that has been shown to be useful for solving problems other than sorting. In contrast, the shuffle-exchange and the cube-connected cycles have been proven useful in solving a wide range of problems.

Another important related result was obtained by Leighton. Leighton has recently shown that his algorithm called columnsort [8] can be used to obtain an efficient algorithm for sorting $N = P^{1+1/k}$ items on a P processor shuffle-exchange [9]. He obtains an $O(k^T P^{1/k} \log P)$ time algorithm, where $T = 1/\log_4 1.5$ (T is approximately 3.419). The algorithm is based on calling columnsort in a nested manner so that the N items are sorted by repeatedly sorting groups of $P^{1/k}$ items each. Furthermore, there is a possibility that the value of the exponent T can be reduced to less than 1 by using Leighton's concept of closesorting [8],[9]. Finally, a similar result using columnsort was obtained by Aggarwal [1]. More research into the applications of columnsort is clearly needed.

The paper is divided as follows. Section 2 presents an abstract description of the new sorting algorithm and proves its correctness. Section 3 shows how this sorting algorithm can be implemented efficiently on a number of parallel computers and it presents an algorithm for simulating a Priority-CRCW PRAM with a shuffle-exchange computer. Throughout this paper, N will be the number of items to be sorted and P will be the number of processors available.

2. Cubesort

This section contains a description of a new parallel sorting algorithm that the authors call *cubesort*. The description of cubesort given in this section is independent of the architecture that is used to implement it. Cubesort works by repeatedly partitioning the N items to be sorted into small groups and sorting these groups separately and in parallel. In particular, let

$N = M^D$ where M and D are integers. Each step of cubesort partitions the M^D items into either M^{D-1} groups of M items each or M^{D-2} groups of M^2 items each, and sorts the groups in parallel.

The M^D items to be sorted can be viewed as occupying a D-dimensional cube, where each side of the cube is of length M. Each location L in the cube has an address of the form $L = (L_D, L_{D-1}, ..., L_1)$, where $(L_D, L_{D-1}, ..., L_1)$ is a D-digit base-M number and L_i is the projection of location L along the i-th dimension. This numbering of the locations in the cube corresponds to an ordering of the locations that will be called *row-major order*. Cubesort will sort the items in the cube into row-major order.

In addition to viewing the items as forming a single D-dimensional cube, they can be viewed as forming a number of cubes of smaller dimension. A *j-cube*, where $0 \le j \le D$, is a set of M^j items with base-M addresses that differ only in the j least significant digits. That is, $A = (A_D, A_{D-1}, ..., A_1)$ and $B = (B_D, B_{D-1}, ..., B_1)$ are in the same j-cube if and only if $A_i = B_i$ for all i, $j+1 \le i \le D$. Each j-cube, where $0 \le j \le D$, is classified as being either *even* or *odd*. A j-cube, where $0 \le j \le D-1$, is even if it contains a location L where $L_{j+1} \bmod 2 = 0$, and it is odd otherwise. The D-cube that contains all N items is defined to be even.

There are D different partitions, represented as P_j where $1 \le j \le D$, that are used by cubesort. A group in partition P_j consists of a set of items with base-M addresses that differ only in digits j and j-1. Note that each group in P_1 contains M items, while each group in the remaining partitions contains M^2 items.

Finally, it is sometimes useful to view the items in a j-cube as forming a 2-dimensional array. A *j-array*, where $2 \le j \le D$, is an $M^2 \times M^{j-2}$ array of the items in a j-cube, where the items are placed in the array in row-major order. Thus each (j-2)-cube forms a row in a j-array, and each (j-1)-cube forms a band of M consecutive rows in a j-array. Also, each column in a j-array is a group in P_j.

In order for cubesort to work correctly, it is assumed that $D \ge 3$ and that $M \ge (D-1)(D-2)$. Cubesort makes use of two subroutines, Sort_Ascending and Sort_Mixed. The subroutine Sort_Ascending(i) sorts the groups in partition P_i in ascending row-major order. The subroutine Sort_Mixed(i, j) sorts the groups in partition P_j that are in even i-cubes in ascending order, while it sort the groups in P_j that are in odd i-cubes in descending order. Cubesort is called by first setting the global variables M and D and then calling Cubesort(D). The pseudo-code of cubesort is given below.

```
Cubesort(S)                           /* Abstract Description of Cubesort */
integer S;
{
   if S = 3 then
   {
                                      /* PHASE 1: */
      Limit_Dirty_Cubes(S);


                                      /* PHASE 2: */
      Sort_Mixed(S-1, S-1);


                                      /* PHASE 3: */
      Merge_Dirty_Cubes(S, S);
   }
   else
   {
                                      /* PHASE 1: */
      Limit_Dirty_Cubes(S);
      Limit_Dirty_Cubes(S);


                                      /* PHASE 2: */
      Cubesort(S-1);


                                      /* PHASE 3: */
      Merge_Dirty_Cubes(S, S);
   }
}


Limit_Dirty_Cubes(S)
integer S;
{
   if S > 2 then
      Limit_Dirty_Cubes(S-1);
   Sort_Ascending(S);
}


Merge_Dirty_Cubes(S, T)
integer S, T;
{
   Sort_Mixed(S, T);
   if T > 2 then
      Merge_Dirty_Cubes(S, T-2);
}
```

The call Cubesort(S) sorts each even S-cube in ascending row-major order and each odd S-cube in descending row-major order. In order to prove that cubesort works correctly, it is necessary to use the zero-one principle [7], which states that "if a network with n input lines sorts all 2^n sequences of 0's and 1's into nondecreasing order, it will sort any arbitrary sequence of n numbers into nondecreasing order". In keeping with the zero-one principle, the following discussion will assume that the input consists entirely of 0's and 1's.

The following definitions will be needed in the proof of correctness. A set of items is *dirty* if it contains both 0's and 1's, and it is *clean* otherwise. A sequence of 0's and 1's is *ascending* if it is of the form $0^a 1^b$, where $a,b \geq 0$, and it is *descending* if it is of the form $1^a 0^b$, where $a,b \geq 0$. A sequence is *monotonic* if it is either ascending or descending. A sequence of 0's and 1's is *bitonic* if it is of the form $0^a 1^b 0^c$ or of the form $1^a 0^b 1^c$, where $a,b,c \geq 0$. A j-array is *cross-sorted* if all of its rows are monotonic and if it has at most 1 ascending dirty row and at most 1 descending dirty row. A j-array is *semi-sorted* if all of its rows are bitonic and if it has at most 1 dirty row. A j-array is *block-sorted in ascending (descending) order* if it consists of A rows containing only 0's (1's), followed by B dirty rows, followed by C rows containing only 1's (0's), where $A,B,C \geq 0$. A j-cube is cross-sorted (or semi-sorted or block-sorted) if its corresponding j-array is cross-sorted (or semi-sorted or block-sorted).

The correctness of Cubesort(S) is established next. In order to save space, the proofs have been omitted.

LEMMA 1: If a j-array originally has B dirty rows, and if the columns of the j-array are then sorted in ascending (descending) order, the resulting j-array will be block-sorted in ascending (descending) order and will contain no more than B dirty rows.

LEMMA 2: After calling Limit_Dirty_Cubes(i), where $i \geq 2$, there are at most i-1 dirty (i-1)-cubes in each i-cube, and the dirty (i-1)-cubes are consecutive within each i-array.

LEMMA 3: If a j-array is originally semi-sorted or cross-sorted, and if the columns of the j-array are then sorted in ascending (descending) order, the resulting j-array will be semi-sorted and it will be block-sorted in ascending (descending) order.

THEOREM 1: After calling Cubesort(3), each even 3-cube is sorted in ascending row-major order and each odd 3-cube is sorted in descending row-major order.

LEMMA 4: When $S > 3$, after Phase 1 there are at most 2 dirty (S-1)-cubes in each S-cube, and these dirty cubes are adjacent to one another in the S-array.

LEMMA 5: For any values of S and T, where $1 \leq T \leq S \leq D$, if originally each T-cube is either semi-sorted or cross-sorted, and if Merge_Dirty_Cubes(S, T) is then called, the resulting

T-cubes will all be sorted. Furthermore, the T-cubes that are in even S-cubes will be sorted in ascending order and the T-cubes that are in odd S-cubes will be sorted in descending order.

THEOREM 2: Cubesort(S), where $3 \leq S \leq D$, sorts each even S-cube in ascending order and each odd S-cube in descending order.

3. Implementing Cubesort

The cubesort algorithm given in the previous section sorts $N = M^D$ numbers by performing $O(D^2)$ stages, where each stage consists of sorting, in parallel, groups containing $O(M^2)$ items. This section will show how cubesort can be implemented on a variety of parallel models.

First, the implementation of cubesort on a shuffle-exchange will be presented. It will be assumed that there are N items to be sorted and that P processors are available, where $N = P^{1+1/k}$. The items to be sorted are stored in an N item array A, where A_i is located in processor $j = \text{floor}(i/P^{1/k})$, for $0 \leq i \leq N-1$. In order to use the algorithm from the previous section, let $D = 2k+2$, let $M = P^{1/2k}$, and let location $L = (L_D, L_{D-1}, ..., L_1)$ in the D-dimensional cube correspond to A_L.

Before the groups of a partition are sorted, the data are rearranged so that each group lies within a single processor. There are 2 permutations that are used to perform this rearrangement, namely the M-Shuffle and the M-Unshuffle. The definition of the M-Shuffle of N items is that M-Shuffle(X, N) = MX mod (N-1). The M-Unshuffle is the inverse of the M-Shuffle, so M-Unshuffle(Z, N) = X iff M-Shuffle(X, N) = Z.

The $P^{1/k}$ items that are local to each processor can be sorted in $O((1/k)\, P^{1/k} \log P)$ time. Also, the M-Shuffle and M-Unshuffle of the items to be sorted can each be accomplished in $O((1/k)\, P^{1/k} \log P)$ time. Because the shuffle-exchange implementation of cubesort consists of $O(D^2) = O(k^2)$ applications of sorts that are local to processors and $O(k^2)$ applications of the M-Shuffle and M-Unshuffle routines, the entire algorithm requires $O(kP^{1/k} \log P)$ time.

The implementations of cubesort on the hypercube and the cube-connected cycles are similar to the implementation on the shuffle-exchange. Because of space limitations, only the result will be stated. Cubesort can be implemented in $O(k^2 P^{1/k} \log P)$ time on a hypercube or cube-connected cycles.

In the above discussion, it was assumed that $N = P^{1+1/k}$. However, cubesort can also be used when the number of items per processor grows more slowly. In particular, when $N = P \log P$ cubesort yields an $O(\log^3 P/\log\log P)$ time sorting algorithm for a P processor shuffle-exchange. Again, space limitations prevent including that algorithm.

Finally, cubesort can be used to simulate a Priority-CRCW PRAM with a shuffle-exchange computer. Because of space limitations, only the result will be stated. A single operation of a Priority-CRCW PRAM with N processors and N memory locations can be implemented in $O(kP^{1/k} \log P)$ time on a P processor shuffle-exchange, where $N = P^{1+1/k}$.

Acknowledgments

The authors would like to express their gratitude to Prof. T. Leighton for bringing to their attention his extensions of columnsort for optimal sorting, and for many useful discussions. Also, the authors feel indebted to Prof. S. Hambrusch, Dr. M. Snir, Prof. L. Snyder and Dr. E. Upfal for their helpful comments on the contents of this paper. The work of R. Cypher was supported in part by an NSF graduate fellowship.

Bibliography

1. A. Aggarwal, Unpublished manuscript, 1986.

2. M. Ajtai, J. Komlos, E. Szemeredi, "An O(n log n) Sorting Network", Proc. 15th Annual Symposium on Theory of Computing, 1983, pp. 1-9.

3. K.E. Batcher, "Sorting Networks and their Applications", 1968 AFIPS Conference Proceedings, pp. 307-314.

4. G. Baudet, D. Stevenson, "Optimal Sorting Algorithms for Parallel Computers", IEEE Transactions on Computers, vol. c-27, no. 1, January 1978, pp. 84-87.

5. A. Borodin, J.E. Hopcroft, "Routing, Merging and Sorting on Parallel Models of Computation", Proc. 14th Annual Symposium on Theory of Computing, 1982, pp. 338-344.

6. A. Gottlieb, C.P. Kruskal, "Complexity Results for Permuting Data and Other Computations on Parallel Processors", Journal of the ACM, vol. 31, no. 2, April 1984, pp. 193-209.

7. D.E. Knuth, "The Art of Computer Programming, Vol. 3: Sorting and Searching", Addison-Wesley, Reading, MA, 1973.

8. T. Leighton, "Tight Bounds on the Complexity of Parallel Sorting", IEEE Transactions on Computers, vol. c-34, no. 4, April 1985, pp. 344-354.

9. T. Leighton, Personal communication..

10. D. Nassimi, S. Sahni, "Data Broadcasting in SIMD Computers", IEEE Transactions on Computers, vol. c-30, no. 2, February 1981, pp. 101-107.

11. F.P. Preparata, J. Vuillemin, "The Cube-Connected Cycles: A Versatile Network for Parallel Computation", Communications of the ACM, vol. 24, no. 5, May 1981, pp. 300-309.

12. H.S. Stone, "Parallel Processing with the Perfect Shuffle", IEEE Transactions on Computers, vol. c-20, no. 2, February 1971, pp. 153-161.

A 4d CHANNEL ROUTER FOR A TWO LAYER DIAGONAL MODEL*

E. Lodi, F. Luccio,
Dipartimento di Informatica, Università di Pisa
L. Pagli
Dipartimento di Informatica e Applicazioni,
Università di Salerno

Abstract We consider a variation of the channel routing model called diagonal model (DM), where the connections are laid in two layers, along tracks at +45° (first layer) and -45° (second layer). Based on some theoretical results, we present an algorithm that solves any two-terminal routing problem in DM, in a channel of width 4d. No similar result is known in the Manhattan model, where the best algorithm completely analyzed attains a channel width of $d+O(f)$, $f \leq \sqrt{n}$.

1. Introduction The channel routing problem crucially arises in VLSI design [1]. Two sets of n terminals are displayed on the integer points of two horizontal lines, called *entry* and *exit* lines. The terminals are identified according to their coordinates along these lines (an entry and an exit terminal with the same coordinate are aligned on a vertical line). The stripe bounded by the entry and exit lines is called the *channel*. It has an integer height, called *channel width*.

Each entry terminal must be connected to one exit terminal, and vice-versa (two-terminal problem), and the routing is to be realized inside the channel. Each pair p, q of terminals to be connected is called a *net* $N=(p,q)$. A channel routing problem (CRP) is assigned by a set of nets $(p_1,q_1), \dots , (p_n,q_n)$, such that $p_i \neq p_j$, $q_i \neq q_j$, $\forall$ i≠j. For 1≤i≤n we also have: $1 \leq p_i \leq m$, $1 \leq q_i \leq m$, m≥n, where m is the maximum value among all p_i's and q_i's. A *complete* CRP (or an *incomplete* one) is such that n=m (or

---
*This work has been supported in part by Ministero della Pubblica Istruzione of Italy.

n<m), that is all the integer points 1, ... , m (or, not all such points) correspond to terminals.

CRP has been thoroughly studied in different connection models, with two or more connection layers (see for example [2,3,4,5]). In practice, however, the elementary Manhattan Model is used in most applications. In this model, each net is realized as a sequence of alternating vertical and horizontal segments, routed along the lines of a square grid of unit side, superimposed to the channel. Two layers of wiring are used, to support the horizontal and the vertical lines of the grid, respectively, and contact cuts are introduced between the two layers to connect consecutive segments of the same net.

In this paper we consider a variation of the Manhattan model (shortly MM), called Diagonal model (shortly DM). DM was introduced in [6].The grid lines lay on the two layers with an orientation of $\pm 45°$, with the lines at $+45°$ (*right tracks*) on one layer, and the lines at $-45°$ (*left tracks)* on the other layer. All the tracks meet the entry and the exit lines at integer points, therefore each terminal lies on a right and on a left track. A net is now realized as a sequence of alternating right and left segments, and vias are used to connect consecutive segments.

A classical parameter of a CRP is the *density* d, defined as the maximum number of nets (p,q) such that, for an arbitrary non integer value v, 1<v<m, we have p<v<q or q<v<p. As usual, we represent the nets with arrows. If $p \neq q$ the arrow is drawn horizontally and the starting and ending points have abscissae p and q, respectively (fig.1.a). If p=q we draw a vertical arrow of unit length. Two or more arrows may be drawn on the same horizontal line, as far as they do not overlap. An arrow is called *right, left* or *trivial* if it goes to the right, to the left, or is vertical. The density d is then the maximum number of arrows crossing an arbitrary vertical line v, which intersects the entry

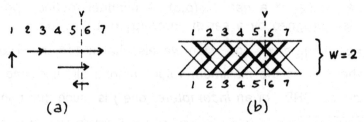

(a)   (b)

Fig.1 (a) Representation of the CRP (1,1),(2,3),(3,7),(4,6),(6,5).
(b) Layout in the diagonal grid of DM.

line in a non integer point. In the example of fig.1 we have d=3.

We also define the *right density* d_r and the *left density* d_l, as the densities evaluated on the right arrows, and on the left arrows, respectively. In the example of fig.1(a) we have $d_r = 2$ and $d_l = 1$. Clearly, $d \le d_r + d_l \le 2d$. It can easily proved that, in all complete problems we have $d_r = d_l = d/2$.

In MM d constitutes a lower bound to the channel width w (at least d horizontal tracks must be used to route the d nets crossing v). In DM, instead, an immediate lower bound for w is $\lceil d/2 \rceil$, since 2w tracks (right or left) cross v. See fig. 1(b), where the diagonal grid is shown on a channel of $w = \lceil d/2 \rceil = 2$ together with a layout for the given CRP.

The goal of a CRP is to construct a layout in a channel of minimum width. The best known algorithm for MM [7] attains an upper bound for w of d+O(f), with $f \le \sqrt{n}$ (f is called the *flux*). In particular, the worst case for f occurs for complete problems (e.g. , shift problems) [5]. The only known algorithm in DM [6], instead, attains an upper bound for w of |p-q|, where (p,q) is the longest net (p,q). The worst case occurs for h of O(n), however, it has been shown in [6] that several problems which are difficult and require a large channel in MM, such as the shift problems, are much better solved in DM. In fig. 2(a), for example, the combination of shift-by-1 with shift-by-(-1) is realized in DM in a channel of width 1 (this is a difficult problem in MM). The difficulties arises in DM if long nets are present. In fig.2.b, a channel of width 3 accomodates a layout for the previous problem, with the inclusion of the long net (1,12). This a good sense solution, while the algorithm of [6] would build a channel of width w=12-1=11.

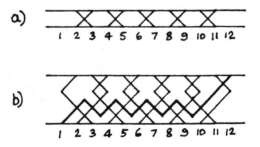

Fig.2 Two layouts in DM. (a) A combination of shift-by-1 with shift-by-(-1). (b) The same, with inclusion of a long net.

In this paper we present a general algorithm to solve any two-terminal CRP in DM, with channel width w=4d for complete problems, and w=4(d_r+d_l) for incomplete ones. The extension of the algorithm to attain w=4d in both cases is rather natural.

2. The routing algorithm The "4d channel" layout will be constructed as a composition of partial layouts for specific families of arrows. In fact, for a section of the channel between abscissae a, b, 1≤a≤b≤m, consider the families of arrows:

F1: (a,a), (a+1, a+1),... ,(b,b),   with a≤b,
   that is, all the arrows are trivial;

F2: (a,b), (a+1, a+1), (a+2, a+2), ... ,(b-1,b-1), with a<b,
   that is, a right arrow crosses all vertical arrows.

F1 and F2 will be routed on a channel of width 4, according to the layouts shown in fig.3. Different cases must be considered, depending on the *size* s=b-a+1 of the family. The layouts will be denoted G1.s and G2.s, for F1 and F2, respectively. The layouts G2.s, s>3, will be realized as generalizations of G2.8 (s even) and G2.9 (s odd).

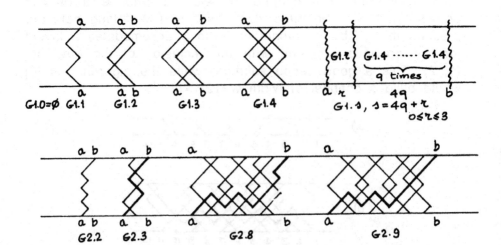

Fig.3   Layouts G1.s, G2.s for the families F1, F2 of size s=b-a+1, in a channel of width w = 4.

G1.s and G2.s have been chosen to be contained, for any value s, within the shaded areas shown in fig. 4 (a). This allows to pack the layouts of several consecutive families in consecutive sections of the channel, as shown in fig. 4 (b), with the rule that two families of type F1 do not occur in adjacent positions. In fact, is two such families $F1'=(a_1,a_1)$, ... ,(b_1,b_1), and $F1''=(a_2,a_2)$, ... ,(b_2,b_2), with $b_1+1=a_2$, can be considered as the unique family $F1'''=(a_1,a_1)$, ... ,(b_2,b_2), and a layout of type G2 can be built for it.

We will show that any CRP can be solved by constructing, one on top of the other, the layouts of d_r+d_l sequences of families, such that each of these layouts can be realized in a channel of width 4 as shown in fig.4 (b). For this purpose, we need some additional results. Let **P1** be an arbitrary CRP:

Definition 1. The *first right chain* R1 of **P1** is a sequence of right arrows of **P1**:

$R1= (p_1, q_1), (p_2, q_2), ... , (p_u, q_u)$, $u\geq1$,

such that, for $1\leq i\leq u-1$ we have $q_i\leq p_{i+1}$, and **P1** does not contain a right arrow (p,q) with $p<p_1$, or $q_i\leq p<p_{i+1}$, or $q_u\leq p$.

Observation. **P1** - R1 is a CRP.

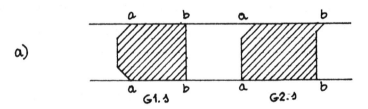

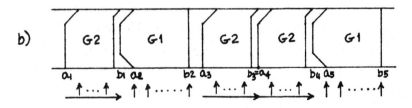

Fig.4 (a) Channel portions occupied by the layouts G1.s and G2.s.
(b) Five consecutive families packed in the channel.

Let **P2**=**P1**-R1, and R2 be the 1st right chain of **P2**. We denote **P2**-R2 by **P3**, and iterate this process until a problem **P(h+1)** without right arrows is generated.

Definition 2. The *right decomposition* D_r of **P1** is the sequence of chains D_r = R1, R2, ... , Rh, such that R2 is the first right chain of **P2** = **P1**-R1, ..., Ri is the first right chain of **Pi**=**P(i-1)**-R(i-1), ... , and **P(h+1)** has no right arrows.

R1, R2, ... , Rh are called the first, second, ... , h-th right chains of **P1.** Recalling that d_r is the right density of the given problem, we have:

Theorem 1. #D_r=d_r.

This theorem is somehow contained among the previous results of channel routing [8]. The proof is carried out by showing that, passing from **Pi** to **P(i+1)**, the right density decreases by one [9]. We have: D_r= R1, ... , Rd_r. In the sample CRP of fig. 5, the chains of the right decomposition are drawn on the upper three horizontal lines.

Consider an arrow (p,q) of an arbitrary right chain Ri of **P1**:

Definition 3. The *target point* t of (p, q)∈ Ri is the minimum integer t≥q such that **P1** does not contain a left or trivial arrow (t, z) (i. e. with t≥z, and t is not a target point for any arrow of the right chains Rj, j<i.

In the example of fig. 5, each target point is indicated by an encircled dot, relative to the arrow on its left. For example, the target point of (3,8)∈ R3 is 11, since position 8 and 9 are target points of arrows in R2 and R1, and 10 is the starting point of the left arrow (10, 10). Note that a target point may be greater than m (e. g., for the arrows (8,15) and (12,17)).

For a chain Ri = (p_1, q_1), (p_2, q_2), ... , (p_u, q_u), let $t_1, t_2, ... ,$ t_u be the corresponding target points. The sequence **Ri**=(p_1, t_1), (p_2, t_2), ... , (p_u, t_u) will be called *extended chain* of Ri. For each **Ri** = (p_1, t_1), ..., (p_u, t_u), consider the families of arrows: for 1≤j≤u+1 (with the limit values t_0=0, p_{u+1}=m+1):

$$F1^j = \begin{cases} (t_{j-1}+1, t_{j-1}+1), \ (t_{j-1}+2, t_{j-1}+2), \ \ldots \ (p_j-1, p_j-1), & \text{if } p_j > t_{j-1}+1 \\ \varnothing, & \text{otherwise;} \end{cases}$$

for $i \leq j \leq u$:

$$F2^j = (p_j, t_j), \ (p_j+1, p_j+1), \ \ldots, \ (t_j-1, t_j-1);$$

where all the trivial arrows (x, x) inserted in $F1^j$ and $F2^j$ are such that x is either an end point (target) of an arrow in **Rj**, j<i, or a start point of an arrow in **P(i+1)** (recall that **P(i+1)** is the CRP **P1** without the arrows of all Rj, j≤i).

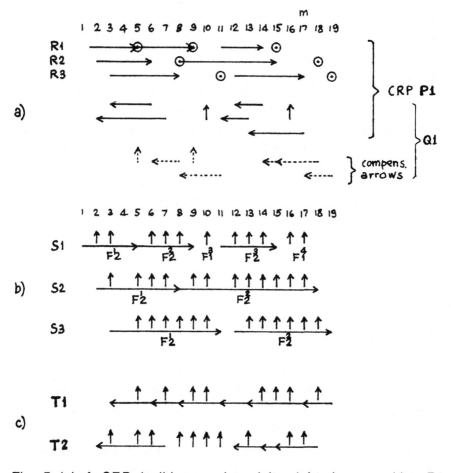

Fig. 5 (a) A CRP (solid arrows) and its right decomposition R1, R2, R3. (Target points are encircled, compensation arrows are dotted). (b), (c) Sequences **S1, S2, S3**, and **T1, T2**.

Consider now the sequence:

$$Si = F1^1 \ F2^1 \ ... \ F1^u \ F2^u \ F1^{u+1}$$

(in fact, any void family $F1^j$ will not be inserted in **Si**). In fig. 5(b) we show the sequence **S1, S2, S3** for the given CRP. For example, the family $F2^2 \in$ **S2**, corresponding to the arrow (8,18) $\in$ **R2**, consists of the arrows (8,18), (9, 9), ... , (17, 17), except for (11, 11), since the CRP arrow starting in 11 belongs to chain R1 which preceeds (not follows) R2. We are now ready to present our algorithm. The sequences **S1, S2, ... , Sd$_r$** are laid out in this order, one on top of the other, with the technique shown in figs. 3,4. This requires a portion of channel of width $4d_r$, starting at the entry line at the bottom (see fig. 6, lower portion). Left and trivial arrows are now to be routed. However, a modification of the original set must take place to compensate for the arrows of each Ri wich have been extended in **Ri**. For each arrow (p,q) $\in$ Ri, transformed into (p,t) $\in$ **Ri**, a new *compensation arrow* (t,q) is inserted among the left or trivial arrows of the CRP (fig.5(a)). That is, (p, q) is actually realized as (p, t) followed by (t, q). Note that, if t=q, the compensation arrow is a trivial one. Letting **C** be the set of all compensation arrows, the residue problem to be solved is then **Q1**= $P(d_r+1) \cup$ **C**. That is, **Q1** consists of all the left and trivial arrows of the original problem, plus the compensation arrows. We have (for a proof see [9]):

Theorem 2. A layout of an arbitrary CRP can be obtained as the layout of **S1, S2, ... , Sd$_r$, Q1** built one on top of the other starting at the entry line.

In our example, we have to build the layout for **Q1** on top of the layout of **S1, S2, S3** of fig. 6. To solve **Q1**, its left arrows are groupped in *left chains* , L1, L2, ..., and a *left decomposition* D_l is considered as an immediate extension of Definition 2. Furthermore, we can state that $\#D_l = d_l(Q1)$, by reverting Theorem 1 ($d_l(Q1)$ is the left density of **Q1**). The following important result holds (for a proof see [9]):

Theorem 3. $d_l(Q1) = d_l(P1)$.

Theorem 3 shows that the left density of a CRP is not increased by the insertion of the compensation arrows. The theorem is proved by showing that, if the compensation arrows relative to the chains R1, R2, ... are inserted in this order, the left density does not increase at any step.

The concept of extended chains is not relevant for the left chains. The target points simply are the end points of the arrows, hence the compensation arrows are all trivial. A set of sequences **T1, T2, ... , Td$_l$** is built for the Li's in the same manner of the **S1, S2, ... , Sd$_r$** (fig. 5(c)), and the layout for each **Ti** is similarly performed in a portion of channel of width $4d_l$ (fig. 6, upper portion). This layout is realized on top of the one for the **Si**, thus solving the whole problem.

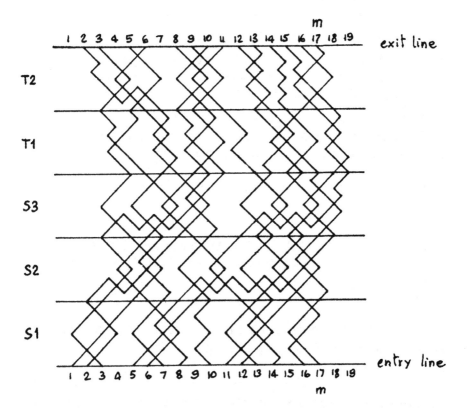

Fig. 6 Layout for the given CRP.

Formally, the routing algorithm differs from the one just described, because the compensation arrows for each R_i are inserted when R_i is built. Note that the successive modifications of the CRP **P** will be indicated by $Z(1), Z(2), \ldots , Z(d_r+d_l)$.

ROUTING ALGORITHM FOR CRP P
```
  Z(1):= P;
  baseline:=0;          {0 is the height of the entry line}
  for i:=1 to d_r do
     begin
         build  Ri → Ri → Si for Z(i);
         draw the layout for Si (with the technique of
                  figs.3,4) starting on baseline;
             Z(i+1):= Z(i) − Ri ∪ Ci, where Ci is the set of
                      compensation arrows for Ri;
             baseline := baseline + 4
     end
  for i:= d_r +1 to d_r + d_l do
     begin
         build Li → Ti for Z(i);
         draw the layout for Ti starting on the baseline;
             Z(i+1):=Z(i)  − Li ∪ Ci, where Ci is the set
                      of compensation arrows for Li;
         baseline := baseline + 4
     end.
```

From all the arguments above, and with the use of obvious data structures, we can easily conclude that:

Corollary 1. The time complexity of ROUTING ALGORITHM is of $O((d_r+d_l)m)$.

Corollary 2. ROUTING ALGORITHM solves any CRP problem in a channel of width $w=4(d_r+d_l)$. If the problem is complete, $d_r=d_l=d/2$, hence $w=4d$.

In fact, it can be shown that a different chain decomposition allows to attain $w=4d$ in any case [9].

3. Concluding remarks In this paper we have shown how to build a layout in a channel of width 4d, under a diagonal channel routing model (DM). This result may be important, since no similar result holds in the standard Manhattan model (MM). However, the use of DM has the disadvantage of bringing the conductive tracks of each layer closer by a factor of $\sqrt{2}$ than in MM. If, for technological reasons, the distance between tracks must be maintaned, then the distance between adjacent terminals, hence the channel width must be increased by $\sqrt{2}$.

Note also that our layout may invade a side portion of the channel of size $\leq d_r$, to the right of the rightmost terminal (e.g., see fig.6). This is generally allowed, and is usually done in MM.

A more important remark is that the simple algorithm presented in [6] allows to build a DM layout in a channel of width h, that is the length of the longest net. If h<4d, such a layout is superior, and our new algorithm shoud not be used. In the tutorial example studied in this paper (figs. 5,6) we have h=7 (length of (8,15)) and 4d=20, hence the algorithm of [6] would yield a much better result. In the worst case, however, we have h$\in$O(n), while the algorithm presented here is provably "good" in all cases.

Finally, we note that our study has been directed to two-terminal problems. An algorithm for multi-terminal nets is presented in [10].

References

[1]   Hu T.C., Kuh E.S. *Theory and Concepts of Circuits Layout.* In T.C. Hu and K.S. Kuh, Eds.: VLSI Circuit Layout: Theory and Design, IEEE Press Selected Rep. Series, New York 1985.

[2]   Leighton T. *A Survey of Problems and Results for Channel Routing,* A.W.O.C., Loutraki, Greece, July 1986.

[3]   Preparata F.P., Lipski W. *Optimal Three-layer Wireable Layouts,* IEEE Trans. on Comp. C-33 (1984), 427-437.

[4]   Sarrafzadeh M. *Hierarchical Approaches to VLSI Circuit Layout,* Coordinated Science Lab., Univ. of Illinois, Tech. Rep. ACT 72, 1986.

[5] Brown D. and Rivest R. L. *New Lower Bounds on Channel Width*, Proc. CMU Conference on VLSI Systems and Computations, Computer Science Press 1981, 178-185.

[6] Lodi E., Luccio F. and Pagli L. *A preliminary study of a Diagonal Channel Routing Model*, Algorithmica, to appear.

[7] Baker B.S., Bhatt S.N., Leighton T. *An Approximation Algorithm for Manhattan Routing,* in F.P.Preparata, Ed., Advances in Computing Research, Vol.2, JAI Press 1984.

[8] Rivest R.L., Baratz A.E. and Miller G. *Provably good channel routing algorithms*, Proc. CMU Conference on VLSI Systems and computations, 1981.

[9] Lodi E., Luccio F. and Pagli L. *Channel Routing in the Diagonal Model*, Dipartimento di Informatica, Univ. of Pisa, Int. Rep. 1988.

[10] Lodi E. *Routing Multiterminal Nets in a Diagonal Model*, Proc. 22nd CISS, Princeton 1988.